D1368573

Natural Compounds
Part 3: Steroids, Terpenes and Alkaloids

Methodicum Chimicum

A Critical Survey of Proven Methods
and Their Application in Chemistry,
Natural Science, and Medicine

Editor-in-Chief

Friedhelm Korte

Volume 11
Natural Compounds

Part 3
Steroids, Terpenes
and Alkaloids

Academic Press New York · San Francisco · London 1978

Georg Thieme Publishers Stuttgart

Maruzen Co., Ltd. Tokyo

Volume 11
Part 3

Steroids, Terpenes and Alkaloids

Edited by F. Korte and M. Goto

Contributions from

Ch. Baumann, Giessen
J.W. Buchler, Aachen
R. Crouch, New York
A. Gossauer, Braunschweig
T. Goto, Nagoya
S. Hayashi, Hiroshima
Y. Hirata, Nagoya
M. Ihara, Sendai
E. R. H. Jones, Oxford
T. Kametani, Sendai
T. Kato, Sendai
A. Kawaguchi, Sendai
T. Kawasaki, Fukuoka
Y. Kitahara, Sendai
T. Kubota, Osaka
K. Nakanishi, New York
R. A. Nicolaus, Naples
S. Nozoe, Sendai
S. Shibata, Tokyo
P. H. Solomon, New York
V. Thaller, Oxford
R.H. Thomson, Old Aberdeen
T. Tokoroyama, Osaka
K. Tsukida, Kobe
S. Yamamura, Nagoya

Academic Press New York · San Francisco · London 1978

Georg Thieme Publishers Stuttgart

Maruzen Co., Ltd. Tokyo

Library of Congress Catalog Card Number: 74-21580
ISBN 0-12-460747-0 (Academic Press)
ISBN 3-13-506401-8 (Thieme)

Preface of the Series

The METHODICUM CHIMICUM is a short critical description of chemical methods applied in scientific research and practice. It is particularly aimed at chemists as well as scientists working in associated areas including medicine who make use of chemical methods to solve their 'interrelated' problems.
Considering the present development of science and the necessity for concise and unambiguous information, the series provides a guide to rapid and reliable detection of the method suitable for the solution of the problem concerned. Thus, particular emphasis is placed on the description of proved procedure whereby a complete and exhaustive compilation of all reported methods and also a detailed description of experimental techniques have been deliberately omitted. Newer methods as well as those which have not yet been reported in review articles are treated more extensively, whereas conventional methods are dealt with concisely. Biological procedures which, in specific cases, are more useful for characterizing substances than chemical or physical methods, will be discussed in the analytical volume. The interrelated methods and concepts which are constantly gaining importance will be fully discussed in the third 'Specific Part'.
The METHODICUM CHIMICUM is comprised of three parts. The first, the 'General Part' consists of Volumes 1, 2 and 3. Volume 1 (Analytical Methods) is concerned with chemical, physical, and biological analytical methods including those necessary for the elucidation of structures of compounds.
Volume 2 (Planning of Syntheses) contains a review on fundamentals, principles, and models with particular respect to the concepts and applications of theoretical chemistry essential to the practically working scientist.
Volume 3 (Types of Reactions) is designed to illustrate the scope and utility of proved working techniques and syntheses.
The second part (Vols. 4–8), which is particularly devoted to 'Systematic Syntheses', deals with proved methods for syntheses of specific compounds. These procedures are classified according to functional groups linked together in the last step of reaction.
Volume 4 (Syntheses of Skeletons) describes the construction of hydrocarbons and heterocyclic compounds.
Volume 5 the formation of C–O-bonds, Volume 6 the formation of C–N-bonds, Volume 7 the syntheses of compounds containing main group elements, and Volume 8 compounds containing transition metal elements.
The third 'Special Part' (Volume 9–11) is concerned with the chemical aspects connected with the formulation of a question or problem.
Volume 9 deals with nonmetallic synthetic fibers and synthetic materials as well as their additives, Volume 10 with synthetic compounds and Volume 11 with natural products.
All Volumes should not contain more than 900 printed pages. They are intended to give the chemist and any person working in fields related to chemistry a sufficient answer to his problem. Selected review articles or important original works are cited for the sake of detailed information.

Friedhelm Korte

Preface of Volume 11

Volume 11 of METHODICUM CHIMICUM is devoted to the methods of structural determinations and syntheses of natural products which are of interest not only to the chemist but also to the scientist who works in associated areas, including medicine. This compilation contains a short discussion of the principles of well-proved procedures. This volume comprised of three parts. Part 1 is concerned with the chemistry and biochemistry of nucleic acids, proteins, and carbohydrates. Part 2 describes the different aspects of antibiotics, vitamins, hormones and other compounds with biochemical significance. Part 3 is designed to illustrate the scope and utility of proved working techniques and syntheses in the fields of steroids, terpenes, alkaloids, and natural pigments.

Separation and determination of main classes of compounds were already discussed in Volumes 1A and 1B of this series, and, to avoid duplication, some important discussions on purification and isolation of classes of compounds are intentionally omitted in Volume 11. Synthetic compounds with industrial importance or with physiological significance are to be discussed in Volumes 9 and 10. In this volume, special emphasis was layed on the structures and syntheses of compounds, and physiological and biochemical functions of compounds are discussed secondarily.

In this volume, the literature has been surveyed up to 1972. Thanks to the close collaboration of the authors, it has been possible to include new significant results recently published. We are grateful to the authors for their cooperation and their many valuable suggestions.

Munich, July 1976 Friedhelm Korte
Tokyo, July 1976 Miki Goto

Table of Contents for Volume 11, Part 3 (Chapter 1–6)

Preface of the Series . . . v
Preface of Volume 11 . . . vii

1 Steroids . . . 1

1 Steroids . . . 2
(*Rosalie Crouch, Koji Nakanishi, Philippa H. Solomon*)

2 Terpenoids . . . 43

2.1 Mono- and Sesquiterpenoids . . . 44
(*Shuichi Hayashi*)
2.2 Diterpenoids . . . 66
(*Tadahiro Kato, Yoshio Kitahara*)

3 Saponins . . . 87

3 Saponins . . . 88
(*Toshio Kawasaki*)

4 Alkaloids . . . 101

4 Alkaloids . . . 102
(*Yoshimasa Hirata, Tetsuji Kametani, Masataka Ihara*)

5 Natural Pigments . . . 133

5.1 Pyran Compounds . . . 134
(*Toshio Goto, Shosuke Yamamura*)
5.2 Pyrrol Compounds . . . 142
5.2.1 The Chemistry of Naturally Occurring Bile Pigments . . . 142
(*Abert Gossauer*)
5.2.2 Macrocyclic Tetrapyrrole Pigments . . 153
(*Johann W. Buchler*)
5.3 Quinonoid Compounds . . . 159
(*Ronald H. Thomson*)
5.4 Polyene Compounds . . . 168
(*Kiyoshi Tsukida*)
5.5 Polyacetylene Compounds . . . 175
(*Sir Ewart R. H. Jones, Viktor Thaller*)
5.6 Polyphenolic Compounds . . . 179
(*Takashi Tokoroyama, Takashi Kubota*)
5.7 Melanins . . . 190
(*Rodolfo A. Nicolaus*)
5.8 Visual Pigments . . . 200
(*Christian Baumann*)

6 Miscellaneous . . . 207

6.1 Lichen Substances . . . 208
(*Shoji Shibata*)
6.2 Biosynthesis of Isoprenoids . . . 223
(*Shigeo Nozoe, Akihiko Kawaguchi*)

Subject Index . . . 236

1 Steroids

Contribution by

R. Crouch, New York
K. Nakanishi, New York
P.H. Solomon, New York

1 Steroids

Rosalie Crouch, Koji Nakanishi and Philippa Heggs Solomon

Department of Chemistry, Columbia University, New York, New York 10027, USA

1.1 Introduction

The steroids, the basic skeleton of which comprises the 17 carbon atoms of the perhydrocyclopentenophenanthrene system (see p. 2, cholestane) are probably one of the largest classes of natural products. They are distributed widely in both the animal and plant kingdoms and include compounds of vital importance to the maintenance of life, such as cholesterol, the bile acids, the D vitamins, sex hormones, corticoid hormones, cardiac aglycones, antibiotics and insect moulting hormones.

Because of their varying bioactivities, the steroids have been the focus of a prodigious quantity of synthetic work, including both total synthesis and modification of natural compounds. Furthermore, the relative ease in making modifications and derivatives, and the rigid and planar skeleton are important attributes that have made the steroids an ideal group of reference compounds for developing new chemical reactions, spectroscopic techniques, microbial reactions, etc.

In the following section are listed some typical steroids, as well as some recently isolated steroids with unusual structural modifications.

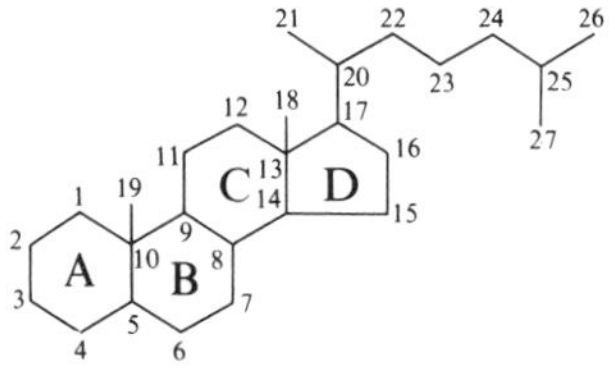

Steroid ring system: cholestane

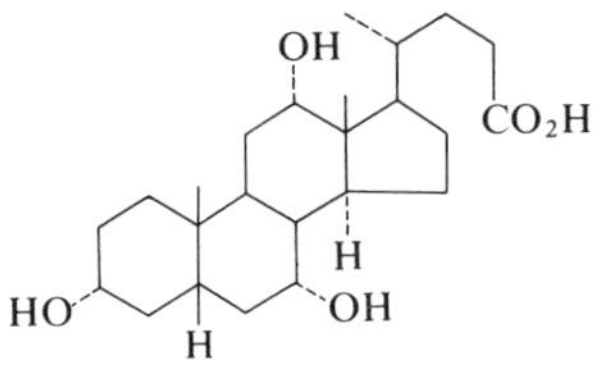

Cholic acid

—isolated from bile, where it occurs as the conjugate acid in which the steroid is joined by a peptide bond to glycine or taurine; emulsifying agent which aids in the digestion of fats.

—other members of group—desoxycholic, chenodesoxycholic and lithocholic acids, lack 12-, 7-, and 7-/12-hydroxyl groups, respectively.

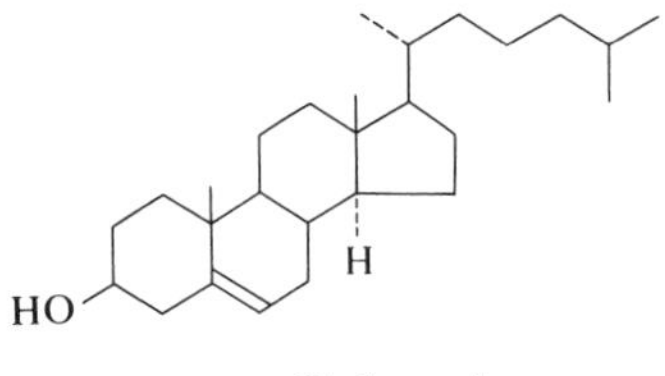

Cholesterol

—major sterol from body tissues.

—side chain alkylation at C-24 leads to phytosterols.

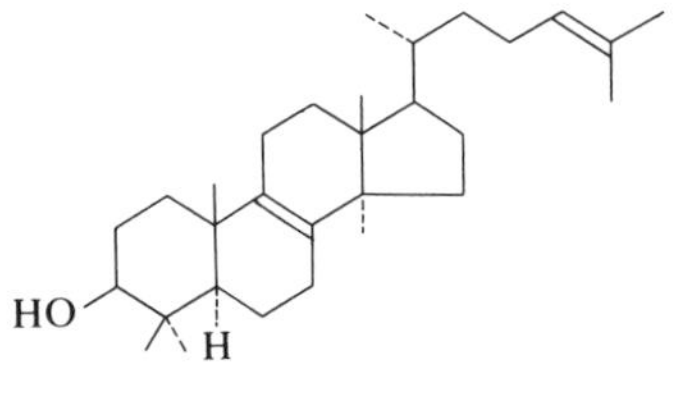

Lanosterol

—widely distributed in plants; major component of sheep's wool fat.

—key intermediate in steroid biosynthesis from squalene.

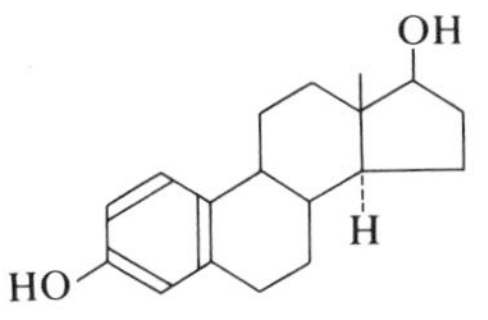

Estradiol

—primary female sex hormone.
—estrone (C-20 oxo-analogue) and estriol (16α-OH-estradiol) show less activity.
—synthetic analogues used in cases of estrogen deficiency.
—equilenin (equine female sex hormone) is A,B-naphthalenic analogue of estrone.

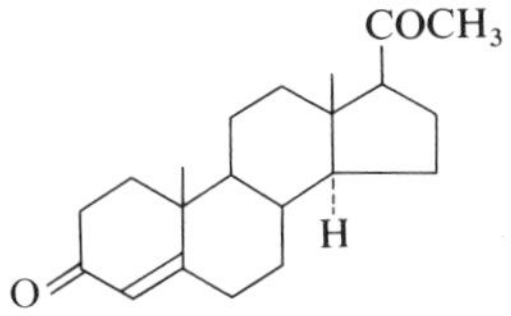

Progesterone

—female sex hormone required for maintenance of pregnancy.
—synthetic analogues used to prevent pregnancy.

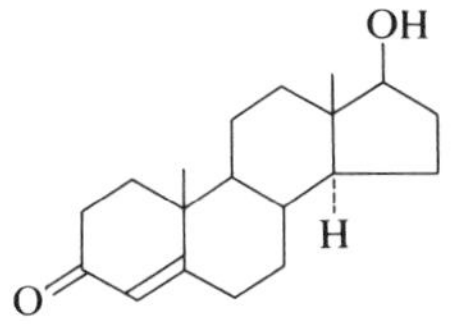

Testosterone

—most active of androgenic hormones.
—other androgenic hormones differ in oxidation state.

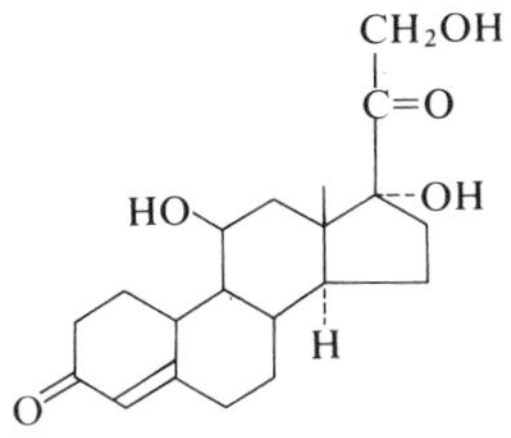

Cortisol

—most important of adrenalcorticosteroid hormones, which are vital to the maintenance of life; all natural compounds possess 3-keto-4-ene and oxygenated two carbon side chain.
—cortisone (C-11-oxo-analogue) potent anti-inflammatory.
—synthetic analogues widely used in medicine.

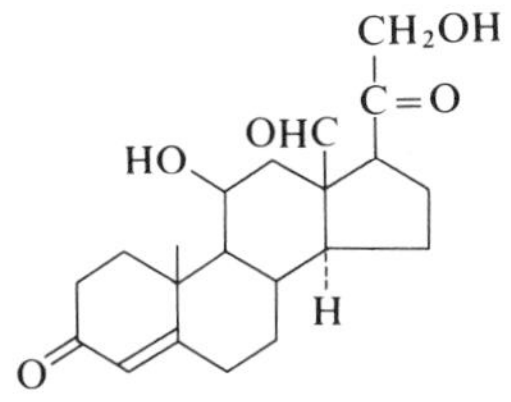

Aldosterone

—controls electrolyte balance; used in treatment of Addison's disease.
—unique structural feature is C-18 oxo-function.

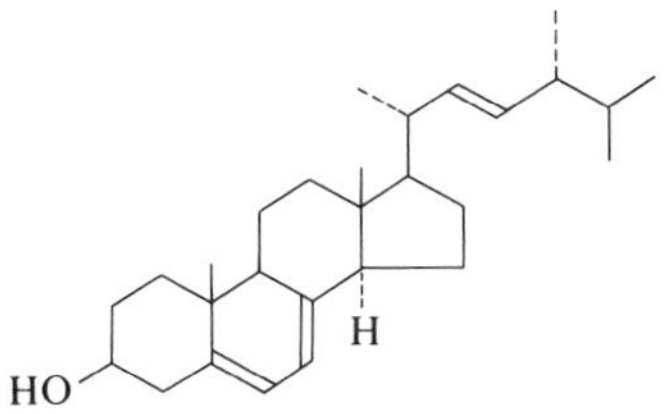

Ergosterol

—major sterol from yeast; widely distributed in nature.
—ultraviolet irradiation produces lumisterol, tachysterol and vitamin D_2 which differs from vitamin D_3 in the side chain.

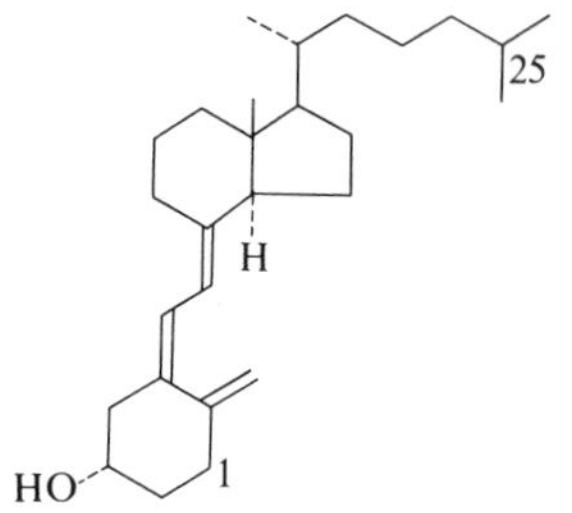

Vitamin D_3

—most physiologically active compound of the D vitamin complex[1]; produced by ultraviolet irradiation of 7-dehydrocholesterol.
—vitamin D_3 is transformed into a number of more active of metabolites[2], of which the most active is 1α,25-dihydroxyvitamin D_3, a hormone involved in Ca^{++} and phosphate transport[3-5]; synthetic analogues are being introduced for the treatment of bone diseases[6].

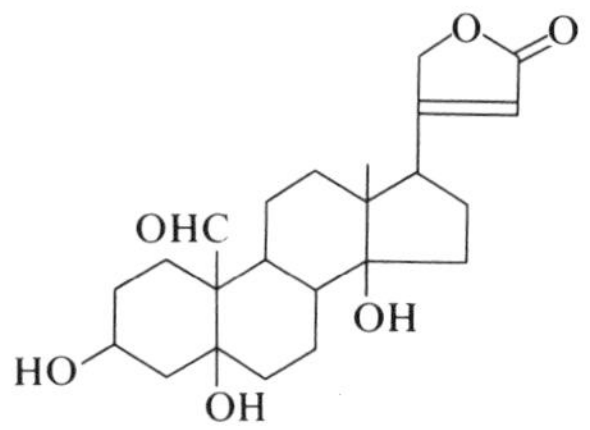

Strophanthidin

—one of the cardiac active principles occurring as the 3-*O*-glycosides in a number of plants.
—strophanthidin has unusual C-19 oxo function.
—cardiac aglycones are characterized by α,β-unsaturated side chain lactone (butenolide) and a 14β-OH group.

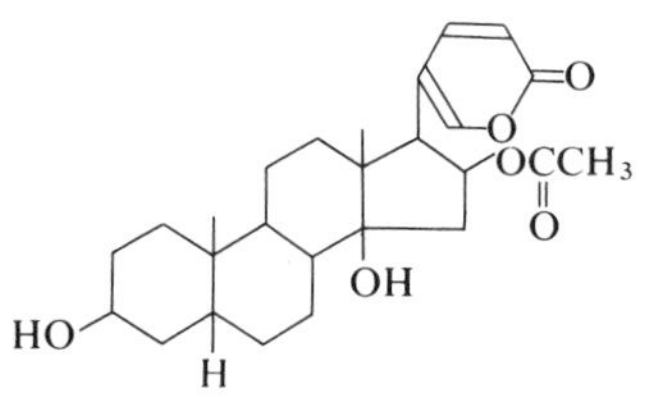

Bufotalin

—genin isolated from toads as the poisonous C-14 suberylarginine ester bufatoxin.
—characteristic bufadienolide ring is found in many glycosides of plant origin, all with 5β, 14β-configuration, many highly oxygenated.
—both genins and conjugates are cardiac active.

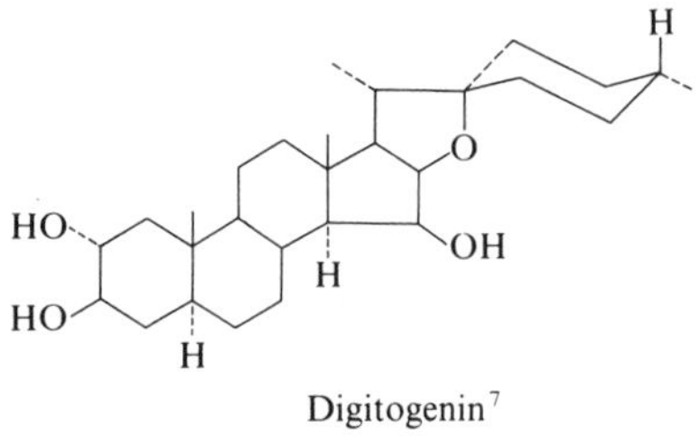

Digitogenin[7]

—the sapogenins are widely distributed in plants as saponins (*i.e.*, glycosides of digitogenin and other sapogenins).
—sapogenins all have a characteristic spiro-ketal side chain; configuration is α at C-14, variable at C-5 and C-25.
—the glycoside digitonin forms highly insoluble complexes with cholesterol and other 3β-sterols.

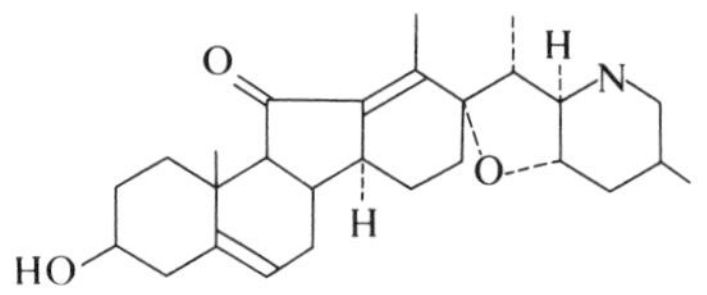

Jervine[8]

—rearranged (C-nor/D-homo) steroid skeleton added to the challenge of structure determination.
—steroid alkaloids exhibit a wide variety of structural features.

---

1 *H. F. DeLuca*, Vitamins and Hormones *25*, 315 (1967).
2 *H. F. DeLuca*, Fed. Proc. *33*, 2211 (1974).
3 *J. W. Blunt, H. F. DeLuca, H. K. Schnoes*, Biochemistry *7*, 3317 (1968); *J. W. Blunt, H. F. DeLuca, H. K. Schnoes*, Chem. Commun. 801 (1968).
4 *M. F. Holick, H. K. Schnoes, H. F. DeLuca, R. W. Gray, I. T. Boyle, T. Suda*, Biochemistry *11*, 4251 (1972).
5 *H. Lam, H. K. Schnoes, H. F. DeLuca, T. C. Chen*, Biochemistry *12*, 4851 (1973).
6 *R. G. Harrison, B. Lythgoe, P. W. Wright*, J. Chem. Soc. Perkin I 2654 (1974).
7 *C. Djerassi, T. T. Grossnickel, L. B. High*, J. Am. Chem. Soc. *78*, 3166 (1956).
8 *H. Mitsuhashi, Y. Shimizu*, Tetrahedron *19*, 1027 (1963); *T. Masamune, M. Takasugi, Y. Mori*, Tetrahedron Lett. 489 (1965).

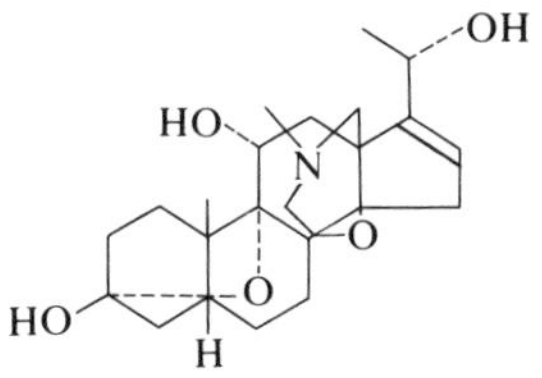

Batrachotoxinin A (see p. 11)

—potent frog venom, isolated as the 21-*O*-2,4-dimethylpyrrole-3-carboxylate.

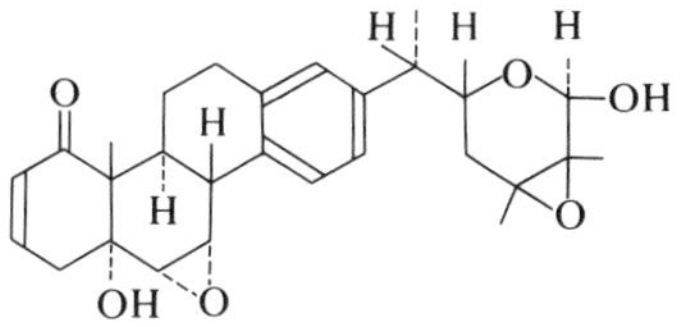

Nicandrenone (see p. 18)

—unusual D-homo aromatic structure is a feature of several recently isolated plant products.
—unusual structural features are also exhibited by the physalins (seco C/D ring junctions) and fukujusonorones (C-18 norsteroid).

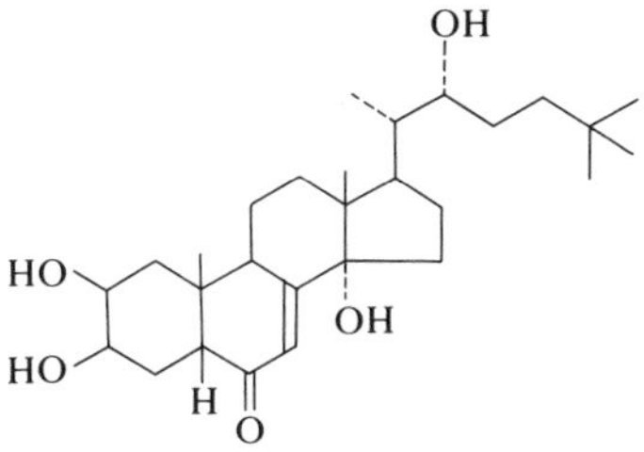

α-Ecdysone[9a,9b]

—insect moulting hormone; also isolated with similar compounds from plant sources.
—β-ecdysone (20-hydroxyecdysone; 20R, 22R)[10] is the moulting hormone of crustaceans[11] as well as insects[12].
—a total of about 45 ecdysones have been isolated from plants and animals.

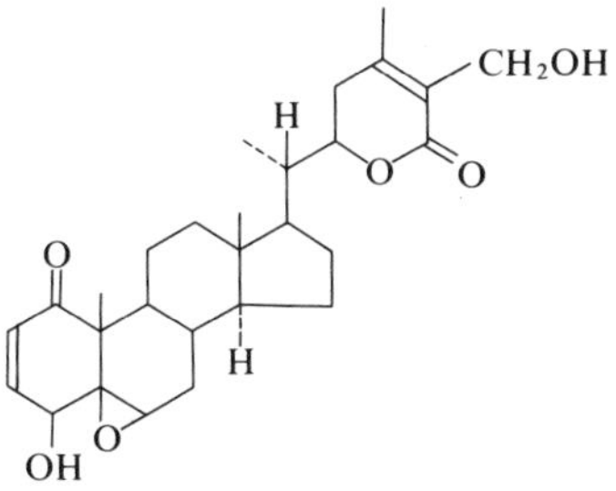

Withaferin A (see p. 19)

—*in vitro* tumor inhibitor.
—isolated with many similar compounds from plant sources.
—the C_9 side-chain with α,β-unsaturated lactone is characteristic of the class.

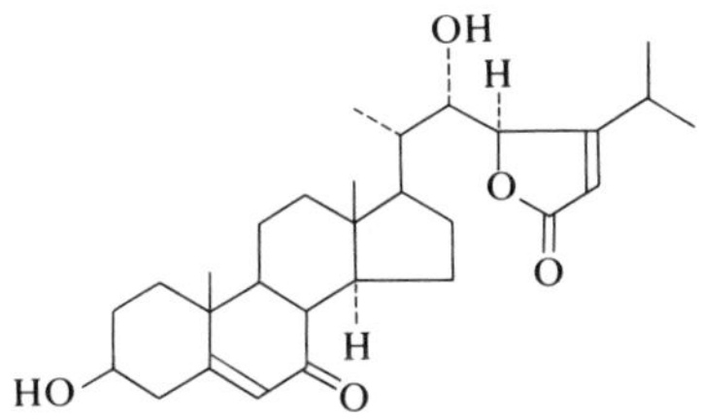

Antheridiol (see p. 14)

—sex hormone of aquatic fungus, *Achlya bisexualis.*

[9a] *D. H. S. Horn*, The Ecdysones *in* Naturally Occurring Insecticides, *M. Jacobson, D. G. Crosby* (Eds.), Marcel Dekker, New York 1971.

[9b] *C. Rufer, H. Hoffmeister, W. Hoppe, R. Huber*, Chem. Ber. *98*, 2353 (1965); *R. Huber, W. Hoppe*, Chem. Ber. *98*, 2403 (1965).

[10] *M. Koreeda, D. A. Schooley, K. Nakanishi, M. Hagiwara*, J. Am. Chem. Soc. *93*, 4084 (1971).

[11] *F. Hampshire, D. H. S. Horn*, Chem. Commun. 37 (1966).

[12] *D. H. S. Horn, E. J. Middleton, J. A. Wunderlich, F. Hampshire*, Chem. Commun. 339 (1966).

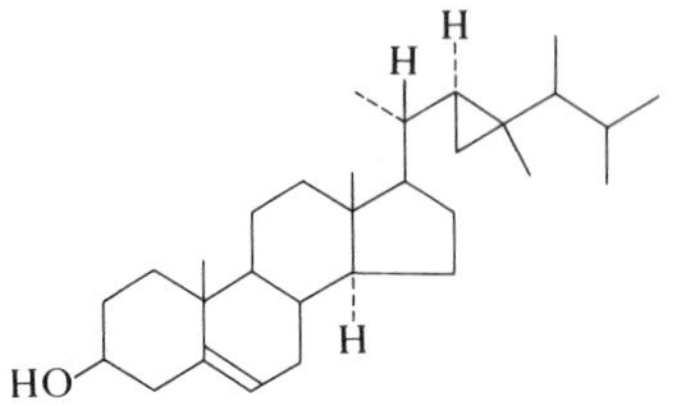

Gorgosterol (see p.12)

—isolated from marine coelenterates.
—unique structural feature is C-22/C-23 cyclopropane ring; this moiety is also present in the marine sterols 23-demethylgorgosterol[13] and acanthasterol[14].

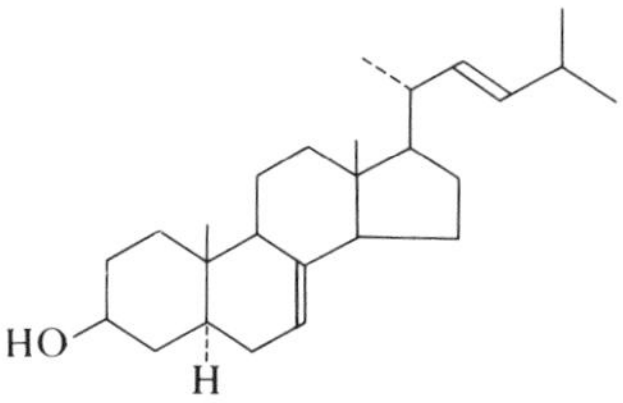

Asterosterol[15]

—one of a large class of marine sterols with normal tetracyclic nucleus and side chains ranging from C_7 (a C_{26} steroid, *e.g.*, asterosterol) to C_{11} (C_{30} steroids).
—the Δ-7 unsaturation is a common feature of marine sterols.

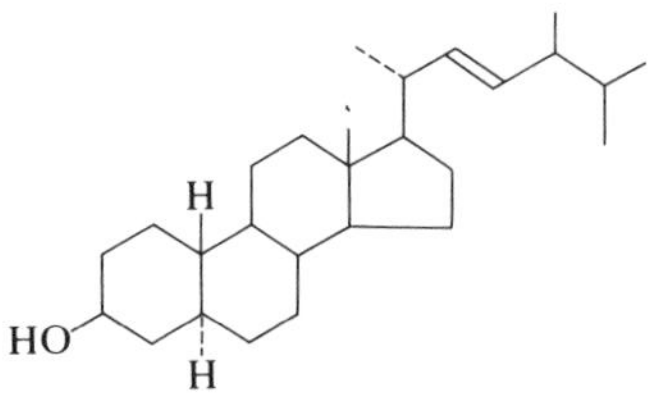

19-nor-5α,10β-ergost-trans-22-en-3β-ol[16]

—absence of 19-Me group is unique among ring A/B non-aromatic steroids.
—a homologous series of 19-norcholestanols, differing in side chain alkylation, has been isolated from the sponge, *axinella polyploides*.

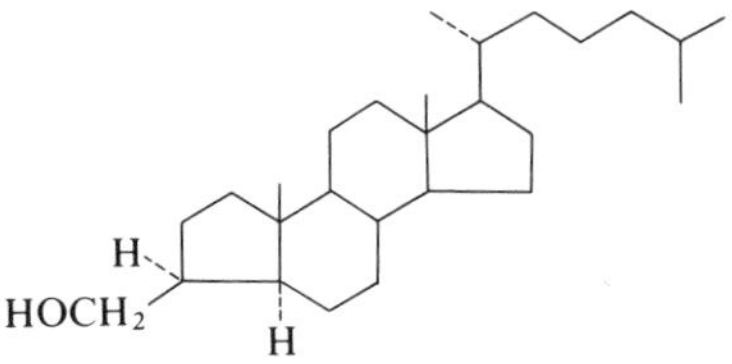

3-hydroxymethyl-A-nor-5α-cholestane[17]

—major sterol component of sponge *axinella nerrucosa*.
—contracted A ring is unusual feature of homologous series of cholestanols from the same source.

General References
[18] *L. F. Fieser, M. Fieser*, Steroids, Reinhold Publishing Corp., New York 1959.
[19] *R. H. F. Manske* (ed.), The Alkaloids, Vols. VII, IX, X, XIV, Academic Press, New York.
[20] *P. J. Scheuer*, Chemistry of Marine Natural Products, Academic Press, New York 1973.
[21] Specialist Periodical Reports, Terpenes and Steroids, The Chemical Society, London.
[22] *K. Nakanishi, T. Goto, S. Ito, S. Natori, S. Nozoe* (Eds.), Natural Products Chemistry, Vol. I, p. 421–545, Kodansha, Tokyo/Academic Press, New York 1975

[13] *E. L. Enwall, D. van der Helm, I. Nan Hsu, T. Pattabhiraman, F. J. Schmitz, R. L. Spraggins, A. J. Weinheimer*, Chem. Commun. 215 (1972).
[14] *K. C. Gupta, P. J. Scheuer*, Tetrahedron *24*, 5831 (1968).
[15] *M. Kobayashi, R. Tsuru, K. Todo, H. Mitsuhashi*, Tetrahedron Lett. 2935 (1972); *M. Kobayashi, R. Tsuru, K. Todo, H. Mitsuhashi*, Tetrahedron *29*, 1193 (1973).
[16] *L. Minale, G. Sodano*, J. Chem. Soc., Perkin I 1888 (1974).
[17] *L. Minale, G. Sodano*, J. Chem. Soc., Perkin I 2380 (1974).

2.1 Biosynthetic Outline of Steroids

Mammalians and plants synthesize steroids from acetic acid, whereas insects lack the enzymes for steroid biosynthesis and thus depend on plant C_{28} and C_{29} steroids (phytosteroids) as a cholesterol source. The overall picture is shown in Fig. 1.

2.2 Biosynthesis of Cholesterol and Cycloartenol

The mechanism of cholesterol biosynthesis in mammalians has been clarified in detail through numerous elegant studies employing precursors labeled with ^{14}C and/or ^{3}H[1]. Mevalonic acid **1**, first isolated and characterized from dried distiller solubles[2] and shown to possess the R configuration at C-3[3] (Fig. 2), is the key intermediate for all terpenoid compounds including cholesterol. Mevalonic acid (MVA) has six prochiral hydrogens at C-2, C-4 and C-5; all six hydrogens have now been stereospecifically replaced by deuterium or tritium, and these labeled MVA's (in some cases containing an additional ^{14}C label at C-2 or other carbons) have played a vital role in defining the biosynthetic fate of MVA prochiral hydrogens[1d] as well as the construction of the steroidal skeleton (the prochiral hydrogens are not shown in Fig. 2).

MVA is pyrophosphorylated in two steps by the action of kinase to give **2**, which then decarboxylates to isopentenyl pyrophosphate (PP) **3**. An isomerase converts **3** to γ,γ-dimethylallyl (PP) **4**. The PP group of **4** serves as a leaving group when reacted by **3a**, thus giving the C_{10} geranyl PP **5**, the monoterpenoid precursor. The

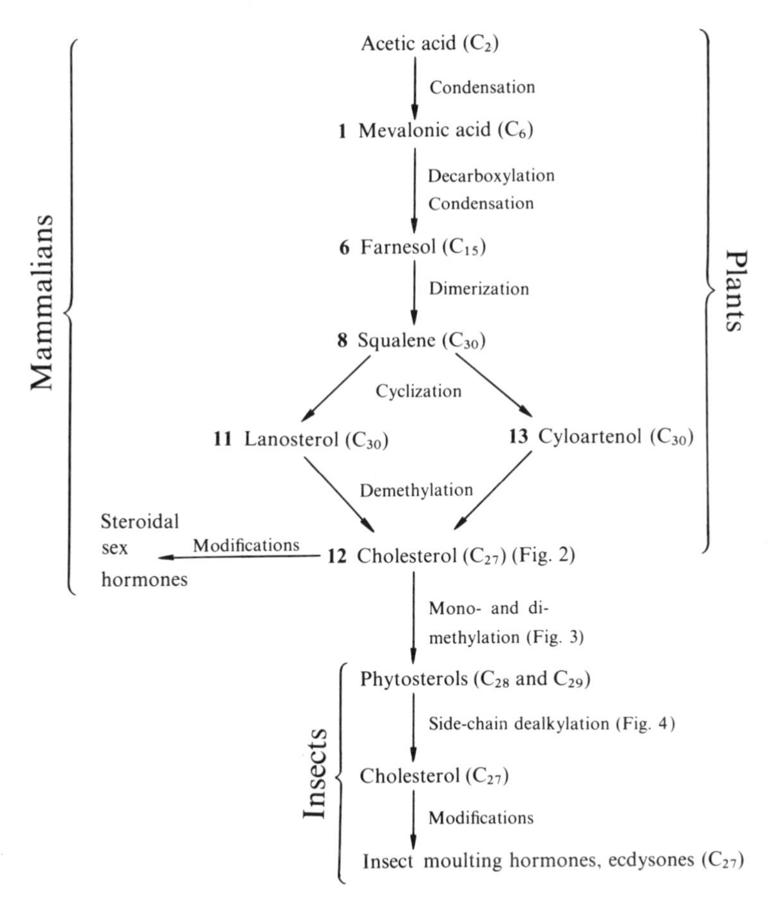

Fig. 1. Outline of steroid biosyntheses.

[1] Reviews: a) *R. B. Clayton*, Quart. Rev. *19*, 168, 201 (1965); b) *J. W. Cornforth*, Quart. Rev. *23*, 125 (1969); c) *L. J. Mulheirn, P. J. Ramm*, Chem. Soc. Rev. *1*, 259 (1972); d) *K. Nakanishi in* Natural Products Chemistry, Vol. I, p. 505, Kodansha, Tokyo/Academic Press, New York 1974.

[2] *D. E. Wolff, C. H. Hoffman, P. E. Aldrich, H. R. Skegg, L. D. Wright, K. Folkers*, J. Am. Chem. Soc. *78*, 4499 (1956); *79*, 1486 (1959).

[3] *M. Eberle, D. Arigoni*, Helv. Chim. Acta *43*, 1508 (1960).

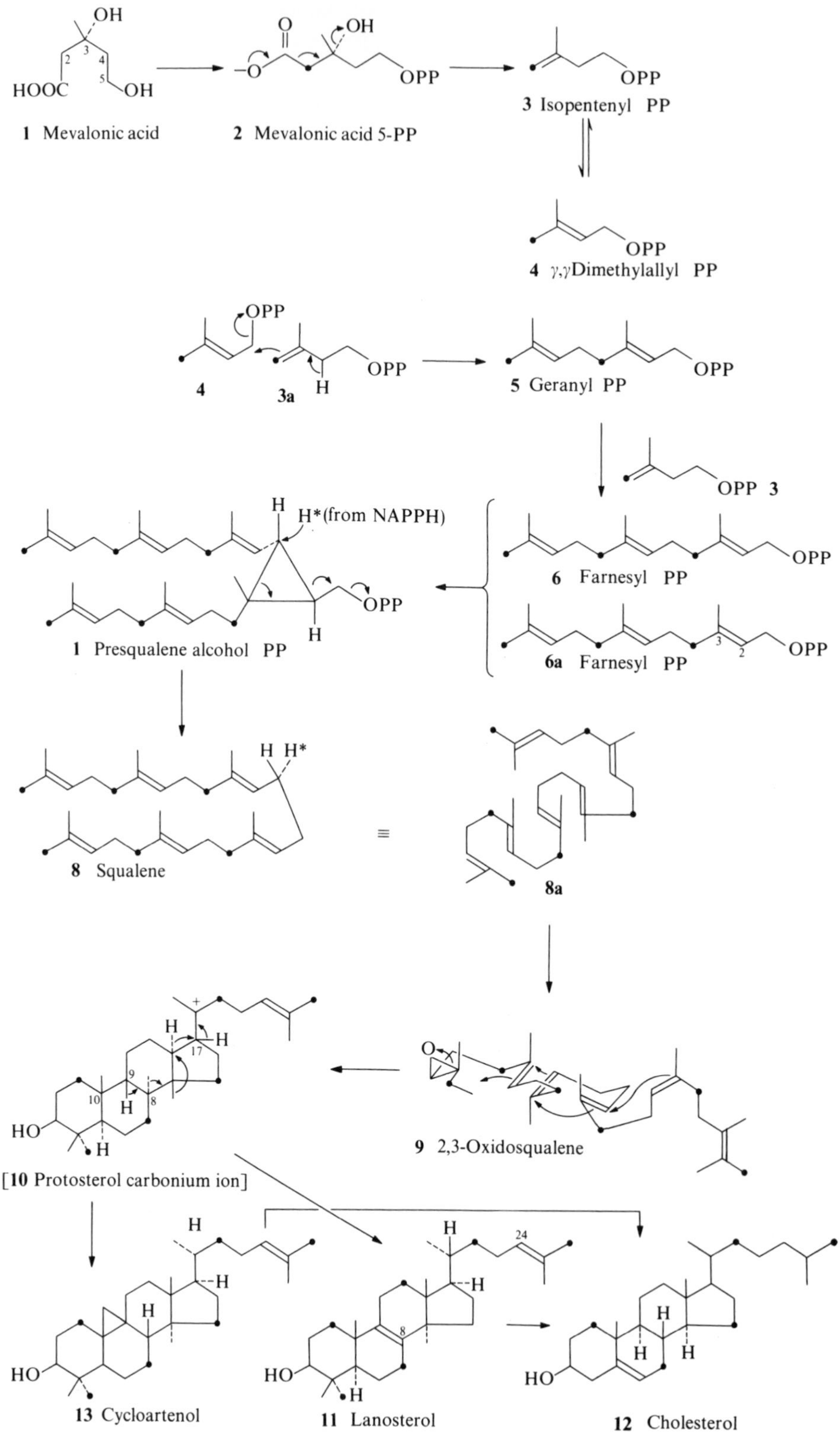

OH
2 3 4 5
HOOC OH
1 Mevalonic acid

O OH
$-O$ OPP
2 Mevalonic acid 5-PP

OPP
3 Isopentenyl PP

OPP
4 γ,γDimethylallyl PP

OPP
4 **3a** H OPP

OPP
5 Geranyl PP

OPP **3**

H H*(from NAPPH)
OPP
H
1 Presqualene alcohol PP

OPP
6 Farnesyl PP

3 2 OPP
6a Farnesyl PP

H H*
8 Squalene

≡

8a

H + H
17
9
10 H 8
HO H
[**10** Protosterol carbonium ion]

O
9 2,3-Oxidosqualene

H
H
H
HO
13 Cycloartenol

H 24
H
8
HO H
11 Lanosterol

H
H H
HO
12 Cholesterol

Fig. 2. Biosynthesis of cholesterol and cycloartenol.

Carbon atoms originating from C–2 of mevalonic acid are denoted by black circles; PP stands for pyrophosphate group.

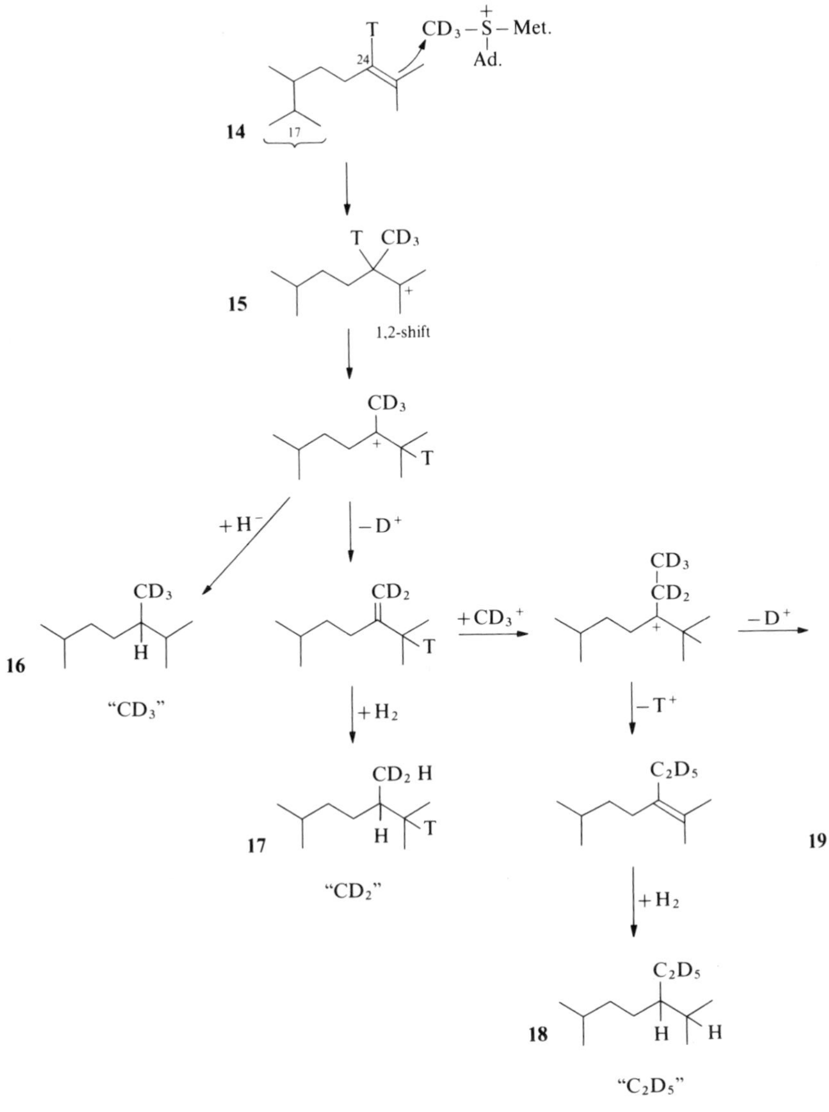

Fig. 3. Side-chain alkylation mechanisms.

condensation between **5** and another molecule of isopentenyl PP **3** affords the precursor for sesquiterpenoids (C_{15}), farnesyl PP **6**. In all these condensations involving the addition of a C_5 unit, the hydrogen which is eliminated *i.e.*, the H shown in structure **3a**, corresponds to the 4-prochiral-S hydrogen in MVA **1**.

Two farnesyl PP units **6** and **6a** condense in a manner to form a cyclopropyl group between C-1 of one unit and C-2/C-3 of the other; this yields the elusive and unique intermediate presqualene alcohol pyrophosphate **7**[4]. Again the hydrogen lost from C-1 of one of the farnesyl pyrophosphate units **6** in the cyclopropyl formation is the 4-pro-*S* hydrogen of MVA **1**. The cyclopropyl ring is cleaved by uptake of 4-pro-*S* hydrogen of the reduced coenzyme NADPH[5] as depicted by arrows in formula **7**, and this yields squalene **8** or **8a**.

[4] *H. C. Rilling, W. W. Epstein*, J. Am. Chem. Soc. *91*, 1041 (1971); *L. J. Altman, R. C. Kowerski, H. C. Rilling*, J. Am. Chem. Soc. *93*, 1782 (1971); *H. C. Rilling, C. D. Poulter, W. W. Epstein, B. Larsen*, J. Am. Chem. Soc. *93*, 1783 (1971); *R. M. Coates, W. H. Robinson*, J. Am. Chem. Soc. *93*, 1785 (1971); *G. Popják, J. Edmond, S.-M. Wong*, J. Am. Chem. Soc. *95*, 2173 (1973).

[5] *G. Popják, DeW. S. Goodman, J. W. Cornforth, R. H. Cornforth, R. Ryhage*, J. Biol. Chem. *236*, 1934 (1961).

The next key intermediate is 2,3-oxidosqualene **9**[6] which undergoes a series of cyclizations triggered by epoxide cleavage leading to the hypothetical protosterol carbonium ion (loss of 17-H to form a 17,20-ene and loss of 4β-Me gives the skeleton of fusidic acid[7], an antibiotic). The 1,2-migrations and loss of 9-H shown in **10** lead to the tetracyclic triterpene lanosterol, which in turn is converted into cholesterol after loss of three methyl groups[8], saturation of 8-ene and 24-ene, and formation of 5-ene[1].

If instead of a 9-H loss in **10**, the 9-H migrates to C-8 and the 10-Me forms a cyclopropane ring, then the product is cycloartenol **13**, which has been shown to replace lanosterol as the principal triterpene in some plants and algae[1c,9].

2.3 Side-Chain Alkylation Leading to Phytosteroids

Although the C_{27} cholesterol was found to be present in plants (algae) first in 1958[10], the majority of phytosteroids contain a 24-Me or 24-Et branching. Intensive biosynthetic studies with deuterated and tritiated precursors have disclosed that the extra carbons are derived from *S*-adenosylmethionine and that four mechanisms "CD_3" "CD_2" "C_2D_5" and "C_2D_4" are operating as summarized in Fig. 3[11]. For example it has been demonstrated that the ^3H in the 24-ene **14** had migrated to C-25 in sterols **17** (C_{28}) and **19** (C_{29})[11], whereas it was absent in another C_{29} sterol **18**[12].

2.4 Side-Chain Dealkylation in Insects

20 α-Ecdysone R = H
21 β-Ecdysone R = OH

22 Cortexone

The insect moulting hormones α- and β-ecdysones **20**, **21** (C_{27}) and the defense secretion of water beetles exemplified by cortexone **22** (C_{21})[13] are steroids. Nevertheless, insects cannot biosynthesize the steroid skeleton and rely on plants as the major source for steroids. The dealkylation mechanisms of C_{29} or C_{28} steroids to C_{27}, C_{21}[14] or C_{19} steroids thus play a significant role in the insect life. The path for the metabolic removal of C-28 and C-29 in phytosterols leading to cholesterol is shown in Fig. 4. Desmosterol **28** appears to be a common intermediate in the conversion of β-sitosterol and campesterol **24** into cholesterol by various insects[15]. The intermediacy of epoxide **27** was shown by trapping experiments using silk-worms (*Bombyx mori*) fed with tritiated steroids[16]; in another experiment with 25-T isofucosterol **30** it was found that ^3H was retained in the final cholesterol, and therefore a mechanism via **31** and **32** was postulated (in the yellow mealworm, *Tenebrio molitor*)[17]. In view of the intermediacy of epoxide **27**, the X in **31**/**32** is presumably oxygen. The epoxide configuration in **27** has been shown to be 24S and 28S[18].

[6] *E. J. Corey, W. E. Russey, P. R. Ortiz de Montellano*, J. Am. Chem. Soc. *88*, 4750 (1966); *E. E. van Tamelen, J. D. Willett, R. B. Clayton, K. E. Lord*, J. Am. Chem. Soc. *88*, 4752 (1966); *E. E. van Tamelen*, Acc. Chem. Res. *1*, 111 (1968).

[7] *W. O. Godtfredsen, H. Lorck, E. E. van Tamelen, J. D. Willett, R. B. Clayton*, J. Am. Chem. Soc. *90*, 208 (1968).

[8] *K. Alexander, M. Akhtar, R. B. Boar, J. F. McGhie, D. H. R. Barton*, Chem. Commun. 383 (1972).

[9] *L. J. Goad, T. W. Goodwin*, Eur. J. Biochem. *7*, 502 (1969); *H. H. Rees, L. J. Goad, T. W. Goodwin*, Biochem. Biophys. Acta *176*, 892 (1969); *G. Ponsinet, G. Ourisson*, Phytochemistry *7*, 757 (1968); *R. Heinz, P. Benveniste*, Phytochemistry *9*, 1499 (1970).

[10] *K. Tsuda, S. Akagi, Y. Kishida, R. Hayatsu, K. Sakai*, Pharm. Bull. Japan *6*, 724 (1958).

[11] Review: *E. Lederer*, Quart. Rev. *23*, 453 (1969).

[12] *P. J. Randall, H. H. Rees, T. W. Goodwin*, Chem. Commun. 1295 (1972).

[13] *H. Schildknecht*, Angew. Chem. Int. Ed. *9*, 1 (1970).

[14] *F. J. Ritter, W. H. J. M. Wientjens*, Steroid Metabolism of Insects, p. 67, Central Laboratorium TNA, Delft 1967.

[15] *J. A. Svoboda, M. J. Thompson, W. E. Robbins*, Lipids *7*, 156 (1972); *F. J. Ritter, W. H. J. M. Wientjens*, TNO-Nieuws *22*, 381 (1967); *J. P. Allais, M. Barbier*, Experientia *27*, 507 (1971).

[16] *M. Morisaki, H. Ohtake, M. Okubayashi, N. Ikekawa*, Chem. Commun. 1275 (1972).

[17] *P. J. Randall, J. G. Lloyd-Jones, I. F. Cook, H. H. Rees, T. W. Goodwin*, Chem. Commun. 1296 (1972).

[18] *S. L. Chen, K. Nakanishi, N. Awata, M. Morisaki, N. Ikekawa, Y. Shimizu*, J. Am. Chem. Soc. **97**, 5297 (1975).

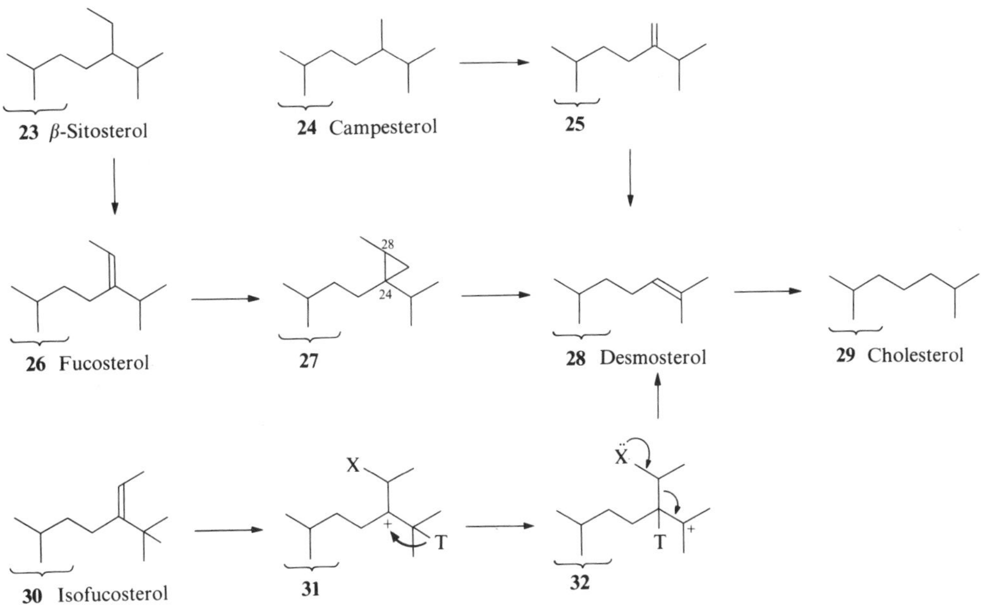

23 β-Sitosterol
24 Campesterol
25
26 Fucosterol
27
28 Desmosterol
29 Cholesterol
30 Isofucosterol
31
32

Fig. 4. Side-chain dealkylation in insects.

3 Some Unusual Structures

The fascinating array of steroidal natural products continues to unfold. Recent discoveries and structural elucidations encompass several novel classes of steroids as well as compounds exhibiting interesting biological properties. The trend continues towards structure studies on increasingly small amounts of material, especially in the case of biologically active compounds.

A wide range of methods has been employed in recent structure determinations, ranging from classical selenium dehydrogenation to a widespread use of X-ray analysis and further sophistication of spectroscopic methods such as carbon magnetic resonance.

3.1 Batrachotoxin

The potent venom of the Columbian arrow poison is extracted from the skins of an endemic species of frogs, *Phyllobates lobates aurotaenia*. Difficulties associated with the structure determination were compounded by the inaccessability of the natural material, in quest of which four expeditions were sent into the dense jungle of the Choco region of western Columbia. Despite the relatively small amount of labile material, four steroidal alkaloids were isolated from the crude venom, batrachotoxin **2**, batrachotoxinin A **1**, pseudobatrachotoxin **3** (which readily isomerized to batrachotoxinin A) and

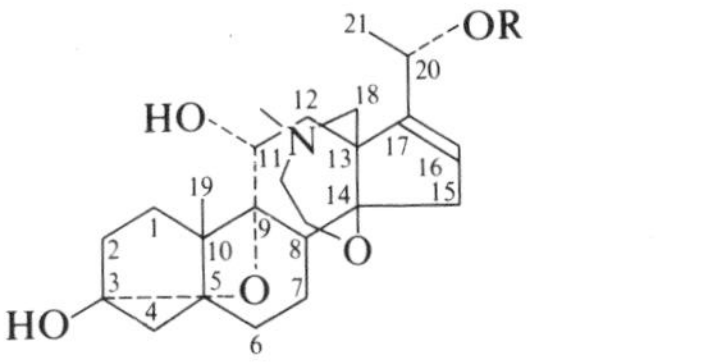

1 Batrachotoxinin A, R = H, $C_{24}H_{35}NO_5$

2 Batrachotoxin, R =

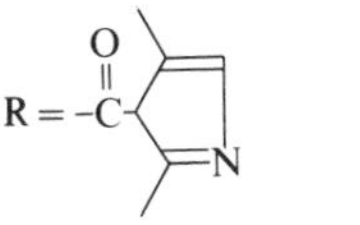

4 Homobatrachotoxin, R =

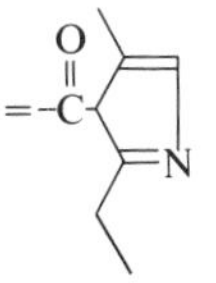

homobatrachotoxin **4** (previously designated isobatrachotoxin)[1]. The structure of pseudobatrachotoxin has not yet been established.

The structure of batrachotoxinin A **1** was elucidated by X-ray diffraction analysis of the *p*-bromobenzoate ester[2]. As the heavy atom method was not applicable, the structure was solved by locating the bromine and oxygen atoms and then using the *p*-bromobenzoate moiety as a known partial structure.

[1] *T. Tokuyama, J. Daly, B. Witkop, I. L. Karle, J. Karle*, J. Am. Chem. Soc. *90*, 1918 (1968).

[2] *T. Tokuyama, J. Daly, B. Witkop*, J. Am. Chem. Soc. *91*, 3931 (1969).

The known structure of batrachotoxinin A **1** led to the elucidation of the structures of batrachotoxin **2** and homobatrachotoxin **4**. The mass spectra of both substances showed peaks which were not present in the spectra of batrachotoxinin A and its *p*-bromobenzoate and these suggested the presence of an additional moiety in **2** and **4**. The largest fragment attributed to this moiety in the spectrum of **2** was *m/e* 139 ($C_7H_9NO_2$) and in that of **4** was *m/e* 153 ($C_8H_{11}NO_2$). Together with the infra-red absorption at 1690 cm^{-1}, ultraviolet absorption at 234 and 264 nm and a strong positive Ehrlich reaction[3], the mass spectral evidence suggested that **2** and **4** were dimethylpyrrolecarboxylate and ethylmethylpyrrolecarboxylate esters of batrachotoxinin A **1**. Careful examination of the NMR spectra for **2** and **4** supported the conclusion that the skeleton was the same as that of batrachotoxinin A. The C-20 proton of the latter (δ 4.57, quartet) is missing in **2** and **4**, and additional signals appear in the region δ 5–7, which are assigned to the aromatic proton of the dialkylpyrrolecarboxylate (δ *ca.* 6.35) and the C-20 proton (δ *ca.* 5.8).

Batrachotoxin was prepared by the reaction of batrachotoxinin A and the mixed ethylchloroformate anhydride of 2,4-dimethylpyrrole-3-carboxylic acid, and was identical with the natural material by comparison of TLC, color reaction with modified Ehrlich's reagent (*p*-dimethylaminocinnaldehyde)[3], toxicity, NMR ($CDCl_3$ and C_6D_6) and mass spectrometry.

These potent venoms (LD_{50} for batrachotoxin is 1 mg/kg in mice) appear to act by interference in the control of ion transport across membranes, in addition to being extremely active cardiotoxins. The toxicity of batrachotoxinin A, which is probably a secondary product arising from decomposition of pseudobatrachotoxin, is about 1/500 of the toxicity of the original venom. Structurally, these substances exhibit several features not encountered previously in naturally occurring pregnane derivatives; these include the $3\alpha,9\alpha$ oxido bridge and 3β hemiacetal, the seven-membered $14\beta,18\beta$ nitrogen containing ring, the Δ^{16} site of unsaturation and the dialkylpyrrole ester moiety.

A formal total synthesis of bachtrachotoxinin A has been achieved by the partial synthesis of the complex alkaloid from 3,20-dioxo-$7\alpha,9\alpha$-dihydroxy-11α,18-diacetoxy-Δ^4-pregnane[4].

3.2 Gorgosterol

Gorgosterol has been isolated from several marine coelenterates[1]. It is the first sterol isolated with a side-chain cyclopropane ring and also the first sterol with carbon substitution at positions 22 and 23.

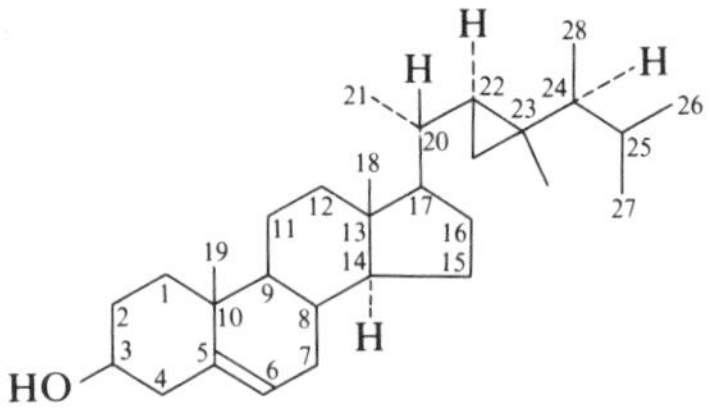

$C_{30}H_{50}O$, gorgosterol **1**
mp. 186.5–188° $[\alpha]_D$ −450°

Although mass spectrometry showed gorgosterol to be a C_{30} sterol, it became apparent that the extra methyl group was not situated at C-4 or C-14 as with other C_{30} sterols. Moffatt oxidation of **1** gave a non-UV-absorbing ketone **2**, which was readily isomerized with base to a ketone **3** (λ_{max} 242 nm, ε 17,000) whose ultraviolet absorption spectrum precluded substitution at C-4. The ORD spectrum of **3** was characteristic of a Δ^4-3-keto steroid. The possibility of substitution at C-14 was eliminated by examination of the mass spectrum of gorgostane **4** which was prepared from **1** by a sequence of hydrogenation, oxidation and Wolff-Kishner reduction. The mass spectrum of **4** showed peaks associated with the normal steroid nucleus; C-14 methyl substitution would have resulted in 14 mass unit shifts of these peaks.

The presence of the cyclopropane ring was suggested by high field absorption (δ −0.13, 0.06–0.37, 0.44) in the NMR spectrum. Opening of the cyclopropane ring in dihydrogorgosterol acetate **5** by treatment with hydrochloric acid in

[3] *L. R. Morgan, R. Schunior,* J. Org. Chem. *27*, 3696 (1962).

[4] *R. Imhof, E. Gössinger, W. Graf, L. Berner-Fenz, H. Berner, R. Schaufelberger, H. Wehrli,* Helv. Chim. Acta *56*, 139 (1973).

[1] *W. Bergmann, M. J. McLean, D. J. Lester,* J. Org. Chem. *8*, 271 (1943); *L. S. Ciereszko, M. A. Johnson, R. W. Schmidt, C. B. Koons,* Comp. Biochem. Physiol. *24*, 899 (1968); *K. C. Gupta, P. Scheuer,* Steroids *13*, 343 (1969).

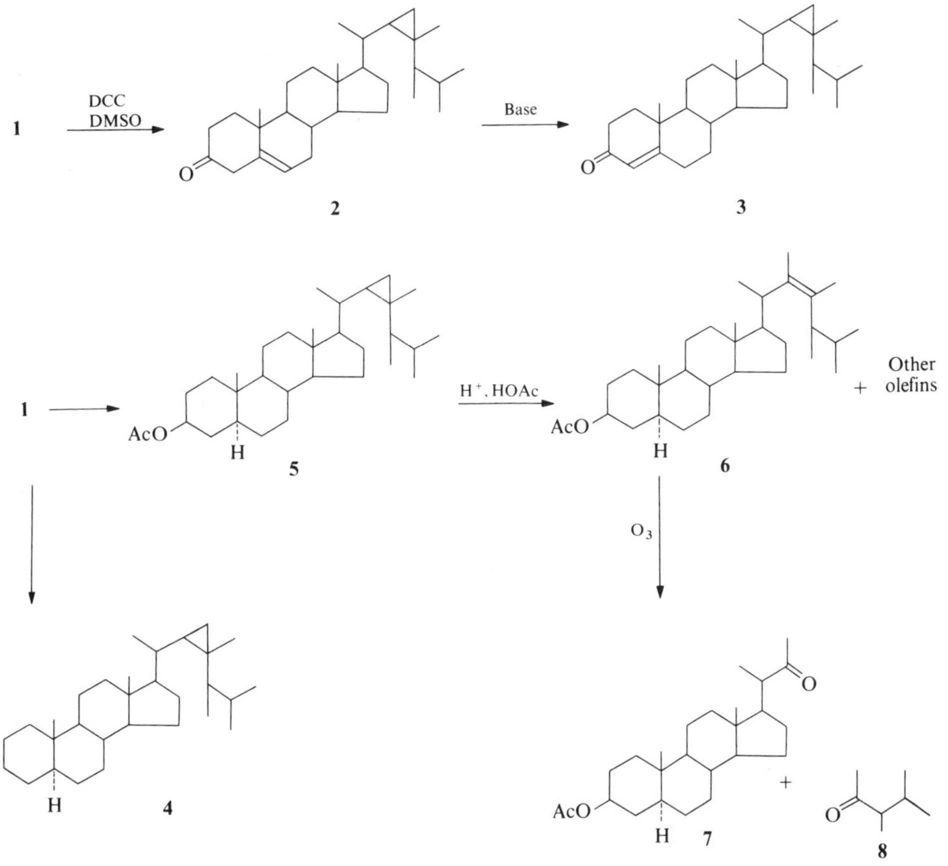

1 DCC DMSO → 2 Base → 3

1 → 5 H⁺, HOAc → 6 + Other olefins

AcO H 5 AcO H 6

O_3

H 4 AcO H 7 + O 8

acetic acid gave a mixture of tetra-substituted olefins. Subsequent ozonolysis yielded **7** and **8** and thus showed one of the olefins to be **6**. At this point no distinction could be made between the possible side-chain structures **9** and **10**[2].

The position of the side-chain cyclopropane ring and the side-chain stereochemistry was de-

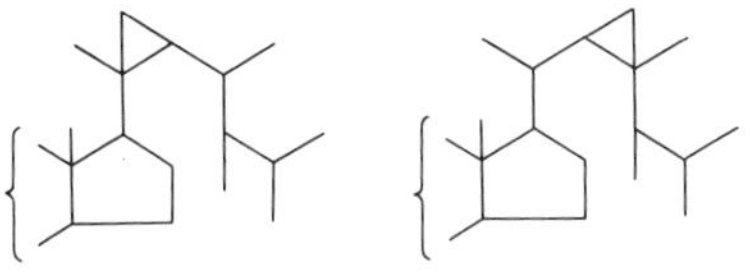

9 10

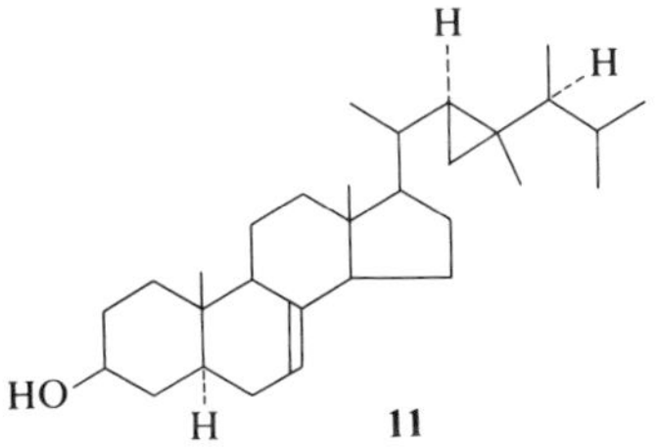

H H

HO H 11

termined by X-ray diffraction analysis of 3*β*-bromogorgostene prepared by treatment of gorgosterol with aluminum bromide in ether. The absolute configuration was derived from the sign of the Cotton effect (+) of the CD curve of 5α-gorgostan-3-one and of the ORD curve of Δ^4-gorgostenone **3**[3].

More recently, acansterol **11**, another C-30 cyclopropane containing sterol isolated from *Acanthaster planci*, has been found to have structure **11**, differing from gorgosterol in the placement of the double bond. The assignment was confirmed by the conversion of **11** to gorgostane **4**, identical with that derived from gorgosterol[4].

A C-29 steroid, calysterol, whose side-chain contains the rare cyclopropene group, has been isolated from the sponge *calyx nicaensis*[5].

---

[2] *R. L. Hale, J. Leclerq, B. Tursch, C. Djerassi, R. A. Gross, Jr., A. J. Weinheimer, K. Gupta, P. J. Scheuer*, J. Am. Chem. Soc. *92*, 2179 (1970).

[3] *N. C. Ling, R. L. Hale, C. Djerassi*, J. Am. Chem. Soc. *92*, 5281 (1970).

[4] *Y. M. Sheikh, C. Djerassi, B. M. Tursch*, Chem. Commun. 217 (1971).

[5] *E. Fattorusso, S. Magno, L. Mayol, C. Santacroce, D, Sica*, Tetrahedron *31*, 1715 (1975).

3.3 Antheridiol

Antheridiol is one of the substances governing sexual reproduction in the aquatic fungus *Achlya bisexualis*[1]. Secreted by the female plant, it shows an extremely high and specific activity, inducing formation of antheridial hyphae on the male plant in concentrations as low as 2×10^{-8} mg/ml[2]. (For a related compound see Ref.[8]).

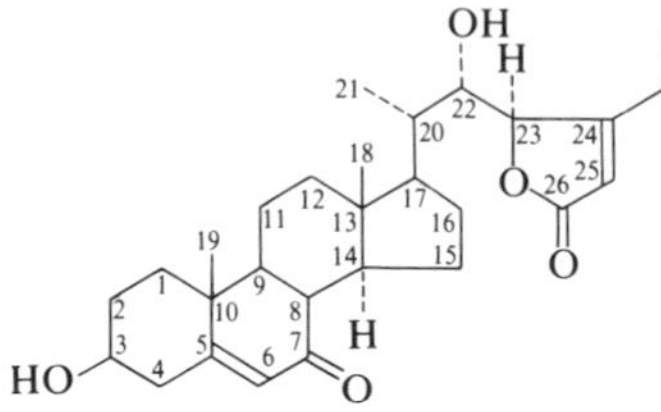

1

Antheridiol, $C_{29}H_{42}O_5$
mp. 250–255° (decomp.)

Because of the scarcity of material early investigation relied heavily on mass spectra using microgram quantities. High resolution mass spectra indicated a molecular formula $C_{29}H_{42}O_5$ and the abundance and distribution of ions corresponding to fragments $C_{19}H_{23}$, $C_{19}H_{25}O$, $C_{19}H_{27}O_2$, $C_{21}H_{30}O$, and $C_{22}H_{30}O_2$ suggested a steroid with two oxygens in the nucleus. Two nuclear double bonds were suggested by the abundant $C_{19}H_{27}O_2$ ion, assumed to result from simple C-17, C-20 cleavage of a side-chain of ten carbons and two oxygens; the position of an oxygen atom at C-22 was suggested by the base peak corresponding to the frament $C_{22}H_{32}O_3$.

Hydrogenation showed two carbon-carbon double bonds, one each in the nucleus and the side-chain. An infra-red spectrum of the tetrahydro derivative **2** showed that the carbonyl bands had shifted from 1672 and 1742 cm^{-1} in **1** to 1709 and 1770 cm^{-1} in **2** and led to the formulation of an α,β unsaturated ketone in the nucleus and an α,β unsaturated δ-lactone in the side-chain. Acetylation showed two hydroxyl groups, again one in the nucleus and one in the side-chain at C-22. The homoallylic relationship of the C-22 hydroxyl was also deduced from mass spectral fragmentation.

The relationship of the nuclear hydroxy and enone functions is indicated from the transformation of **1** to **3**, evidenced by ultraviolet and mass spectral data. The positions of hydroxyl and carbonyl groups at C-3 and C-7 was shown by the fragmentation pattern of the ethylene ketal of **2**. At this point a distinction between the possible side-chain structures **4** and **5** was made on the basis of the abundant $C_7H_{11}O_2$ ion (*m/e* 127) resulting from C-22, C-23 cleavage in **2**. This would not be expected to occur in tetrahydro **5** since it would result in an unstabilized primary carbonium ion.

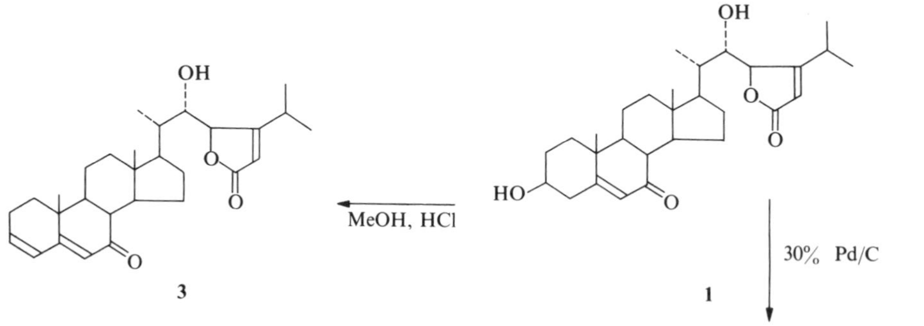

3 ← MeOH, HCl — **1** — 30% Pd/C → **2**

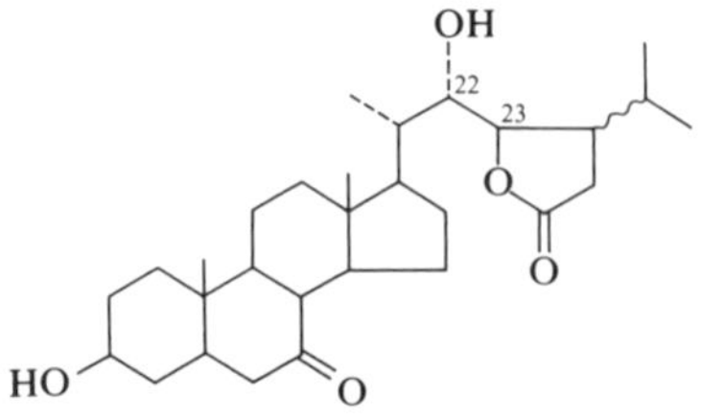

[1] *T. C. McMorris, A. W. Barksdale*, Nature *215*, 320 (1967).
[2] *G. P. Arsenault, K. Biemann, A. W. Barksdale, T. C. McMorris*, J. Am. Chem. Soc. *90*, 5635 (1968).

An NMR spectrum was taken on 4 mg of **1** when more material became available and was in agreement with the structure chosen. In particular, spin decoupling showed the C-22 proton, appearing as a broad doublet at 3.60 δ (spacing of 8 Hz), to be coupled with the single C-23 proton at 4.94 δ.

Confirmation of the structure, and elucidation of C-22, C-23 stereochemistry was achieved through synthesis. The first synthesis from **6** gave a small amount of antheridiol with its 22,23-isomer[3]. A second synthesis from the same starting material provided the four possible 22, 23 epimers **11**→**14**, of which **11** was converted to antheridiol. The *threo* isomers **12** and **14** were also formed from **15** by oxidation with osmium tetroxide followed by removal of the 5,6-epoxide by zinc reduction. The configurations at C-23 of **11**→**14** were determined by comparison of the CD curves with those for cyclogranisolide **16**, of known absolute configuration[5], and those at C-22 followed from the *cis*-hydroxylation to give **10** and **12**. Conversion of the 22, 23R butenolide **11** to antheridiol established the side-chain stereochemistry. An alternate synthesis of antheridiol has been published[6,7].

$C_{19}H_{27}O_2$ **4** $C_{19}H_{27}O_2$ **5**

6 → 1. (2-lithio-3-isopropylfuran) 2. Ac_2O, Py → **7** → $PhCO_3H$(3eq) → **8** → $NaBH_4$ → **9** → 1. Zn, NaI, HOAc 2. 5% H_2SO_4 → **10** → 1. O_2, Py, hematoporphorin 2. $Cu(OAc)_2$ → **1**

[3] *J. A. Edwards, J. S. Mills, J. Sundeen, J. H. Fried*, J. Am. Chem. Soc. *91*, 1248 (1969).

[4] *J. A. Edwards, J. Sundeen, W. Salmond, T. Iwadeen, J. H. Fried*, Tetrahedron Lett. 791 (1972).

[5] *F. H. Allen, J. P. Kutney, J. Tratter, N. D. Westeatt*, Tetrahedron Lett. 283 (1971).

[6] *T. C. McMorris, R. Seshadri*, Chem. Commun. 1646 (1971).

[7] *T. C. McMorris, T. Arundachalam, R. Seshadri*, Tetrahedron Lett. 2677 (1972).

[8] *D. M. Green, J. A. Edwards, A. W. Backsdale, T. C. McMorris*, Tetrahedron *27, 1199 (1971)*.

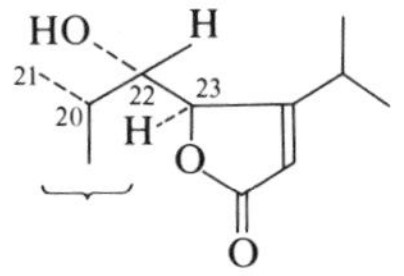

11 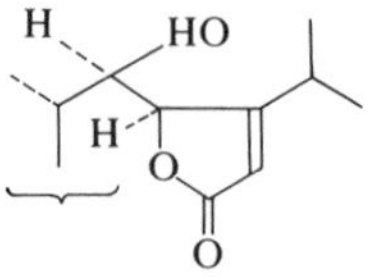 12 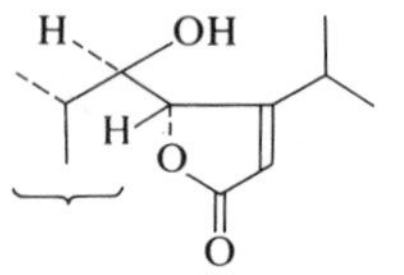 13 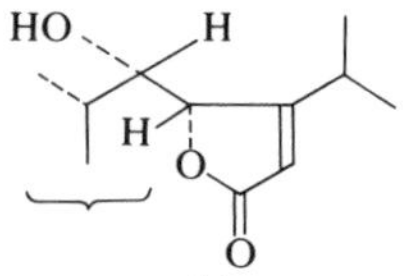 14

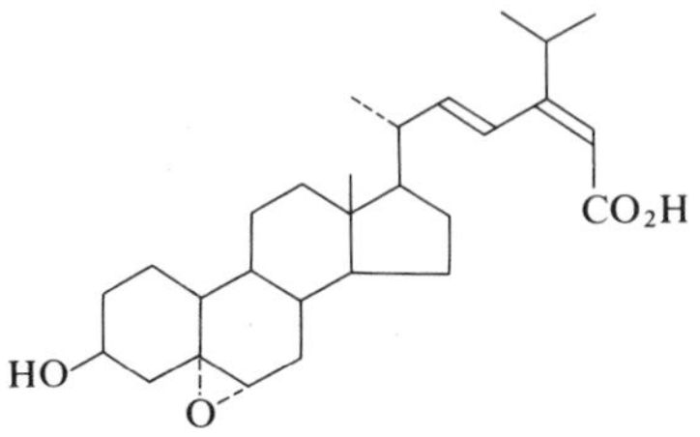

15 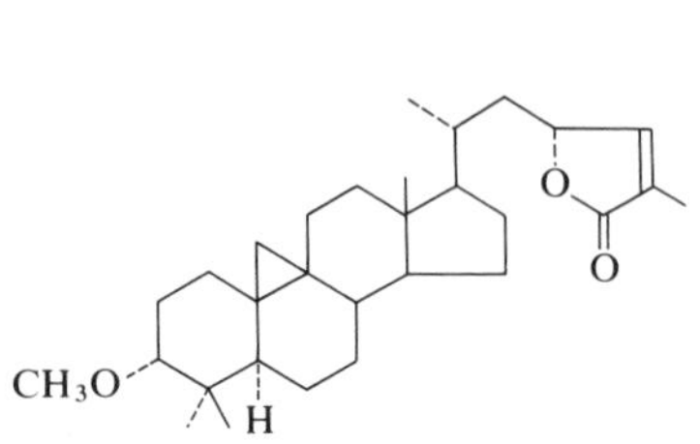 16

3.4 Physalin A

Extraction of the fresh herb *Ph. Alkekengi var Francheti* yielded the very bitter Physalin A together with a less bitter counterpart Physalin B[1]. These highly oxygenated steroids exhibited the unique structural feature of a seco C/D steroid nucleus to which an additional carbocyclic ring (E) has been added.

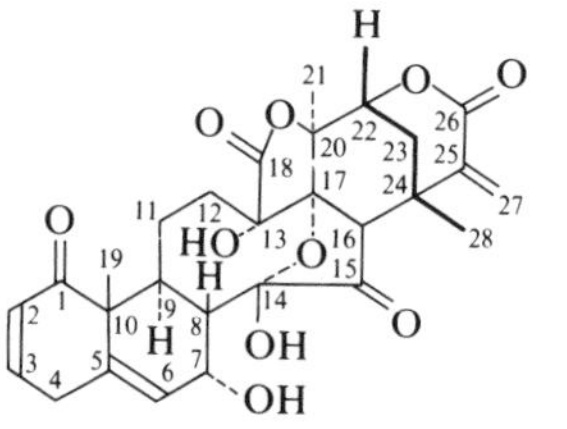

1

Physalin A
$C_{28}H_{30}O_{10}{\cdot}CH_3COCH_3$,
mp. 266° from acetone

Mass spectrometry gave M^+ at *m/e* 526, corresponding to the formula $C_{28}H_{30}O_{10}$. Degradation of **1** by selenium dehydrogenation gave only alkyl-naphthalenes and no phenanthrene derivatives, while pyrolysis of **2** gave **3**. These results suggested that **1** had neither the normal steroid nor the normal triterpenoid skeleton.

Spectroscopic examination of Physalin A showed several carbonyl functions (infrared absorption from 1600–1800 cm^{-1}, including a γ-lactone (1780 cm^{-1}). Conjugation of a ketonic function and possible further conjugated functionality was indicated by the ultra-violet spectrum (λ_{infl} 218 nm, ε 10,000). The NMR spectrum showed five olefinic protons and three methyl singlets. The appearance of an additional secondary methyl group (δ 1.70, $J=8$ Hz) in the NMR spectrum of tetrahydrophysalin **2** suggests the presence of an exo-methylene group in **1**, and

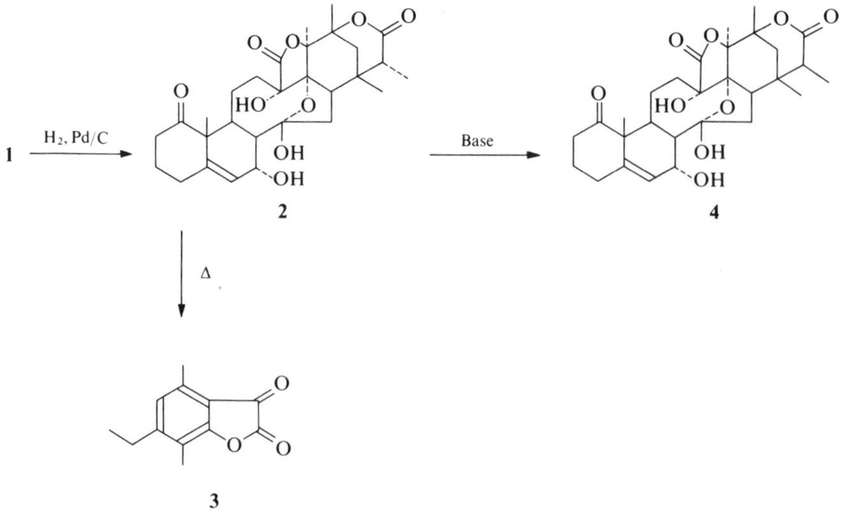

1 → (H_2, Pd/C) → 2 → (Base) → 4

2 → (Δ) → 3

its location adjacent to a carbonyl function was evinced by the ready isomerization of **2** to **4**. The structure of the bromo acetate **5** was established by X-ray crystallographic analysis and the anomalous dispersion of Ca-K_{α} radiation by the bromine atom was used to determine the absolute configuration[2]. Deacetoxybromination of **5** to **4** using Zn-Cu couple showed that no skeletal transformation had occurred in the preparation of **5**. The location of the two double bonds was assigned as C_{25}–C_{27} (terminal methylene) and C_2–C_3. The position of the latter accounts for the broad intense ultraviolet absorption due to superposition of the conjugated ketone and lactone, and the NMR signals which appear at δ 5.84 (J=10 Hz) and δ 6.94 (J=10 Hz) (conjugated enone system) in the spectrum of **1**.

The closely related structure physalin B **6** was elucidated by the conversions shown below[3].

A number of compounds have been isolated which represent intermediate stages in the biogenetic pathway from the withanolides to the physalins[4]; these include withaphysalins A and B[5] and the 13,14-seco-steroid withaphysalin C[6]. Together with nicandrenone and related compounds, the physalins and withanolides form a biogenetic group within the *Solanaceae*[4].

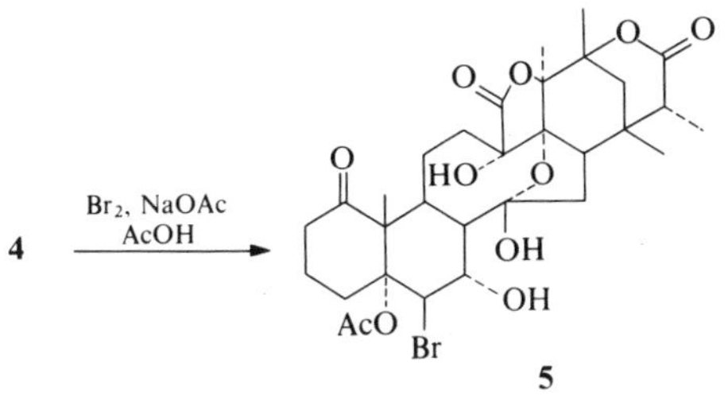

4 —(Br_2, NaOAc, AcOH)→ 5

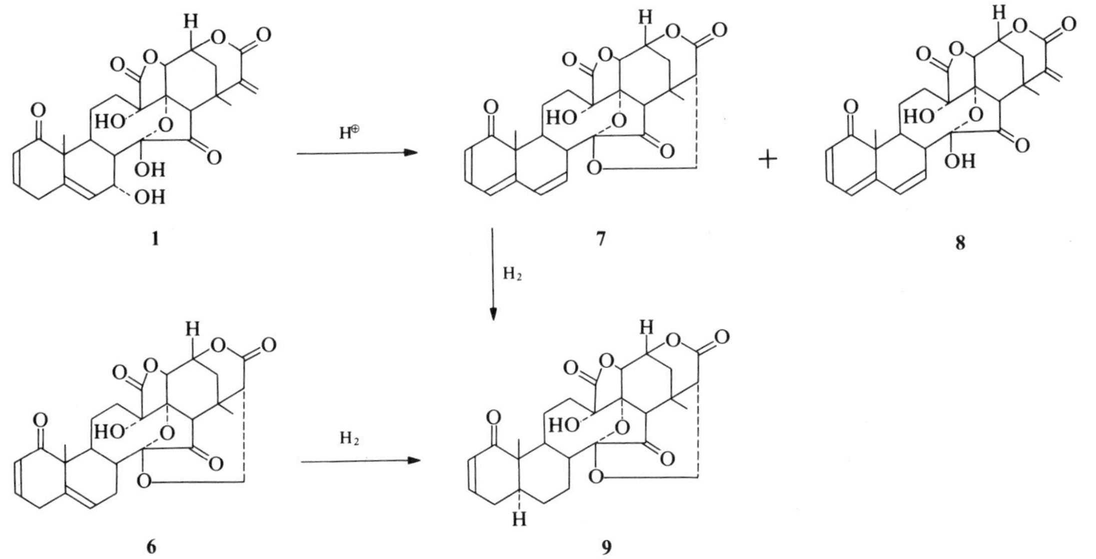

1 —($H^{\oplus}$)→ 7 + 8

7 —(H_2)→ 9

6 —(H_2)→ 9

[1] *T. Matsuura, M. Kawai, R. Nakashima, Y. Butsugan*, Tetrahedron Lett. 1083 (1969).

[2] *M. Kawai, T. Taga, K. Osaki, T. Matsuura*, Tetrahedron Lett. 1087 (1969).

[3] *T. Matsuura. M. Kawai*, Tetrahedron Lett. 1765 (1969).

[4] *M. J. Begley, L. Crombie, P. J. Ham, D. A. Whiting*, J. Chem. Soc. Perkin I 296 (1076).

[5] *E. Glottner, I. Kirson, A. Abraham, P. Sethi, S. Subramanian*, J. Chem. Soc. Perkin I 1370 (1975).

[6] *I. Kirson, V. I. Zaretskii, E. Glottner*, J. Chem. Soc. Perkin I 1244 (1976).

3.5 Nicandrenone: D-Ring Aromatic Steroids

Nicandrenone **1**, a steroid with strong insect repellant and mild insecticidal properties, has been isolated from *Nicandra physaloides* (solanaceae), a Peruvian weed and reputed fly repellant[1]. From the same source, Crombie, Whiting, *et al.*, have isolated four closely related compounds[2], one of which appears to be identical to nicandrenone **1**. These compounds are the first naturally occurring D aromatic steroids to be isolated, and their side chain structure suggests a biogenetic relationship to the withanolides.

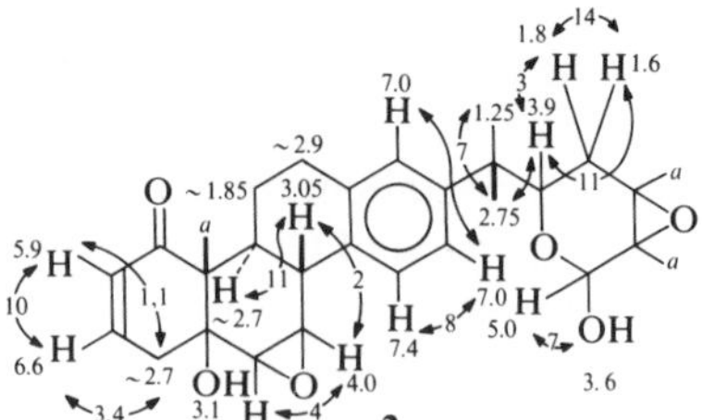

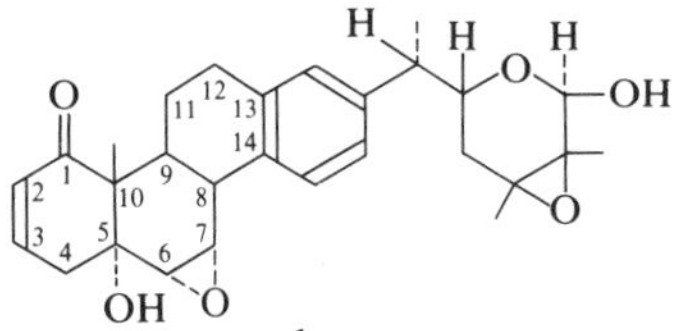

1

Nicandrenone, $C_{28}H_{34}O_6$
mp. 138°

2

$a = 1.25, 1.35, \text{or } 1.35$

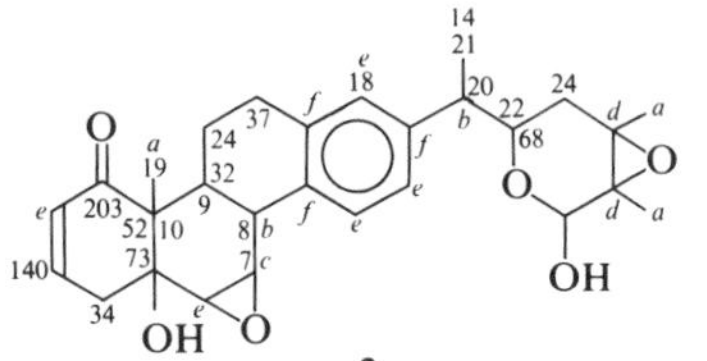

3

$a = 17, 17, \text{ or } 19$
$b = 39 \text{ or } 43$
$c = 56 \text{ or } 57$
$d = 64 \text{ or } 65$
$e = 124, 125, 129, \text{ or } 129$
$f = 135, 137 \text{ or } 142$

Bates and Eckert[3] assigned the planar structure of **1** largely on the basis of extensive NMR studies (CMR and PMR). Assignments of the CMR and PMR signals are indicated in structures **2** and **3**. The proton decoupled CMR spectrum showed 28 carbons, and off-resonance decoupling showed the presence of 4 methyls, 4 methylenes and 12 methines for a total of 32 protons attached to carbon. Two hydroxyl groups and 4 oxygens complete the molecular formula $C_{28}H_{34}O_6$ derived from the parent peak at *m/e* 466 in the mass spectrum. The steroidal nature of **1** was suggested by some previous color reactions and no non-steroidal structure could be constructed which was consistent with the spectroscopic data.

From the NMR spectra were derived the enone system (PMR), placed in ring A, and the trialkylbenzene ring (PMR, CMR) with a 1,2,4 pattern (PMR). Seven rings are present since there are no further sp^2 carbons (CMR), and the three remaining oxygens must be in ether linkages. (Observation of the coupling pattern revealed the structure of the side chain and the partial structure of ring B). Further examination of the NMR spectra indicated a tetraalkyl-substituted epoxide, three unsplit methyls, two methylenes and a tertiary hydroxyl leading to structure **1**. An NOE of 20% was observed between the C-7 epoxide proton and the aromatic proton at δ 7.4 (C-15). Confirmation of the proposed structure was provided by the mass spectrum which showed a strong peak at m/*e* 323 corresponding to cleavage of the C-20 to C-22 bond to give a benzyl cation.

The structure proposed for nicandrenone **1** on the basis of NMR studies, is substantiated by the isolation of compounds **4**→**6** from the same source[2]. The structure of **4** was determined by X-ray analysis using direct methods and structures **5** and **6** were derived by comparison of the NMR, IR and UV spectra with that of **4**. A fourth compound was assigned the structure **1** on the same basis. Comparison of the NMR spectra of the two compounds assigned structure **1** is complicated by the fact that different solvents ($CDCl_3$[3] and C_5D_5N[2]) were used for the measurements. However the peak multiplicities and splitting patterns, together with the fact that both were isolated from the same source suggest they are indeed identical. The stereochemistry of the side-chain is that depicted by the English workers[2]. This is opposite at C-22, 24, 25, 26

---

[1] *O. Nalbandor, R. T. Yamamoto, G. S. Fraenkel*, J. Agric. Food Chem. *12*, 55 (1964).

[2] *M. J. Begley, L. Crombie, P. J. Ham. D. A. Whiting*, Chem. Commun. 1250 (1972).

[3] *R. B. Bates, D. J. Eckert*, J. Am. Chem. Soc. *94*, 8259 (1972).

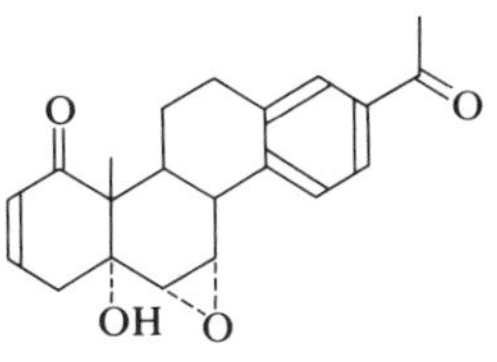
O
O
OH
O

4

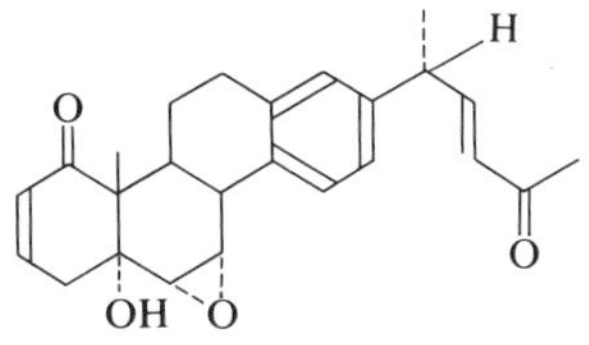
H
O
OH
O
O

5

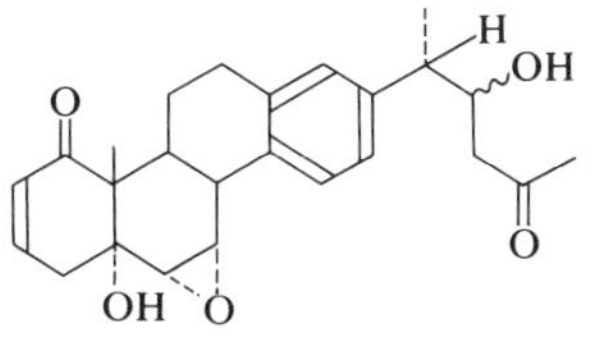
H
OH
O
OH
O
O

6

from that of the two other components of *Nicandra physaloides* whose structures were determined by X-ray analysis[4], and the side-chain stereochemistry bears further investigation.

3.6 Withanolides

The withanolides constitute a new class of steroids characterized by a 9-carbon side-chain containing a δ-lactone. Withaferin A, the first member of the series, was isolated from leaves of *Withania semnifera*[1] (also *Acnistus arborescens*) and exhibits significant *in vitro* tumor inhibitory properties[2].

Elucidation of the structure was initiated by spectroscopic examination of **1** and its hydrogenation products. Reduction with 10% palladium on carbon could be controlled to give dihydrowithaferin **2** (1 mole), dihydrodesoxywithaferin **3** (2 moles) and tetrahydrodesoxywithaferin **4** (3 moles). The 1,4-oxo-hydroxy-ene system was indicated in the downfield region of the NMR spectrum. The olefinic signals at δ 6.18 and 6.97 were absent in the spectrum of **2**, and acetylation of **1** shifted the δ 3.75 doublet to

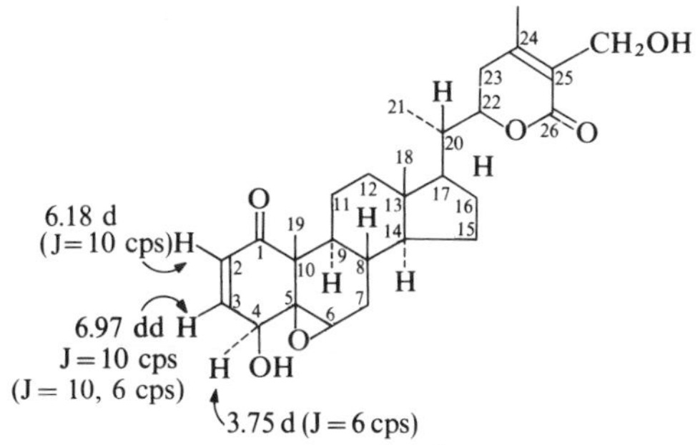
6.18 d
(J=10 cps)H
6.97 dd H
J=10 cps
(J=10, 6 cps)
H OH
3.75 d (J=6 cps)
CH2OH
O

1

Withaferin A, $C_{28}H_{38}O_6$
mp. 243–245°, $[\alpha]_D + 114°$

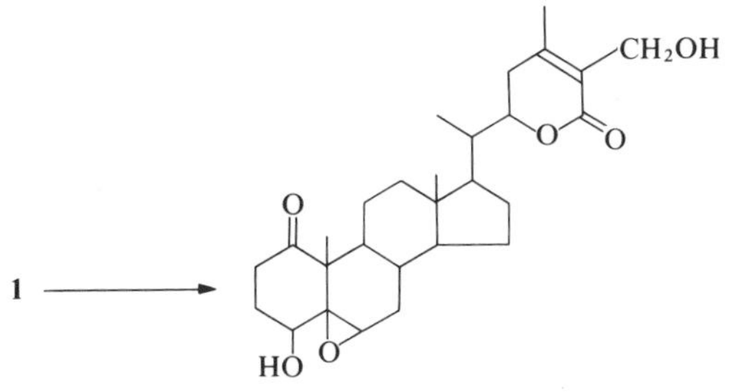
1
CH2OH
O
HO

2

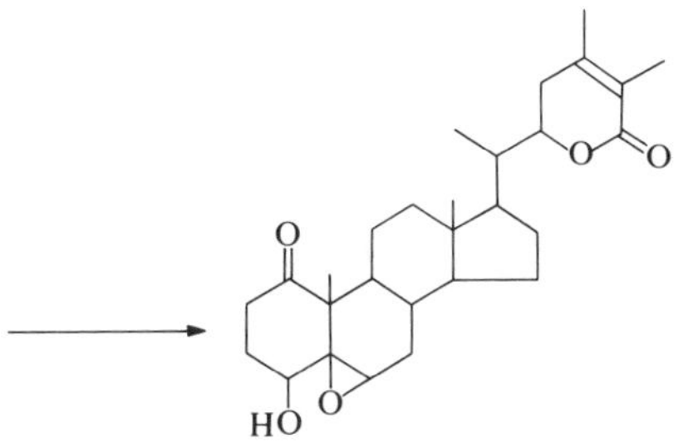
O
HO

3

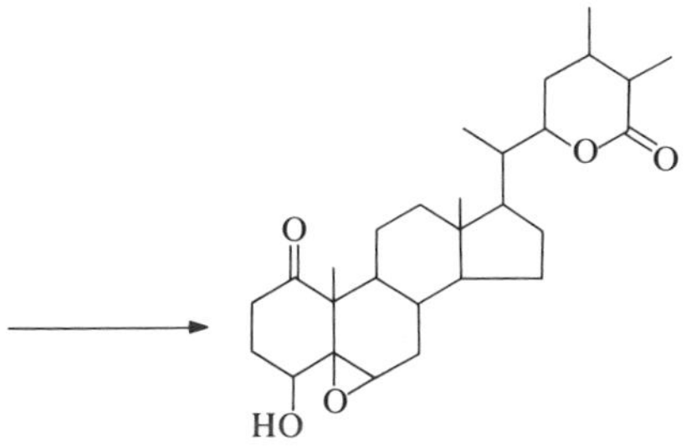
O
HO

4

[4] *M. J. Begley, L. Crombie, P. J. Ham. D. A. Whiting*, Chem. Commun. 1108 (1972).

[1] *D. Lavie, E. Glottner, Y. Shuo*, J. Org. Chem. *30*, 1774 (1965).

[2] *S. M. Kupchan, R. W. Koskotch, P. Bollinger, A. J. McPhail, G. A. Sim. J. A. Saenz Renauld*, J. Am. Chem. Soc. *87*, 5805 (1965).

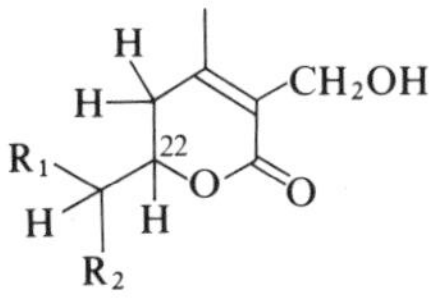

5

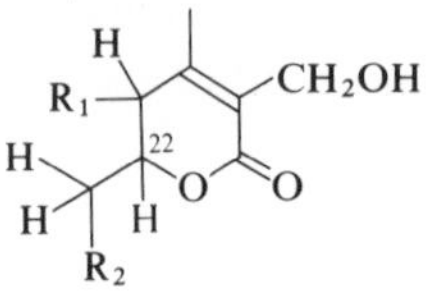

6

δ 4.66. The hydrogenolysis of **2** to **3** was accompanied by the disappearance of the C_{27} methylene singlet and appearance of a vinylic methyl signal at δ 1.93. Conversion of the two vinylic methyl signals of **3** to doublets at δ 0.92 and 1.13 in **4** showed the location of the second double bond, and from changes in the ultraviolet and infrared spectra through the series **1**→**4** the α,β-unsaturated carbonyl functions were inferred. A signal at δ 3.20 which remained unchanged during the transformation **1**→**4** was suggestive of an epoxide functionality. By assuming a signal (double triplet) which appeared at δ 4.40 throughout the **1**→**4** series to be that at C_{22}, two partial structures, **5** and **6**, were possible for Withaferin A. Deuteration of **3** pointed to structure **5** since 5 deuteriums were incorporated and the C-22 proton collapsed to a doublet. Deuteration of **4** resulted in no change in the C-22 proton signal, showing the allylic protons at C-23 to be responsible for the coupling. Mass spectral data established the secondary methyl group adjacent to the δ-lactone (**5**, $R_1 = CH_3$).

The steroidal nuclear structure was revealed by the formation of an ethyl 1,2-cyclopentanophenanthrene and a trialkylnapthalene (both identified from ultraviolet spectra) on selenium dehydrogenation of a Withaferin A derivative[3]. Further examination of unassigned methyl signals in the NMR spectra of several derivatives of **1** revealed that the chemical shift values were those expected for steroidal 10 and 13 methyl groups.

Chemical degradations substantiated the deductions from spectroscopic data. The sequence

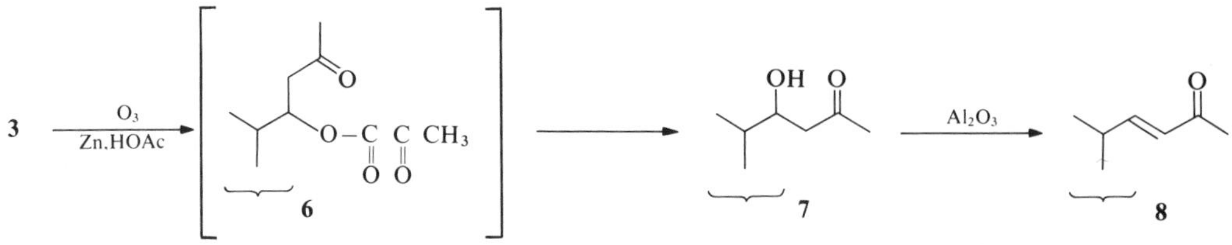

3 → (O_3, Zn.HOAc) → [6] → 7 → (Al_2O_3) → 8

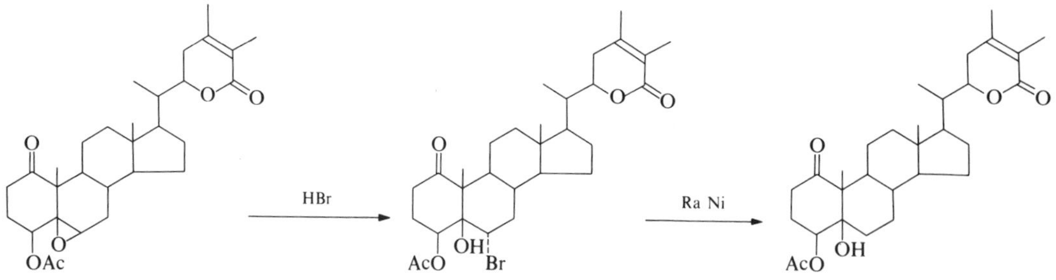

9 → (HBr) → 10 → (Ra Ni) → 11

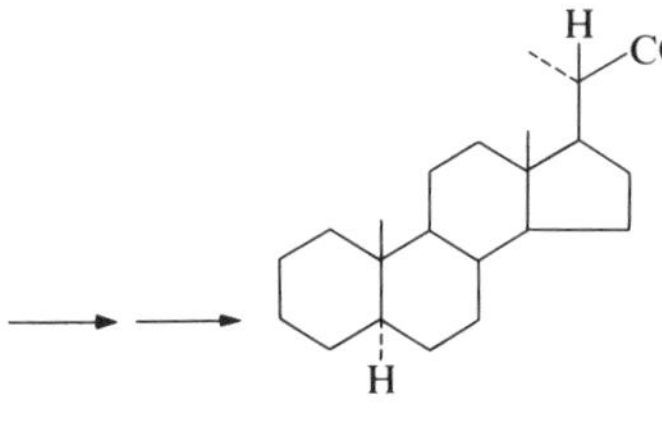

→ → 12

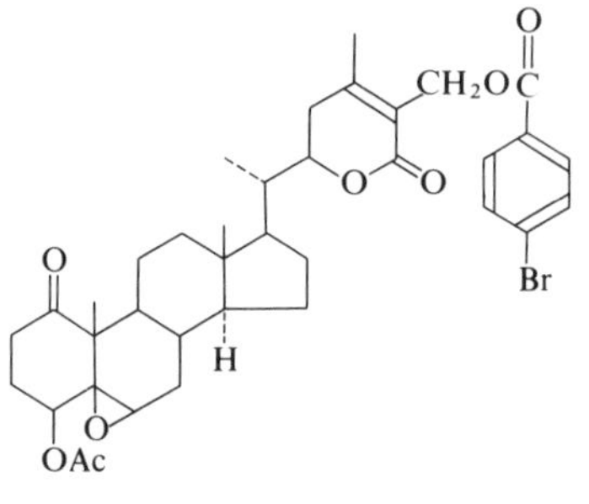

13

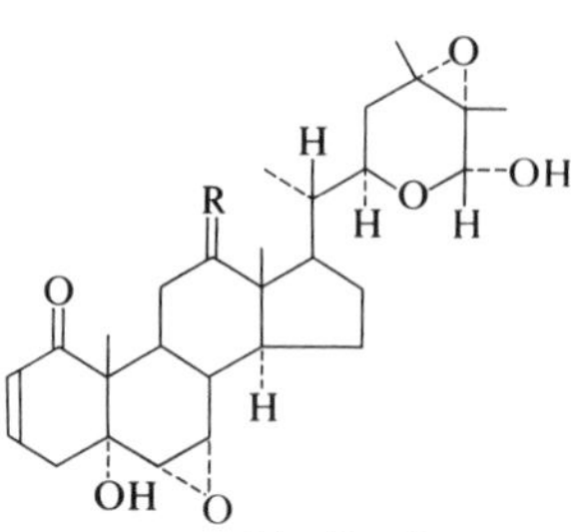

14 R=O

15 R=H, H

3→8 showed the presence of the α,β-unsaturated δ-lactone; and the twelve step degradation of dihydrodesoxywithaferin A acetate **9** to bisnor-(5α)-cholanic acid **12**, (stereochemistry is inverted at C-5 during degradation) not only confirmed the structure of **1** but also determined the stereochemistry. The orientation of the 5,6-epoxide was determined by model studies on 4β-acetoxy-5,6β-epoxy-(5β)-cholestane[4]. The ORD curve for **11** was identical to that of 17β-acetoxy-1-oxo-(5β)-androstane, confirming the β-epoxide structure.

The structure deduced from chemical and spectral studies was confirmed by X-ray crystallographic analysis of the Withaferin A derivative **13**.

Many additional withanolides have recently been isolated[5], among them withanolides D[6] and E[7] whose side chains possess the unusual 17α configuration, and withanolides G, H, I, J, K, all having the unusual feature of a $\Delta^{8(14)}$ double bond. In addition the insect repellant plant *Nicandra physaloides* contains two closely related compounds **14** and **15** (X-ray structures)[8,9].

3.7 Fukujusonorones

In 1969 the structure shown was tentatively assigned to "fukujusonorone" isolated from *Adonis amurensis* Regel et Radd (Ranunculaceae); it was the first 18-norsteroid

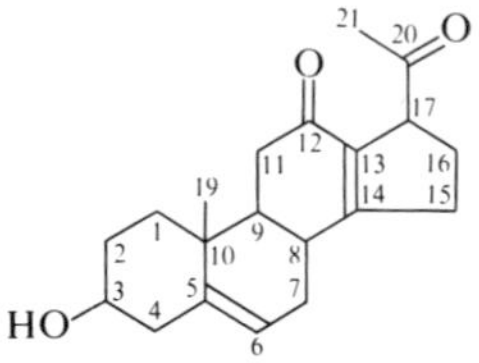

Fukujusonorones (mixture of 17α and 17β)
$C_{20}H_{26}O_3$, mp. 88–90°
UV (EtOH): 249 nm (ε 8,600)
CD (EtOH): 250 nm (Δε −2.83), 285 (+1.37), 330 (−0.34)

[3] *D. Lavie, E. Glottner, Y. Shvo*, J. Chem. Soc. 7517 (1965).

[4] *D. Lavie, S. Greenfield, E. Glottner*, J. Chem. Soc. C 1753 (1966).

[5] *I. Kirson, E. Glottner, D. Lavie, A. Abraham*, J. Chem. Soc. C 2032 (1971).

[6] *D. Lavie, I. Kirson, E. Glottner*, Israel J. Chem. *6*, 671 (1968).

[7] *D. Lavie, I. Kirson, E. Glottner*, J. Chem. Soc. C, 877 (1972).

[8] *E. Glottner, I. Kirson, A. Abraham, D. Lavie*, Tetrahedron *29*, 1353 (1973).

[9] *M. J. Begley, L. Crombie, P. J. Ham, D. A. Whiting*, Chem. Commun. 1108 (1972).

isolated from a natural source[1]. The sharp melting point, chromatographic properties, and singlet nature of the NMR methyl peaks at 1.05 (19-H) and 2.26 ppm (21-H) indicated it to be homogeneous. Subsequent carbon magnetic resonance (CMR) studies confirmed the proposed structure but also showed that it was in fact a mixture of two epimers at C-17.

The following studies[2] were carried out as an application of partially relaxed Fourier transform (PRFT) CMR measurements[3] to simplify assignment of peaks. The proton noise decoupled spectrum is shown in Fig. 1c. Some peaks at low field are clearly doublets but peaks in the high field region are narrowly spaced and no concrete conculsion can be drawn. This also applied to the continuous wave (off-resonance) decoupled spectrum (Fig. 1a) where the 20–40 ppm region is too congested to allow differentiation between a quartet (methyl), triplet (methylene), doublet (methine) and a singlet (quaternary).

The T_1 relaxation times of ^{13}C contained in alicyclic systems, *e.g.*, cholesteryl chloride[4], usually decrease in the sequence CH_2, CH, CH_3 and C with no H; when this general tendency is considered in conjunction with the partially relaxed and continuous wave decoupled spectrum, the assignment of CMR peaks is greatly simplified; for such purposes it is usually unnecessary to measure accurate T_1 relaxation times. In Fig. 1b, which was taken at an interval time τ of 0.4 sec in the $(180–\tau–90–T)_n$ sequence[5] of measurement, all CH_2 peaks are positive, the CH peaks are weakly negative, whereas the CH_3, quaternary C and carbonyl C peaks are strongly negative. The separation of closely spaced peaks due to C-1/C-10 and C-15/C-21 into positive and

[1] *Y. Shimizu, Y. Sato, H. Mitsuhashi*, Experientia *25*, 1129 (1969).

[2] *P. H. Solomon, K. Nakanishi, W. E. Fallon, Y. Shimizu*, Chem. Pharm. Bull. *22*, 1671 (1974).

[3] *K. Nakanishi, R. Crouch, I. Miura, X. Dominguez, A. Zamudio, R. Villareal*, J. Am. Chem. Soc. *96*, 609 (1974); *P. R. Zanno, I. Miura, K. Nakanishi, D. L. Elder*, J. Am. Chem. Soc. *97*, 1975 (1975), and references cited therein.

[4] *A. Allerhand, D. Doddrell, R. Komoroski*, J. Chem. Phys. *55*, 189 (1971).

[5] *G. C. Levy, G. L. Nelson*, Carbon-13 Nuclear Magnetic Resonance for Organic Chemists, p. 182, Wiley-Interscience, New York, N.Y. 1972.

[6] *H. J. Reich, M. Jautelat, M. T. Messe, F. J. Weigert, J. D. Roberts*, J. Am. Chem. Soc. *91*, 7445 (1969).

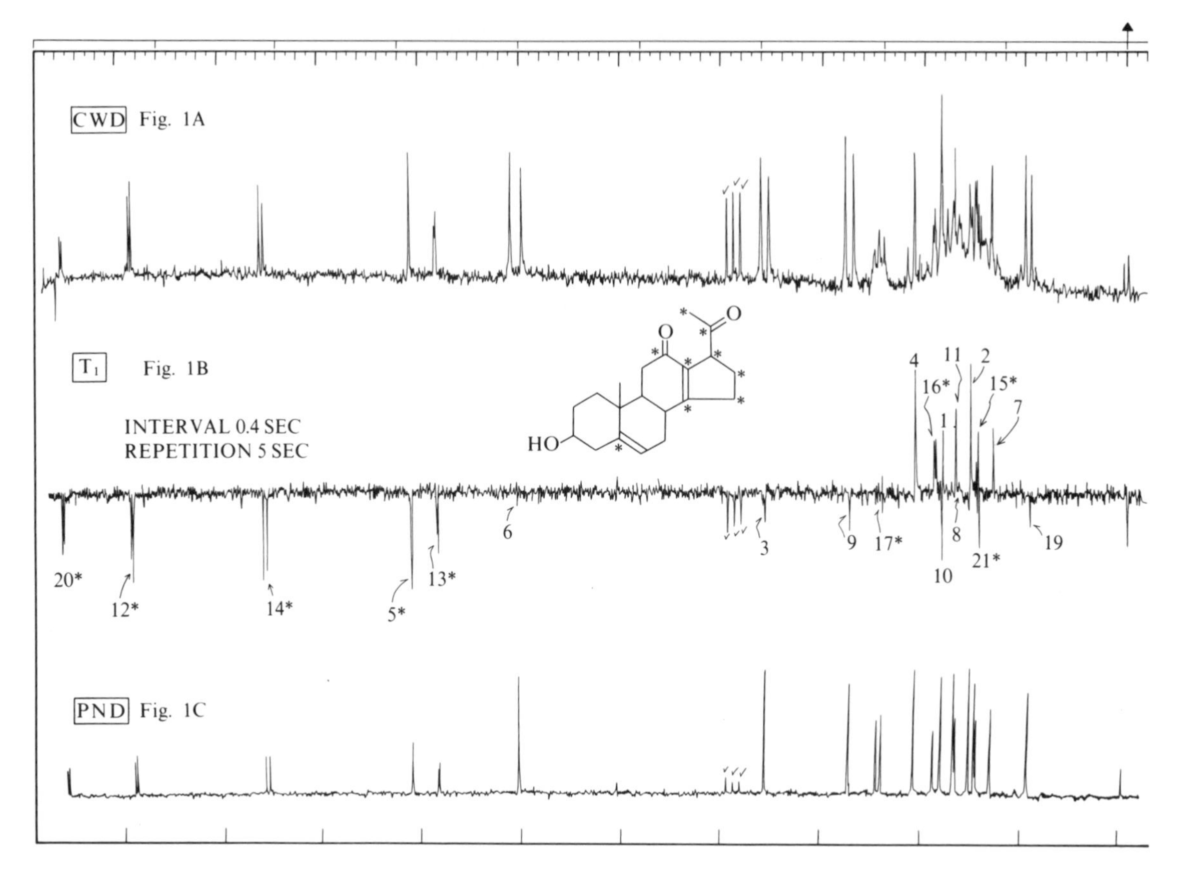

negative signals should be noted. Thus Figs. 1a–1c and consideration of steroidal CMR chemical shifts[6] made it possible to assign all peaks without comparison with other spectra of fukujusonorone derivatives.

The proposed structure is confirmed from the CMR data. However, it is also clear that peaks due to carbons surrounding C-17 are doublets (marked by asterisks in Fig. 1b); *i.e.*, C-20, -12, -14, -13, -17, -16, -15 and -21. This showed that the starting material was a mixture (approximately 1:1) of 17-epimers, a result which was corroborated by mass spectroscopic studies (with deuteration) and high pressure liquid chromatography. The doublet nature of the C-5 CMR peak cannot be explained.

4 Total Synthesis

Much effort has been expended on the total synthesis of various steroids. The aromatic steroids (equilenin and estrone), containing the least number of asymmetric centers, were the first to be synthesized (late 1930's and 1940's). Synthesis of the nonaromatic steroid came in 1951 with both Woodward and Robinson reporting their respective routes in that year. Since then almost every major class of steroids has been synthesized. As most of this work is reviewed elsewhere[1], this discussion will cover only the more recent work in this field.

In the field of aromatic steroids a vastly improved synthesis of estrone and equilenin arose from investigations of the D-homo analogues by Torgov in 1960[2], and simultaneous studies by five groups on the extension of this synthesis to the five-membered D ring compounds[3]. Thus syntheses of *dl* equilenin methyl ether **8** and estrone methyl ether **9** with overall yields of 10.5% and 27% respectively from **1** were evolved (Scheme 1). This route differed from earlier schemes in that cyclization of the D ring, which

[1] *A. A. Akhrem, Y. A. Titov,* Total Steroid Synthesis, Plenum Press, New York 1970.

[2] *S. Ananchenko, I. V. Torgov,* Dokl. Akad. Nauk SSSR *127*, 553 (1959).

[3] *D. Crirpin, J. Whitehurst,* Proc. Chem. Soc. 22 (1963); *T. Miki, K. Hiraga, T. Asako,* Proc. Chem. Soc. 139 (1963); *T. Windholz, J. Fried, A. Patchett,* J. Org. Chem. *28*, 1092 (1963); *S. N. Ananchenko, I. Torgov,* Tetrahedron Lett. 1553 (1963).

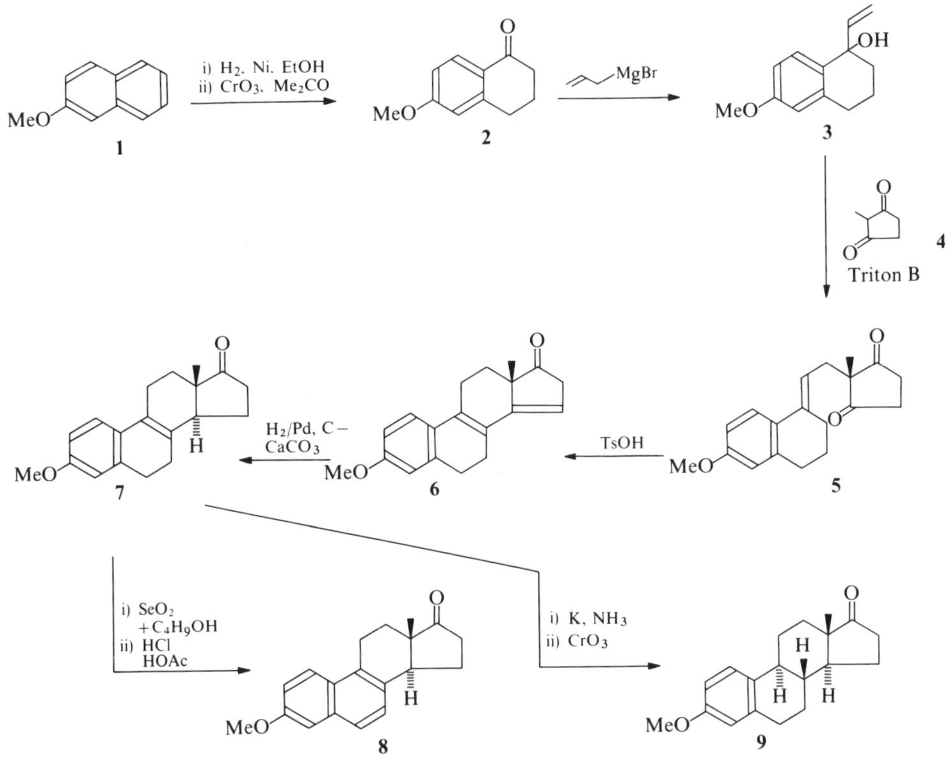

Scheme 1.

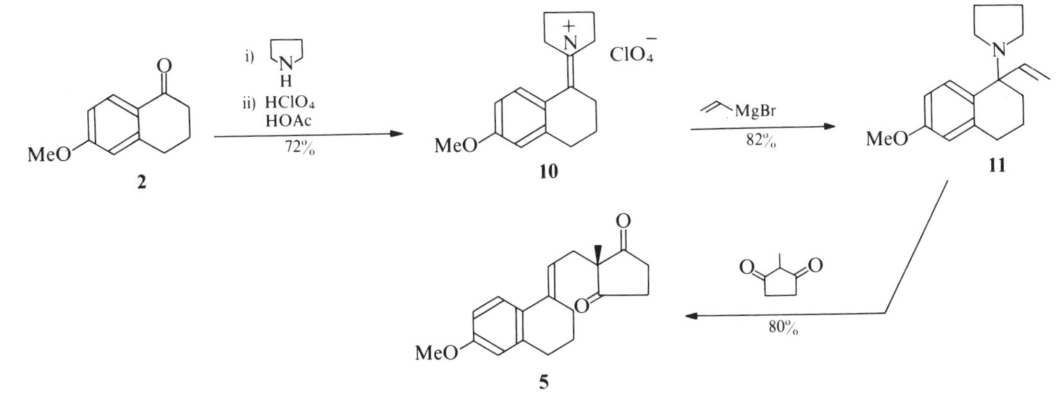

Scheme 2.

was not amenable to stereochemical control, was avoided, and the trans C/D junction was formed by the stereospecific hydrogenation of **6** to **7**. A novel modification of the Torgov synthesis has been developed in which Pandit *et al.*[4] utilized the allylic amine **11**, which is readily prepared from the enamine of *m*-methoxytetralone (Scheme 2). A yield of 80% was obtained on condensation of **11** with the cyclopentadione. In a recent estrone synthesis reported by Valenta and coworkers[5], the *dl* steroid is obtained in a 22% overall yield based on **12** (Scheme 3). In a key step, use of boron trifluoride as a catalyst in the Diels-Alder reaction of **12** and **13** reverses the

[4] *U. K. Pandit, F. A. van der Vlugt, A. C. van Dalen,* Tetrahedron Lett. 3697 (1969).

[5] *R. A. Dickinson, R. Kubela, G. A. MacAlpine, Z. Stojanac, Z. Valenta,* Can. J. Chem. *50*, 2377 (1972).

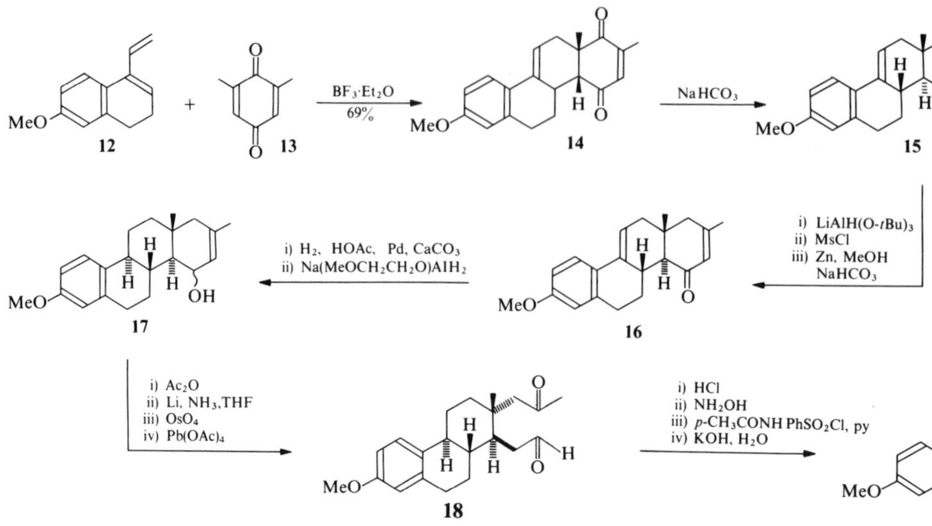

12 + 13 → ($BF_3 \cdot Et_2O$, 69%) → 14 → ($NaHCO_3$) → 15

i) $LiAlH(O\text{-}tBu)_3$
ii) MsCl
iii) Zn, MeOH, $NaHCO_3$
→ 16

i) H_2, HOAc, Pd, $CaCO_3$
ii) $Na(MeOCH_2CH_2O)AlH_2$
→ 17

i) Ac_2O
ii) Li, NH_3, THF
iii) OsO_4
iv) $Pb(OAc)_4$
→ 18

i) HCl
ii) NH_2OH
iii) $p\text{-}CH_3CONH\,PhSO_2Cl$, py
iv) KOH, H_2O
→ 9

Scheme 3.

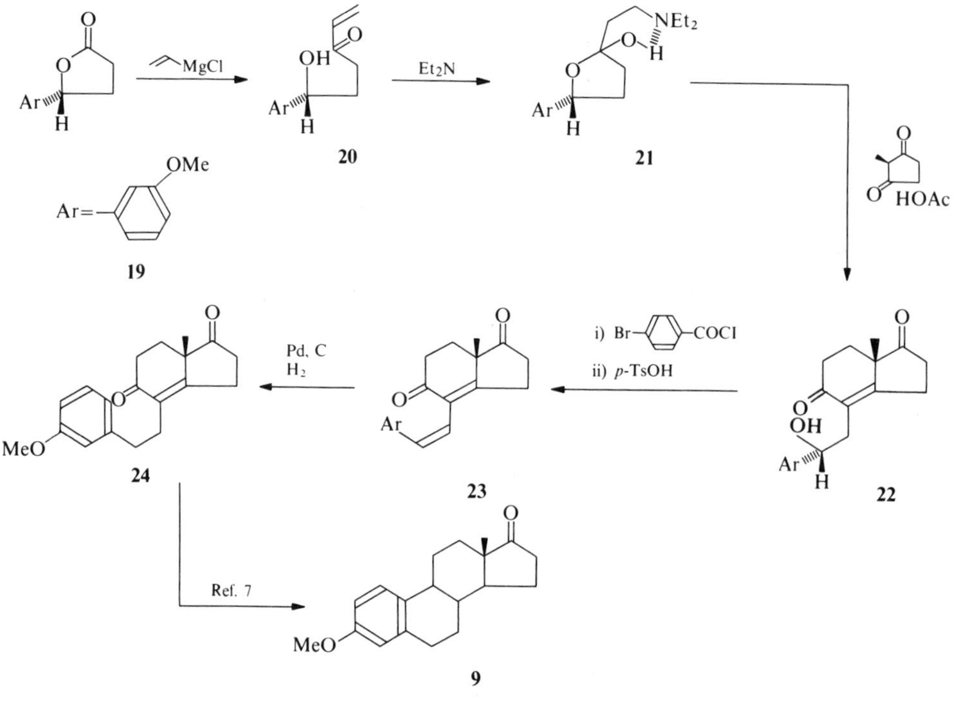

Ar = 3-methoxyphenyl (19)

MgCl (vinyl) → 20 → (Et_2N) → 21 → (2-methylcyclopentane-1,3-dione, HOAc) → 22

i) Br–C_6H_4–COCl
ii) p-TsOH
→ 23

Pd, C, H_2 → 24

Ref. 7 → 9

Scheme 4.

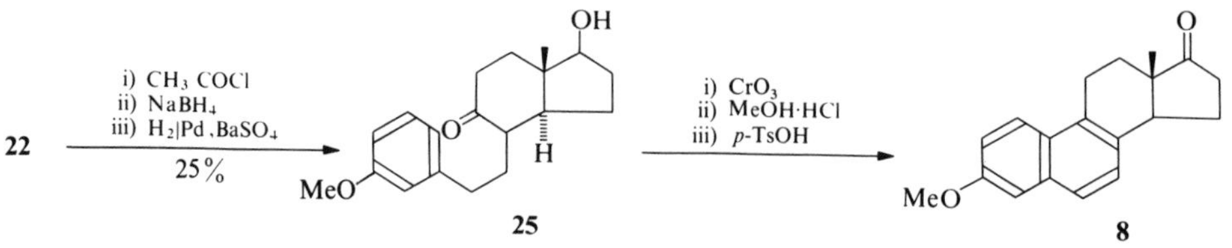

22 →
i) CH_3COCl
ii) $NaBH_4$
iii) H_2|Pd, $BaSO_4$
25%
→ 25

i) CrO_3
ii) MeOH·HCl
iii) p-TsOH
→ 8

Scheme 5.

25A + 25B —CuCl→ 25C

25D —i) PhLi, CH_2O; ii) $-H_2O$→ 25E → 25F —i) 2-methylcyclopentane-1,3-dione; ii) L-Phenylalanine, $HClO_4$→

25G (93% 13S) —i) $NaBH_4$; ii) H_2, 10%Pd/C, H^+; iii) HO OH, p-TsOH→ 25H —i) Na, NH_3; ii) NaOH; iii) H^+→

25I → → (+) 25J

25K → [25L] → 25M

normal thermal orientation to give the methyl at the desired C-13 rather than C-14. Facile isomerization of **14** to the *trans* isomer was followed by conversion of the six-membered enedione ring D to the five membered estrone D ring.

Saucy and coworkers have recently extended their method for synthesizing optically active 19-norsteroids (see Scheme 4) utilizing the optically active Mannich base **21** for the synthesis of aromatic steroids[6]. Crystallization of the *p*-bromobenzoate of **22** allowed separation of the optically pure material. After the selective reduction of **23**, the enedione **24** was transformed by known methods[7] to optically active estrone. Racemic equilenin methylether was obtained in a similar manner (Scheme 5).

A new synthesis of estrone and equilenin methyl ethers has been reported by Saucy and coworkers[22] employing the addition of Grignard reagent **25A** to the enone **25B** to form the key intermediate **25C** in 80% yield. Conversion to the final product was achieved by known methods[7].

[6] *N. Cohen, B. Banner, J. Blount, M. Tasi, G. Saucy*, J. Org. Chem. *38*, 3229 (1973).

[7] *G. Douglas, J. Graves, D. Hartley, G. Hughes, B. McLoughlin, J. Siddall, H. Smith*, J. Chem. Soc. 5072 (1963).

By developing the novel conversion of 6-substituted α-picolines to cyclohexenones[23], Danishefsky and Cain[24] have synthesized optically active estrone **25J** in 9.7% overall yield from 2,6-lutidine **25D**.

Michael addition to **25E** readily gave **25F** which reacted smoothly with 2-methylcyclopentane-1,3-dione; subsequent cyclization to **25G** gave material of 84% optical purity when the reaction was catalyzed by perchloric acid and L-phenylalanine. Birch reduction of the heteroaromatic ring resulted in **25I** which was transformed to estrone **25J**.

A formal synthesis of estrone has been completed by Kametani *et al.*[25], employing the thermal ring opening of the intermediate **25K** to the triene **25L**. This *o*-quinodimethane undergoes cycloaddition regio- and stereo-specifically to form methyl D-homoestrone **25M** in 97% yield from **25K**. Conversion of **25M** to estrone has been described previously[31].

In the field of nonaromatic steroids, the routes involve many steps and the overall yields are still not high. However, significant improvements have been made and several new interesting synthetic approaches have recently been reported.

Velluz (1960) carried out several syntheses of steroids from a tricyclic compound embodying the crucial C/D *trans* ring juncture[8]. This ketone **29** prepared from 6-methoxy-1-tetralone, can be obtained optically active by the resolution of the acid **28**. After reduction of the aromatic B ring, alkylation gives the thermodynamically preferred isomer **34** with the 8β proton. Unsaturation at the 9,10 position gave versatility to the intermediates **32**→**34** which could be alkylated and reduced, while facile migration of unsaturation to the 9,11 position provided entry, via oxidation, to the corticosteroids[9,10].

These workers' synthesis of cortisone[11] (Scheme 6) is highlighted by improved yield and a decrease in number of steps compared to earlier routes (1% yield in 27 steps from **2**). Key features include the stereospecific methylation of **34** to **35**, introduction of the 11-oxo group via the 9α,11β-bromohydrin and introduction of the side chain by regiospecific addition of sodium acetylide to the 17-keto group.

The synthesis of *dl*-progesterone **52** developed by Stork and coworkers[12] utilized both a new annulation reaction involving an isoxazole derivative for the construction of the A and B rings, and a stereospecific introduction of the C-10 methyl by reductive alkylation. All of the nuclear carbon atoms except for C-19 were assembled by the alkylation of the Wieland-Mischler ketone **44** with the isoxazole **43**. The tricyclic enone **47** was obtained in 60% yield from **45** by reduction of the 17-oxo group and $\Delta^{8,14}$ bond, followed by hydrogenolysis of the isoxazole, and cyclization. Again the crucial C/D *trans* ring juncture was obtained by catalytic hydrogenation. Reductive methylation by trapping of the intermediate enolate gave an almost quantitative yield of the β-isomer **48**. Deketalization and closure of ring A gave *dl*-D-homotestosterone **49** in 26% yield from **40**, and manipulation on the D-homo ring gave *dl*-progesterone **52** (Scheme 7)[13].

A new approach to steroid synthesis was demonstrated by Johnson *et al.*, (1968) who employed a biogenetic-type olefin cyclization which generated five asymmetric centers stereospecifically[13]. The key step is the treatment of the carbinol **62** with excess trifluoroacetic acid at −78° followed by reduction with lithium aluminum hydride to give the tetracyclic diene in 30% yield. The hydrocarbon **63** was accompanied by a hydroxyl containing fraction which on dehydration with phosphorous oxychloride gave more **63**. The carbinol **62** is rapidly dehydrated prior to cyclization and it appears that the resulting pentaene, having no asymmetric centers, undergoes a one step stereospecific cyclization to form the five asymmetric centers present in **63**. Rings A and D are simultaneously cleaved and then recyclized to give *dl*-16-dehydroprogesterone (Scheme 8).

---

[8] *L. Velluz, J. Mathieu, G. Nomine*, Tetrahedron, Suppl. *8, II*, 495 (1966).

[9] *Y. Kurosawa, H. Shimojima, U. Osawa*, Steroids, Suppl. *I*, 185 (1965).

[10] *L. Velluz, G. Amiard, R. Heymes*, Bull. Soc. Chim. France 1015 (1954).

[11] *L. Velluz, G. Nomine, J. Mathieu*, Angew. Chem. *72*, 725 (1960); *L. Velluz, G. Nomine, J. Mathieu, E. Toromanoff, D. Bertin, R. Bucourt, J. Tessier*, Compt. Rend. *250*, 1293 (1960).

[12] *G. Stork, S. Danishefsky, M. Ohashi*, J. Am. Chem. Soc. *89*, 5459 (1967); *G. Stork, J. E. McMurry*, J. Am. Chem. Soc. *89*, 5463, 5464 (1967).

[13] *W. S. Johnson, M. F. Semmelhack, M. U. S. Sultanbawa, L. A. Dolak*, J. Am. Chem. Soc. *90*, 2994 (1968); *S. F. Brady, M. A. Ilton, W. S. Johnson*, J. Am. Chem. Soc. *90*, 2882 (1968).

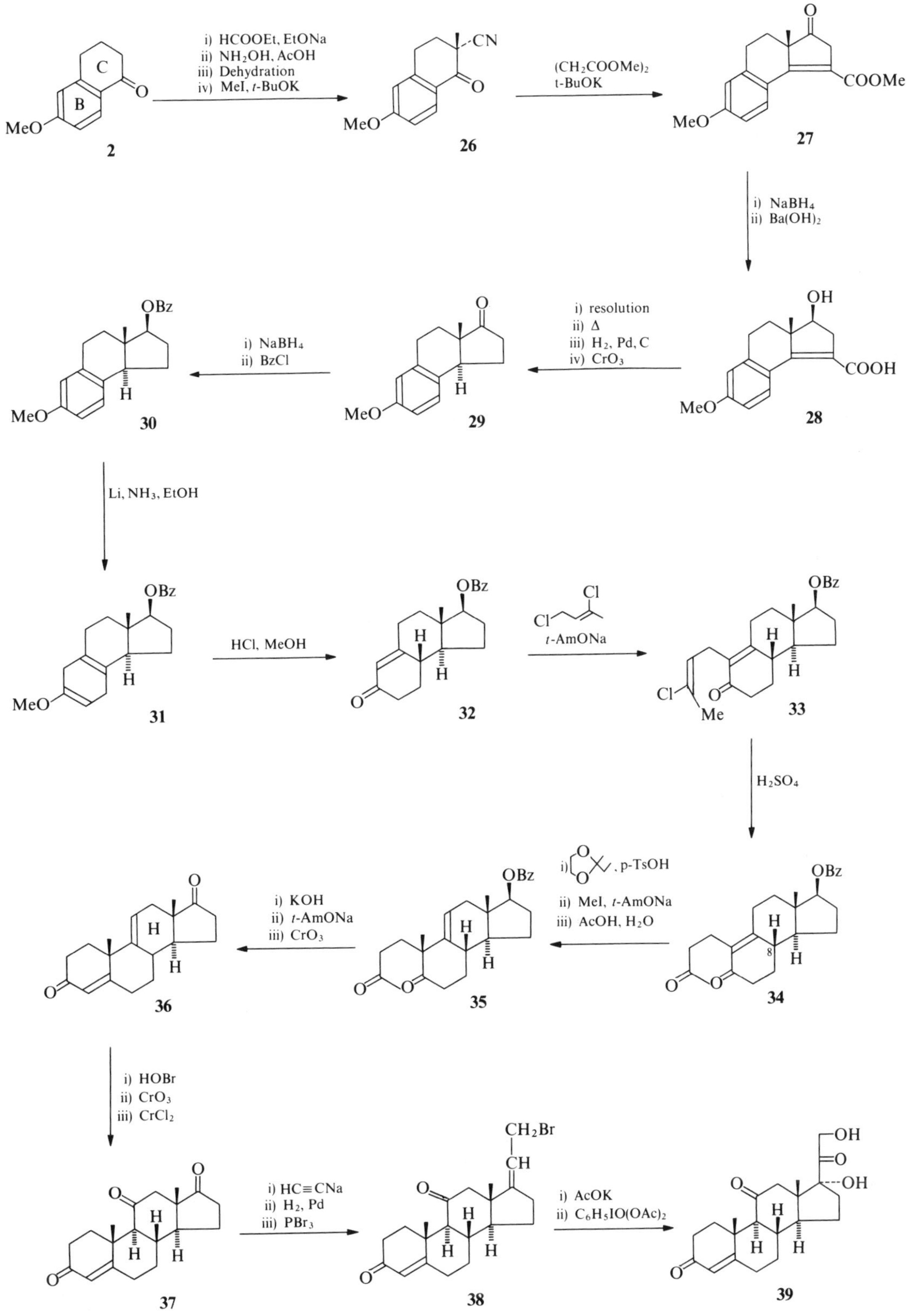

MeO C B O **2**

i) HCOOEt, EtONa
ii) NH_2OH, AcOH
iii) Dehydration
iv) MeI, *t*-BuOK

MeO CN O **26**

$(CH_2COOMe)_2$
t-BuOK

MeO O COOMe **27**

i) $NaBH_4$
ii) $Ba(OH)_2$

MeO OH COOH **28**

i) resolution
ii) Δ
iii) H_2, Pd, C
iv) CrO_3

MeO O H **29**

i) $NaBH_4$
ii) BzCl

MeO OBz H **30**

Li, NH_3, EtOH

MeO OBz H **31**

HCl, MeOH

O OBz H H **32**

Cl Cl
t-AmONa

Cl Me O OBz H H **33**

H_2SO_4

O O OBz H 8 H **34**

i) (dioxolane), p-TsOH
ii) MeI, *t*-AmONa
iii) AcOH, H_2O

O O OBz H H **35**

i) KOH
ii) *t*-AmONa
iii) CrO_3

O O H H **36**

i) HOBr
ii) CrO_3
iii) $CrCl_2$

O O O H H H **37**

i) HC≡CNa
ii) H_2, Pd
iii) PBr_3

CH_2Br CH O O H H H **38**

i) AcOK
ii) $C_6H_5IO(OAc)_2$

OH O OH O O H H H **39**

Scheme 6.

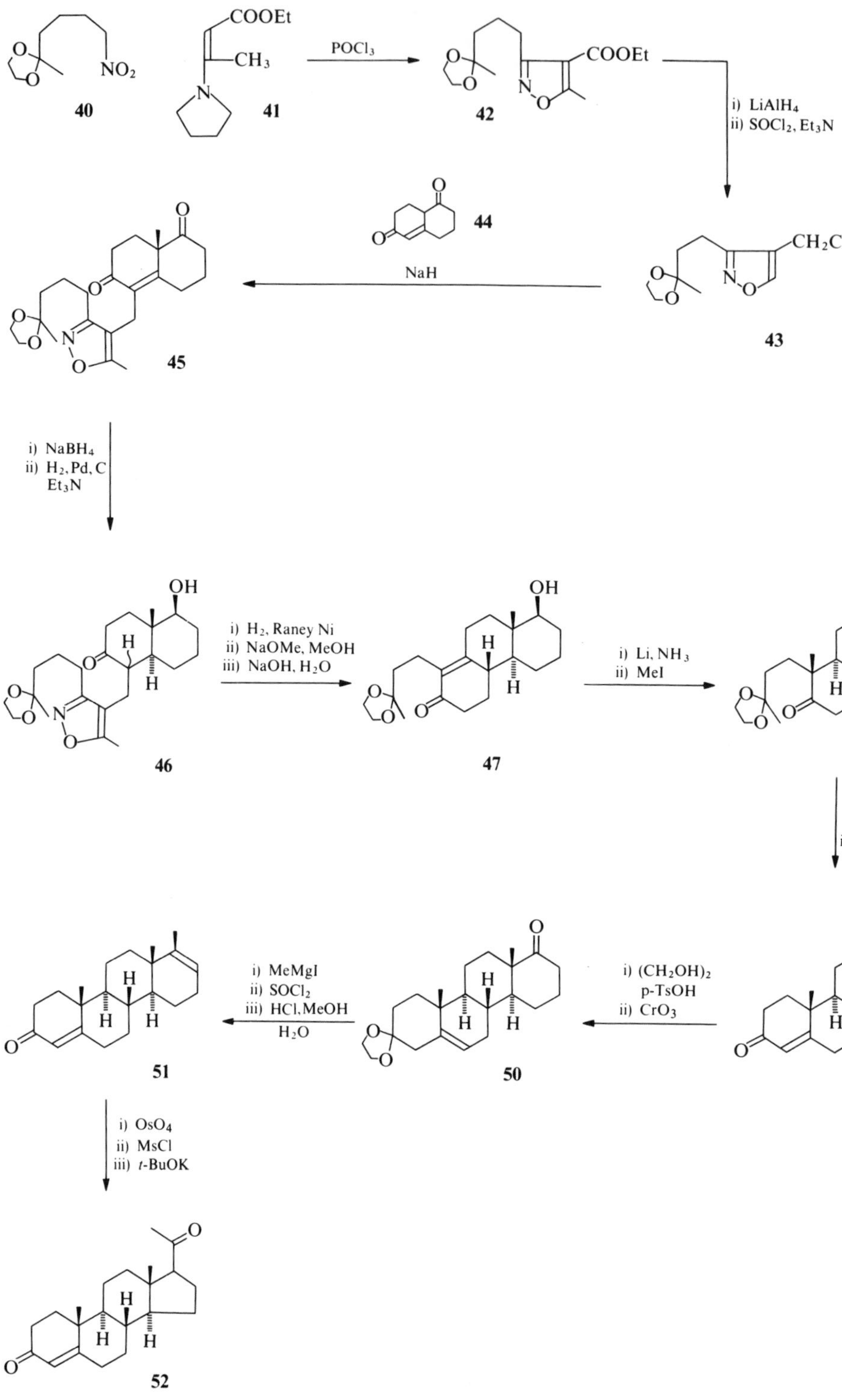
POCl3
COOEt
CH3
NO2
40
41
42
i) LiAlH4
ii) SOCl2, Et3N
44
CH2Cl
NaH
43
45
i) NaBH4
ii) H2, Pd, C
Et3N
46
i) H2, Raney Ni
ii) NaOMe, MeOH
iii) NaOH, H2O
47
i) Li, NH3
ii) MeI
48
i) H+
ii) NaOH
49
i) (CH2OH)2
p-TsOH
ii) CrO3
50
i) MeMgI
ii) SOCl2
iii) HCl, MeOH
H2O
51
i) OsO4
ii) MsCl
iii) t-BuOK
52

Scheme 7.

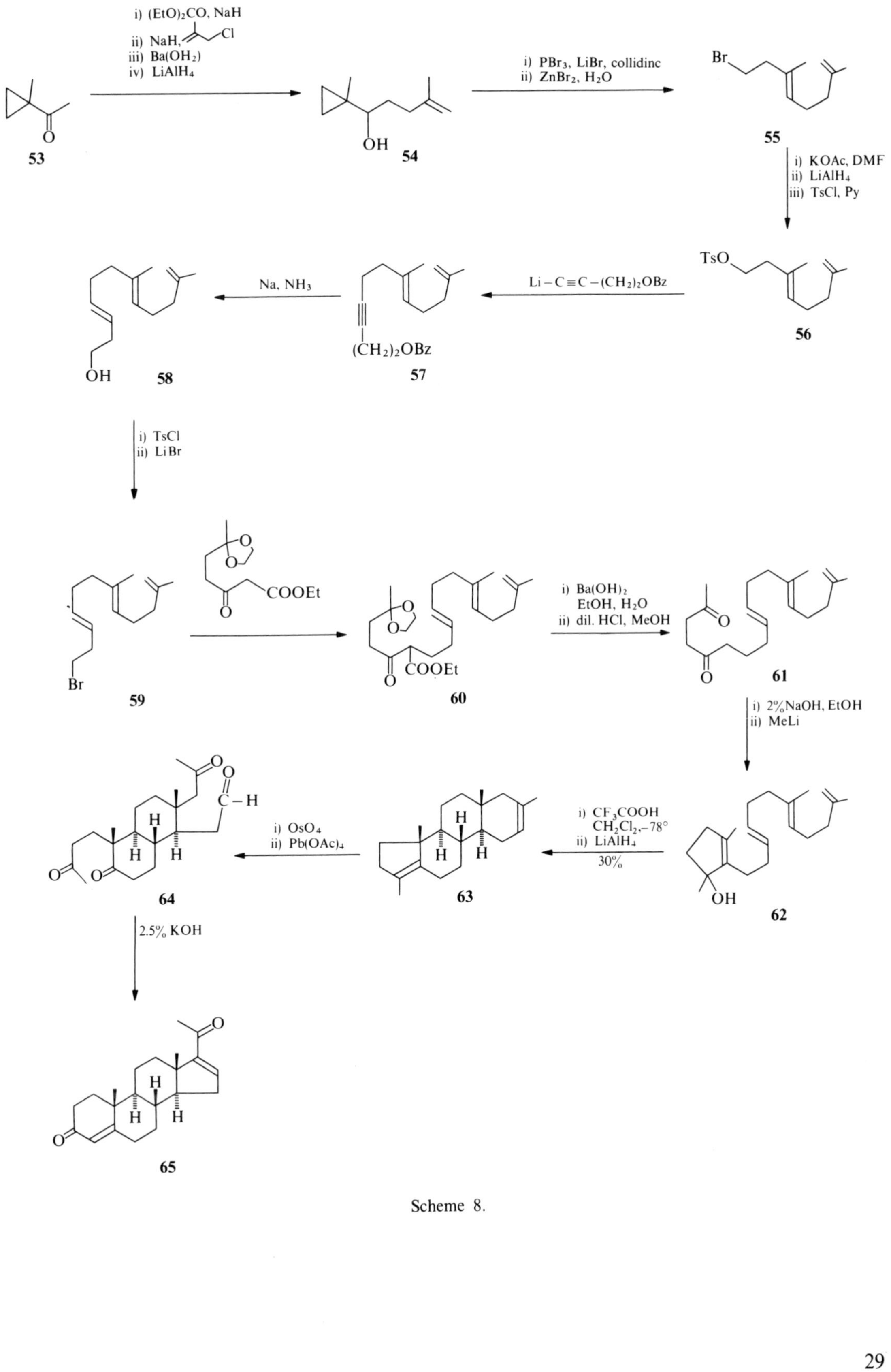

i) $(EtO)_2CO$, NaH
ii) NaH, (2-methylallyl chloride)Cl
iii) $Ba(OH_2)$
iv) $LiAlH_4$

53

OH
54

i) PBr_3, LiBr, collidine
ii) $ZnBr_2$, H_2O

Br
55

i) KOAc, DMF
ii) $LiAlH_4$
iii) TsCl, Py

TsO
56

$Li-C\equiv C-(CH_2)_2OBz$

$(CH_2)_2OBz$
57

Na, NH_3

OH
58

i) TsCl
ii) LiBr

Br
59

COOEt

COOEt
60

i) $Ba(OH)_2$, EtOH, H_2O
ii) dil. HCl, MeOH

61

i) 2% NaOH, EtOH
ii) MeLi

OH
62

i) CF_3COOH, CH_2Cl_2, $-78°$
ii) $LiAlH_4$
30%

63

i) OsO_4
ii) $Pb(OAc)_4$

C–H
64

2.5% KOH

65

Scheme 8.

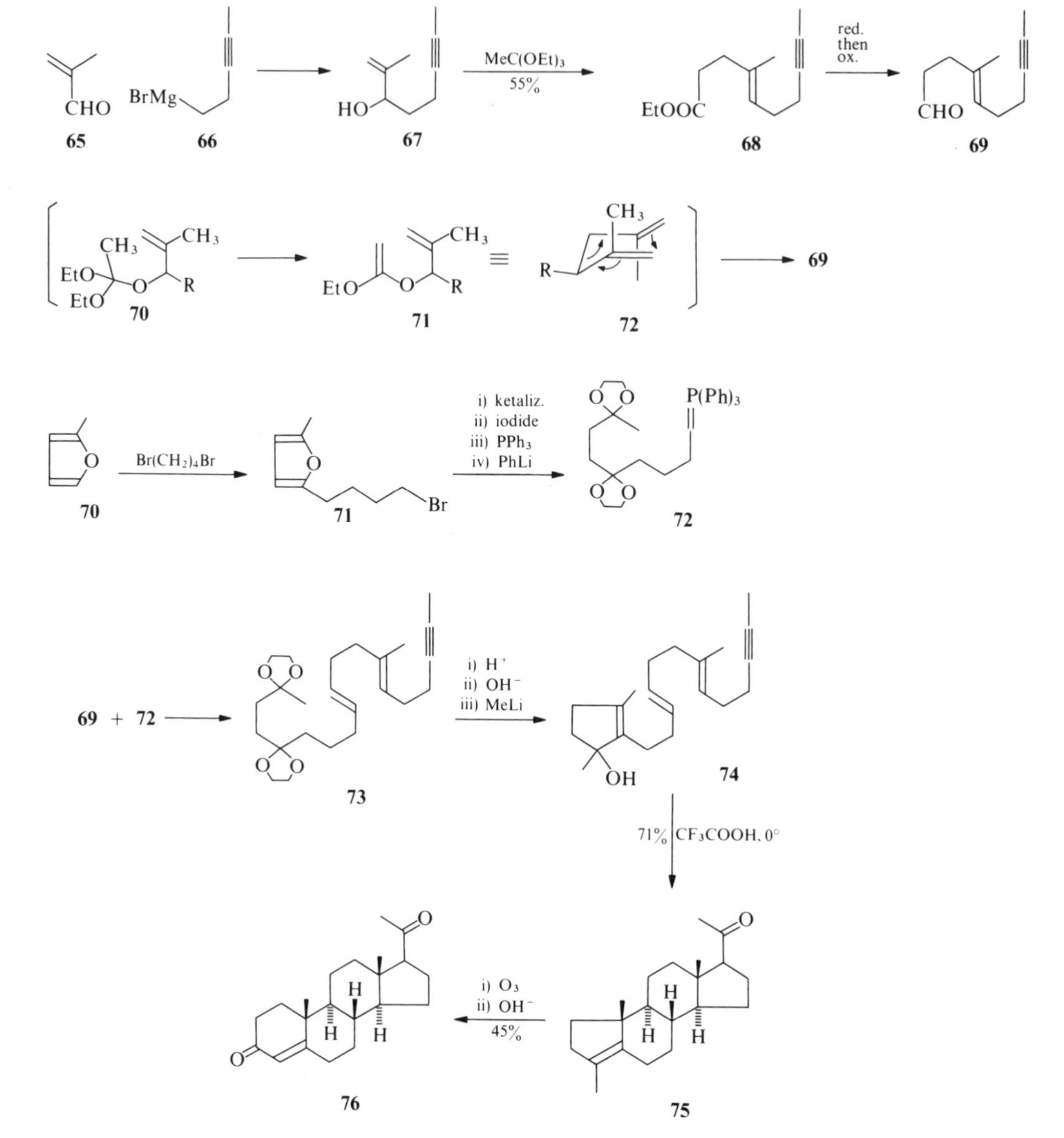

Scheme 9.

An ingenious modification[14] of scheme 8 involved application of the orthoester Claisen rearrangement[15] and use of acetylene bond participation in the cyclization. The Claisen rearrangement is highly stereoselective leading to almost exclusive formation of trans-enyne aldehyde **69**; the stereoselectivity is attributed to nonbonded interaction between the OEt and R groups that forces R to adopt an equatorial-like conformation, and hence insure *trans* backbone arrangement as in **68**. A Wittig reaction between aldehyde **69** and ylide **72** gave diketal dienyne **73**, which was cyclized to **74**. Modification of ring A yielded crude **76** (85:15 mixture of 17α and 17β) from which pure *dl*-progesterone was obtained by crystallization.

Recently Johnson reported the nonenzymatic cyclization of the tetraenic acetal **77** in which six asymmetric centers are produced and the pro-

[14] *W. S. Johnson, M. B. Gravestock, B. E. McCarry*, J. Am. Chem. Soc. *93*, 4332 (1971); *W. S. Johnson, M. B. Gravestock. R. J. Parry, R. F. Myers, T. A. Bryson, D. H. Miles*, J. Am. Chem. Soc. *93*, 4330 (1971).

[15] *W. S. Johnson, L. Werthemann, W. R. Bartlett, T. J. Brocksom, T.-t Li, D. J. Faulkner, M. R. Peterson*, J. Am. Chem. Soc. *92*, 741 (1970).

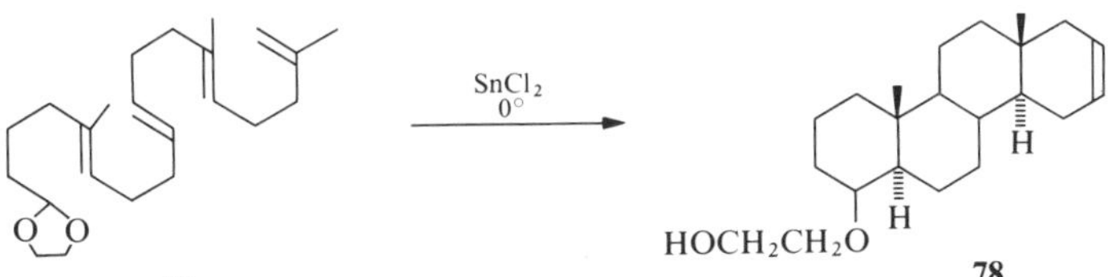

Scheme 10.

duct **78** has the D-homosteroid nucleus (Scheme 10).

Saucy and coworkers (1972)[17] have developed an elegant total synthesis of 19-norsteroids, utilizing the isoxazole moiety, and a similar approach has been used to prepare the unnatural retrotesterone[18] and retroprogesterone.[19] Resolution was accomplished on intermediate **83** in which the A, B and C rings were already assembled, and the D ring was then introduced by Michael addition of methylcyclopentanedione. Once more the stereochemistry of the CD ring junction was controlled by catalytic hydrogenation, and closure to give, as expected, the thermodynamically favored product. The overall yield was high, **92** being obtained from **93** in a yield of 10.2% (Scheme 11).

Precalciferol$_3$, the first isolable product of irradiation of the provitamins, which may in turn be reversibly transformed by mild heating in inert solvent into the corresponding vitamin D, has now been synthesized by Lythgoe and coworkers (Scheme 12)[20]. By utilization of the optically active intermediates **94** and **95**, the absolute configuration of C-13, -14 and -20 were predetermined. Carbonyl addition of **101a** to **101** completed the carbon skeleton, the stereochemistry at C-1 and the double bond position in ring A being established by the optically active cyclohexene adduct **102**. An overall yield of 0.34% from **93** was obtained. The work constitutes the first total synthesis of a vitamin D by nonphotochemical methods. In a previous formal synthesis the 5-*trans* vitamin was first prepared and then converted by irradiation to the 5-*cis* isomer[21].

A number of syntheses of ecdysones, which involve conversions from other naturally occurring steroids, have been reported. Ecdysones have been obtained by conversion of ergosterol derivatives[26,27], stigmasterol[28], and 3β-hydroxybisnorchol-5-enic acid[29]. Nakanishi *et al.*[30] have converted diosgenin **105** to α-ecdysone **109** by a process commencing with the stereospecific reduction of the diosgenin spiro-ketal group. Key steps were the reduction of **105** to **106** (92% yield) and the reductive cleavage of the C_{16}-O bond in **107** to yield **108**.

---

[16] *W. S. Johnson, K. Wiedhaup, S. Brady, G. Olson*, J. Am. Chem. Soc. *96*, 3979 (1974).

[17] *J. W. Scott, G. Saucy*, J. Org. Chem. *37*, 1652 (1972); *J. W. Scott, R. Borer, G. Saucy*, J. Org. Chem. *37*, 1659 (1972).

[18] *G. Saucy, R. Borer*, Helv. Chim. Acta *54*, 2517 (1971).

[19] *A. M. Krubiner, G. Saucy, E. P. Oliveto*, J. Org. Chem. *33*, 3548 (1968).

[20] *T. M. Dawson, J. Dixon, P. S. Littlewood, B. Lythgoe, A. K. Saksena*, J. Chem. Soc. C 2960 (1971); *I. J. Bolton, R. G. Harrison, B. Lythgoe, R. S. Manwaring*, J. Chem. Soc. C 2944 (1971); *I. J. Bolton, R. G. Harrison, B. Lythgoe*, J. Chem. Soc. C 2950 (1971); *T. M. Dawson, J. Dixon, P. S. Littlewood, B. Lythgoe*, J. Chem. Soc. C 2352 (1971).

[21] *G. K. Kochi, D. Singleton, L. Andrews*, Tetrahedron *24*, 3503 (1968); *J. K. Kochi, D. Singleton*, J. Am. Chem. Soc. *90*, 1582 (1968).

[22] *N. Cohen, B. Banner, W. Eichel, D. Parrish, G. Saucy, J. Cassal, W. Meinter, A. Fuerst*, J. Org. Chem. *40*, 681 (1975).

[23] *S. Danishefsky, P. Cain, A. Nagel*, J. Am. Chem. Soc. *97*, 380 (1975).

[24] *S. Danishefsky, P. Cain*, J. Am. Chem. Soc. *98*, 4975 (1976).

[25] *T. Kametani, H. Nemoto, H. Ishikawa, K. Schiroyama, K. Fukumoto*, J. Am. Chem. Soc. *98*, 3378 (1976).

[26] *A. Furlenmeier, A. Furst, A. Langemann, G. Waldvogel, P. Hocks, R. Weichert*, Experientia *22*, 573 (1966).

[27] *D. H. R. Barton, P. G. Feakins, J. P. Poyser, P. G. Sammes*, J. Chem. Soc. *C* 1584 (1970).

[28] *H. Mori, K. Shibata, K. Tsuneda, S. Sawai*, Chem. Pharm. Bull. *16*, 563, 2416 (1968).

[29] *J. B. Siddall, A. D. Cross, J. H. Fried*, J. Am. Chem. Soc. *88*, 862 (1966).

[30] *E. Lee, Y.-T. Liu, P. Solomon, K. Nakanishi*, J. Am. Chem. Soc. *98*, 1634 (1976).

[31] *W. S. Johnson, D. K. Banerjee, W. P. Schneider, C. D. Gutsche, W. E. Shelberg, L. J. Chinn*, J. Am. Chem. Soc. *74*, 2832 (1952).

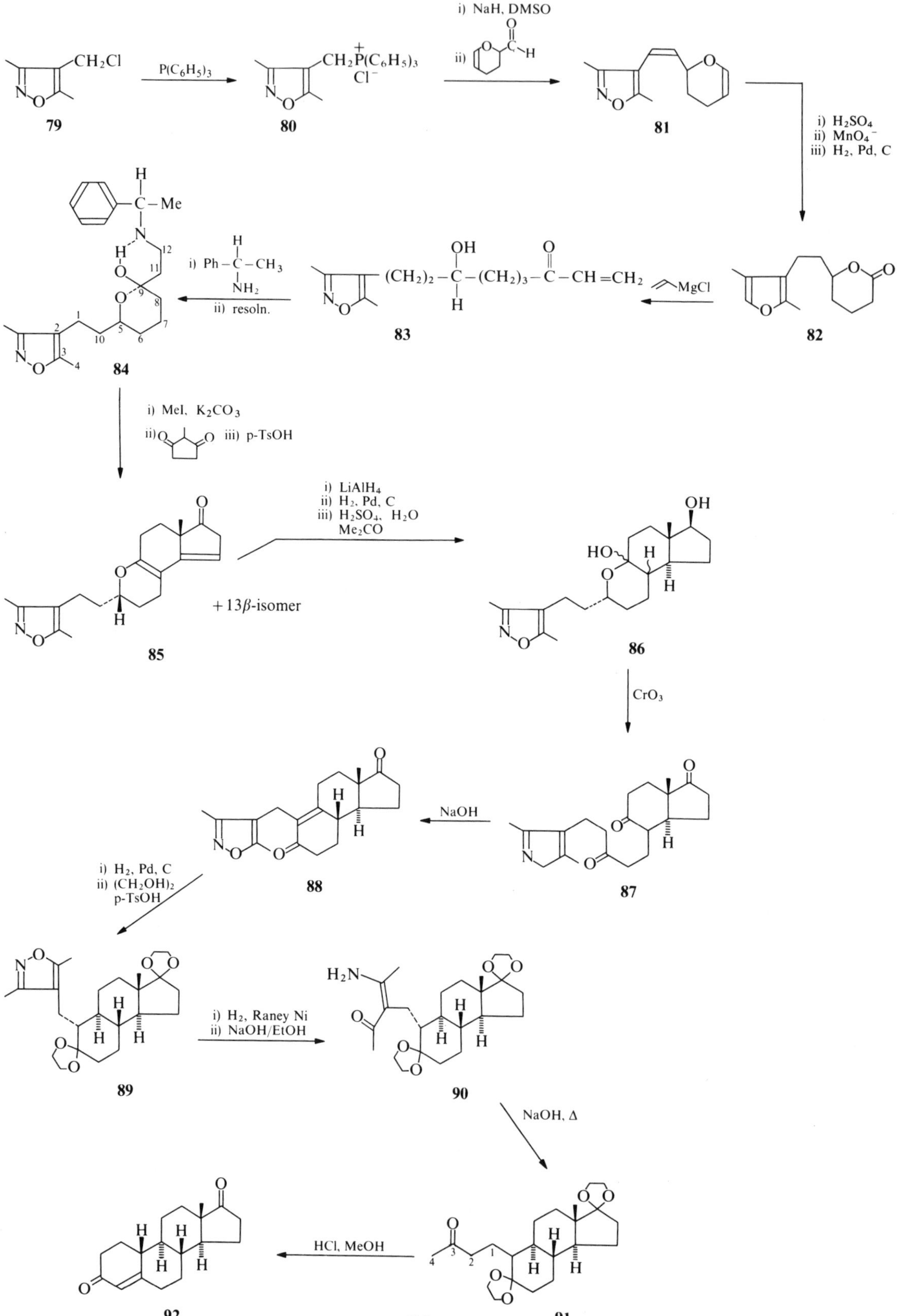

i) NaH, DMSO
ii)
$P(C_6H_5)_3$
CH_2Cl
$CH_2\overset{+}{P}(C_6H_5)_3$ Cl^-
79
80
81
i) H_2SO_4
ii) MnO_4^-
iii) H_2, Pd, C
$(CH_2)_2-\underset{H}{\overset{OH}{C}}-(CH_2)_3-\overset{O}{\overset{\|}{C}}-CH=CH_2$
MgCl
i) Ph–C(H)(NH₂)–CH₃
ii) resoln.
84
83
82
i) MeI, K_2CO_3
ii)
iii) p-TsOH
i) $LiAlH_4$
ii) H_2, Pd, C
iii) H_2SO_4, H_2O
Me_2CO
+ 13β-isomer
85
86
CrO_3
NaOH
88
87
i) H_2, Pd, C
ii) $(CH_2OH)_2$
p-TsOH
i) H_2, Raney Ni
ii) NaOH/EtOH
89
90
NaOH, Δ
HCl, MeOH
92
91

Scheme 11.

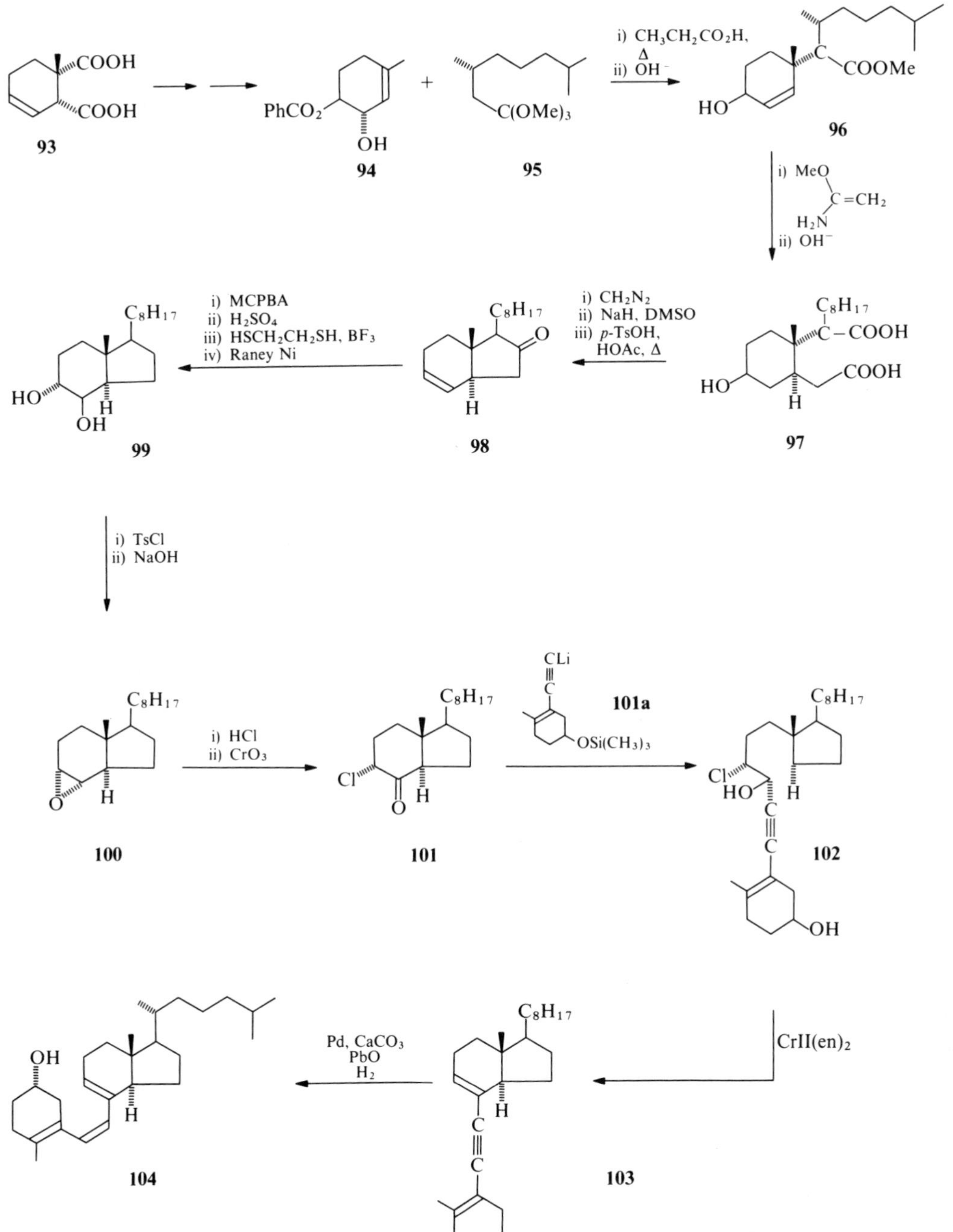

93 → → 94 + 95

i) $CH_3CH_2CO_2H$, Δ
ii) OH^-

96

i) MeO(H$_2$N)C=CH$_2$
ii) OH^-

97

i) CH_2N_2
ii) NaH, DMSO
iii) *p*-TsOH, HOAc, Δ

98

i) MCPBA
ii) H_2SO_4
iii) $HSCH_2CH_2SH$, BF_3
iv) Raney Ni

99

i) TsCl
ii) NaOH

100

i) HCl
ii) CrO_3

101

101a

102

CrII(en)$_2$

103

Pd, $CaCO_3$
PbO
H_2

104

Scheme 12.

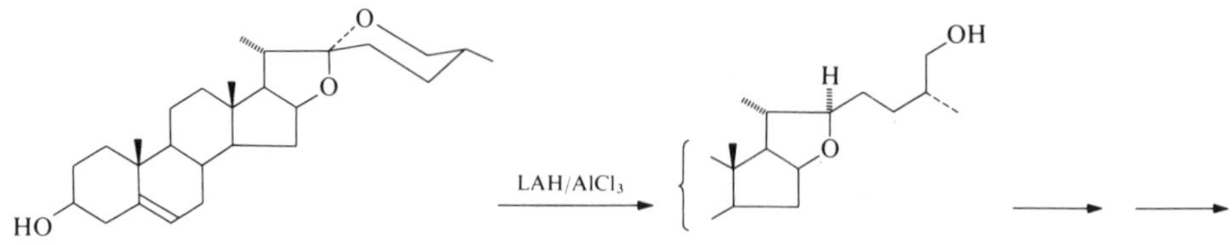

LAH/AlCl₃

105

106

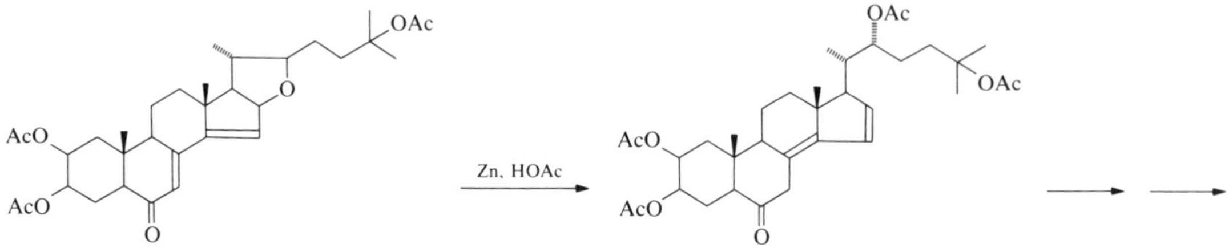

Zn, HOAc

107

108

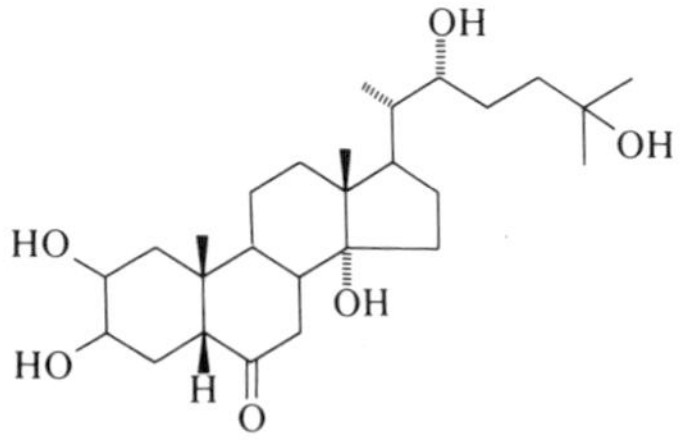

109

5 Interconversions

In addition to work in total synthesis, interest still continues in the interconversion of steroid classes and functionalities. Due to the biological importance of various derivatives much research has been carried out on the transformation of the readily available steroids to their more active forms. Many of these transformations are of particular interest as they possess wide application in other areas in addition to the steroid field.

Breslow, *et al.*, have developed a process of remote oxidation by which unactivated carbons can be functionalized by intramolecular hydrogen abstraction. Irradiation of the hemisuccinate of 3α,5α-androstane **1** with benzophenone-4-carboxylic acid gives the 17-keto steroid **2** in a 21% yield[1]. Furthermore, by attachment of a rigid benzophenone reagent to 3α-cholestanol, these workers have functionalized the steroid selectively at C-7, -12 and -14[2].

Remote oxidation is also achieved by attachment of a benzophenone reagent to ring D of androstan-17β-ol **3** or the side chain C-24 of cholestan-24-ol which permits the functionalization of C-14, -15 and -9. The latter is particularly important as it provides a route to the functionalization necessary for the 11-oxygenated corticosteroids. Although the yields are low in this process, the recovery of starting material is high[3].

Remote oxidation can also be achieved by the use of aryliodine reagents, either in solution or attached to the steroid skeleton by an ester linkage[30]. This methodology has been applied to a synthesis of cortisone[31].

A convenient conversion of primary and secondary α-ketols to α-diketones has been developed utilizing cupric acetate in methanol[4]. This reagent is particularly applicable to the preparation of 21-aldehydes from 21-hydroxy-20-ketones (for example **6** to **7**).

[1] *R. Breslow, P. Scholl,* J. Am. Chem. Soc. *93*, 2331 (1971).

[2] *R. Breslow, S. W. Baldwin,* J. Am. Chem. Soc. *92*, 732 (1970).

[3] *R. Breslow, P. Kalicky,* J. Am. Chem. Soc. *93*, 3540 (1971); *R. Breslow, S. Baldwin, T. Flechtner, P. Kalicky, S. Liu, W. Washburn,* J. Am. Chem. Soc. *95*, 3251 (1973).

[4] *M. Lewbart, V. R. Mattox,* J. Org. Chem. *28*, 2001 (1963).

hν
21%

1 → 2

i) $h\nu Cl_2FC-CF_2Cl$
ii) $Pb(OAc)_4$
iii) OH^-

3 → 4 (16%) + 5 (4%)

$Cu(OAc)_2$, MeOH
97%

6 → 7

$Cu(OAc)_2$
2,2′-bipyridyl
1,4-diazabicyclo[2,2,2]octane

8 → 9

DMSO
DCC
90%

10 → 11

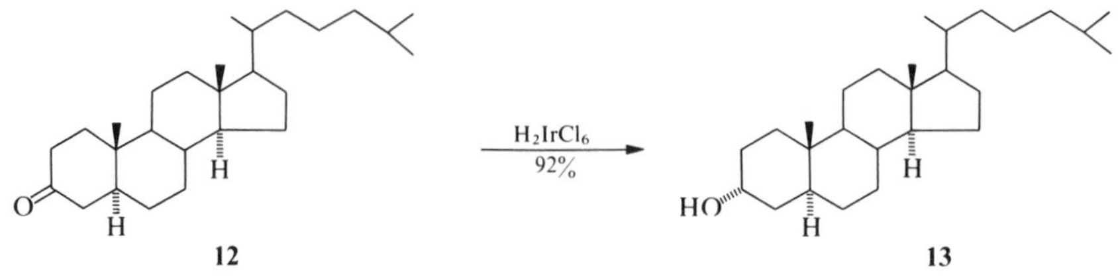

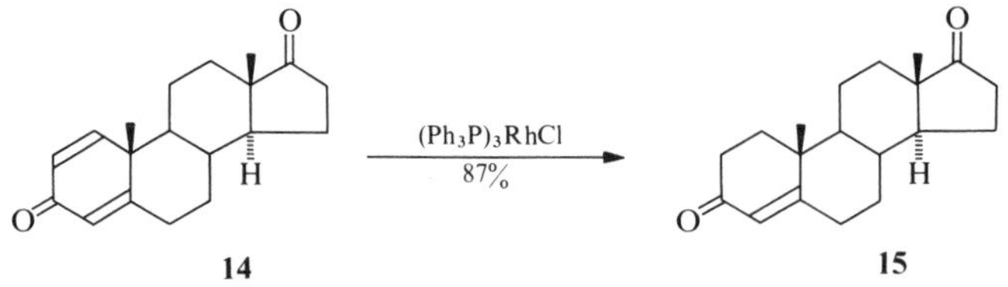

H_2IrCl_6
92%

12 → 13

$(Ph_3P)_3RhCl$
87%

14 → 15

An additional use for cupric acetate is the oxygenation of branched aldehydes to ketones. When complexed with an amine, cupric acetate has been used as catalyst for the oxidation of **8** to **9** in high yield[5].

The use of dimethylsulfoxide (DMSO) as an oxidizing agent has found extensive application in the steroid field, as the reactions involve mild conditions and give high yields. The system utilizing dicyclohexylcarbodiimide (DCC) and a weak acid oxidizes both primary and unhindered secondary alcohols to the corresponding ketone or aldehyde, while sulfonates, tertiary alcohols, olefins, esters and epimerizable centers are stable to the reagent[6]. Furthermore there is no difficulty with allylic oxidation (for example, **10** to **11**)[7]. The system has been modified by the use of an alkynyl amine or diphenylketene-*p*-tolylimine in place of the diimide[8].

An excellent method for reduction of an unhindered steroidal ketone (especially at the C-3 position) to the corresponding axial alcohol utilizes chloroiridic acid (or sodium chloroiridate) and trimethyl phosphite in 2-propanol[9].

Tris (triphenylphosphine) chlororhodium is a highly selective homogeneous hydrogenation catalyst which reduces the Δ^1, Δ^2 and Δ^3 double bonds but not isolated Δ^4 or Δ^5 olefins[10].

Steroids containing ene-diol or triol substituents distributed over C-3 through C-6 may be aromatized in ring A by treatment with HBr[11]. There is some evidence that this rearrangement occurs through a spiro-diene cation **17**[12].

It has been found that selective addition of an electrophilic agent to a side chain olefinic bond can be achieved in a stereospecific manner[13]. The approach of the electrophile is from the less hindered side and attack of the conjugate ion occurs away from the steroid nucleus. The stereochemistry of **21** was confirmed by X-ray.

A short route from ergosterol **22** to the biosynthetically important ergosteroldiene **24**[14] involves the dehydrobromination of the dibromo derivative **23** forming the less substituted diene system double bond[15]. Protection of the ring B double bond was achieved by formation of an adduct with a triazole.

The same triazole has been used to protect the B

[5] *V. Van Rhennen,* Tetrahedron Lett. 985 (1969).

[6] *K. E. Pfitzner, J. G. Moffatt,* J. Am. Chem. Soc. *87*, 5661, 5670 (1965).

[7] *D. C. Wigfield, S. Feiner, D. J. Phelps,* Steroids *20*, 435 (1972).

[8] *R. E. Harmon, C. V. Zenarosa, S. K. Gupta,* Chem. Commun. 327, 537 (1969).

[9] *Y. M. Y. Haddad, H. B. Henbest, Mrs. J. Husbands, T. R. B. Mitchell,* Proc. Chem. Soc. 361 (1964).

[10] *C. Djerassi, J. Gutzwiller,* J. Am. Chem. Soc. *88*, 4537 (1966).

[11] *J. R. Hanson,* Tetrahedron Lett. 4501 (1972).

[12] *J. Libmann, Y. Mazer,* Chem. Commun. 729 (1971); *J. R. Hanson,* Chem. Commun. 1119, 1343 (1971).

[13] *D. H. R. Barton, J. P. Poyser, P. G. Sammes, M. B. Hursthouse, S. Neidle,* Chem. Commun. 715 (1971).

[14] *D. H. R. Barton, T. Shioire, D. A. Widdowson,* J. Chem. Soc. C 1968 (1971).

[15] *A. B. Garry, J. M. Midgley, W. B. Whalley, B. J. Wilkens,* Chem. Commun. 167 (1972).

OAc

HBr

HO OH H

16 17 18

I $_2$/AgOAc
52%

H H I H
H OAc

OAc$^{\ominus}$

19 20 21

i) N N O, N–Ph, O
ii) Br_2

AcO 22

Br
Br
AcO N O N–Ph 23

1, 5-diazabicyclo-5-[4,3,0]non-5-ene
52%

AcO 24

OAc
NOF
Al_2O_3

AcO 25

AcO F N–NO_2 26

AcO F O 27

28 → 29: PbF_4, 27%

29 → 30: i) KOH, MeOH; ii) CrO_3; iii) NaOAc

31 → 32: $Ph_3C^+BF_4^-$

31: *a* R = CH_2Ph; *b* = THP; *c* = $OC(O)C_6H_4OCH_3$

32: *a*-60%; *b*-67%; *c*-90%

33 → 34: Ph_3CLi

34 → 35: $LiAlH_4$ *in situ*, 40–75%

ring diene system during the synthesis of vitamin D_3 metabolites[32]. For alternative methods of protecting the 5,7-diene system see Ref. 33.

As it has been shown that the introduction of a fluorine into a steroidal compound modifies and enhances its biological activity, great attention has been devoted to the development of new fluorination procedures. Among the more recent advances is the application of nitrosyl fluoride to an olefin to generate an α-fluoro ketone[16] (**25→27**).

A general laboratory procedure for addition of fluorine to an olefin utilizes lead tetrafluoride, generated *in situ* from lead dioxide and hydrogen fluoride[17]. Although the yields are low the starting material may be recovered.

Good yields of fluoro-steroids may be derived from the corresponding alcohol by the action of phenyl tetrafluorophosphorane on the trimethylsilyl ether[34]. Direct fluorination at saturated carbon, at positions 9, 14 and 17, can be achieved in a very selective manner using elemental fluorine or fluoroxy reagents (eg CF_3OF) due to the directive effects of polar substituents suitably situated on the steroid nucleus[35].

A new method for the deprotection of masked steroidal alcohols has been developed utilizing hydride transfer to the trityl cation. This method involves extremely mild and neutral conditions. Benzyl ethers, tetrahydropyranyl ethers, and benzyloxycarbonyl esters were successfully cleaved[18].

A novel method for the protection of steroidal ketones during a metal hydride reduction has been reported[19]. The α,β unsaturated 3-ketosteroid may be selectively converted to the metal enolate with subsequent reduction of the 11-keto function by lithium aluminum hydride. It has also been shown that a keto function on the 17-side-chain may likewise by protected.

Vanadium (IV) chloride has been found to react with the steroidal alcohol **36** to form the rearranged allylic chloride **37**, a useful intermediate for the production of the corticoid side chain[20].

[16] *G. A. Boswell, Jr.*, J. Org. Chem. *33*, 3699 (1968).

[17] *A. Bowers, P. G. Holton, E. Denot, M. C. Loza, R. Urquiza*, J. Am. Chem. Soc. *84*, 1050 (1962).

[18] *D. H. R. Barton, P. D. Magnus, G. Streckert, D. Zurr*, Chem. Commun. 1109 (1971).

[19] *D. H. R. Barton, R. H. Hesse, M. M. Pecket, C. Wiltshire*, Chem. Commun. 1017 (1972).

[20] *A. Krubiner, A. Perrotta, H. Lucais, E. P. Oliveto*, Steroids *19*, 649 (1972).

OH
CH=CH$_2$
CH$_3$O
H
36

VCl_4
NaOAc
70%

CH–CH$_2$Cl
CH$_3$O
H
+ $VOCl_2$
37

OAc
C≡CH
H
H
38

Zn
86%

CH$_2$
C
H
H
39

OAc
CH=CH$_2$
H
H
40

R–Cu$^{\ominus}$–RLi$^{\oplus}$
25–40%

CH$_2$–R
H
H
41

O
$^{\ominus}$O
$^{\oplus}$N
CH$_3$
H
42

p-TsCl
pyridine
52%

O
O
N
CH$_3$
H
43

R
O
H
44
a R=βMeO,H
b R=O

hν

R
O
45

R
O
46
a=20%
b=40%

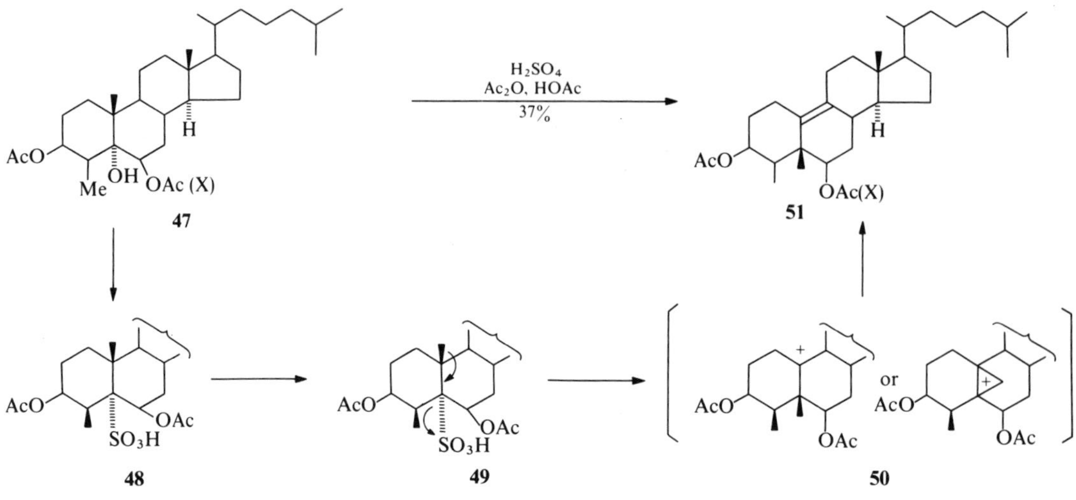

AcO Me OH OAc (X) H
47

H_2SO_4
Ac_2O, HOAc
37%

AcO OAc(X) H
51

AcO OAc SO_3H
48

AcO OAc SO_3H
49

AcO OAc or AcO OAc
50

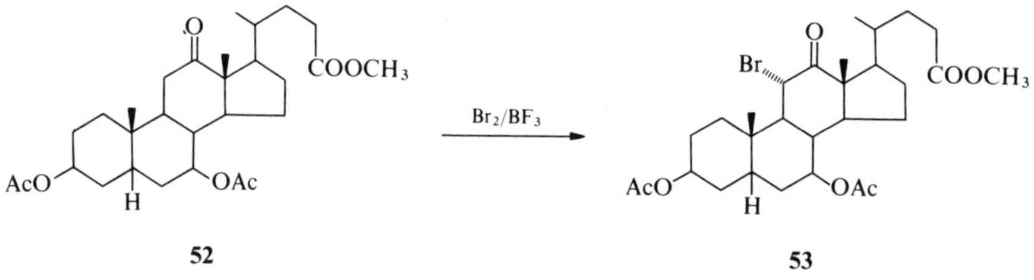

O COOCH$_3$ AcO H OAc
52

Br_2/BF_3

Br O COOCH$_3$ AcO H OAc
53

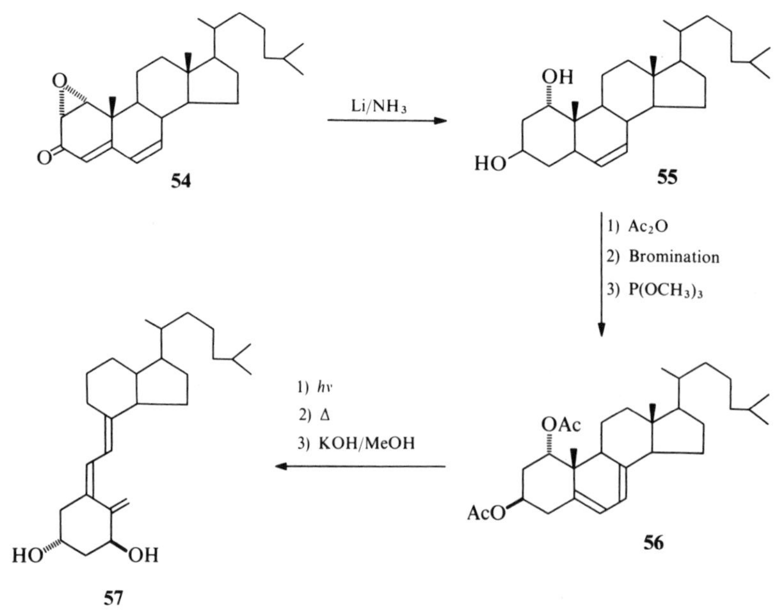

O O
54

Li/NH_3

OH HO
55

1) Ac_2O
2) Bromination
3) $P(OCH_3)_3$

OAc AcO
56

1) $h\nu$
2) Δ
3) KOH/MeOH

HO OH
57

Zinc dust has been employed as a convenient reagent for the conversion of the 17α-ethynyl carbinol acetate **38** to the allenyl steroid **39**, another intermediate suitable for the preparation of the corticoid side chain[21].

Lithium dialkylcopper has become increasingly important in the steroid field as an alkylating agent. It is particularly useful for the formation of the 21-alkyl side chains[22] (**40**→**41**).

A convenient alternative to the Beckman rearrangement has been developed for the formation of A-aza-A homosteroid analogues from a Δ^4-13-ketosteroid. Nitrone **42**, formed by conventional methods, on treatment with *p*-toluene-sulfonyl chloride in pyridine rearranges smoothly to the amide. The direction of rearrangement does not appear to depend on the stereochemistry of the nitrone as mixtures of isomeric nitrones gave a single amide product[23].

The *β*,*γ* unsaturated ketones **44a** and **44b** were found to undergo photoepimerization to the 5α compounds **46a** and **46b**[24]. The mechanism of epimerization proposed by the authors involved a diradical intermediate.

Much study has been devoted to the mechanism of the Westphalen rearrangement[25]. In a recent study on the rearrangements of the 6-acetoxy-5α-hydroxy-4-methyl steroids, numerous isomers were studied and kinetic data obtained. For these compounds, a synchronous methyl migration and C(5)–*O*–cleavage were proposed[26]. The reaction proceeds most effectively when the substituent X is electronegative and relatively inefficient as a participating neighboring group (*e.g.*, X=F or CN)[27].

A stereospecific bromination of **52** catalyzed by boron trifuoride has been reported[28]. No reaction was obtained with bromine in the presence of hydrobromic acid or sodium acetate. An intermediate complex with the boron trifluoride is proposed.

Barton and coworkers have obtained 1α-hydroxyvitamin D_3 **57** from the cholestadienone **54**. The key step is the reductive conversion of **54** with metal in liquid ammonia, in which the rate of reduction is faster than the rate of enolization. An alternative route to the B ring diene employs allylic oxidation of a Δ^5-steroid to the 5-en-7-one, followed by modified Bamford-Stevens reaction of the corresponding tosylhydrazone[36].

---

[21] *M. Biollaz, W. Haefliger, E. Velarde, P. Crabbé, J. H. Fried,* Chem. Commun. 1322 (1971).

[22] *P. Rona, L. Tokes, J. Tremble, P. Crabbé,* Chem. Commun. 43 (1969).

[23] *D. H. R. Barton, M. J. Day, R. H. Hesse, M. M. Pechet,* Chem. Commun. 945 (1971).

[24] *R. J. Chambers, B. A. Marples,* Tetrahedron Lett. 3747 (1971).

[25] *J. Westphalen,* Chem. Ber., *48*, 1064 (1915).

[26] *T. G. Ll. Jones, B. A. Marples,* J. Chem. Soc. C 572 (1971).

[27] *B. Marples, B. O'Callaghan, J. Scotlow,* J. Chem. Soc., Perkin I 1026 (1974).

[28] *Y. Yanuka, G. Halperin,* J. Org. Chem. *38*, 2587 (1973).

[29] *D. H. R. Barton, R. H. Hesse, M. M. Pechet, E. Rizzardo,* J. Am. Chem. Soc. *95*, 2748 (1973); *M. Galbraith, B. Horn, E. Middleton, J. Thompson,* Chem. Commun. 203 (1973).

[30] *R. Breslow, R. Corcoran, J. A. Dale, S. Liu, P. Kalicky,* J. Am. Chem. Soc. *96*, 1973 (1974); *R. Corcoran,* Tetrahedron Lett. 317 (1976).

[31] *R. Breslow, B. B. Snider, R. J. Corcoran,* J. Am. Chem. Soc. *96*, 6791 (1974).

[32] *S. C. Eyley, D. H. Williams,* J. Chem. Soc. Perkin I 727, 731 (1976).

[33] *D. H. R. Barton, A. A. Leslie Gunatilaka, T. Nakanishi, H. Patin, D. A. Widdowson, B. R. Worth,* J. Chem. Soc. Perkin I 821 (1976).

[34] *N. Bautin, D. Robert, A. Combon,* Bull. Soc. Chim. Fr. *12*, 2861 (1974).

[35] *D. H. R. Barton, R. H. Hesse, R. E. Mackwell, M. M. Pecket, S. Rosen,* J. Am. Chem. Soc. *98*, 3036 (1976).

[36] *R. Moriarity, H. Paaren, J. Gilmore,* Chem. Commun. 927 (1974).

2 Terpenoids

Contributions by

S. Hayashi, Hiroshima
T. Kato, Sendai
Y. Kitahara, Sendai

In this chapter sesterterpenoids and triterpenoids will not be handled; they are discussed in Chapter 3 and in Chapter 6.2.

2.1 Mono- and Sesquiterpenoids

Shuichi Hayashi
Faculty of Science, Hiroshima University, Hiroshima, 730 Japan.

1 Introduction

When leaves, stems, flowers or fruits of a plant are distilled with steam, an oil is obtained which is called essential oil because it has the characteristic odor of the origin. The lower boiling fraction of the oil is known in a great portion to be a mixture of acyclic and alicyclic isomers of $C_{10}H_{16}$ hydrocarbons and their oxygen derivatives[1].

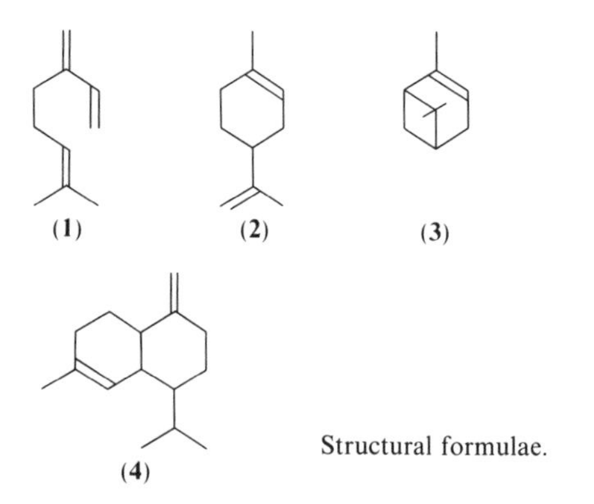

Structural formulae.

Myrcene (**1**) which was first isolated from bay oil, the essential oil of *Myrcia acris* D.C, has the molecular formula of $C_{10}H_{16}$ and is formally a head-to-tail dimer of isoprene molecules. Limonene (**2**), a major compound of the essential oil from the rind of citrus, is a Diels-Alder addition product of one molecule of isoprene to the other molecule, and α-pinene (**3**), contained in turpentine oil in a large amount, is formed in theory by further cyclization of limonene. In addition, many compounds combining three isoprene units in head-to-tail fashion such as γ-cadinene (**4**) are isolated from the higher boiling fraction of essential oils. This mode of construction of the carbon skeleton has been rationalized by the term, isoprene rule. The compounds obeying the isoprene rule are defined as terpenoids, and members of higher classes of terpenoids also occur in the vegetable and animal kingdoms.

Terpenoids are classified according to the number of isoprene units in the molecule.

Number of carbon atoms	Class
10	Monoterpenoids
15	Sesquiterpenoids
20	Diterpenoids
25	Sesterterpenoids
30	Triterpenoids
40	Tetraterpenoids
>40	Polyterpenoids

The simplest members contain two units. Sesquiterpenoids are being found increasingly. Tetraterpenoids are plant pigments known as carotenoids and polyterpenoids include rubber, gutta-percha and balata.

2 Structure Elucidation of Monoterpenoids

The fundamental features of the chemistry of monoterpenoids, including acyclic and cyclohexano-monoterpenic compounds, were delineated before 1920. In recent years, newer members of the monoterpenoids named the iridoids, which are characterized by the cyclopentanopyran ring system, have been detected in nature. Because a large number of these members shows remarkable physiological activity, the chemistry of this area has rapidly expanded[2]. Nepetalactone[3] was isolated as the attractive principle for many species of the cat family from the essential oil of the catmint plant, *Nepeta cataria* L. Nepetalactone (**5**) which has the molecular formula of $C_{10}H_{14}O_2$ is an unusual enol-lactone; hydrolysis with dilute sodium hydroxide solution furnished the epimeric (*) nepetalic acids (**6**) containing a methyl and carboxyl group together with a –CH(Me)CHO side chain. The structure of this side chain was established by alkaline hydrogen peroxide oxidation which gave the carboxy methyl ketone (**7**). Further oxidation of **7** with sodium hypoiodite yielded the dibasic acid (**8**), which could readily form an

[1] *J. L. Simonsen*, The Terpenes, Vols. I–V, The University Press, Cambridge 1947–1957; *E. Guenther*, The Essential Oils, Vols. I–VI, Van Nostrand, New York 1948–1952; *G. Gildemeister, Fr. Hoffmann*, Die Ätherischen Öle, Vols. I–V, Akademie-Verlag, Berlin 1956–1959.

[2] *W. I. Taylor*, Cyclopentanoid Terpene Derivatives, Marcel Dekker, Inc., New York 1969.

[3] *S. M. McElvain, R. D. Bright, P. R. Johnson*, J. Am. Chem. Soc. *63*, 1558 (1941); *S. M. McElvain, E. J. Eisenbraut*, J. Am. Chem. Soc. *77*, 1599 (1955).

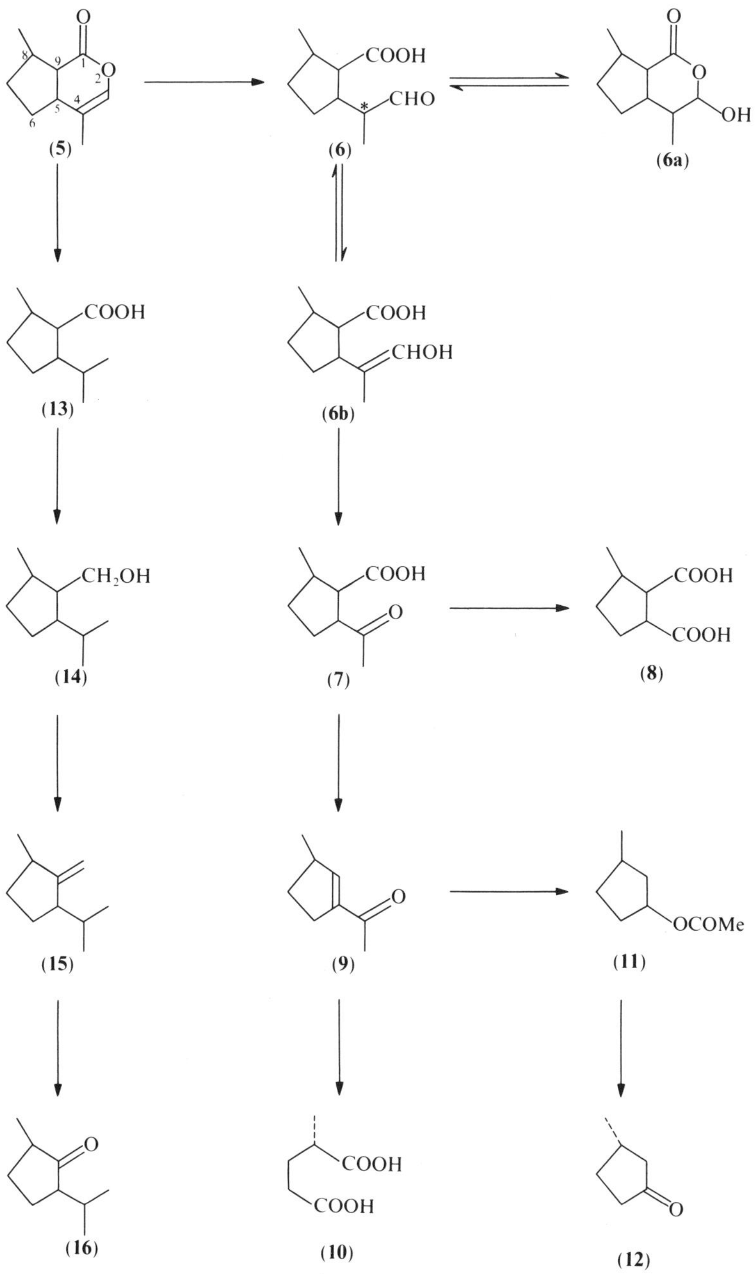

Scheme 1.

anhydride on treatment with acetic anhydride. Based on this easy formation of the anhydride and the fact that the original nepetalic acid (**6**) showed aldehyde–lactol tautomerism, the carboxyl group and the –CH(Me)CHO substituents can be assigned adjacent positions on a methyl cyclopentane nucleus. When **7** was subjected to the action of lead dioxide, oxidative decarboxylation occurred to yield the cyclopentenyl methyl ketone (**9**). The ketone was oxidized with potassium permanganate to give acetic acid and known (+)-α-methylglutaric acid (**10**) and it was also, after being converted to the saturated ketone, degraded to known (+)-3-methylcyclopentanone (**12**) by a reaction sequence of Baeyer-Villiger reaction, saponification and oxidation with chromic acid. The formation of (+)-α-methylglutaric acid and (+)-3-methylcyclopentanone reveals the cyclopentane nucleus and 1,2,3-relationship of the methyl, carboxyl and –CH(Me)CHO substituents in nepetalic acid. Furthermore, the structure of nepetalactone

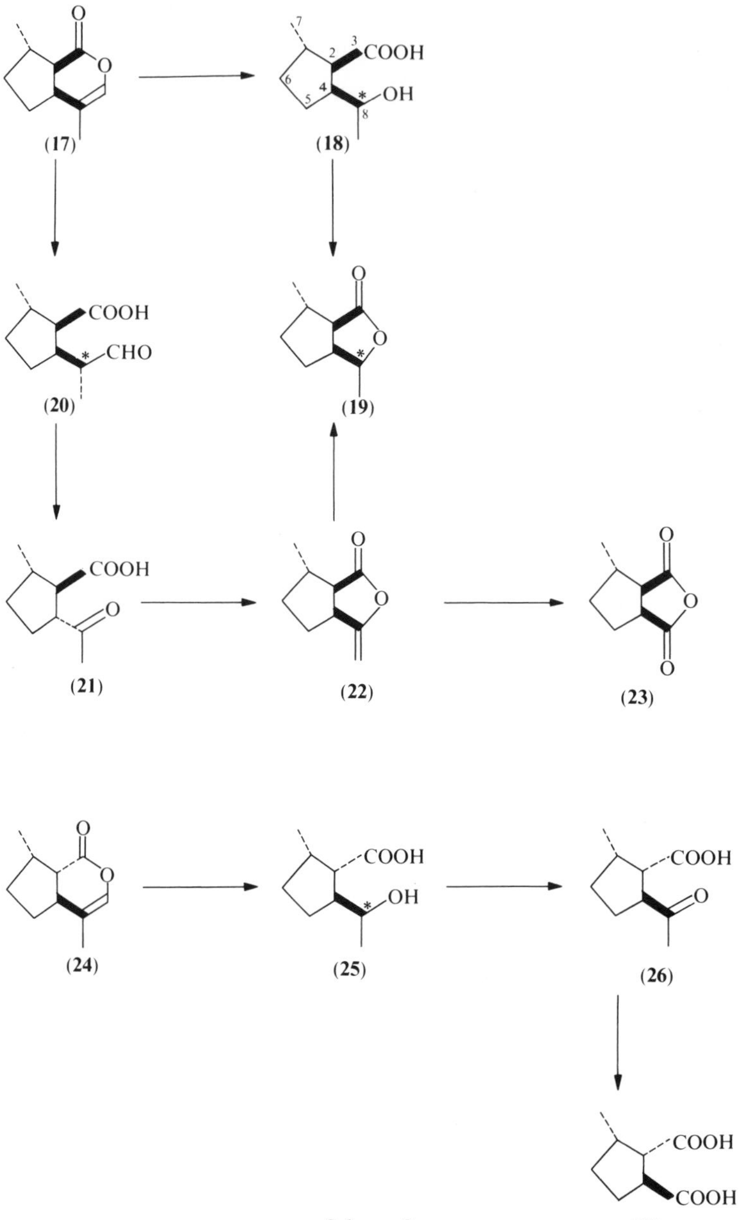

Scheme 2.

was confirmed by an alternative reaction sequence[4]. Hydrogenolysis of the enol-ether gave 2-methyl-5-isopropylcyclopentane carboxylic acid (**13**). The acid was pyrolyzed, via the alcohol (**14**) and its acetate, to give the olefin (**15**), the ozonolysis of which afforded the known 2-methyl-5-isopropylcyclopentanone (**16**).

The major constituent of oil of catnip was established to be *cis, trans*-nepetalactone (**17**) in the following way[5]. When the nepetalactone was ozonolyzed and the product was reduced with sodium borohydride at $-70°$, the epimeric (*) hydroxy acids (**18**) were obtained, which readily formed the γ-lactone (**19**). Since such mild conditions should not cause an inversion of configuration at C-2 or C-4, both substituents at C-2 and C-4 take the *cis* configuration. Furthermore, the carboxy methyl ketone (**21**) which was ob-

[4] *J. Meinwald,* J. Am. Chem. Soc. *76*, 4571 (1954).

[5] *R. B. Bates, E. J. Eisenbraun, S. M. McElvain,* J. Am. Chem. Soc. *80*, 3420 (1958).

tained as mentioned above from the nepetalactone by hydrolysis and oxidation was converted into the enolic lactone (**22**) by treatment with acetic anhydride, which gave one of the epimers of the *cis*-γ-lactone (**19**) by catalytic hydrogenation. Ozonolysis of **22** also furnished the anhydride (**23**) showing absorption bands at 1852 and 1776 cm^{-1} characteristic of a succinic anhydride. The identification of this anhydride was carried out by synthesis of the ($\pm$)-*cis, trans*-anhydride (**23**). The stereochemistry of the three substituents in the nepetalactone was thus established to be the *cis, trans*-configuration. Since the formation of (+)-α-methylglutaric acid (**10**) in the above-mentioned degradation of nepetalic acid correlates the *cis, trans*-nepetalactone with L-glyceraldehyde, the absolute configuration of *cis, trans*-nepetalactone is given by the formula (**17**).

In the ozonolysis–sodium borohydride treatment of natural nepetalactone, another nepetalic acid (**25**) was isolated together with the above *cis, trans*-nepetalactones (**18**). This acid was not lactonized by the action of chromic acid, but was converted into the carboxy methyl ketone (**26**), the structure of which was identified, after oxidation to the dibasic acid (**27**) and by comparison of the IR spectrum with that of the synthetic compound. The result indicates *trans,cis*-nepetalactone (**24**) to be a minor component of catnip oil. Gas chromatography of natural nepetalactone, in fact, exhibited 75%of the *cis, trans*-isomer and 25%of the *trans, cis*-compound.

3 Synthesis of Monoterpenoids

Various compounds have been identified, similar to *cis,trans*-nepetalactone and its derivatives, including some chemical defenses of the insects (iridomyrmecin, isoiridomyrmecin and iridodial) isolated from the secretion of the anal glands of the Argentine ant, *Iridomyrmex humilis*, and of Australian *Iridomyrmex* ants[6], as well as the attracting and exciting principles against the cat family (iridomyrmecin, isoiridomyrmecin, dihydronepetalactone, isodihydronepetalactone and neonepetalactone) and strong attracting substances against *Chrysopa septempunctata Wesmael* (neomatatabiol, matatabiether, allomatatabiol and others) in essential oils of galls and leaves of *Actinidia polygama* Mig[7]. Many syntheses of these compounds have been reported starting from citronellal[8], *e.g. trans*-pulegenic acid prepared by the ring contraction of (+)-pulegone[9] and other compounds[10].

The synthesis using 2,6-dimethylbicyclo[3,3,0] octan-3-one as a key intermediate is herein described[7,11]. The key compound has ideal stereochemistry for the synthesis of these cyclopentano-monoterpenoids, $H_{(c-1)}$–$H_{(c-5)}$–$H_{(c-6)}$: *cis-trans*, and was prepared stereospecifically from ethyl 3-methyl-2-oxocyclopentane carboxylate. The condensation product of the ethyl ester with 3-bromobutyne was, after hydration, hydrolyzed to compound (**30**), which was also synthesized by an alternative route of condensation of the ester with 1-bromo-ethylmethylketone directly. The compound (**30**) was treated with dilute hydrochloric acid to yield the diketone (**31**), which was converted to the key compound (**33**) through intramolecular condensation by alkali and catalytic hydrogenation.

The benzilidene derivative (**34**) of the key compound was ozonolyzed to give the dicarboxylic acid (**35**) which was identical with racemic *cis, trans*-nepetalinic acid in the melting point and IR spectrum.

Then the benzilidene derivative (**34**) was reduced with sodium borohydride and treated with acetic anhydride to give a crystalline acetate and an oily one (**36**), the latter predominating much over the former. The ozonolysis of the predominant oily acetate furnished the α-acetoxyketone (**37**). Although the treatment of the α-acetoxyketone with dilute alkali afforded the α-ketol, this product was too unstable to be distilled and the acetoxy ketone was immediately oxidized with periodate to the aldehyde carboxylic acid (**38**). The product, with some character of lactol (**39**),

---

[6] *G. W. K. Cavill, D. L. Ford, H. J. Locksley*, Aust. J. Chem. *9*, 288 (1956); *10*, 352 (1957).

[7] *T. Sakan, F. Murai, S. Isoe, S. B. Hyeon, Y. Hayashi*, Nippon Kagaku Zasshi *90*, 507 (1967).

[8] *K. J. Clark, G. I. Fray, R. H. Jager, R. Robinson*, Tetrahedron *6*, 217 (1959).

[9] *S. A. Achmad, G. W. K. Cavill*, Aust. J. Chem. *16*, 858 (1963); *J. Wolinsky, D. Chan*, J. Org. Chem. *30*, 41 (1965); *S. A. Achmad, G. W. Cavill*, Aust. J. Chem. *18*, 1989 (1965); Proc. Chem. Soc. 166 (1963).

[10] *J. Wolinsky, T. Gibson, D. Chan, H. Wolf*, Tetrahedron *21*, 1247 (1965).

[11] *T. Sakan, A. Fujino, F. Murai, A. Suzui, Y. Butsgan*, Bull. Chem. Soc. Japan *33*, 1737 (1960).

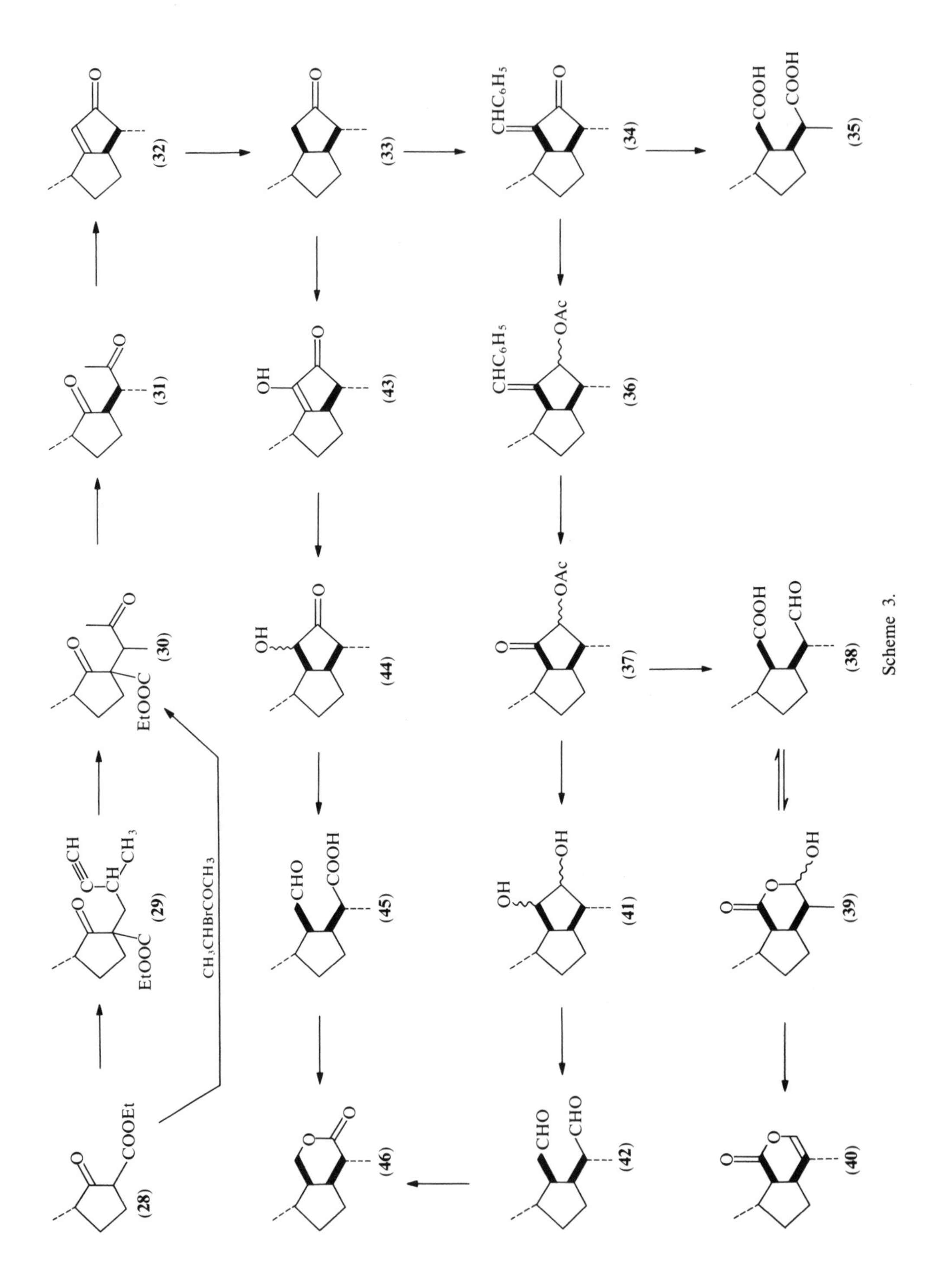

COOEt
(28)
EtOOC
C≡CH
CH
CH3
(29)
CH3CHBrCOCH3
EtOOC
(30)
(31)
(32)
(33)
CHO
COOH
(45)
(46)
OH
(44)
(43)
CHC6H5
(34)
OAc
(36)
(37)
OH
(41)
CHO
(42)
COOH
(35)
(38)
(39)
(40)

Scheme 3.

was dehydrated by heating at about 200° to give (±)-*cis,trans*-nepetalactone. The α-acetoxyketone (**37**) also, after conversion to the glycol (**41**), was oxidized with periodate to yield (±)-iridodial (**42**).

On the other hand, the key compound was oxidized with selenium dioxide and then subjected to catalytic hydrogenation followed by lead tetraacetate oxidation to give the aldehyde carboxylic acid (**45**). The sodium salt of the aldehyde carboxylic acid was reduced with sodium borohydride. The hydroxy acid thus obtained was finally converted to (±)-isoiridomyrmecin (**46**) by treatment with dilute hydrochloric acid.

4 Structure Elucidation of Sesquiterpenoids

The liverworts (*Hepaticae*), placed phylogenetically between the vascular plants and the algae, form a unique division in the plant kingdom. Their haploidal plant bodies (gametophytes) which grow from the spores contain characteristic oil bodies in the cell. From these liverworts several novel sesquiterpenoids have been isolated.

Taylorione (**47**)[12] was isolated from *Mylia taylori* (Hock.) Gray belonging to the *Jungermanniaceae*, by adsorbing the hexane extract onto silica gel and eluting with a mixed solvent of hexane and ethyl acetate (4:1). Elemental analysis of the ketone yielded $C_{15}H_{22}O$ (M^+ 218) and the UV, IR and NMR spectral data indicated that it contained a *gem*-dimethyl, three active methylenes connected to a C=O and/or C=C bond and an exo double bond conjugated with a tri-substituted double bond. When the ketone was hydrogenated over PtO_2 in AcOH, a saturated tetrahydro ketone (**48**) was produced, containing a newly formed secondary methyl and active methylene together with the two original groups, *gem*-dimethyl and acetyl. Thus, it is certain that taylorione is a bicyclic sesquiterpene ketone containing a conjugated diene system.

The NMR spectrum of tetrahydrotaylorione showed clearly the signal due to two cyclopropane protons (δ 0.2–0.5, 2H), but the IR spectrum did not show an absorption band attributable to cyclopropane methylene. This

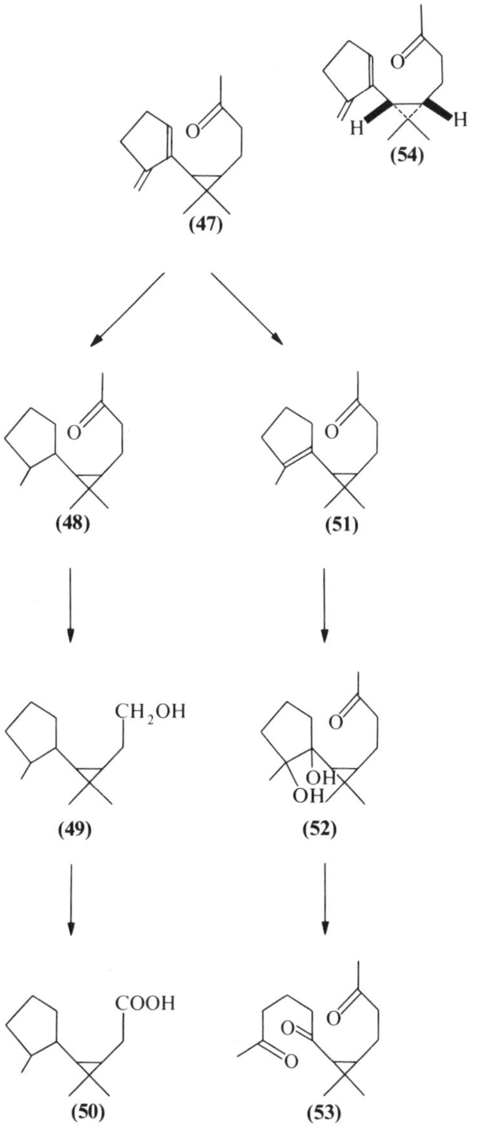

(47) (54) (48) (51) (49) CH₂OH (52) (50) COOH (53)

Scheme 4.

fact indicates that taylorione and tetrahydrotaylorione have in common, as a partial structure, a tetra-substituted cyclopropane system which consists of two methines and a quarternary carbon atom bearing the *gem*-dimethyl group. In tetrahydrotaylorione, also, it is certain that the acetyl group connects to a saturated carbon chain, because the active methylene adjacent to the carbonyl group showed a triplet ($J = 8.0$) centering at δ 2.40. The length of this connecting carbon chain was determined by a series of degradation reactions: tetrahydrotaylorione was submitted to the Baeyer-Villiger reaction with CF_3CO_3H and the ester thus obtained was then hydrolyzed with KOH to give

[12] A. Matsuo, S. Sato, M. Nakayama, S. Hayashi, Tetrahedron Lett. 3681 (1974).

a bis-nor primary alcohol (**49**) which was converted into an acid (**50**) by a Jones oxidation. As to the degradation products, the carbinol methylene of the bis-nor alcohol (**49**) showed a triplet ($J=7.0$) and the active methylene of the acid (**50**) a doublet ($J=7.0$). These facts indicate that, in taylorione and tetrahydrotaylorione, the acetyl group connects to a methine group with a distance of two methylene units.

The residual portion of the taylorione molecule, on the basis of the molecular formula and the above-mentioned spectroscopic evidence, was deduced to contain the exo double bond conjugated with a tri-substituted double bond and two active methylenes adjacent to the double bonds. For such partial structure a 2-alkyl-1-methylen-cyclopent-2-ene system was selected from three possible structures by a series of the following reactions; taylorione took up one molar equivalent of hydrogen during catalytic hydrogenation over Pd–C in EtOH to give a dihydro ketone (**51**) which contained a tetra-substituted double bond bearing a methyl group. The formation of such a vinylic methyl is expected in the 1,4-addition of hydrogen to the conjugated diene system of the taylorione molecule and the value of the UV absorption maximum (λ_{max}^{EtOH} 212 nm, ε 5900) suggests that a tetra-substituted double bond is in conjugation with the cyclopropane ring. The dihydro ketone was then oxidized with osmic acid to give a glycol which, by means of glycol fission with periodic acid, was lastly converted into a tri-ketone (**53**). The tri-ketone thus obtained possessed a ketone group conjugating a cyclopropane ring and two acetyl groups together with a *gem*-dimethyl, three active methylenes and two aliphatic methylenes.

From the above evidence, the gross structure of taylorione can be expressed by formula **47**. This novel carbon skeleton may be biosynthesized from an aromadendrene-type precursor, which is common to co-occurring myliol, *via* the oxidative cleavage of the C_1–C_{10} bond.

For the conformation of the proposed structure and the stereochemistry, taylorione was ozonolyzed at $-15°$ to give a cyclopentyl carboxylic acid, $C_{10}H_{18}O_3$; $[\alpha]_D$ $-20.6°$. The ozonolysis product was identical in every respect with *cis*-1-methoxycarbonyl-2,2-dimethyl-3(3′-oxo-*n*-butyl)-cyclopropane which was obtained by the similar ozonolysis of (+)-car-2-ene and whose absolute configuration has been established, except the negative sign of the optical rotation was opposite to the positive sign of that of the decomposition product from (+)-carene. The absolute configuration of taylorione, hence, should be represented by formula **54**.

5.1 General Remarks of Sesquiterpenoids*

Although the sesquiterpenoids appear to be composed of three isoprene units linked head-to-tail, their carbon skeleton is very much varied owing to secondary C–C bond formation and molecular rearrangement during the biogenetic course.

In this chapter, the description is limited to germacrane, elemane, guaiane and ambrosane classes, because in these classes of the sesquiterpenoids the most remarkable development have been made.

5.2 Germacrane Class

So far about 80 compounds of this class have been isolated. Although these compounds include hydrocarbons, alcohols, epoxides, ketones (the above are nonlactonic), γ-lactones (germacranolides) and furanolides, the members of this class are thermally unstable, owing to the fact that they have an unsaturated ten-membered ring containing two endo-cyclic double bonds or epoxides at the 1(10)- and 4-positions as a common structure of the molecule. Especially, the sesqui-hydrocarbons of this class contain an additional double bond in the interior or exterior of the ring and are senstive to heating. The detection of hydrocarbons such as germacrene A[184], C[146] and D[158] was therefore delayed until 1970, when they were isolated by employing a low temperature procedure.

5.2.1 Nonlactonic Compounds

Among ten members of non-lactonic compounds of this class, shiromool and shiromodiol mono- and di-acetate were found to be insect anti-feedant[130,183].

* Hereafter, Ref. in chronological order.

13 *Y. Asahina, C. Ukita*, J. Pharm. Soc. Japan *61*, 376 (1941).

14 *S. Naito*, J. Pharm. Soc. Japan *75*, 93 (1955).

15 *G. Büchi, D. Rosenthal*, J. Am. Chem. Soc. *78*, 3860 (1956).

16 *Z. Cěkan, V. Herout, F. Sŏrm*, Chem. Ind. 1234 (1956).

17 *V. Sykora, V. Herout, F. Sŏrm*, Coll. Czech. Chem. Commun. *21*, 267 (1956).

18 *D. H. R. Barton, C. R. Narayanan*, J. Chem. Soc. 963 (1958).

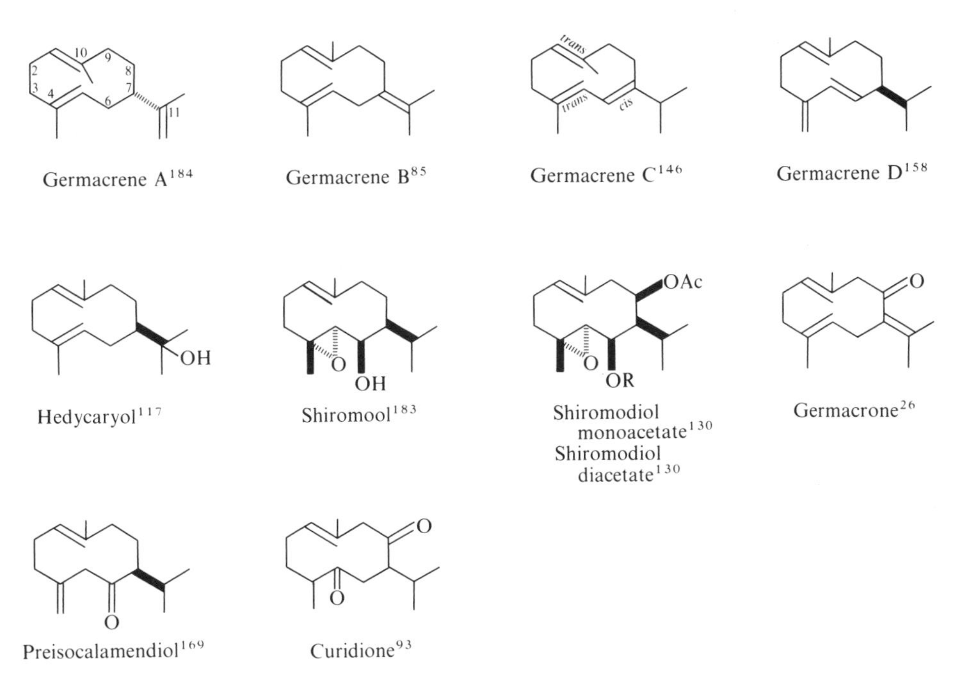

Fig. 1. Nonlactonic sesquiterpenoids of the germacrane class.

5.2.2 Germacranolides

This group of germacrane class includes monolactones and dilactones, and all their members have one lactone functional group at C-12, with the exception of aristolactone[55], which carries this function between C-6 and C-15. The germacranolides, accordingly, are classified into two subgroups, *i.e.* C-6 closing and C-8 closing types depending upon direction of the lactone formation involving the C-12 functional group. As far as the stereochemistry is concerned, both closing types uniformly have a stereostructure of C-12 lactone function attaching to C-6 or C-8 in *trans*-fashion; H-7 always takes α-orientation and H-6 or H-8, β. Pyrethrosin[22] is the only exception with a *cis*-fused lactone.

In the dilactones of the C-6 closing type, the additional lactone ring is oriented between C-14 and C-2 accompanied by α-oriented H-2, and in those of the C-8 closing type, between C-15 and C-6 with β-oriented H-6.

Urospermal was reported to occur in two conformers; one conformer has a hydrogen bond between the aldehydic and secondary hydroxyl groups, and the other one holds the H-bond between the primary and secondary hydroxyl groups[160].

Recently, relactonization of some germacranolides which contain two α-oriented lactonizable oxygen functions on C-6 and C-8 was examined using strong alkaline treatment followed by acidification[185]. Salonitenolide, cnicin and chihuahuin, which are C-6 closing germacranolides having a lactonizable oxygen function at C-8, after being correspondingly hydrolyzed with excess aqueous sodium hydroxide, were acidified to afford only artemisiifolin or chamissonin, which have a C-8 closing γ-lactone and a hydroxyl group at C-6. In reverse, chamissonin did not give rise to relactonization in a similar treatment. These results indicate that the C-8 closing type is in a more stable form compared with the C-6 closing type in the germacranolides. Heliangin, isolated from the leaves of *Helianthus tuberosus*, was shown to have antiauxin activity[75]; elephantin and elephantopin, which were isolated from *Elephantopus elatus*, exhibited cytotoxicity[141]. Jurineolide was isolated as the bitter principle of *Jurinea cyanoides*[153].

---

19 *I. Ogujanov, D. Ivanov*, C. R. Acad. Bulgare. Sci. *11*, 469 (1958).

20 *I. Ogujanov, V. Herout, M. Horak, F. Sŏrm*, Coll. Czech. Chem. Commun. *24*, 2371 (1959).

21 *M. Suchý, V. Herout, F. Sŏrm*, Coll. Czech. Chem. Commun. *25*, 507 (1959).

22 *D. H. R. Barton, O. C. Böchman, P. de Mayo*, J. Chem. Soc. 2263 (1960).

23 *W. Herz, G. Högenauer*, J. Org. Chem. *26*, 5011 (1961).

24 *W. Herz, K. Ueda*, J. Am. Chem. Soc. *83*, 1139 (1961).

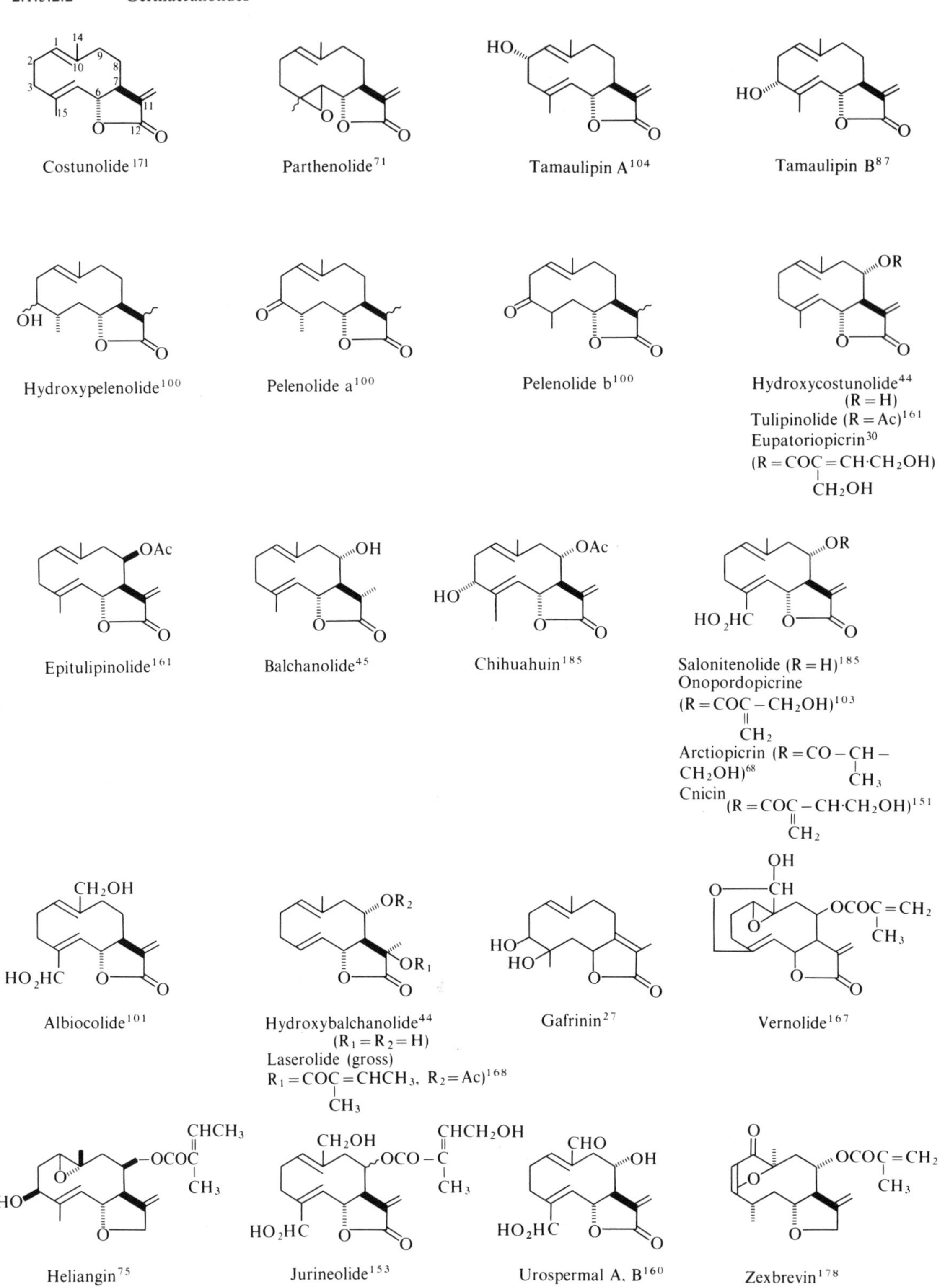
Costunolide[171]
Parthenolide[71]
Tamaulipin A[104]
Tamaulipin B[87]
Hydroxypelenolide[100]
Pelenolide a[100]
Pelenolide b[100]
Hydroxycostunolide[44] (R = H)
Tulipinolide (R = Ac)[161]
Eupatoriopicrin[30]
$(R = COC(CH_2OH)=CH\cdot CH_2OH)$
Epitulipinolide[161]
Balchanolide[45]
Chihuahuin[185]
Salonitenolide (R = H)[185]
Onopordopicrine
$(R = COC(=CH_2)-CH_2OH)$[103]
Arctiopicrin $(R = CO-CH(CH_3)-CH_2OH)$[68]
Cnicin $(R = COC(=CH_2)-CH\cdot CH_2OH)$[151]
Albiocolide[101]
Hydroxybalchanolide[44]
$(R_1 = R_2 = H)$
Laserolide (gross)
$R_1 = COC(CH_3)=CHCH_3,\ R_2 = Ac)$[168]
Gafrinin[27]
Vernolide[167]
Heliangin[75]
Jurineolide[153]
Urospermal A, B[160]
Zexbrevin[178]

Fig. 2.

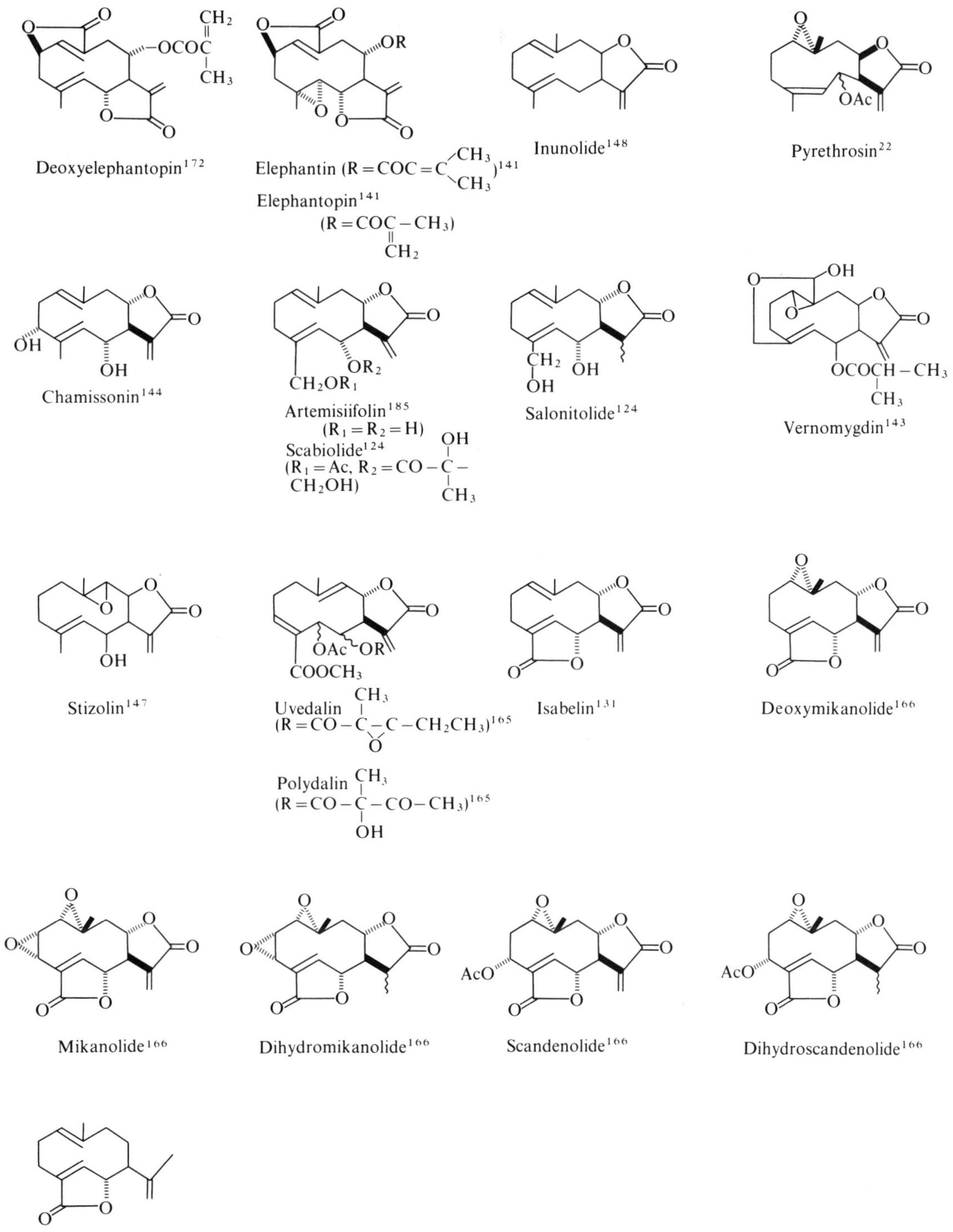

Deoxyelephantopin[172]

Elephantin ($R = COC = C(CH_3)_2$)[141]

Elephantopin[141] ($R = COC(=CH_2) - CH_3$)

Inunolide[148]

Pyrethrosin[22]

Chamissonin[144]

Artemisiifolin[185] ($R_1 = R_2 = H$)

Scabiolide[124] ($R_1 = Ac$, $R_2 = CO - C(OH)(CH_3) - CH_2OH$)

Salonitolide[124]

Vernomygdin[143]

Stizolin[147]

Uvedalin ($R = CO - C(CH_3) - C - CH_2CH_3$, epoxide)[165]

Polydalin ($R = CO - C(CH_3)(OH) - CO - CH_3$)[165]

Isabelin[131]

Deoxymikanolide[166]

Mikanolide[166]

Dihydromikanolide[166]

Scandenolide[166]

Dihydroscandenolide[166]

Aristolactone[55]

Fig. 2. Germacranolides.

---

[25] *H. Minato*, Tetrahedron Lett. *8*, 280 (1961).

[26] *M. Suchý, V. Herout, F. Šorm*, Coll. Czech. Chem. Commun. *26*, 1358 (1961).

[27] *J. P. de Villiers*, J. Chem. Soc. 2049 (1961).

[28] *R. B. Bates, R. C. Slagel*, J. Am. Chem. Soc. *84*, 1307 (1962).

[29] *R. B. Bates, R. C. Slagel*, Chem. Ind. 1715 (1962).

[30] *L. Dolejs, V. Herout*, Coll. Czch. Chem. Commun. *27*, 2654 (1962).

[31] *J. A. Hamilton, A. T. McPhail, G. A. Sim*, J. Chem. Soc. 708 (1962).

[32] *J. B. Hendrickson, R. Rees*, Chem. Ind. 1424 (1962).

[33] *W. Herz, H. Watanabe, M. Miyazaki, Y. Kishida*, J. Am. Chem. Soc. *84*, 2601 (1962).

5.2.3 Germafuranolides

About 20 members of this group, which have the isopropenyl side chain of the germacrane carbon skeleton as a part of a furan ring, have been mainly detected from plants of the *Lauraceae* family. In all the members the oxide bond of the furan ring is oriented between the C-8 and C-12 positions; the alternative 6,12-oxide form has never been found. Among them, also, several members have an additional lactone ring between C-15 and C-6 with the above-mentioned furan ring. Thus, sericenic acid[105] isolated from leaves of *Neolitsea sericea* is thought to be an intermediary compound showing biogenetic transformation from the simple germafuranolide to the lactonized germafuranolide. Besides, some pairs of geometric isomers such as furanodienone and isofuranodienone[138], neosericenine[175] etc., have been isolated regarding the cyclodeca-1,5-diene system of the molecule. The pairs were respectively characterized by NOE examination to comprize one isomer having two endo-cyclic double bonds in *trans-trans* fashion and a second isomer containing two *trans-cis* oriented double bonds.

The conformational investigation of a germacra-1(10),4-diene system was first carried out by means of X-ray diffraction analysis with a silver

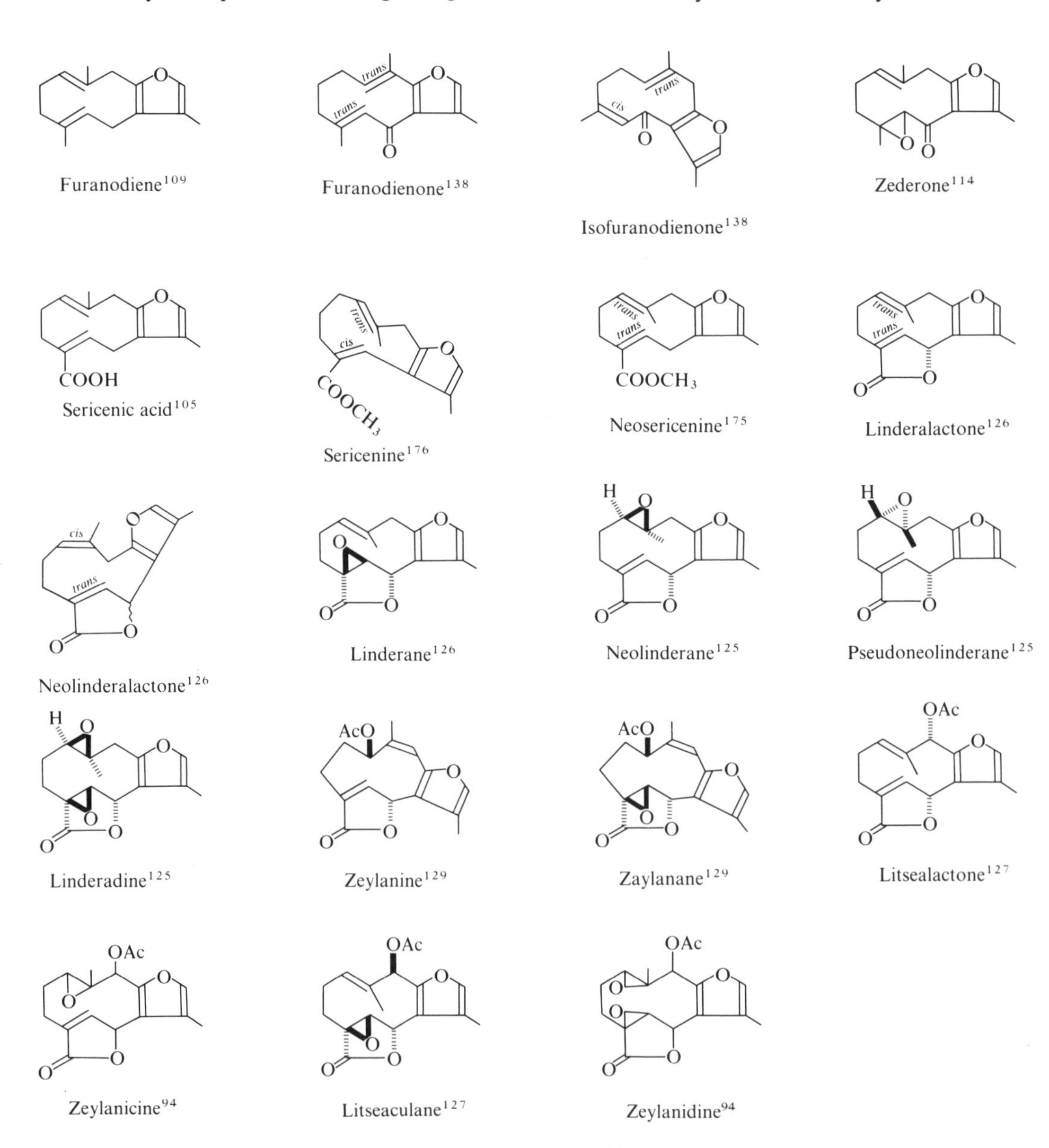

Fig. 3. Germafuranolides.

nitrate adduct of germacrene B prepared by the reduction of germacrone[84]. It was thereby found that the germacrene system has two endo-cyclic double bonds in a crossed-orientation. Recently, conformational analysis by means of NOE examination has further been tried with some germacranolides in solution and it was suggested that isabelin[131] and neolinderalactone[154], which have two γ-lactone rings or a γ-lactone ring and a furan ring fixed to the ten-membered system, each exists as two conformers; one conformer contains two endo-cyclic double bonds in crossed-orientation and the other in the parallel orientation. The equilibrium ratio of these two conformers was established to be 10:7 for isabelin and 8:2 for neolinderalactone.

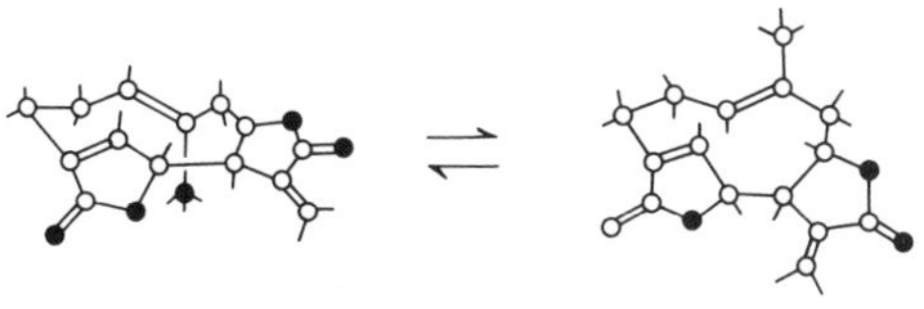

Isabelin

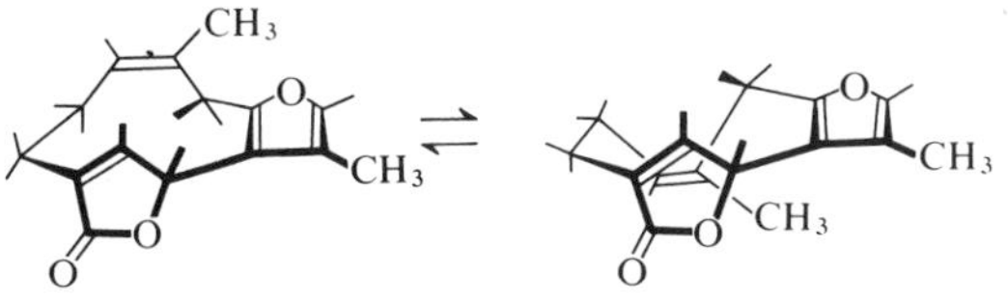

CH_3 CH_3 CH_3 CH_3

Neolinderalactone

Fig. 4. Conformations of isabelin and neolinderalactone.

5.3 Elemane Class

This class also contains γ-lactones and furanolides together with hydrocarbons, alcohols and carbonyl compounds like the germacrane class.

As to the stereochemistry at the C-5, C-7 and C-10 centers, it can be said, in general, that two isopropenyl groups on C-5 and C-7 have β and a vinyl group on C-10 has α-configuration in the absolute configuration of the members of this class.

Thermal instability of germacrene hydrocarbons has been mentioned in the foregoing section. A large number of germacrane sesquiterpenoids containing a cyclodeca-1,5-diene system such as germacrene A[184], hedycaryol[117], germacrone[26], chamissonin[144], furanodiene[109], furanodienone[138], isofuranodienone[138], sericenine[176], linderalactone[126], and litsealactone[127] easily underwent thermal cleavage of the ten-membered ring at the allylic position for the two endo-double bonds to give elemane type compounds. This thermal reaction is known as the Cope rearrangement[40]. Accordingly, it is to be questioned whether compounds of this elemane class collected here are true natural products or not.

When linderalactone[126], litsealactone[127] and dihydrotamaulipin B acetate[87] were heated, mixtures of the original compound and a rearrangement product were obtained always in a ratio of 2:3. The rearrangement products themself were reversely heated to give the same mixtures. The results indicate that, in some elemane-type sesquiterpenoids, retro-Cope rearrangement is possible. Similar facts were also observed between some germacranolides and elemane-type sesquiterpenoids, for example costunolide and dehydrosaussurealactone; dihydrocostunolide and saussurealactone[171]. Investigations on the stereospecificity of the Cope rearrangement are now under way.

Recently, some new types of sesquiterpenoids with the elemane carbon skeleton such as vermonenin[143], occidenol[177], etc., have been isolated. The formation of a 4,5-dihydrooxepin ring in occidenol-type compounds is explained by the Cope rearrangement of a hypothetical dihydromikanolide-like precursor, but the δ-lactone formation in the vermonenin-type compounds could not be explained without assuming a biogenetic oxidation process of an elemane-type intermediate.

---

[34] *W. Herz*, J. Org. Chem. *27*, 4043 (1962).

[35] *H. Minato*, Tetrahedron *18*, 365 (1962).

[36] *S. K. Paknikar, S. C. Bhattacharyya*, Tetrahedron *18*, 1509 (1962).

[37] *R. B. Bates, V. Prochazka, Z. Cekan*, Tetrahedron Lett. *14*, 877 (1963).

[38] *W. Herz, A. R. de Vivar, J. Romo, N. Viswanathan*, J. Am. Chem. Soc. *85*, 19 (1963).

[39] *W. Herz, A. R. de Vivar, J. Romo, N. Viswanathan*, Tetrahedron *19*, 1359 (1963).

[40] *S. T. Rhoads, in* Molecular Rearrangements Part 1, p. 655, *P. de Mayo* (Ed.), Interscience, New York 1963; *G. B. Gill*, Quart. Rev. *22*, 386 (1968).

[41] *J. Romo, A. R. de Vivar, W. Herz*, Tetrahedron *19*, 2317 (1963).

[42] *F. Sanchez-Viesca, J. Romo*, Tetrahedron *19*, 1285 (1963).

[43] *D. M. Simonovic, A. S. Rao, S. C. Bhattacharyya*, Tetrahedron *19*, 1061 (1963).

2.1:5.3 Elemane Class

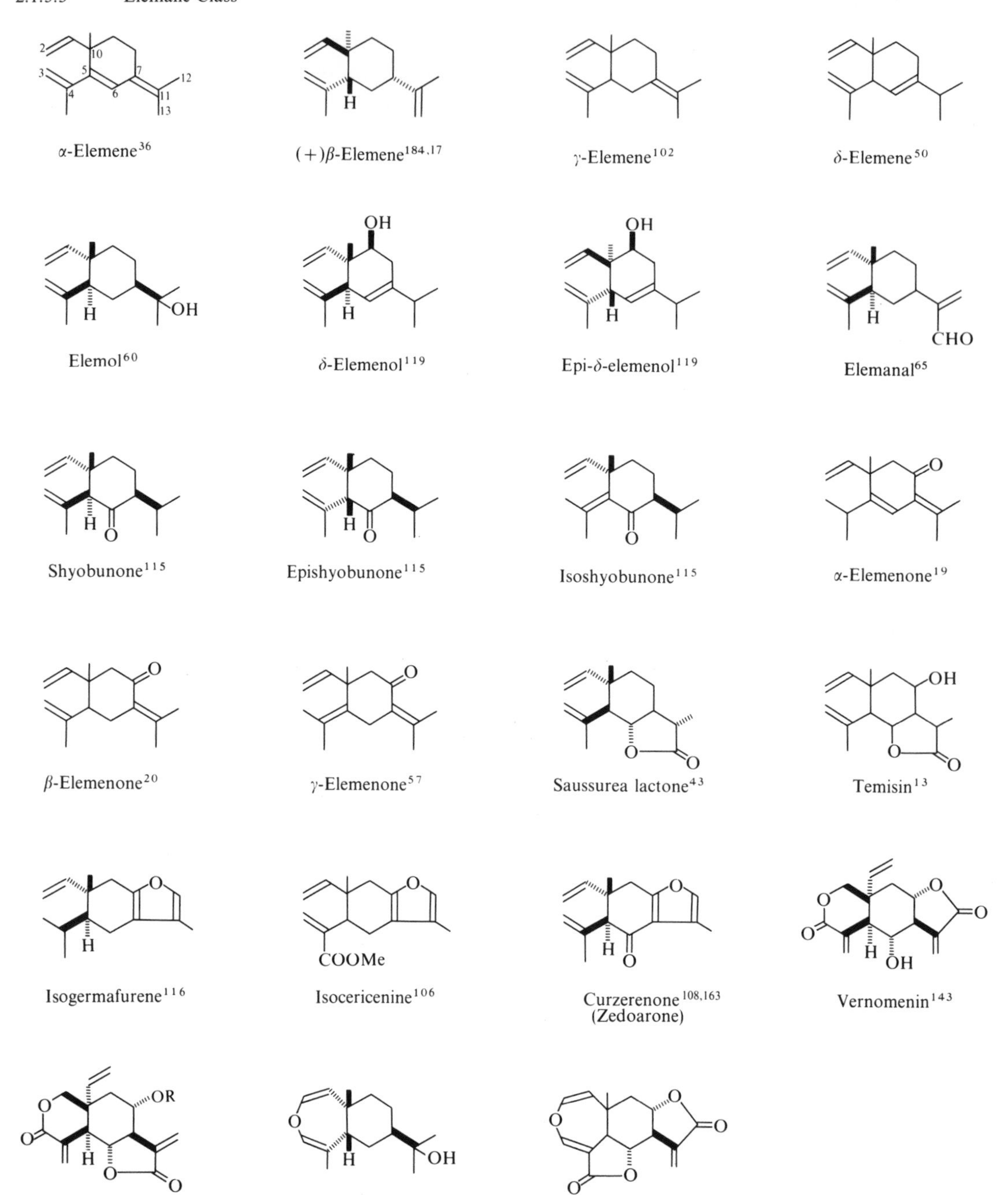

Fig. 5. Sesquiterpenoids of the elmane class.

---

[44] *M. Suchý, V. Herout, F. Sŏrm*, Coll. Czech. Chem. Commun. *28*, 1618 (1963).

[45] *M. Suchý, V. Herout, F. Sŏrm*, Coll. Czech. Chem. Commun. *28*, 1715 (1963).

[46] *M. Suchý, V. Herout, F. Sŏrm*, Coll. Czech. Chem. Commun. *28*, 2257 (1963).

[47] *J. P. de Villiers, K. Pachler*, J. Chem. Soc. 4989 (1963).

[48] *E. H. White, R. E. K. Winter*, Tetrahedron Lett. *3*, 137 (1963).

[49] *V. Benesova, P. N. Chou, V. Herout, Y. R. Naves, V. Lamparsky*, Coll. Czech. Chem. Commun. *29*, 1042 (1964).

5.4 Guaiane Class

5.4.1 Nonlactonic Compounds

Although the treatment of bulnesol with sulfuric acid in glacial acetic acid furnishes α-, β-, and γ-guaiene[28], the last two of these hydrocarbons have not yet been detected in nature. As naturally occurring hydrocarbons having the guaiane carbon skeleton, α-guaiene[35], β-bulnesene[29], guai-3,7-diene[182], and γ-gurjunene[134] are now count up. This class further includes three hemiketals which have a 5,8-epoxy-8-ol system[64,112,137] and some others of two types, 5,11-oxide[170] and 10,11-oxide[66,90,92] together with a large number of γ-lactones (guaianolides) to be described in the following section. The 5,11-oxides isolated from zedoary such as liguloxide, liguloxidol and guaioxide have a *cis*-fused bicyclo[5.3.0]decane system accompanied by an α-oriented H-5, while the 10,11-oxides isolated from japanese valerian root such as kessane, α-kessyl alcohol, kessanol and kessoglycol diacetate possess a *trans*-fused bicyclo [5.3.0] decane system accompanied by the same α-oriented H-5; in both types of ether, the oxide rings are in β-orientation.

---

50 *J. H. Gough, M. D. Sutherland*, Aust. J. Chem. *17*, 1270 (1964).
51 *W. Herz, S. Inayama*, Tetrahedron *20*, 341 (1964).
52 *W. Herz, Y. Kishida, M. Y. Lakshmikantham*, Tetrahedron *20*, 979 (1964).
53 *R. A. Lucas, S. Rovinski, R. J. Kiesel, L. Dorfman, H. B. MacPhillamy*, J. Org. Chem. *29*, 1549 (1964).
54 *R. A. Lucas, R. G. Smith, L. Dorfman*, J. Org. Chem. *29*, 2101 (1964).
55 *M. Martin-Smith, P. de Mayo, S. J. Smith, J. B. Stenlake, W. D. Williams*, Tetrahedron Lett. *35*, 2391 (1964).
56 *M. Suchý, V. Herout, F. Sŏrm*, Coll. Czech. Chem. Commun. *29*, 1829 (1964).
57 *G. V. Pigulevskii, N. V. Belova*, Zh. Obsch. Khim *34*, 1345 (1964).
58 *J. Romo, P. Joseph-Nathan, A. Fernando Diaz*, Tetrahedron *20*, 79 (1964).
59 *K. Takeda, H. Minato, M. Ishikawa*, J. Chem. Soc. 4578 (1964).
60 *A. D. Wagh, S. K. Paknikar, S. C. Bhattacharyya*, Tetrahedron *20*, 2647 (1964).
61 *W. Herz, M. V. Lakshmikantham*, Tetrahedron *21*, 1711 (1965).
62 *W. Herz, A. R. de Vivar, M. V. Lakshmikantham*, J. Org. Chem. *30*, 118 (1965).
63 *W. Herz, P. S. Santhanam*, J. Org. Chem. *30*, 4340 (1965).
64 *H. Hikino, K. Meguro, K. Sakurai, T. Takemoto*, Chem. Pharm. Bull. (Tokyo) *13*, 1484 (1965).
65 *S. Ito, K. Endo, H. Honma, K. Ota*, Tetrahedron Lett. *42*, 3777 (1965).
66 *S. Ito, H. Hikino, M. Kodama, Y. Hikino, T. Takemoto*, J. Pharm. Soc. Japan *68*, 804 (1965).
67 *M. V. Nazarenko*, Zh. Prikl. Khim. *38*, 2372 (1965).
68 *M. Suchý, Z. Samek, V. Herout, F. Sŏrm*, Coll. Czech. Chem. Commun. *30*, 3473 (1965).
69 *J. B. Barrera, J. L. B. Funes, A. G. Gonzalez, J. Chem. Soc. (C) 1298 (1966).*
70 *T. J. Batterham, N. K. Hart, J. A. Lanberton*, Aust. J. Chem. *11*, 143 (1966).
71 *A. S. Bawdekar, G. R. Kelkar, B. S. Bhattachryya*, Tetrahedron Lett. *11*, 1225 (1966).
72 *W. Herz, S. Rajappa, M. V. Lakshmikantham, J. J. Schmid*, Tetrahedron *22*, 693 (1966).
73 *W. Herz, S. Rajappa, S. K. Roy, J. J. Schmid, R. N. Mirrington*, Tetrahedron *22*, 1907 (1966).
74 *W. Herz, V. Sudarsanam, J. J. Schmid*, J. Org. Chem. *31*, 3232 (1966).
75 *S. Iriuchijima, S. Kuyama, N. Takahashi, S. Tamura*, Agr. Biol. Chem. (Tokyo) *30*, 1152 (1966).
76 *P. Joseph-Nathan, J. Romo*, Tetrahedron *22*, 1723 (1966).
77 *T. J. Mabry, W. Renold, H. E. Miller, H. B. Kagan*, J. Org. Chem. *31*, 681 (1966).
78 *M. V. Nazarenko, L. I. Leout'eva*, Khim. Prir. Soedin. *2*, 399 (1966).
79 *M. Nakazaki, H. Chikamatsu, M. Maeda*, Tetrahedron Lett. *37*, 4499 (1966).
80 *S. D. Sastry, M. L. Maheshwari, S. C. Bhattacharyya*, Tetrahedron Lett. *10*, 1035 (1966).
81 *A. R. de Vivar, E. A. Bratoeff, T. Rios*, J. Org. Chem. *31*, 673 (1966).
82 *J. Romo, P. Joseph-Nathan, G. Siade*, Tetrahedron *22*, 1499 (1966).
83 *J. Romo, A. R. de Vivar, P. Joseph-Nathan*, Tetrahedron Lett. *10*, 1029 (1966).
84 *H. Allen, D. Rogers*, Chem. Commun. 588 (1967).
85 *E. D. Brown, M. D. Solmon, J. K. Sutherland, A. Torre*, Chem. Commun. 111 (1967).
86 *N. H. Fischer, T. J. Mabry*, Tetrahedron *23*, 2529 (1967).
87 *N. H. Fisher, T. J. Mabry*, Chem. Commun. 1235 (1967).
88 *W. Herz, S. Rajappa, M. V. Lakshmikantham, D. Raulais, J. J. Schmid*, J. Org. Chem. *32*, 1042 (1967).
89 *W. Herz, P. S. Santhanam*, J. Org. Chem. *32*, 507 (1967).
90 *H. Hikino, Y. Hikino, Y. Takeshita, K. Shirata, T. Takemoto*, Chem. Pharm. Bull. (Tokyo) *15*, 324 (1967).
91 *H. Hikino, K. Meguro, G. Kusano, T. Takemoto*, J. Pharm. Soc. Japan *87*, 70 (1967).
92 *H. Hikino, Y. Hikino, Y. Takeshita, T. Takemoto*, Tetrahedron *23*, 553 (1967).
93 *H. Hikino, Y. Sakurai, H. Takahashi, T. Takemoto*, Chem. Pharm. Bull. (Tokyo) *15*, 1390 (1967).
94 *B. S. Joshi, V. N. Kamat, T. R. Govindachari*, Tetrahedron *23*, 273 (1967).
95 *H. E. Miller, T. J. Mabry*, J. Org. Chem. *32*, 2929 (1967).
96 *T. Rios, A. R. de Vivar, J. Romo*, Tetrahedron *23*, 4265 (1967).
97 *J. Romo, P. Joseph-Nathan, A. R. de Vivar, C. Alvarez*, Tetrahedron *23*, 529 (1967).
98 *J. Romo, A. R. de Vivar, P. Joseph-Nathan*, Tetrahedron *23*, 29 (1967).
99 *A. R. de Vivar, A. Cabrera, A. Ortega, J. Romo*, Tetrahedron *23, 3903 (1967).*

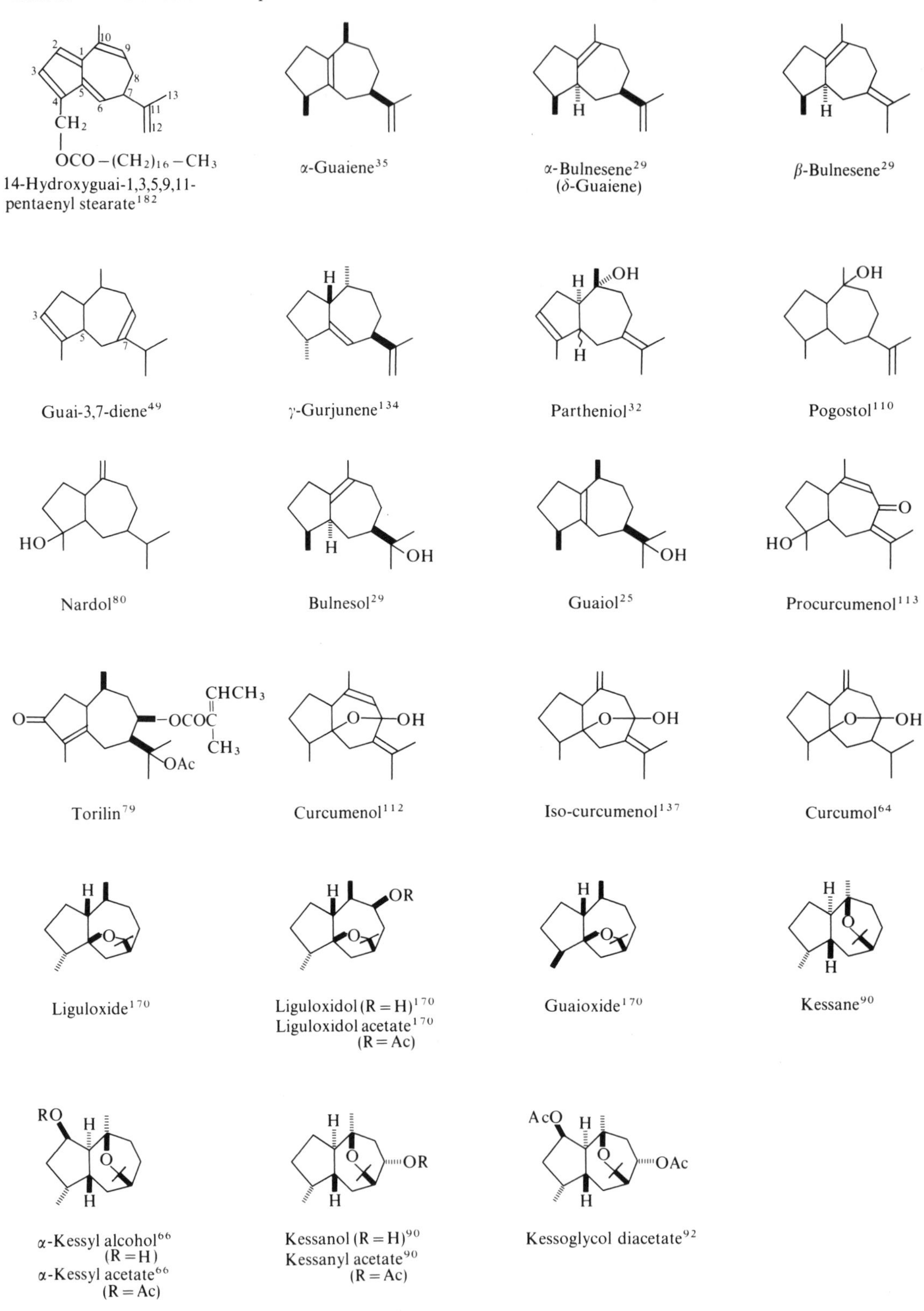

Fig. 6. Nonlactonic sesquiterpenoids of the guaiane class.

5.4.2 Guaianolides

During chemotaxonomic investigations on the *Compositae*, over 100 compounds of sesquiterpene lactones which comprise two main classes, the guaianolides and the ambrosanolides (pseudoguaianolides), have been isolated. Since all the members of both groups have a lactone function on C-12, they are again classified into two subgroups of C-6 closing and C-8 closing types according to the direction of lactone formation. As far as the stereostructure has been established, all the guaianolides of the C-6 linkage type have the configurations of H-5 (α), H-6 (β), and H-7 (β); the C-8 closing guaianolides predominantly possess H-5 (α), H-7 (β), and H-8 (α). Three chlorine-containing guaianolides, eupachlorin, its acetate and eupachloroxin[142], which show tumorinhibitory activity have been isolated from the alcoholic extract of *Eupatorium rotundifolium*, and the stereostructures were confirmed by means of X-ray analysis. These compounds are new type of sesquiterpenoids, together with some bromine-containing sesquiterpenoids[139] which have been isolated from a seahare and sea algae in recent years, but which are not described in this chapter because they belong to a different class. Zaluzanin A, a δ-lactone containing a cyclopropane ring on the C-8, C-9 and C-10, was treated with an alkali solution to be easily converted into a γ-lactone, zalzanin, which occurs as its acetate in nature[98].

Lactucin[18] and cynaropicrin[21] were investigated as bitter principles and matricarin[156] was found to have digitalis-like activity.

[100] *M. Suchý, Z. Samek, V. Herout, R. B. Bates, G. Snatzke, F. Šorm*, Coll. Czech. Chem. Commun 32, 3917 (1967).
[101] *M. Suchý, Z. Samek, V. Herout, F. Šorm*, Coll. Czech. Chem. Commun. 32, 3934 (1967).
[102] *J. A. Wenniger, R. L. Yates, M. Dolinsky*, J. Assoc. Off. Anal. Chem. 50, 1304 (1967).
[103] *B. Drozdz, M. Holub, Z. Samek, V. Herout, F. Šorm*, Coll. Czech. Chem. Commun. 33, 1730 (1968).
[104] *N. H. Fischer, T. J. Mabry, H. B. Kagan*, Tetrahedron 24, 4091 (1968).
[105] *N. Hayashi, S. Hayashi, T. Matsuura*, Tetrahedron Lett. 48, 4957 (1968).
[106] *S. Hayashi, N. Hayashi, T. Matsuura*, Tetrahedron Lett. 16, 1999 (1968).
[107] *W. Herz, C. M. Gast, P. S. Subramaniam*, J. Org. Chem. 33, 2780 (1968).
[108] *H. Hikino, K. Agatsuma, T. Takemoto*, Tetrahedron Lett. 24, 2855 (1968).
[109] *H. Hikino, K. Agatsuma, T. Takemoto*, Tetrahedron Lett. 8, 931 (1968).
[110] *H. Hikino, K. Ito, T. Takemoto*, Chem. Pharm. Bull. (Tokyo) 16, 1608 (1968).
[111] *H. Hikino, D. Kuwano, T. Takemoto*, Chem. Pharm. Bull. (Tokyo) 16, 1601 (1968).
[112] *H. Hikino, Y. Sakurai, S. Numabe, T. Takemoto*, Chem. Pharm. Bull. (Tokyo) 16, 39 (1968).
[113] *H. Hikino, Y. Sakurai, T. Takemoto*, Chem. Pharm. Bull. (Tokyo) 16, 1605 (1968).
[114] *H. Hikino, S. Takahashi, Y. Sakurai, T. Takemoto*, Chem. Pharm. Bull. (Tokyo) 16, 1081 (1968).
[115] *M. Iguchi, A. Nishiyama, H. Koyama, S. Yamamura, Y. Hirata*, Tetrahedron Lett. 51, 5315 (1968).
[116] *H. Ishi, T. Tozyo, M. Nakamura, K. Takeda*, Tetrahedron 24, 625 (1968).
[117] *R. V. H. Jones, M. D. Sutherland*, Chem. Commun. 1229 (1968).
[118] *N. P. Kiryalov, S. V. Serkerov*, Khim. Prir. Soedin. 4, 341 (1968).
[119] *K. Morikawa, Y. Hirose*, Tetrahedron Lett. 24, 2899 (1968).
[120] *A. Ortega, A. R. de Vivar, J. Romo*, Can. J. Chem. 46, 1539 (1968).
[121] *J. Romo, A. R. de Vivar, E. Dioz*, Tetrahedron 24, 5625 (1968).
[122] *J. Romo, T. Rios, L. Quijano*, Tetrahedron 24, 6087 (1968).
[123] *J. Romo, A. R. de Vivar, A. Velez*, Can. J. Chem. 46, 1535 (1968).
[124] *M. Suchý, Z. Samek, V. Herout, F. Šorm*, Coll. Czech. Chem. Commun. 33, 2238 (1968).
[125] *K. Takeda, I. Horibe, H. Minato*, Chem. Commun. 1168 (1968).
[126] *K. Takeda, I. Horibe, M. Teraoka, H. Minato*, Chem. Commun. 637 (1968).
[127] *K. Takeda, I. Horibe, M. Teraoka, H. Minato*, Chem. Commun. 940 (1968).
[128] *F. P. Toribio, T. A. Geisman*, Phytochemistry 7, 1623 (1968).
[129] *K. Tōri, M. Ohtsuru, I. Horibe, K. Takeda*, Chem. Commun. 943 (1968).
[130] *K. Wada, Y. Enomoto, K. Matsui, K. Munakata*, Tetrahedron Lett. 45, 4673 (1968).
[131] *H. Yoshioka, T. J. Mabry, H. E. Miller*, Chem. Commun. 1679 (1968).
[132] *J. B. Barrera, C. Betancor, J. L. B. Funes, A. C. Gonzalez*, An. Quim. 285 (1969).
[133] *T. A. Dullforce, G. A. Sim, D. N. J. White, E. J. Kelsey, S. M. Kupchan*, Tetrahedron Lett. 12, 973 (1969).
[134] *C. Ehret, G. Ourisson*, Tetrahedron 25, 1785 (1969).
[135] *R. I. Evstratova, V. I. Sheichenko, K. S. Rybalko, A. I. Bankovski*, Khim-Farm. Zh. 3, 39 (1969).
[136] *W. Herz, P. S. Subramaniam, N. Dennis*, J. Org. Chem. 34, 2915 (1969).
[137] *H. Hikino, K. Agatsuma, T. Takemoto*, Chem. Pharm. Bull. (Tokyo) 17, 959 (1969).
[138] *H. Hikino, C. Konno, T. Takemoto, K. Tori, M. Ohtsuru, I. Horibe*, Chem. Commun. 662 (1969).
[139] *T. Irie*, Nippon Kagaku Zasshi (Japan) 90, 1179 (1969).
[140] *R. Jesus, A. R. de Vivar, M. Aguilar*, Bol. Inst. Quim. Univ. Nac. Auton. Mex. 21, 66 (1969).

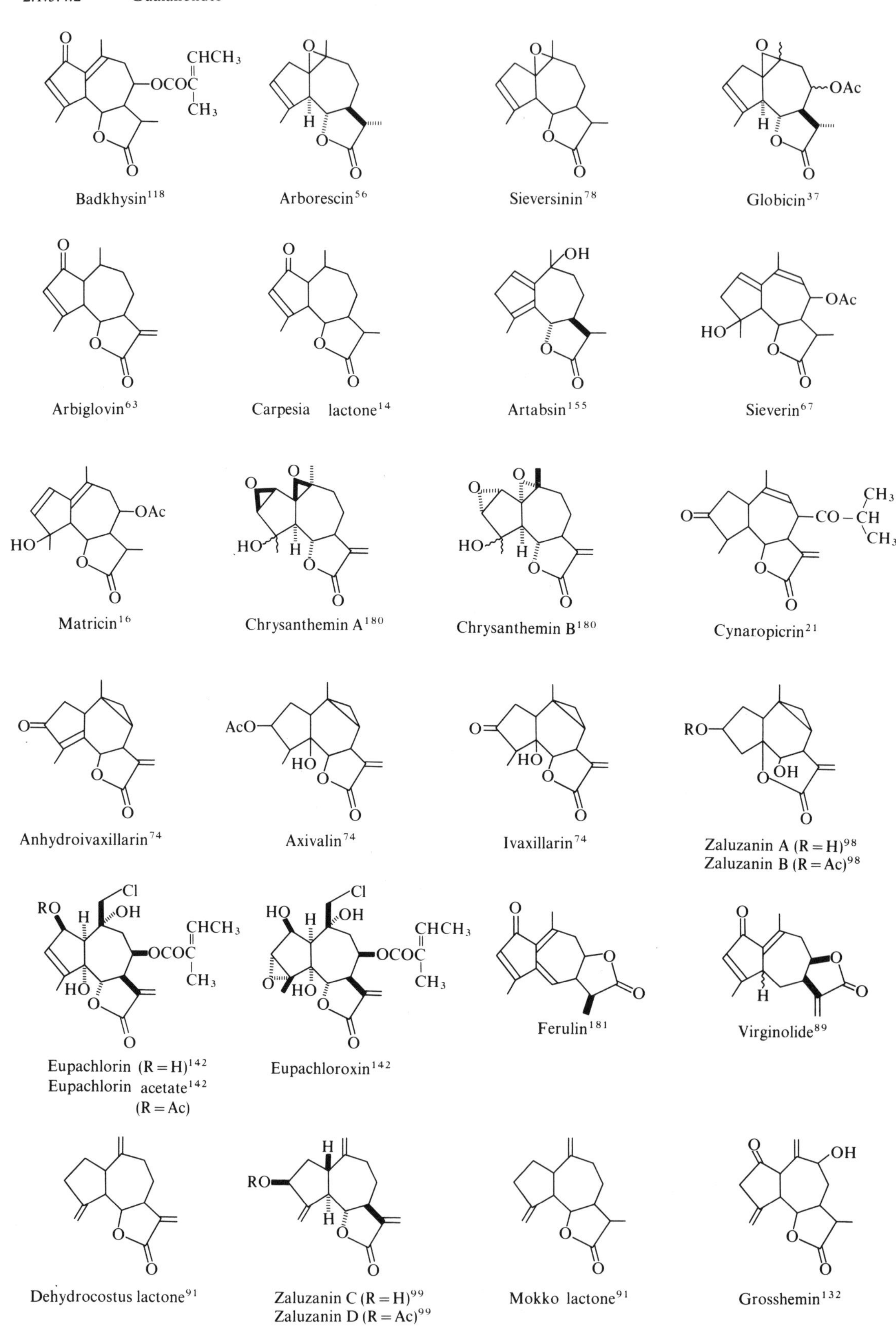

Badkhysin[118]
Arborescin[56]
Sieversinin[78]
Globicin[37]
Arbiglovin[63]
Carpesia lactone[14]
Artabsin[155]
Sieverin[67]
Matricin[16]
Chrysanthemin A[180]
Chrysanthemin B[180]
Cynaropicrin[21]
Anhydroivaxillarin[74]
Axivalin[74]
Ivaxillarin[74]
Zaluzanin A (R = H)[98]
Zaluzanin B (R = Ac)[98]
Eupachlorin (R = H)[142]
Eupachlorin acetate[142] (R = Ac)
Eupachloroxin[142]
Ferulin[181]
Virginolide[89]
Dehydrocostus lactone[91]
Zaluzanin C (R = H)[99]
Zaluzanin D (R = Ac)[99]
Mokko lactone[91]
Grosshemin[132]

Fig. 7.

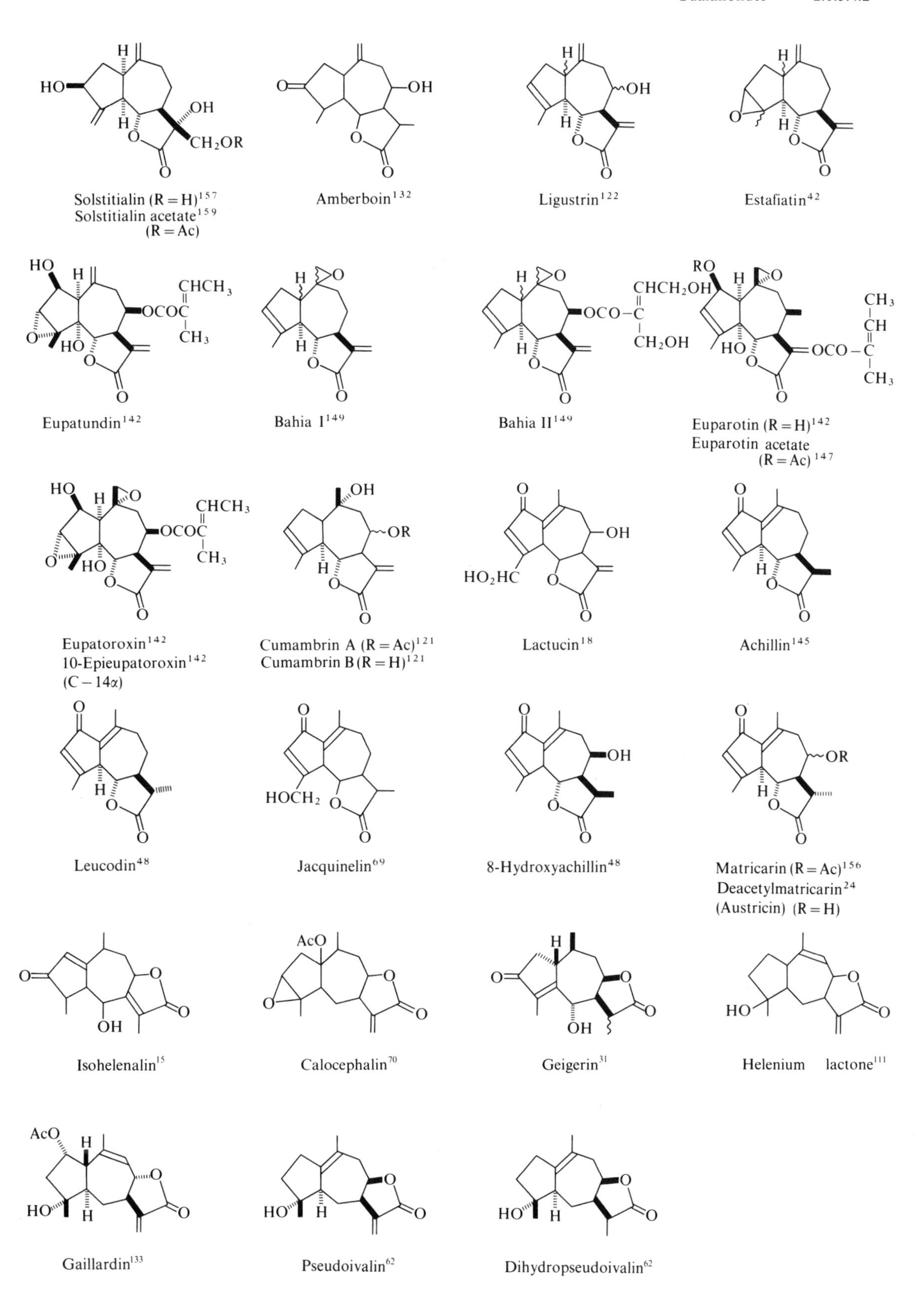

Fig. 7. Lactonic sesquiterpenoids of the guaiane class.

5.5 Ambrosane Class (Pseudoguaiane)

Formation of the ambrosane carbon skeleton is explained by the biogenetic 1,2-shift of a methyl group in the guaiane skeleton from C-4 to C-5. Up to now a large number of the members of this class have been isolated and all the compounds are characterized as γ-lactone (ambrosanolides) with the exception of damsic acid[162] which is the only sesquiterpene acid having the ambrosane carbon skeleton and was isolated from *Ambrosia ambrosioides*.

The ambrosanolides, which include both C-6 and C-8 closing types, have a *trans*-fused bicyclo [5.3.0] decane system along with a β-oriented C-15 methyl group. With respect to the stereochemistry of the lactone ring, the C-6 closing ambrosanolides all have *cis*-fusion (α-oriented H-6), but such common figure has not been recognized in the C-8 closing type.

Damsin[46] is a cytotoxic principle of *Ambrosia ambrosioides* and amaralin[53] was found to have a potent analgesic activity.

Mexicanin E[41] and dihydromexicanin E[54] are nor-sesquiterpene lactone without a C-15 methyl group and mexicanin D[38] is an abnormal guaianolide having a methyl group at the C-2 position instead of C-15.

[141] *S. M. Kupchan, Y. Aynehchi, J. M. Cassady, H. K. Schnos, A. L. Barlingame*, J. Org. Chem. *34*, 3867 (1969).
[142] *S. M. Kupchan, J. E. Kelsey, M. Maruyama, J. M. Cassady, J. C. Hemingway, J. R. Knox*, J. Org. Chem. *34*, 3876 (1969).
[143] *S. M. Kupchan, R. J. Hemingway, A. Karim, D. Werner*, J. Org. Chem. *34*, 3903 (1969).
[144] *M. F. L'Homme, T. A. Geisman, J. Yoshioka, T. H. Porter, W. Renold, T. J. Mabry*, Tetrahedron Lett. *37*, 3161 (1969).
[145] *J. N. Marx, E. H. White*, Tetrahedron *25*, 2117 (1969).
[146] *K. Morikawa, Y. Hirose*, Tetrahedron Lett. *22*, 1799 (1969).
[147] *M. N. Mukhametzhanov, V. I. Sheichenko, A. I. Bankovskii, K. S. Rybalko, K. I. Boryaev*, Khim. Prir. Soedin. *5*, 567 (1969).
[148] *R. Raghavan, K. R. Ravindranath, G. K. Trivedi, S. K. Paknikar, S. C. Bhattacharyya*, Indian J. Chem. *7*, 310 (1969).
[149] *A. R. de Vivar, A. Ortega*, Can. J. Chem. *47*, 2849 (1969).
[150] *H. Rüesch, T. J. Mabry*, Tetrahedron *25*, 805 (1969).
[151] *Z. Samek, M. Holub, V. Herout, F. Sŏrm*, Tetrahedron Lett. *34*, 2931 (1969).
[152] *S. V. Serkerov*, Khim. Prir. Soedin. *5*, 490 (1969).
[153] *M. Suchý, L. Dolejs, V. Herout, F. Sŏrm, G. Snatzke, J. Himmelreich*, Coll. Czech. Chem. Commun. *34*, 229 (1969).
[154] *K. Takeda, in* Chemistry of Natural Products 6, p. 181 IUPAC (Ed.), Butterworths, London 1969.
[155] *K. Vokac, Z. Samek, V. Herout, F. Sŏrm, Coll. Czech. Chem. Commun. 34*, 2288 (1969).
[156] *E. H. White, S. Eguchi, J. N. Marx*, Tetrahedron *25*, 2099 (1969).
[157] *E. T. William, H. Hope, N. Zarghami, D. E. Heinz*, Chem. Ind. 460 (1969).
[158] *K. Yoshihara, Y. Ohta, T. Sakai, Y. Hirose*, Tetrahedron Lett. *27*, 2263 (1969).
[159] *N. Zarghami, D. E. Heinz*, Chem. Ind. 1556 (1969).
[160] *R. K. Bentley, J. G. St. C. Buchanan, T. G. Halsall, V. Thaller*, J. Chem. Soc. (D) *7*, 435 (1970).
[161] *R, W. Doskotch, F. S. El-Feraly*, J. Org. Chem. *35*, 1928 (1970).
[162] *R. W. Doskotch, C. D. Hufford*, J. Org. Chem. *35*, 486 (1970).
[163] *S. Fukushima, M. Kuroyanagi, A. Ueno, Y. Akahori, Y. Saiki*, J. Pharm. Soc. Japan *90*, 863 (1970).
[164] *W. Herz, K. Aota*, J. Org. Chem. *35*, 2611 (1970).
[165] *W. Herz, S. V. Bhat*, J. Org. Chem. *35*, 2605 (1970).
[166] *W. Herz, P. S. Subramanian, P. S. Santhanam, K. Aota, A. L. Hall.* J. Org. Chem. *35*, 1453 (1970).
[167] *C. M. Ho, R. Toubiana*, Tetrahedron *26*, 941 (1970).
[168] *M. Holub, Z. Samek, D. P. Popa, V. Herout, F. Sŏrm*, Coll. Czech. Chem. Commun. *35*, 284 (1970).
[169] *M. Iguchi, A. Nishiyama, S. Yamamura, Y. Hirata*, Tetrahedron Lett. *11*, 855 (1970).
[170] *H. Ishii, T. Tozyo, M. Nakamura, H. Minato*, Tetrahedron *26*, 2911 (1970).
[171] *T. C. Jain, C. M. Banks, J. E. McCloskey*, Tetrahedron Lett. *11*, 841 (1970).
[172] *T. Kurokawa, K. Nakanishi, W. Wu, H. Y. Ysu, M. Maruyama, S. M. Kupchan*, Tetrahedron Lett. *33*, 2863 (1970).
[173] *V. I. Novikov, F. N. Forostyan, D. P. Popa*, Khim. Prir. Soedin. *6*, 29 (1970).
[174] *T. Sekita, S. Inayama, Y. Iitaka*, Tetrahedron Lett. *2*, 135 (1970).
[175] *K. Takeda, I. Horibe, H. Minato*, J. Chem. Soc. (C) 1547 (1970).
[176] *K. Takeda, K. Tori, I. Horibe, H. Minato, N. Hayashi, T. Matsuura*, J. Chem. Soc. (C) 985 (1970).
[177] *B. Tomita, Y. Hirose*, Tetrahedron Lett. *3*, 235 (1970).
[178] *A. R. de Vivar, C. Guerreo, E. Diaz, A. Ortega*, Tetrahedron *26*, 1657 (1970).
[179] *A. R. de Vivar, M. Aguilar*, Tetrahedron *26*, 2775 (1970).
[180] *J. Romo, A. R. de Vivar, R. Trivimo, J.-N. Pedro, E. Diay*, Phytochemistry *9*, 1615 (1970).
[181] *S. V. Serkerov*, Khim. Prir. Soedin. *6*, 371 (1970).
[182] *K. Vokac, Z. Samek, V. Herout, F. Sŏrm, Coll. Czech. Chem. Commun. 35*, 1296 (1970).
[183] *K. Wada, Y. Enomoto, K. Munakata*, Agr. Biol. Chem. (Tokyo) *34*, 946 (1970).
[184] *A. J. Weinheimer, W. W. Youngblood, P. H. Washecheck, T. K. B. Karns, L. S. Ciereszko*, Tetrahedron Lett. *7*, 497 (1970).
[185] *H. Yoshioka, W. Renold, T. J. Mabry*, Chem. Commun. 149 (1970).
[186] *H. Yoshioka, H. Rüesch, E. Rodriguez, A. Higo, J. A. Mears, T. J. Mabry, J. G. Caizada, X. A. Dominguez*, Tetrahedron *26*, 2167 (1970).
[187] *H. Yoshioka, E. Rodriguez, T. J. Mabry*, J. Org. Chem. *35*, 2888 (1970).

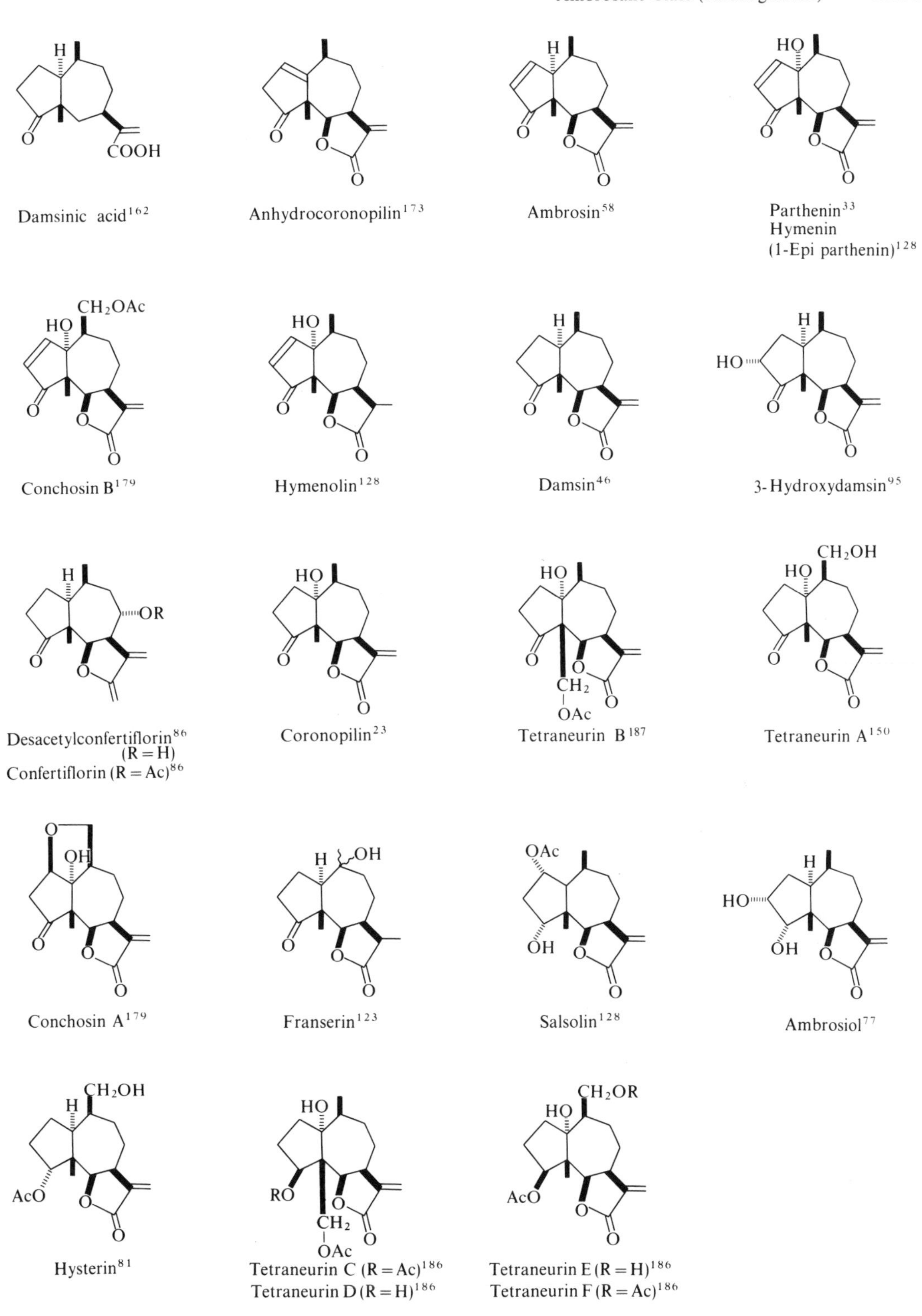
H
O
COOH
Damsinic acid[162]
O
O
O
Anhydrocoronopilin[173]
H
O
O
O
Ambrosin[58]
HO
O
O
O
Parthenin[33]
Hymenin
(1-Epi parthenin)[128]
CH₂OAc
HO
O
O
O
Conchosin B[179]
HO
O
O
O
Hymenolin[128]
H
O
O
O
Damsin[46]
H
HO
O
O
O
3-Hydroxydamsin[95]
H
OR
O
O
Desacetylconfertiflorin[86]
(R = H)
Confertiflorin (R = Ac)[86]
HO
O
O
O
Coronopilin[23]
HO
O
O
CH₂
OAc
O
Tetraneurin B[187]
CH₂OH
HO
O
O
O
Tetraneurin A[150]
O
OH
O
O
O
Conchosin A[179]
H
OH
O
O
O
Franserin[123]
OAc
OH
O
O
Salsolin[128]
H
HO
OH
O
O
Ambrosiol[77]
CH₂OH
H
AcO
O
O
Hysterin[81]
HO
RO
O
CH₂
OAc
O
Tetraneurin C (R = Ac)[186]
Tetraneurin D (R = H)[186]
CH₂OR
HO
AcO
O
O
Tetraneurin E (R = H)[186]
Tetraneurin F (R = Ac)[186]

Fig. 8.

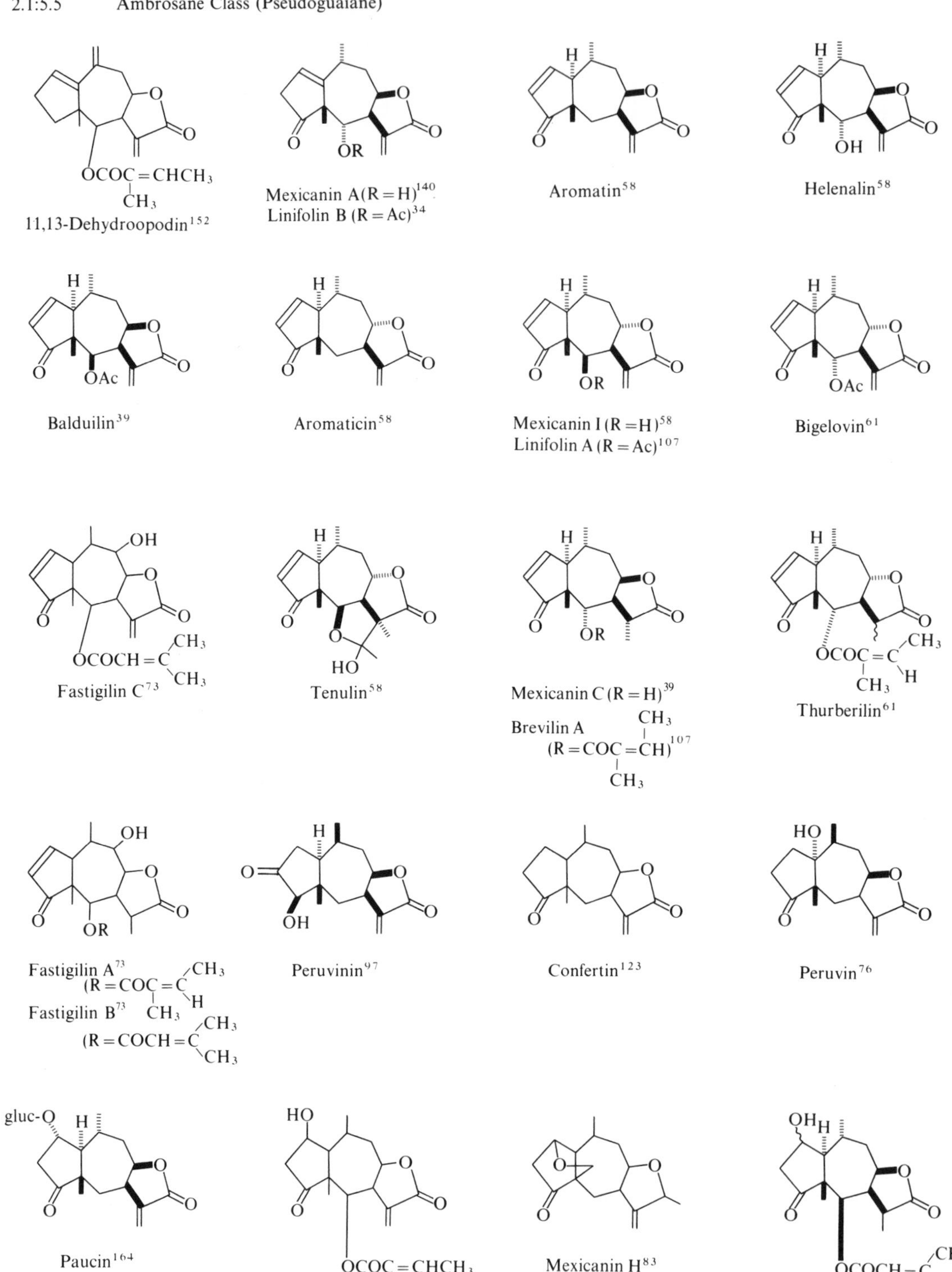
OCOC=CHCH3
CH3
11,13-Dehydroopodin[152]
OR
Mexicanin A(R=H)[140]
Linifolin B (R=Ac)[34]
H
Aromatin[58]
H
OH
Helenalin[58]
H
OAc
Balduilin[39]
H
Aromaticin[58]
H
OR
Mexicanin I (R=H)[58]
Linifolin A (R=Ac)[107]
H
OAc
Bigelovin[61]
OH
OCOCH=C
CH3
CH3
Fastigilin C[73]
H
HO
Tenulin[58]
H
OR
Mexicanin C (R=H)[39]
Brevilin A
(R=COC=CH)[107]
CH3
CH3
H
OCOC=C
CH3
H
CH3
Thurberilin[61]
OH
OR
Fastigilin A[73]
(R=COC=C
CH3
H
CH3
Fastigilin B[73]
(R=COCH=C
CH3
CH3
H
OH
Peruvinin[97]
Confertin[123]
HO
Peruvin[76]
gluc-O
H
Paucin[164]
HO
OCOC=CHCH3
CH3
Arnifolin[135]
Mexicanin H[83]
OH
H
OCOCH=C
CH3
CH3
Flexuosin B[52]

Fig. 8.

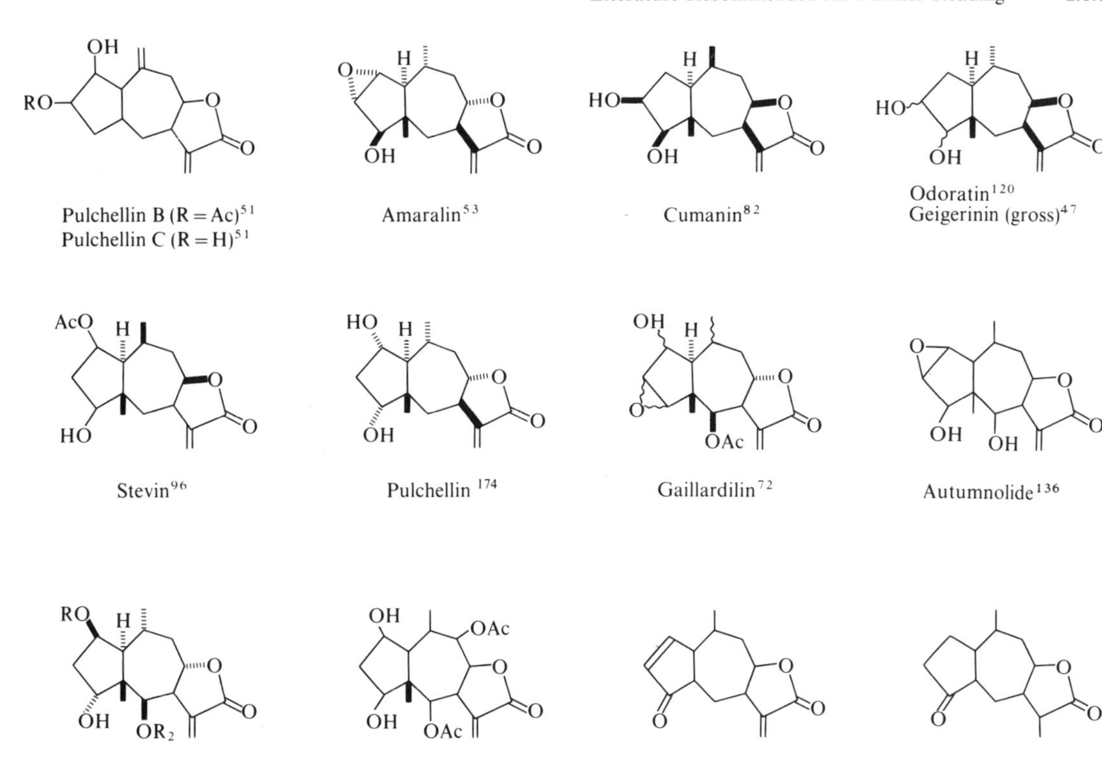

Fig. 8. Sesquiterpenoids of the ambrosane class.

6 Literature Recommended for Further Reading

[188] *A. A. Newman,* Chemistry of Terpenes and Terpenoids, Academic Press, London, New York 1972.

[189] *W. Templeton,* An Introduction to the Chemistry of the Terpenoids and Steroids, Butterworths, London 1969.

[190] *T. W. Goodwin,* Aspect of Terpenoid Chemistry and Biochemistry, Academic Press, London, New York 1971.

[191] *L. A. Porter,* Chem. Rev. *67*, 441 (1967).

[192] *W. I. Taylar, A. R. Battersby,* Cyclopentanoid Terpene Derivatives, Marcel Dekker, New York 1969.

[193] *K. H. Overton,* Terpenoids and Steroids, Vol. 1–5, The Chemical Society, Burington House, London 1971–1975.

[194] *T. K. Devon, A. I. Scott,* "Handbook of Naturally Occurring Compounds, Vol. II. Terpenes," Academic Press, New York, London 1972.

[195] *K. Nakanishi,* "Natural Products Chemistry, Vol. I," Academic Press, New York, London 1974.

2.2 Diterpenoids*

Tadahiro Kato and (the late) Yoshio Kitahara
Department of Chemistry, Faculty of Science, Tohoku University, Sendai, 980 Japan

Since diterpenoids can be regarded as formed from geranyl geraniol or geranyl linalool in accordance with the principles of the biogenetic isoprene rule, it is convenient to discuss the diterpenoids under the appropriate sub-headings based on the biogenesis. Structural variety in the diterpenoids arises from secondary rearrangement, ring cleavage, contraction and expansion. Cyclization of the common precursor in a fundamentally different manner leads to unusual carbon skeletons which are discussed in a separated section "Miscellaneous".

1 Bicyclic Diterpenes

Labdane and Related Types

An increasing number of halogenated terpenoids has been elucidated from marine origins such as seaweeds, sea hares and soft corals[1,2]. Of these compounds aplysin 20[3] and concinndiol[4] belong to brominated labdane type diterpenes, which are presumably biosynthesized in nature by bromonium ion initiated cyclization of geranyl geraniol. It is noteworthy that aplysin 20 has an axial hydroxyl group at the C-8 position and an equatorial bromine atom located at the C-3 position of labdane skeleton.

For the realization of biogenetic type brominative cyclization of polyenes in a laboratory, several reagent systems have been developed using model polyenes and the results are shown in Table 1. Although further studies are necessary to improve the yield, 2,4,4,6-tetrabromocyclohexadienone reacts with polyenes by the liberation of Br^{+}[8b].

Based on biogenetic consideration, α- and β-(*dl*)-levantenolides[9] were synthesized from an acyclic progenitor by the action of $SnCl_4$[10]. On the meanwhile, partial syntheses of labdane type compounds, marrubiin[11], kolavelool[12], hardwickiic acid[12], antipodes of polyalthic acid[13], and levantenolides[14] have been carried out starting from degradation products of natural terpenoids. These synthetic works have established unequivocally the assigned structures.

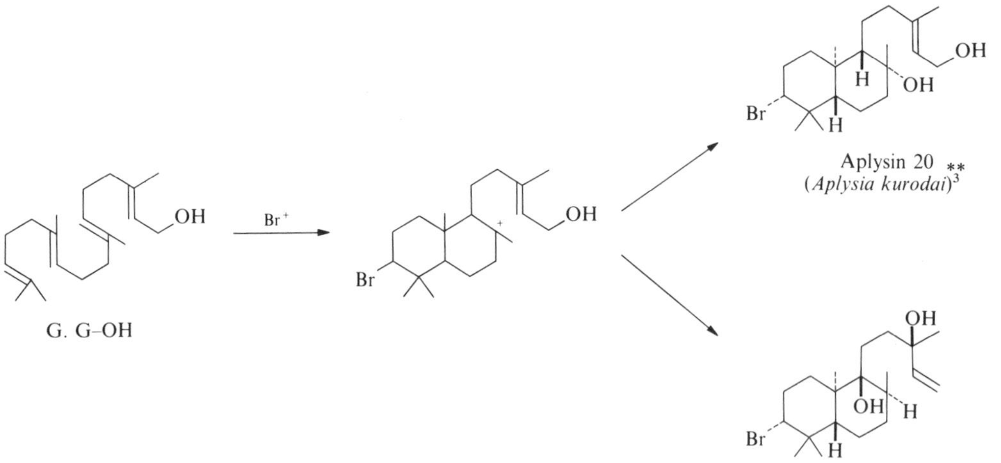

G. G-OH $\xrightarrow{Br^+}$

Aplysin 20 ** (*Aplysia kurodai*)[3]

Concinndiol (Red alga, *Laurencia concinna*)[4]

* The diterpenes discussed in this chapter are those reported after publication of the recent excellent reviews by McCrindle and Overton [Advances in Organic Chemistry, Vol. 5, 47 (1965), Interscience Publishers, and Rodd's Chemistry of Carbon Compounds, IIc, 369 (1969), Elsevier Publishing Co. Ltd.]. The compounds listed in [Handbook of Naturally Occurring Compounds II, 185 (1972)] by Devon and Scott, Academic Press, are also excluded. ** See, p. 85.

[1] *J. F. Siuda, J. F. DeBernardis*, Lloydia *36*, 107 (1973).
[2] *W. Fenical*, J. Phycol. *11*, 245 (1975).
[3] *S. Yamamura, Y. Hirata*, Bull. Chem. Soc. Japan *44*, 2560 (1971).
[4] *J. J. Sims, G. H. Y. Lin, R. M. Wing, W. Fenical*, Chem. Commun. 470 (1973).
[5] *E. E. van Tamelen, E. J. Hessler*, Chem. Commun. 411 (1966).

Table 1. Brominative cyclization of polyenes

Reaction	Ref.
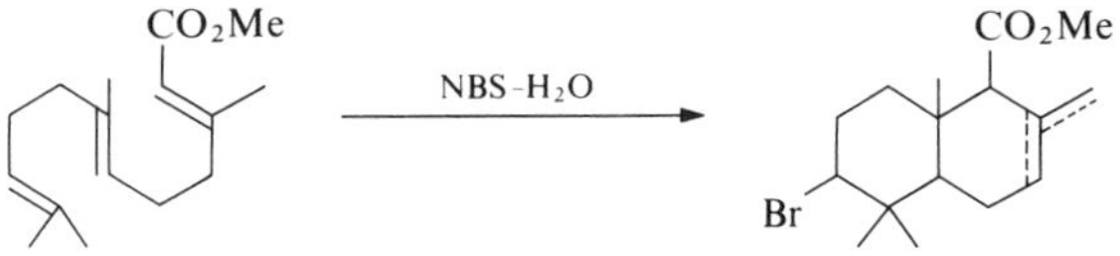 CO_2Me → ($NBS–H_2O$) → CO_2Me, Br	5
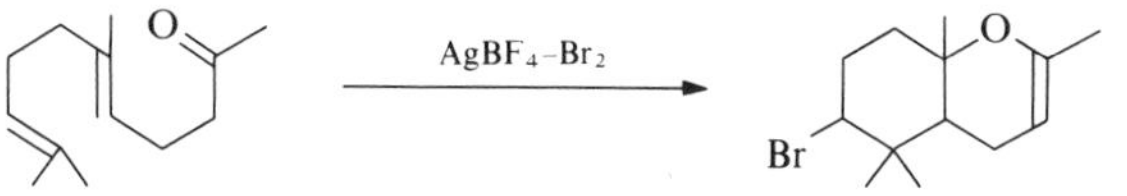 O → ($AgBF_4–Br_2$) → O, Br	6
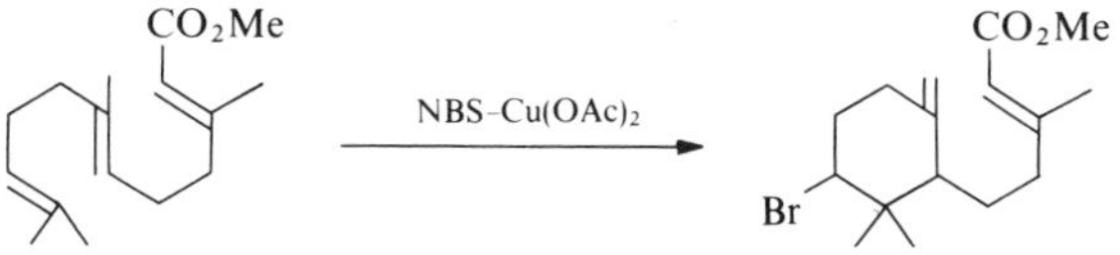 CO_2Me → ($NBS–Cu(OAc)_2$) → CO_2Me, Br	7
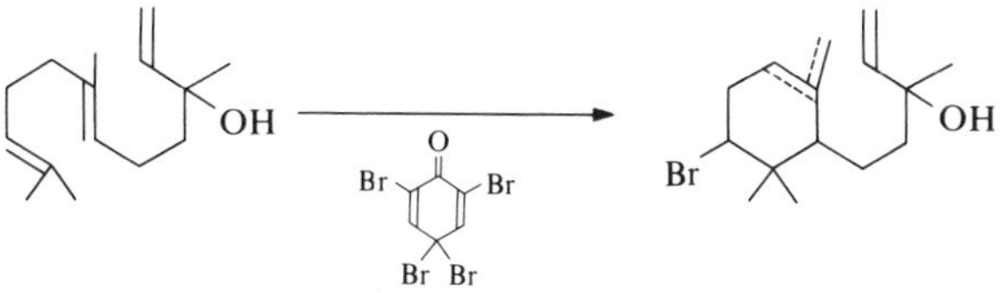 OH → (Br, Br, Br Br, O) → OH, Br	8

O, O, OMe

$SnCl_4$

O, O, O, H, H

α-Levantenolide

O, O, O, H, H

β-Levantenolide

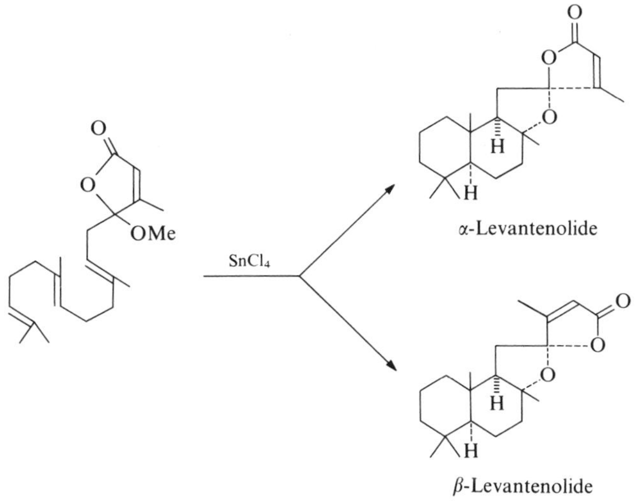

[6] *L. E. Wolinsky, D. J. Faulkner*, J. Org. Chem. *41*, 597 (1976).

[7] *A. G. Gonzalez, J. D. Martin, C. Perez, M. A. Ramirez*, Tetrahedron Lett. 137 (1976).

[8a] *T. Kato, I. Ichinose, Y. Kitahara*, Chem. Commun. 518 (1976).

[8b] *T. Kato, I. Ichinose, S. Kumazawa, Y. Kitahara*, Bioorg. Chem. *4*, 188 (1975).

[9] *A. G. Gonzalez, C. G. Francisco, R. Freire, R. Hernandez, J. A. Salazar, E. Suarez*, Tetrahedron Lett. 1897 (1976).

[10] *T. Kato, M. Tanemura, T. Suzuki, Y. Kitahara*, Bioorg. Chem. *1*, 84 (1971).

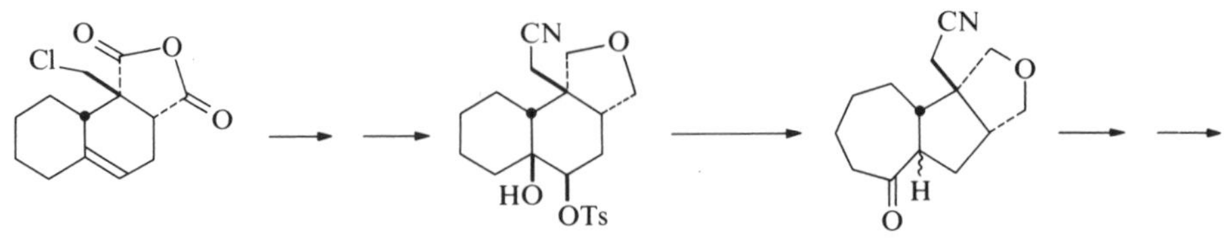

Bicyclic anhydride → → Diol monotosylate → Rearranged ketone → →

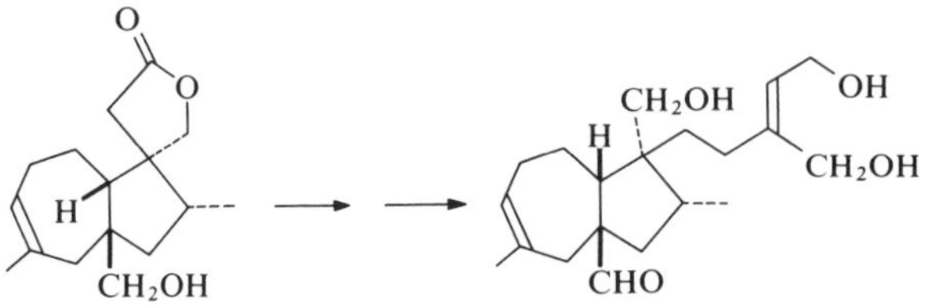

Hydroxy lactone → → Portulal
(*Portula grandiflora*)[15]

Portulal is a diterpene which has been isolated from *Portula grandiflora* as an unique plant growth regulator[15]. Structurally it has a perhydroazulene skeleton with a clerodane type substitution. The structure has been confirmed by its total synthesis starting from bicyclic anhydride. The perhydroazulene skeleton was constructed from diol monotosylate upon treatment with *t*-BuOK in *t*-BuOH to give the rearranged ketone, which was then converted to hydroxy lactone via multi step reactions. The γ-lactone, identical with the degradation product from natural portulal, was used as a relay compound and finally transformed into portulal[16].

Clerodendrin A, a bitter principle of *Clerodendron tricotomum thunb*, has been revealed to possess the antifeeding repellent activity for the larvae of *Spodoptera littoralis* Boisd[17]. The first naturally occurring chloro-diterpene, gutierolide was isolated from the annual herb, *Gutierrezia dracunculoides* (DC) Blake[18]. The stereostructure including absolute configuration has been determined by X-ray analysis.

Typical bicyclic diterpenoids are listed in Table 2.

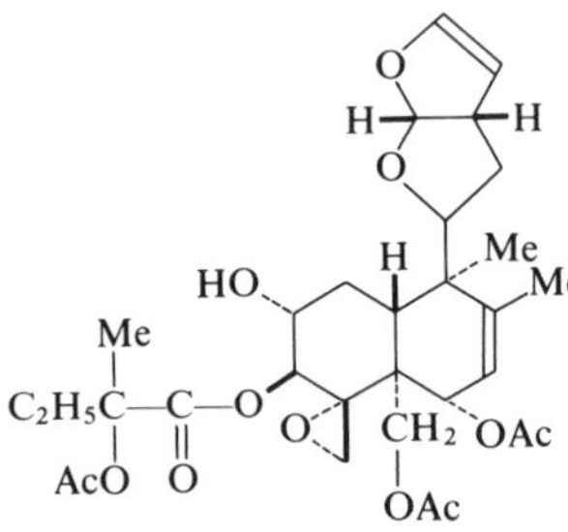

Clerodendrin A
(*Clerodendron tricotomum thunb*)[17]

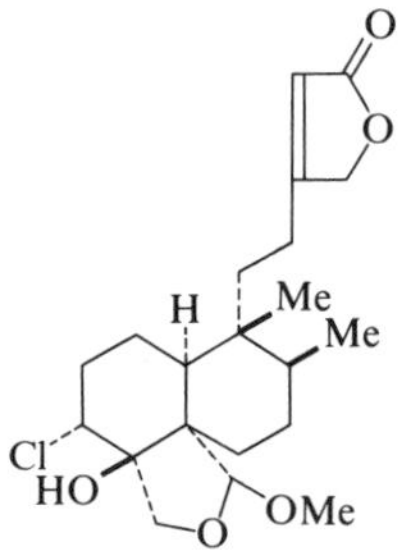

Gutierolide
(*Gutierrezia dracunculoides* (DC) Blake)[18]

[11] *L. Mangoni, M. Adinolfi, G. Laonigro, E. Doria*, Tetrahedron Lett. 4167 (1968).

[12] *R. Misra, Sukh Dev*, Tetrahedron Lett. 2685 (1968).

[13] *S. W. Pelletier, L. B. Hawley, K. W. Gopinath*, Chem. Commun. 96 (1967).

[14] *A. G. Gonzalez, C. G. Francisco, R. Freire, T. Hernandez, J. A. Salazar, E. Suarez*, Tetrahedron Lett. 2725 (1976).

[15] *S. Yamazaki, S. Tamura, F. Marumo, Y. Saito*, Tetrahedron Lett. 359 (1969).

[16a] *T. Tokoroyama, K. Matsuo, R. Kanazawa, H. Kotsuki, T. Kubota*, Tetrahedron Lett. 3093 (1974).

[16b] *R. Kanazawa, H. Kotsuki, T. Tokoroyama*, Tetrahedron Lett. 3651 (1975).

[17] *N. Kato, S. Shibayama, K. Munakata, C. Katayama*, Chem. Commun. 1632 (1971).

[18] *W. B. T. Cruse, M. N. G. James*, Chem. Commun. 1278 (1971).

[19] *G. Gafner, G. J. Kruger, D. E. A. Rivett*, Chem. Commun. 249 (1974).

Table 2. Typical bicyclic diterpenes

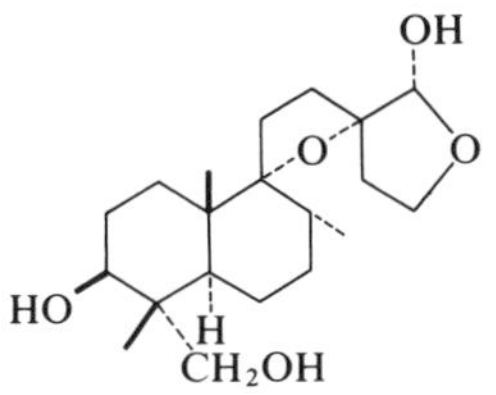

Lasiocoryin
(*Lasiocorys capensis*)[19]

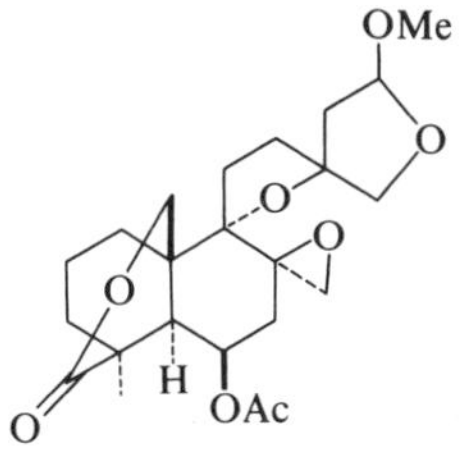

Methoxynepetaefolin
(*Leonotis nepetaefolia*)[20]

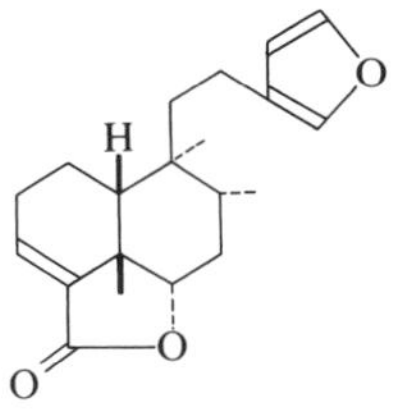

(*Solidago arguta*)[21]

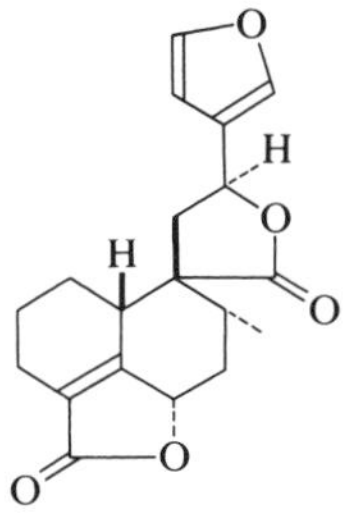

Teucvin
(*Teucrium viscidum* Bl. var. *Miquelianum* Maxim.)[22]

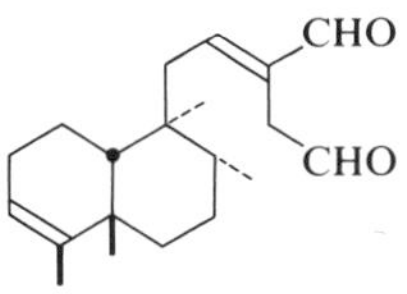

Linaridial
(*Linaria japonica* Miq.)[23]

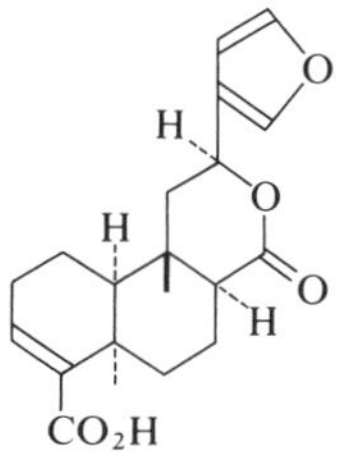

Floribundic acid
(*Evodia floribunda*)[24]

2 Tricyclic Diterpenes

2.1 Pimarane and Rosane Types

From fermentation of the fungus *Aspergillus chevalier* two novel pimarane diterpenes, designated LL-S491β and γ respectively, have been elucidated. The former displays significant antibacterial activity against certain gram positive organisms and the latter exhibits antiviral activity against Herpes simplex. Both compounds possess strong antiprotozoal activity against *Tetrahymena pyriformis*[1]. When LL-S491β was treated with *p*-TsOH–Ac_2O, acid catalyzed migration of C-13 vinylidene group to C-14 was observed[2]. This evidence provides a chemical model for the postulation[3] that the cleistanthane diterpene of the rearranged skeleton is derived biogenetically from a pimarane involving a 1,2 migration of the C-13 vinylidene fragment.

By the aids of germination assay, three active inhibitors of germination, named momilactone-A, B, and C, were isolated from rice husk[4]. All of these compounds have a novel C_9–C_{10} *cis* relation in common pimarane skeleton[5]. Annonalide elaborated from *Annona coriacea*[6] has been chemically correlated with structurally similar momilactone-B[7].

Several pimarane type compounds such as kirenol[8] and thermalol[9] have been characterized.

[20] *P. S. Manchand*, Tetrahedron Lett. 1907 (1973).

[21] *A. B. Anderson, R. McCrindle, E. Nakamura*, Chem. Commun. 453 (1974).

[22] *E. Fujita, I. Uchida, T. Fujita, N. Masaki, K. Osaki*, Chem. Commun. 793 (1973).

[23] *I. Kitagawa, M. Yoshihara, T. Tani, I. Yoshioka*, Tetrahedron Lett. 23 (1975).

[24] *D. Billet, M. Durgeat, S. Heitz, A. Ahond*, Tetrahedron Lett. 3825 (1975).

[1] *G. A. Ellestadt, M. P. Kunstmann, P. Mirando, G. O. Morton*, J. Am. Chem. Soc. *94*, 6206 (1972).

[2] *G. A. Ellestadt, M. P. Kunstmann, G. O. Morton*, Chem. Commun. 312 (1973).

[3] *E. J. McGarry, K. H. Pegel, L. Phillips, E. S. Waight*, J. Chem. Soc. (C) 904 (1971).

[4] *T. Kato, C. Kabuto, N. Sasaki, T. Tsunakawa, H. Aizawa, K. Fujita, Y. Kato, Y. Kitahara, N. Takahashi*, Tetrahedron Lett. 3861 (1973).

[5] *M. Tsunakawa, M. Oba, C. Kabuto, N. Sasaki, T. Kato, Y. Kitahara*, Chemistry Lett. 1157 (1976).

[6] *P. Mussini, F. Orsini, F. Pelizzoni, B. L. Buckwalter, E. Wenkert*, Tetrahedron Lett. 4849 (1973).

[7] *F. Pelizzoni*, private communication.

[8] *T. Murakami, T. Isa, T. Satake*, Tetrahedron Lett. 4990 (1973).

[9] *A. Matsuo, S. Uto, M. Nakayama, S. Hayashi, K. Yamasaki, R. Kasai, O. Tanaka*, Tetrahedron Lett. 2451 (1976).

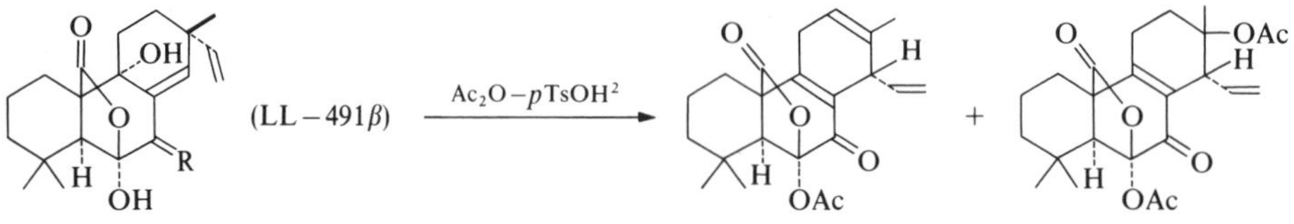

(LL − 491β) $\xrightarrow{Ac_2O-pTsOH^2}$ +

LL − 491β R = O
LL − 491γ R = β-OH
(*Aspergillus chevalier*)[1]

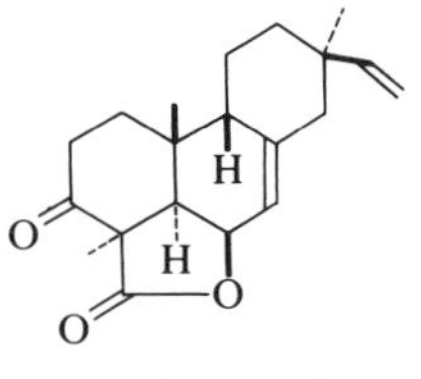

Momilactone-A
(Rice husk)[4]

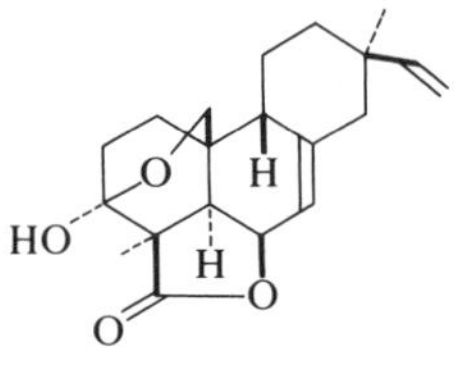

Momilactone-B
(Rice husk)[4]

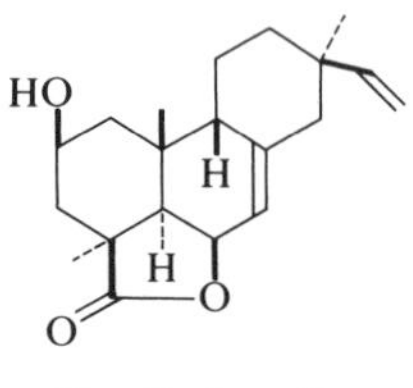

Momilactone-C
(Rice husk)[5]

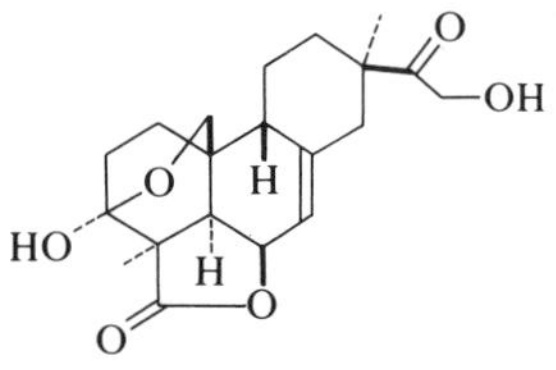

Annonalide
(*Annona coriacea*)[6]

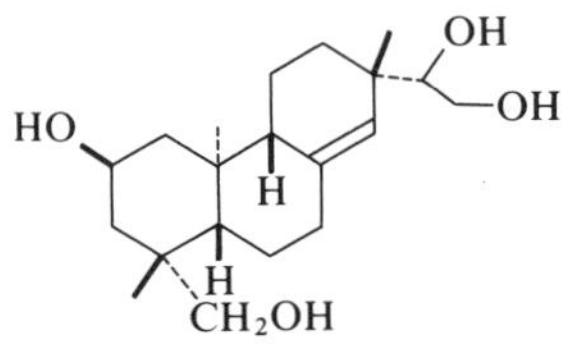

Kirenol
(*Siegesbeckia pubescens* Makino)[8]

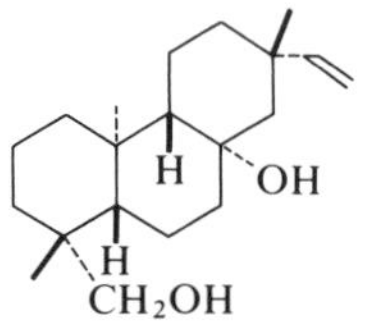

Thermarol
(*Jungermannia thermarum*)[9]

It has been observed[10] that the rearranged pimaradien-3β-ol and isopimaradien-3β-ol are produced directly from an all *trans* 14,15-oxido-geranyl geranyl ester (X = OCOOMe) by means of a specific sequence involving appearance of the exocyclic methylene bicycle, loss of allylic X, and further cyclization to pimaradien-3β-ols, an overall process closely related to that by which the same category of tricycles is believed to be formed enzymically from a geranyl geranyl species or its terminal epoxide[11]. Exposure of the ester to BF_3-ether in nitromethane at 0° for 0.5 hr provided (±)-pimara-8(9),15-dien-3β-ol, (±)-isopimara-8(9), 15-dien-3β-ol, and (±)-isopimara-7(8),15-dien-3β-ol.

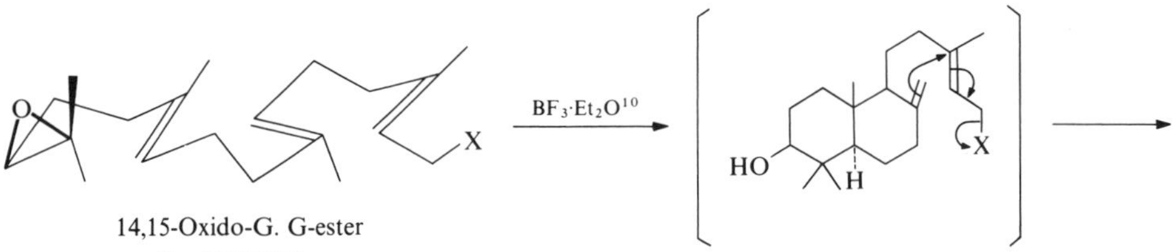

$\xrightarrow{BF_3 \cdot Et_2O^{10}}$

14,15-Oxido-G. G-ester
X = OCOOMe

Exocyclic methylene bicycle

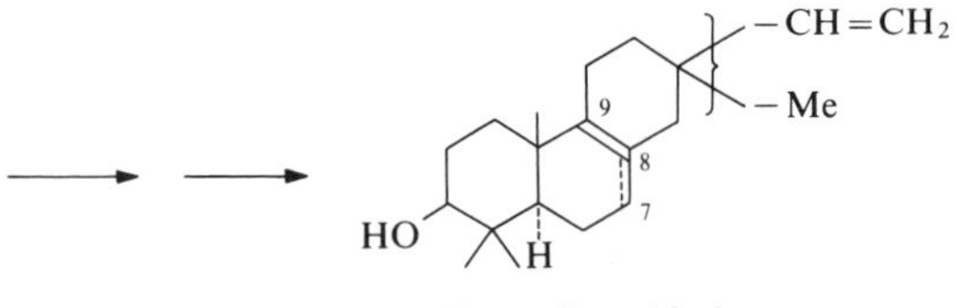

Pimaradiene-3β-ol

H_2SO_4–AcOH

Methyl isocupressate → Rosadienoic ester → → Epoxide

→ → Unsaturated lactone Deoxyrosenonolactone Rosenonolactone

The nonenzymic conversion of bicyclic diterpenes to hydrocarbons of the pimaradiene and other types has been investigated in details[12]. The greater thermodynamic stability of rosane compounds over that of pimarane derivatives[13] has been exploited in a biogenetically modeled synthesis of rosenonolactone from isocupressic acid[14]. Methyl isocupressate has been transformed with AcOH–H_2SO_4 into the rosadienoic ester, which was converted via the epoxide into the unsaturated lactone, and thence into deoxyrosenonolactone and resenonolactone.

2.2 Abietane and Its Variant Types

Several abietane type diterpenes with highly oxidized C rings show significant physiological activities. Taxodione and taxodone from *Taxodium distichum* have a tumor inhibitory activity[1], while callicarpone is a fish killing component of *Callicarpa candicans*[2]. Fractionation of an EtOH extract of *Tripterygium wilfordii* Hook, which shows significant activity against the L-1210 and P-388 leukemias resulted in the isolation of novel natural products, triptolide and tripdiolide, containing the 18(4→3) abeo-abietane skeleton with three epoxide rings[3]. *Stemodia maritima* (Scrophulariaceae) was found to produce a similar compound, stemolide[4]. One possibility on the genesis of the epoxide group in stemolide involves reaction of singlet oxygen with a diene followed by rearrangement of the resulting endo epoxide.

From aereal part of *Salvia ballotaeflora*, conacytone and icetexone were elucidated[5]. Natural 3β-acetate of diterpenoid alkaloid, 3β-acetoxynorerythrosuamine, was isolated from *Erythrophleum chlorostachys* as a cytotoxic material. It has been revealed that the original acetate is a thousand times more active than the parent alcohol and has an ED_{50} value of 0.0003 μg/ml[6]. Biological screening of plants, *Coleus barbathus* which are used in medicine as a stomach aid, led to the isolation of barbatusin[7] and cyclobutatusin[8]. Both compounds have the rare feature containing a methylcyclopropyl ring at the C-13 atom. In view of the reactivity of the cyclopropane molecule, barbatusin may represent an

[10] *E. E. van Tamelen, S. A. Marson*, J. Am. Chem. Soc. *97*, 5614 (1975).

[11] *A. J. Birch, R. W. Richards, H. Smith*, Proc. Chem. Soc., London 192 (1958).

[12] *S. F. Hall, A. C. Oehlschlager*, Tetrahedron *28*, 3155 (1972).

[13] *T. McCreadie, K. H. Overton*, J. Chem. Soc. (C) 312 (1971).

[14] *T. McCreadie, K. H. Overton, A. J. Allison*, J. Chem. Soc. (C) 317 (1971).

[1] *S. M. Kupchan, A. Karim, C. Marcks*, J. Org. Chem. *34*, 3912 (1969).

[2] *K. Kawazu, T. Mitsui*, Tetrahedron Lett. 3519 (1966).

[3] *S. M. Kupchan, W. A. Court, R. G. Dailey, C. J. Gilmore, R. F. Bryan*, J. Am. Chem. Soc. *94*, 7194 (1972).

[4] *P. S. Manchand, J. F. Blount*, Tetrahedron Lett. 2489 (1976).

[5] *W. H. Watson, Z. Taira, X. A. Dominguez, H. Gonzales, M. Guiterrez, R. Aragon*, Tetrahedron Lett. 2501 (1976).

[6] *J. W. Loder, R. H. Nearn*, Tetrahedron Lett. 2497 (1975).

[7] *A. H. J. Wang, I. C. Paul, R. Zelnik, K. Mizuta, D. Lavie*, J. Am. Chem. Soc. *95*, 598 (1973).

[8] *A. H. J. Wang, I. C. Paul, R. Zelnik, D. Lavie, E. C. Levy*, J. Am. Chem. Soc. *96*, 580 (1974).

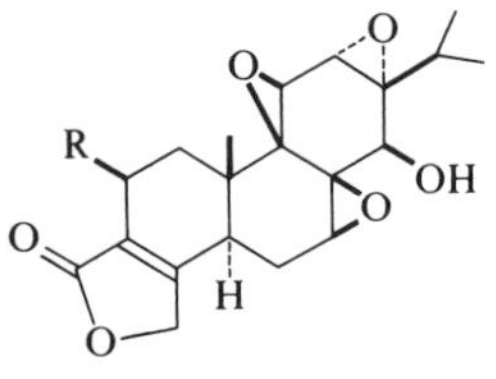

Triptolide R = H
Tripdiolide R = OH
(*Tripterygium wilfordii*)[3]

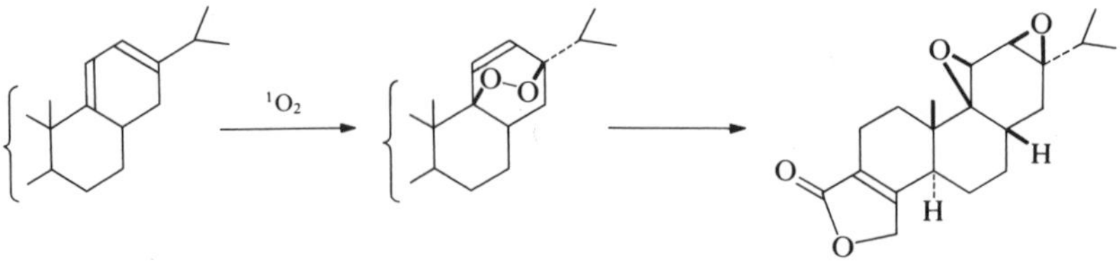

Stemolide
(*Stemodia maritima*)[4]

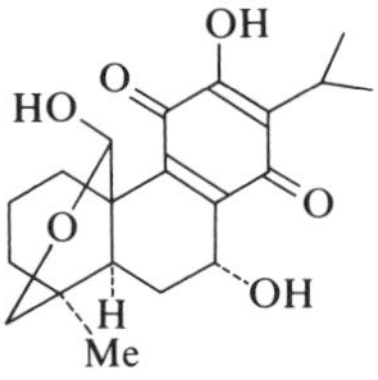

Conacytone
(*Salvia ballotaeflora* Benth)[5]

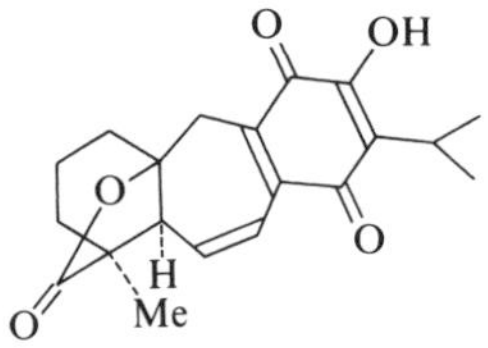

Icetexone
(*Salvia ballotaeflora* Benth)[5]

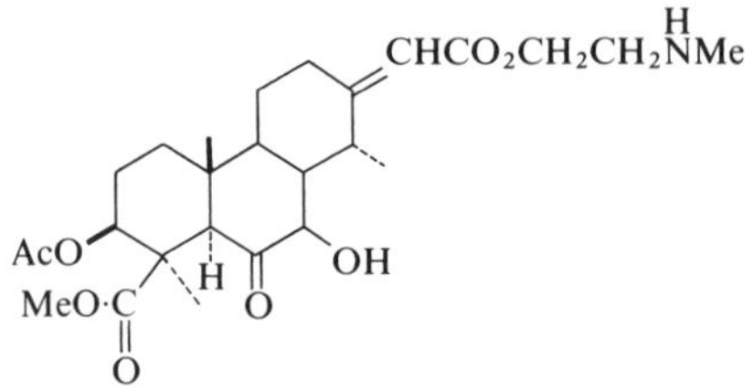

3β-Acetoxynorerythrosuamine
(*Erythrophleum chlorostachys*)[6]

interesting and unusual intermediate in the biosynthesis of some naturally occurring quinonoid diterpene such as coleon E. Cyclobutatusine, which retains the basic ring skeleton of barbatusine, has the outstanding new feature of a four membered ring formed by a bond between C-1 and C-11. Coleon E from the glands on the leaves of *Coleus barbatus* (Benth) Agnew, has fully conjugated unsaturated system extending over A, B and C rings[9]. The methyl group at C-3 and *n*-propyl side chain at C-13 instead of the usual isopropyl group are regarded as arising by modification of an abietane precursor[7]. Reports on the characterization of several abietane type diterpenes have been published as exemplified coleon D[10], lycoxanthol[11], pododacric acid[12], and salviol[13].

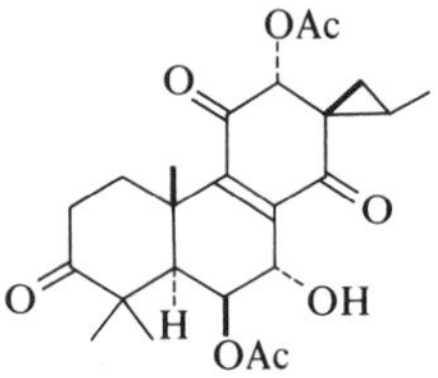

Barbatusin
(*Coleus barbathus*)[7]

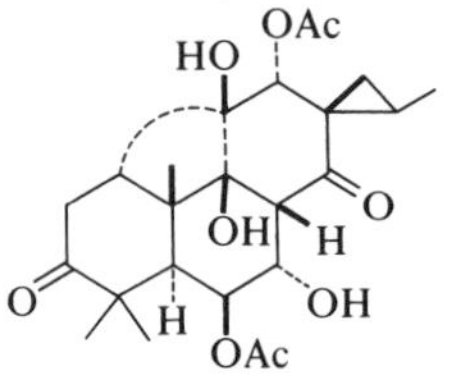

Cyclobutatusin
(*Coleus barbathus*)[8]

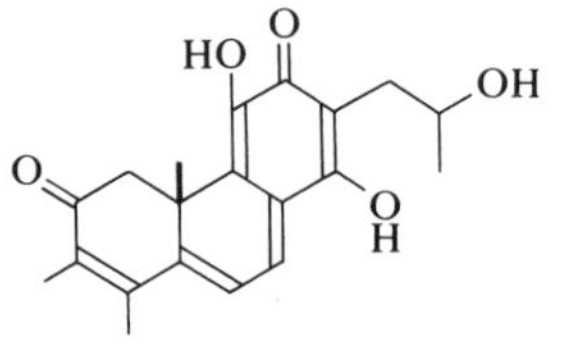

Coleon E
(*Coleus barbatus* (Benth) Agnew)[9]

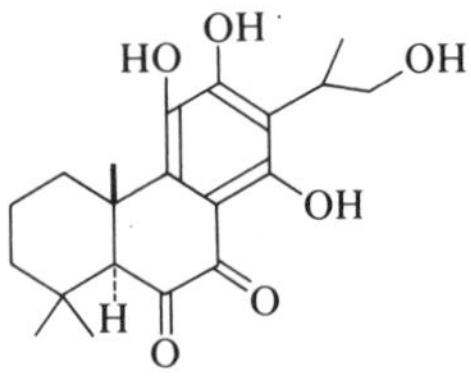

Coleon D
(*Coleus aquaticus* Guercke)[10]

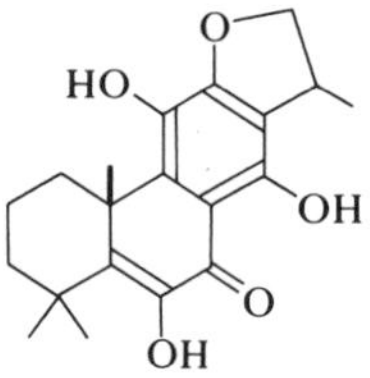

Lycoxanthol
(*Lycopodium lucidulum*)[11]

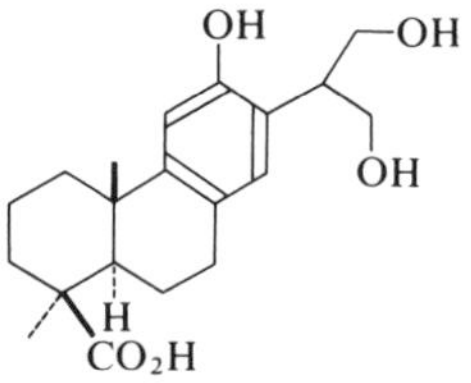

Pododacric acid
(*Podocarpus dacrydioides*)[12]

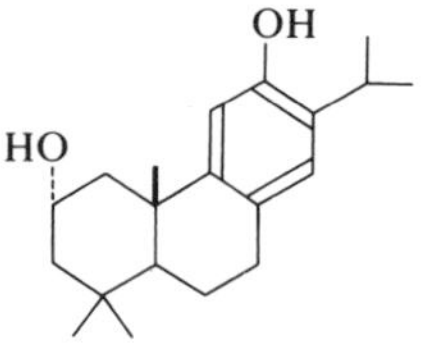

Salviol
(*Salvia miltiorrhiza* Bunge)[13]

---

[9] *P. Rüedi, C. H. Eugster*, Helv. Chim. Acta *55*, 1994 (1972).
[10] *P. Rüedi, C. H. Eugster*, Helv. Chim. Acta *55*, 1736 (1972).
[11] *R. H. Burnell, M. Moinas*, Chem. Commun. 897 (1971).
[12] *R. C. Cambie, K. P. Mathai*, Chem. Commun. 154 (1971).
[13] *T. Hayashi, T. Handa, M. Ohashi, H. Kakisawa*, Chem. Commun. 541 (1971).

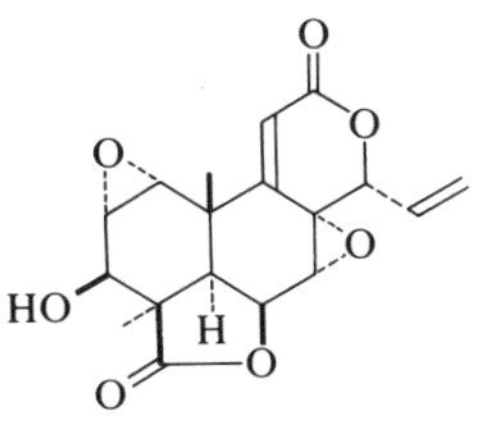

Inumakilactone B
(*P. macrophyllus* D. Don)[14]

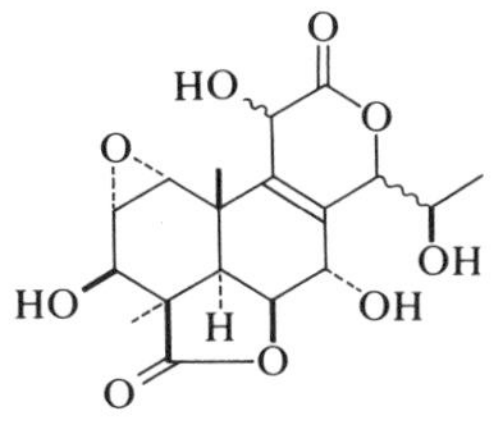

Inumakilactone C
(*P. macrophyllus* D. Don)[14]

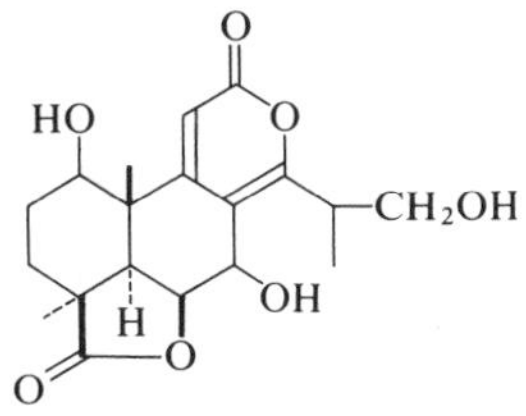

Inumakilactone E
(*P. macrophyllus* D. Don)[15]

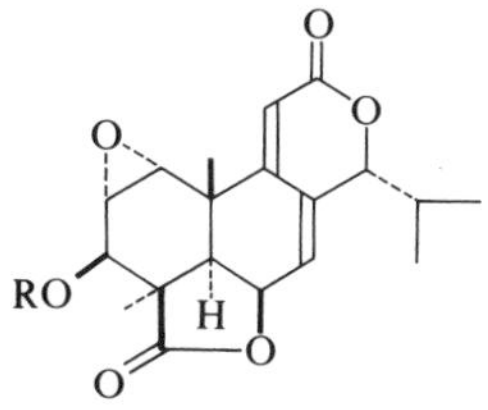

Ponalactone A
R = H
Ponalactone A glucoside
R = glucose
(*P.* nakaii HAY)[16]

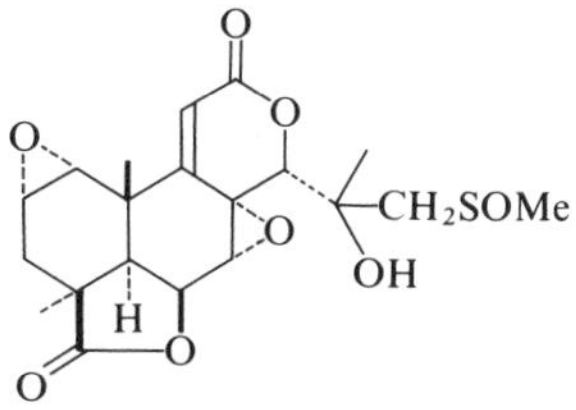

Podolactone C
(*P. neriifolius*)[17]

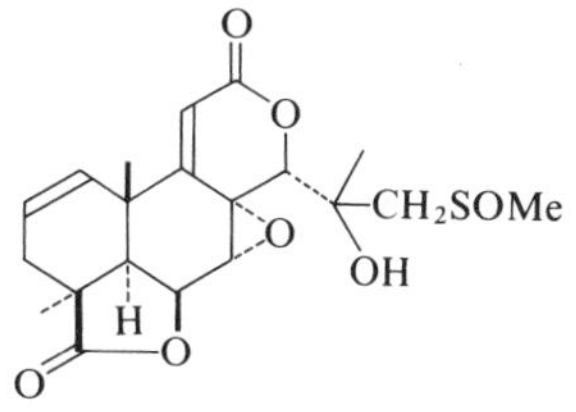

Podolactone D
(*P. neriifolius*)[17]

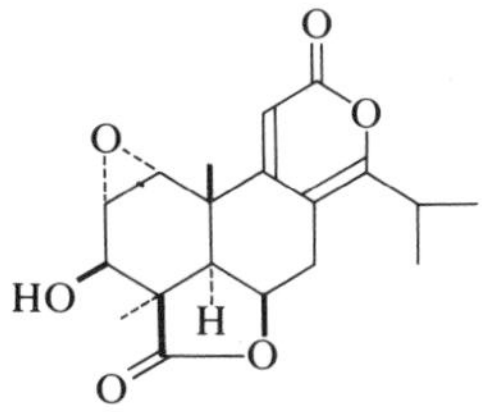

Hallactone A
(*P. hallii*)[18]

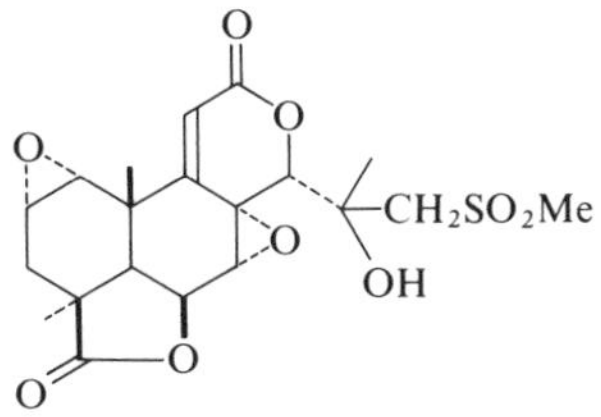

Hallactone B
(*P. hallii*)[18]

A series of nor and bisnorditerpenes was elaborated from *podocarpus* species. Diterpenes include inumakilactones[14,15], ponalactones[16], podolactones[17], and hallactones[18]. Some of these nor- and bisnorditerpenes have been shown to possess the potent inhibitory activities against expansion and mitosis of plant cells and also against plant growth. Podolactones (C and D) and hallactone B are the first terpenes known to contain a sulphoxide grouping. The presence of a sulphoxide group in a terpene is unique and its biogenesis can be expected to take place via the corresponding C-16 thiol. As well as nagilactone C, hallactones are the toxic components responsible for the mortality of house-fly larvae reared on a diet containing the materials from leaves of *Podocarpus hallii* Kirk.

The lactones isolated from the metabolite of a mould, *Acrostalagmus* species have a novel C_{16}-carbon skeleton[19-21]. These are LL-Z1271α, β and γ, acrostalidic acid and isoacrostalidic acid. The carbon skeletone has been demonstrated by incorporation experiment to arise from a diter-

[14] *S. Ito, M. Sunagawa, M. Kodama, H. Honma, T. Takahashi*, Chem. Commun. 91 (1971).

[15] *T. Hayashi, H. Kakisawa, S. Ito, Yuh Pan Chen, Hong-Yen Hsu*, Tetrahedron Lett. 3385 (1972).

[16] *S. Ito, M. Kodama, M. Sunagawa, M. Koreeda, K. Nakanishi*, Chem. Commun. 855 (1971).

[17] *M. N. Galbraith, D. H. S. Horn*, Chem. Commun. 1362 (1971).

[18] *G. B. Russell, P. G. Fenemore, P. Singh*, Chem. Commun. 166 (1973).

[19] *G. A. Ellestad, R. H. Evans, M. P. Kunstmann, J. E. Lancaster, G. O. Morton*, J. Am. Chem. Soc. *92*, 5483 (1970).

[20] *G. A. Ellestad, R. H. Evans, M. P. Kunstmann*, Tetrahedron Lett. 497 (1971).

[21] *M. Sato, T. Ruo, T. Hayashi, H. Kakisawa*, Tetrahedron Lett. 2183 (1974).

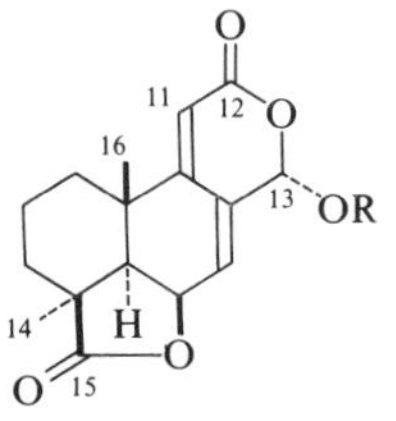

LL-Z1271α
R = Me
LL-Z1271γ
R = H
(*Agrostalagmus*)[19]

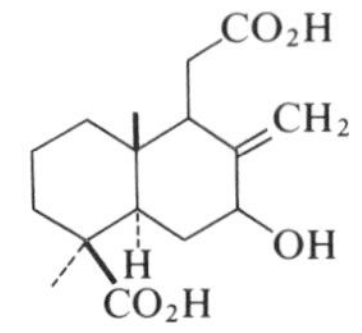

LL-Z1271β
Acrostalic acid
(*Acrostalagmus*)[20,21]

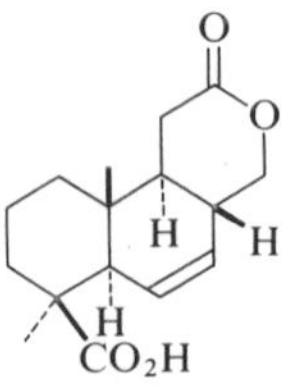

Acrostalidic acid
(*Acrostalagmus* NRRL-3481)[21]

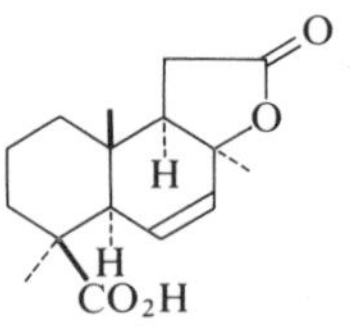

Isoacrostalidic acid
(*Acrostalagmus* NRRL-3481)[21]

pene with oxidative loss of four carbon atoms[22]. The almost equal enhancement of the signal intensities of C-12 and C-15 in the CMR spectra of the labelled lactone (LL-Z1271α), produced by incorporation of [2-^{13}C]AcOH, shows that the lactone is biosynthesized from a diterpene derived from AcOH through mevalonic acid. Administration of [2-^{14}C, 5-$^{3}H_2$]mevalonic acid (^{3}H/^{14}C = 5.4) to the mould afforded the doubly labelled lactone whose ^{3}H/^{14}C ratio was shown to be 2.5. This shows that the lactone is constructed from four molecules of mevalonic acid with loss of four tritium atoms. If the lactone were derived from three molecules of mevalonic acid, the ^{3}H/^{14}C ratio of the lactone would be 3.6. These experiments demonstrate that the C_{16} skeleton of the lactone is derived from microbiological degradation of a normal diterpene such as labdadienol by oxidative cleavage between C-12 and C-13. The lactone (LL-Z1271α) has a strong inhibitory activity on the growth of an *Avena coleoptile* section comparable to those of the inumaki-, nagi-, and podolactones.

Much work on the syntheses of abietane type diterpenes has been carried out including fichtelite[23-25], sugiol[26], ferruginol[27], carnosate[28], royleanone[29], sempervirol[30,31], tanshinone[32], taxodione[32], and LL-Z1271α[34].

3 Tetracyclic Diterpenes

3.1 Kaurane and Related Types

Although many members of the cyclic diterpenes possess a hydroxyl group at C-3 position, two different routes have been demonstrated concerning the biogenetic origin of the hydroxyl group. One is the introduction by an oxygenation reaction subsequent to the cyclization stages[1,2]. The other is a direct outcome of the enzymatic cyclization of (R,S)14,15-oxidogeranyl geranyl pyrophosphate. The latter pathway in kaurane type diterpenes was proven by soluble enzyme preparations obtained from immature *Echinocystis macrocaspa* which contains kaurene synthetase[3].

A number of ring C and D functionalized kaurenoid derivatives have been reported as typified ent-11α-hydroxykauren-15α-yl acetate[4], mascaroside[5], and phlebiakauranols[6]. Phlebiakauranols are antibacterial metabolites from

---

22 *H. Kakisawa, M. Sato, T. Ruo, T. Hayashi*, Chem. Commun. 802 (1973).

23 *N. P. Jensen, W. S. Johnson*, J. Org. Chem. *32*, 2045 (1967).

24 *A. W. Burgstahler, J. N. Marx*, J. Org. Chem. *34*, 1562 (1969).

25 *W. S. Johnson, N. P. Jensen, J. Hooz, E. J. Leopold*, J. Am. Chem. Soc. *90*, 5872 (1968).

26 *W. L. Meyer, G. B. Cleman*, Tetrahedron Lett. 4255 (1966).

27 *M. Ohashi, T. Maruishi, H. Kakisawa*, Tetrahedron Lett. 719 (1968).

28 *W. L. Meyer, E. Schindler*, Tetrahedron Lett. 4261 (1966).

29 *E. Wenkert, J. D. MeChesney, D. J. Watts*, J. Org. Chem. *35*, 2422 (1970).

30 *U. R. Ghatak, N. R. Chattejee*, Tetrahedron Lett. 4783 (1967).

31 *C. Romualdo, M. Lorenzo*, Gazz. Chim. Ital. *97*, 920 (1967).

32 *H. Kakisawa, Y. Inoue*, Chem. Commun. 1327 (1968).

33 *K. Mori, M. Matsui*, Tetrahedron *26*, 3467 (1970).

34 *M. Adinolfi, L. Mangoni, G. Barone*, Tetrahedron Lett. 695 (1972).

---

1 *H. J. Bakker, E. L. Ghisalberti, P. R. Jefferies*, Phytochemistry *11*, 2221 (1972).

2 *A. B. Anderson, T. McCrindle, J. K. Turnbull*, Chem. Commun. 143 (1973).

3 *R. M. Coates, R. A. Conradi, D. A. Ley, A. Akeson, J. Harada, S.-C. Lee, C. A. West*, J. Am. Chem. Soc. *98*, 4659 (1976).

4 *J. D. Connolly, I. M. S. Thornton*, J. Chem. Soc., Perkin I 736 (1973).

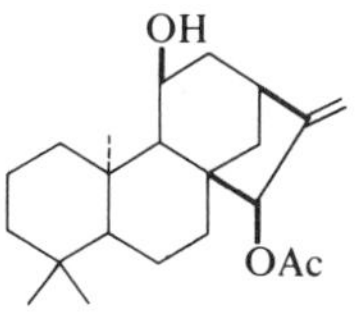

ent-11α-Hydroxy-kauren-15α-yl acetate (*Solenostoma triste*)[4]

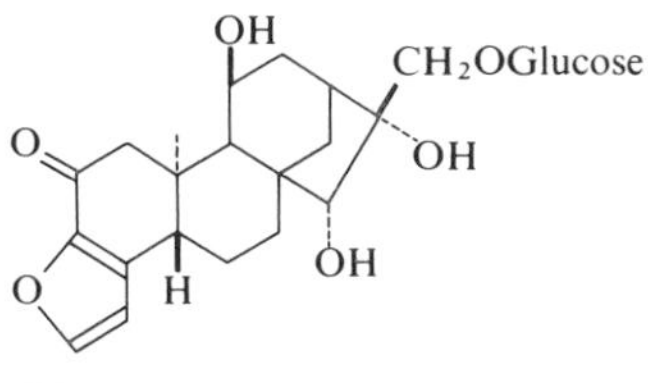

Mascaroside (Malagasy *Coffea vianneyi* Leroy)[5]

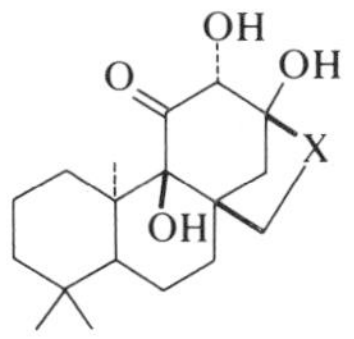

Phlebiakauranol
X = –CH(OH)–CH(OH)OMe

Phlebianorkauranol
X = CO
(*Phlebia strigosozonata*)[6]

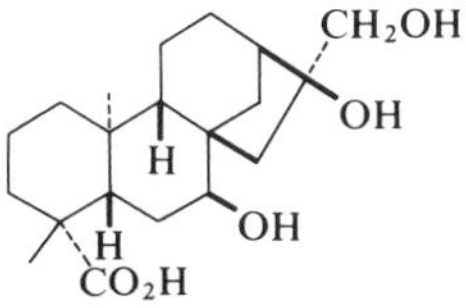

ent-7α,16α,17-Trihydroxy-kauran-19-oic acid (evening glory, *Calonyction aculeatum*)[7]

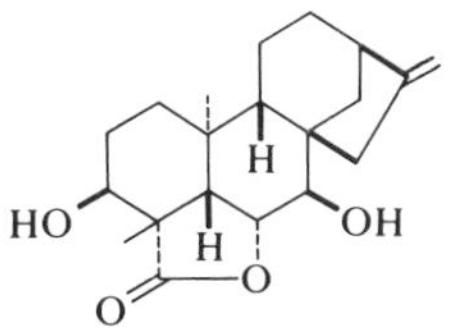

3β,7β-Dihydroxykaurenolide (*Gibberella fujikuroi*)[8]

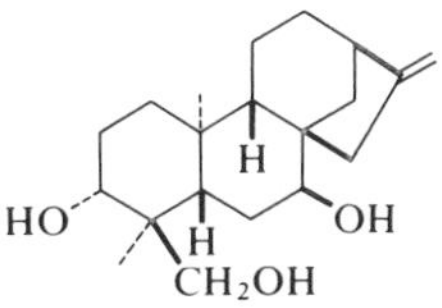

Foliol (*Sideritis leucantha* Cav.)[9]

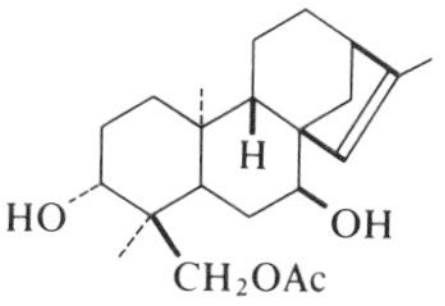

Isolinearol (*Sideritis leucantha* Cav.)[9]

Phlebia strigosozonata and are active against the strain of *Staphylococcus aureus*. Trihydroxy kauranoic acids[7], 3β,7β-dihydroxykaurenolide[8], and six diterpenes[9] isolated from *Sideritis leucantha* belong to the C-7 hydroxylated kaurenoid which is the key intermediate in the biosynthesis of gibberellin skeleton.

New sweet steviol glucosides, named rebaudioside-A and B were discovered from *Stevia rebaudiana*. In determining the structures, the identification of genuine aglycones and the elucidation of location and configuration of glucosyl linkage were achieved by ^{13}C NMR spectroscopy[10]. ^{13}C NMR has been recognized as a useful technique for structure elucidation in the pimaradienes[11], kaurenes[12], hibaenes[13], and gibberellins[14]. Structures of paniculoside I, II, and III from *Stevia paniculata* were determined by ^{13}C NMR spectrum without any chemical degradations[10]. Sideritol[15], isosideritol[15], isolated from *Sideritis angustifolia*, are the first examples of an oxygenated nitrogen free diterpenoid of the ent-atisane class. Vakognavine[17] is the first compound of an N, C(19) secoditerpene alkaloid and may be a plausible intermediate in the biosynthesis of the hetisine type skeleton.

---

[5] *A. Ducruix, C. Pascard-Billy, M. Hamonniere, J. Poisson*, Chem. Commun. 396 (1975).

[6] *J. M. Lisy, J. Clardy, M. Anchel, S. M. Weinreb*, Chem. Commun. 406 (1975).

[7] *N. Merofushi, T. Yokota, N. Takahashi*, Tetrahedron Lett. 789 (1973).

[8] *J. H. Bateson, B. E. Cross*, Tetrahedron Lett. 3407 (1971).

[9] *T. G. de Quesada, B. Rodriguez, S. Valverde*, Tetrahedron Lett. 2187 (1972).

[10] *K. Yamazaki, H. Kohda, T. Kobayashi, R. Kasai, O. Tanaka*, Tetrahedron Lett. 1005 (1976).

[11] *E. Wenkert, B. L. Buckwalter*, J. Am. Chem. Soc. *94*, 4367 (1972).

[12] *J. R. Hanson, G. Savona, M. Siverns*, J. Chem. Soc., Perkin I 2001 (1974).

[13] *C. von Carstenn-Lichterfelde, C. Pasual, J. Pons, R. Ma. Rabanal, B. Rodriguez, S. Valverde*, Tetrahedron Lett. 3569 (1975).

[14] *I. Yamauchi, N. Takahashi, K. Fujita*, J. Chem. Soc., Perkin I 992 (1975).

[15] *W. A. Ayer, J.-A. H. Ball, B. Rodriguez, S. Valverde*, Can. J. Chem. *52*, 2792 (1974).

[16] *I. Carrascal, B. Rodriguez, S. Valverde*, Chem. Commun. 815 (1975).

[17] *S. W. Pelletier, K. N. Iyer, L. H. Wright, M. G. Newton, N. Singh*, J. Am. Chem. Soc. *93*, 5942 (1971).

[18] *E. Wenkert*, Chem. Ind. 282 (1955).

[19] *C. H. Bieskorn, E. Pöhlmann*, Chem. Ind. 5661 (1968).

[20] *R. B. Turner, K. H. Ganshirt, P. E. Shaw, J. D. Tauber*, J. Am. Chem. Soc. *88*, 1776 (1966).

[21] *R. A. Bell, R. E. Ireland, R. A. Partyka*, J. Org. Chem. *31*, 2530 (1966).

[22] *K. Mori, Y. Nakahara, M. Matsui*, Tetrahedron Lett. 2411 (1970).

[23] *D. J. Beames, T. R. Klose, L. M. Mander*, Chem. Commun. 773 (1971).

[24] *U. R. Ghatak, P. C. Chakraborti, B. C. Ranu, B. Sanyal*, Chem. Commun. 548 (1973).

[25] *A. J. Baker, A. C. Goudie*, Chem. Commun. 180 (1971).

[26] *T. Kato, T. Suzuki, N. Ototani, Y. Kitahara*, Chemistry Lett. 887 (1976).

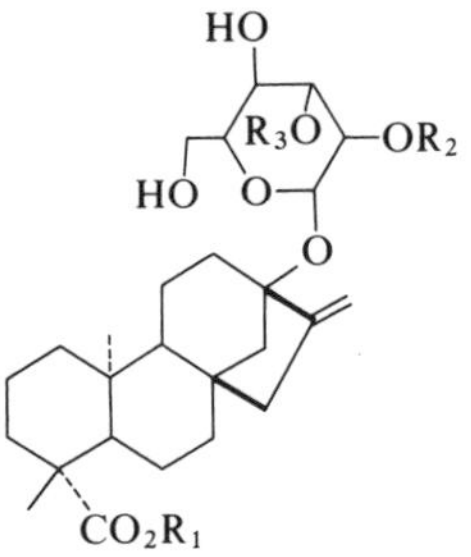

Rebaudioside A
$R_1 = R_2 = R_3 = Glc$
Rebaudioside B
$R_1 = H$, $R_2 = R_3 = Glc$
Glc = β-glucopyranosyl
(*Stevia rebaudiana*)[10]

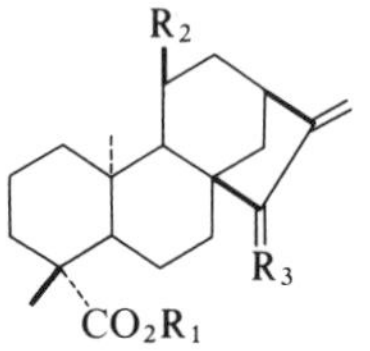

Paniculoside I
$R_1 = Glc$, $R_2 = H$, $R_3 = $ OH, H
Paniculoside II
$R_1 = Glc$, $R_2 = R_3 = OH$
Paniculoside III
$R_1 = Glc$, $R_2 = OH$, $R_3 = O$

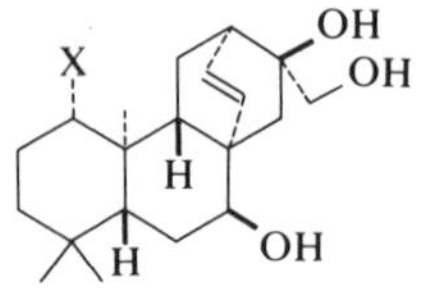

Sideritol
X = OH
Isosideritol
X = H
(*Sideritis angustifolia*)[16]

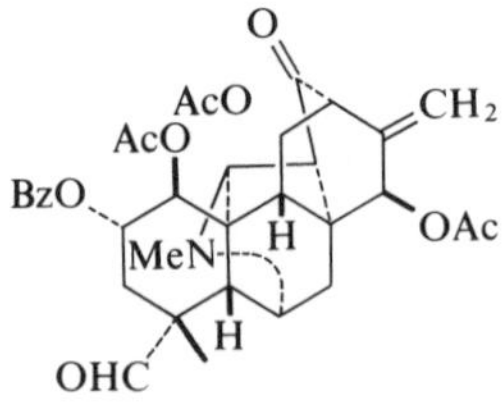

Vakognavine
(*Aconitum palmatum*)[17]

Because of the structural complexity, much effort was expended on the total synthesis as well as on the construction method of various types of C/D ring systems in tetracyclic diterpenoids. Chemical interconversion of C/D rings between cyclic terpenoids have also been examined, based on the Wenkert's hypothesis[18,19] concerning the biogenesis of tetracyclic diterpenoids. Table 3 shows the representative results of these studies.

Table 3. Construction methods of C/D ring systems

Reaction	Ref.
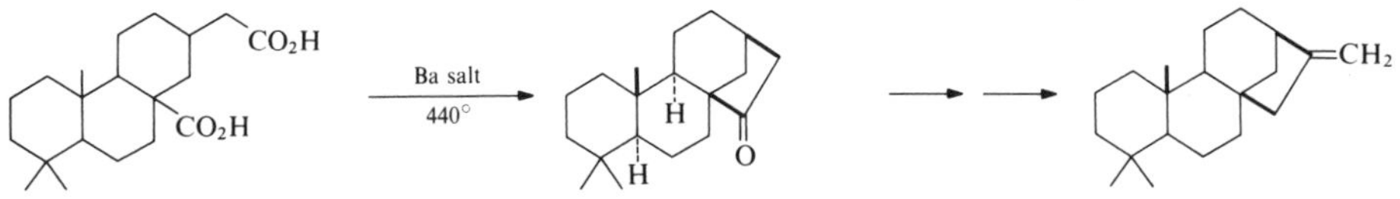 Ba salt, 440° → → → Phyllocladene	20
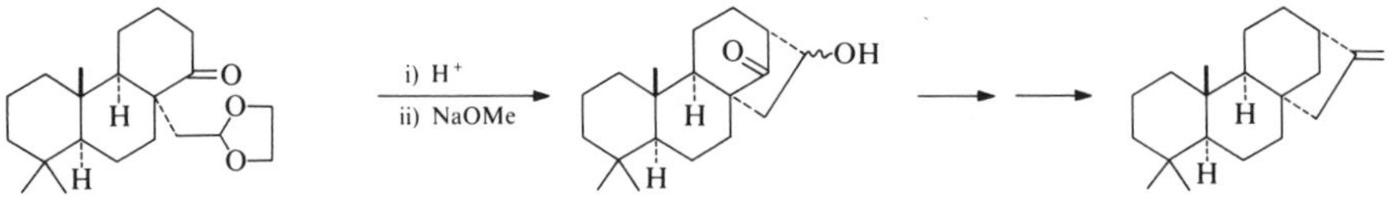 i) H^+ ii) NaOMe → → → Kaurene	21
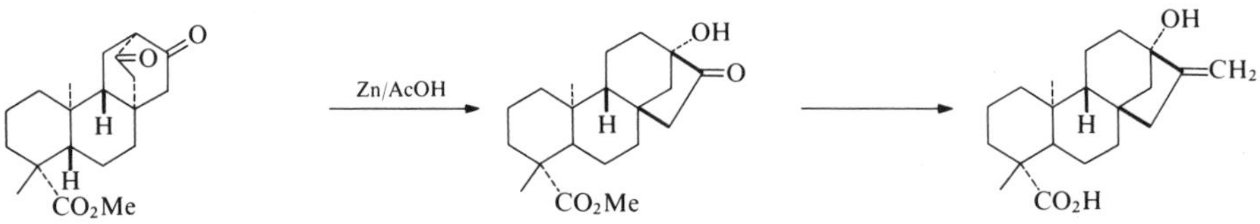 Zn/AcOH → → Steviol	22
 HBF_4 →; n = 1 (97%), n = 2 (90%)	23,24
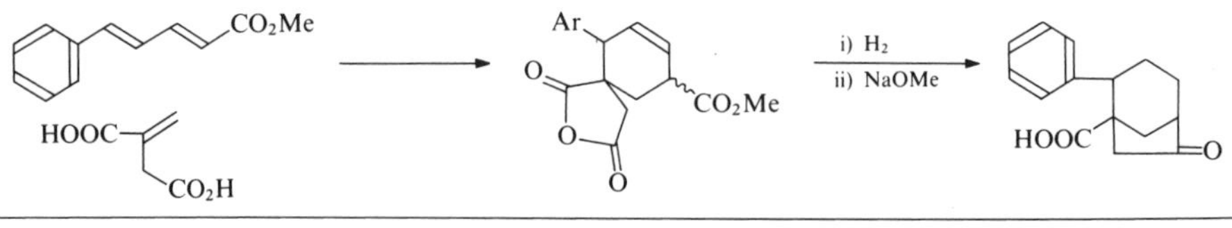 → i) H_2 ii) NaOMe →	25,26

Table 3. (continued)

Reaction (reagents / labels)	Ref.
O; $h\nu$, $CH_2=C=CH_2$; O, H, $=CH_2$; P – TsOH; O, H, H	27
Me, O, MeO, CHSBu; 180°; [MeO, CH_2, Me, O, CH – SBu]; Me, O, MeO, SBu	28
OH, H, CH_2; HCO_2H; [H, H, +]; RO, H, H	29
OH, $=CH_2$, H, H; HCl; Me, O, H, H; Me, H, H; Hibaene	30
N – NHTs, O, HO, H, H; $NaBH_4$; H, Me, HO, HO, H, H	31
OMs, H; Me, NaOS = CH_2/DMSO; Me, Z, H, H; Z = O, H, OH	32
H, H, CH_2OH, O; 95% HCO_2H; OCHO, H, H, CH_2OCHO	33

[27] *D. K. Manh Duc, M. Fetizon, S. Lazare*, Chem. Commun. 282 (1975).

[28] *T. Kametani, H. Nemoto, K. Fukumoto*, Chem. Commun. 400 (1976).

[29] *E. Wenkert, Z. Kumazawa*, Chem. Commun. 140 (1968).

[30] *R. A. Bell, R. E. Ireland, L. N. Mander*, J. Org. Chem. *31*, 2536 (1966).

[31] *K. H. Pegel, L. P. L. Piacenza, L. Phillips, E. S. Waight*, Chem. Commun. 552 (1973).

[32] *R. B. Kelly, J. Eber, Hing-Kwok Hung*, Chem. Commun. 689 (1973).

[33] *R. D. H. Murray, R. W. Mills, J. M. Young*, Tetrahedron Lett. 2393 (1971).

3.2 Gibbane and Related Types

By the development of experimental techniques, it has become possible to characterize a trace component of physiologically active natural products, especially among the gibberellins. The power of combined gas chromatography-mass spectrometry in detecting and identifying plant gibberellins has been demonstrated by the characterization[1] of the known gibberellins (A_1, A_5, A_6, A_8, and A_{19}) and by the detection of an unknown gibberellins in extracts of *Phaseolus multiflorus*[2].

Gibberellins have a characteristic plant growth promoting activity which is only observed with various gibbane derivatives and with diterpenes which have been indicated to be intermediates of normal gibberellin biosynthetic pathways[3]. The detailed information on the gibberellin biosynthetic pathway has been provided by studies with normal *Gibberella fujikuroi* and its UV-induced mutant (Bl-41a)[4,5]. These experiments show that the direct precursor of gibberellins is 7β-hydroxy-(−)-kaur-16-en-19-oic acid, which is effectively converted into the GA_{12} aldehyde and gibberellic acid[6]. Furthermore, it has been disclosed that the ring contraction occurs by a shift of 6β-hydrogen to C-7 as the C_7, C_8 bond migrates[7].

The structure was assigned to the antheridium-inducing factor, antheridiogen-An (AAn), isolated from culture media of the fern *Anemia phyllitidis* (Schizaiaceae). It is the first fern antheridiogen to be characterized, and induces antheridia at 10 g/*l* and also substitutes for light requirement in spore germination. Biosynthetically, a close structural relation between AAn and gibberellins is obvious; i.e., such an intermediate as an epoxide, readily derivable from a gibberellin, could undergo facile rearrangement leading to the AAn skeleton[8].

Pharbitic acid, elaborated from immature seed of Japanese morning glory, is apparently biosynthesized by oxidation of gibberellin A_3 accompanied by addition of mercaptopyruvic acid or cysteine to the α,β-unsaturated ketone. Since pharbitic acid shows no gibberellin like activity and is contained in relatively high quantity, it seems to function physiologically as regulating the content and the activity of GA_3[9].

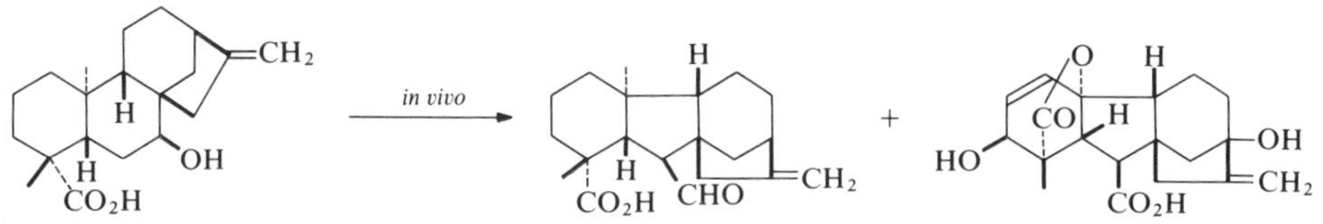

7β-Hydroxy-(−)-kaur-16-en-19-oic acid — *in vivo* → GA_{12}-aldehyde + Gibberellic acid

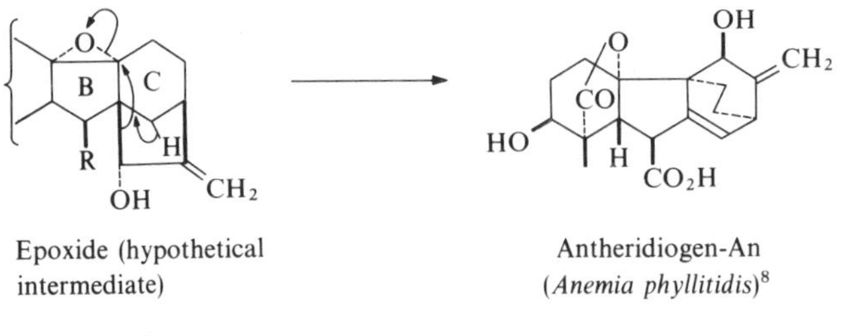

Epoxide (hypothetical intermediate) → Antheridiogen-An (*Anemia phyllitidis*)[8]

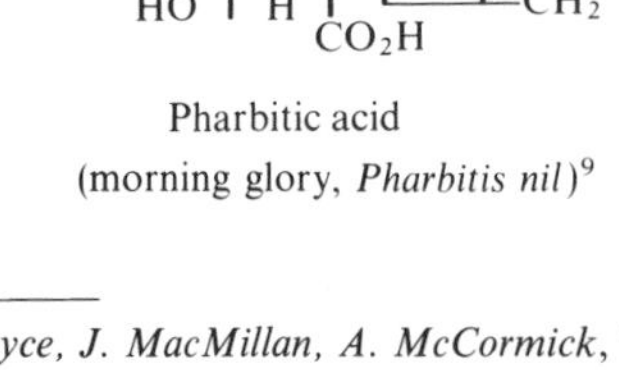

Pharbitic acid (morning glory, *Pharbitis nil*)[9]

[1] *R. J. Pryce, J. MacMillan, A. McCormick*, Tetrahedron Lett. 5009 (1967).

[2] *J. MacMillan, R. J. Pryce*, Tetrahedron Lett. 1537 (1968).

[3] *I. F. Cook, P. R. Jefferies, J. R. Knox*, Tetrahedron Lett. 2157 (1971).

[4] *J. R. Bearder, P. Hedden, J. MacMillan, C. M. Wels, B. O. Phinney*, Chem. Commun. 777 (1973).

[5] *J. R. Bearder, J. MacMillan, C. M. Wels*, Chem. Commun. 778 (1973).

[6] *J. R. Hanson, A. F. White*, Chem. Commun. 410 (1969).

[7] *J. R. Hanson, J. Hawker*, Chem. Commun. 208 (1971).

[8] *K. Nakanishi, M. Endo, U. Näf, L. F. Johnson*, J. Am. Chem. Soc. *93*, 5579 (1971).

[9] *T. Yokota, S. Yamazaki, N. Takahashi, Y. Iitaka*, Tetrahedron Lett. 2957 (1974).

Table 4. Construction methods of A/B ring of gibberellins

Reaction	Ref.
$\xrightarrow{70\%\ HClO_4}$ → i) H_2 ii) Et_3OBF_4	10,11
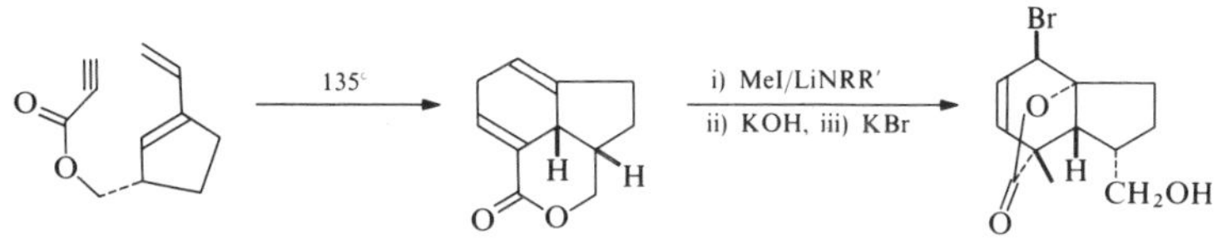 $\xrightarrow{135°}$ → i) MeI/LiNRR′ ii) KOH, iii) KBr	12,13
t-BuOK/t-BuOH	14
$Me_2S{=}CH_2$ → i) BF_3·Ether ii) Oxidn.	15
i) BuLi ii) CO_2 → →	16

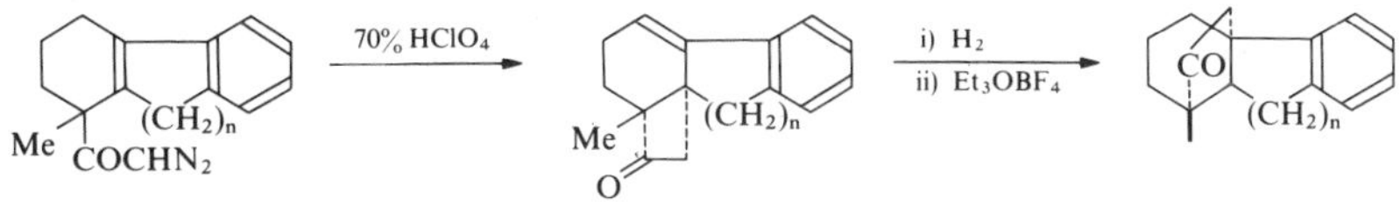

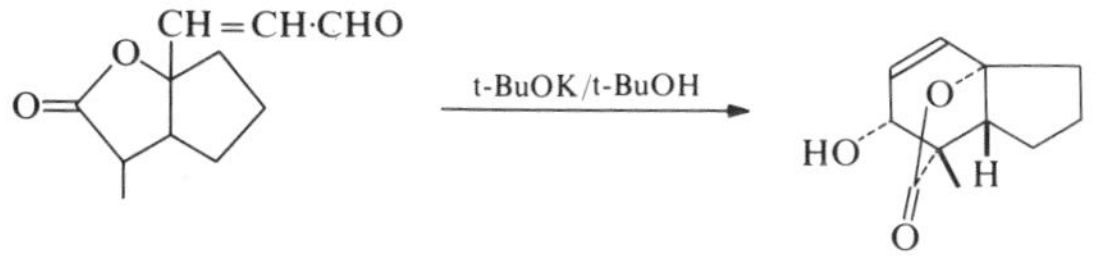

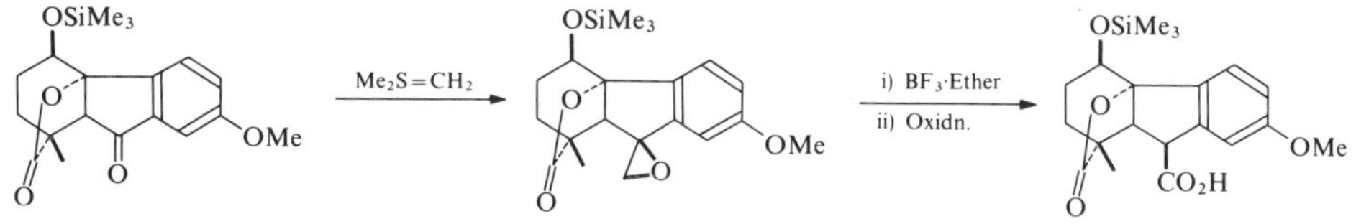

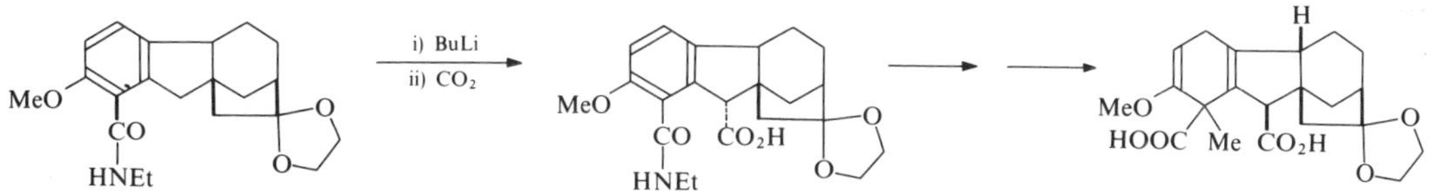

Due to its special hormone activity and structural complexity, many attempts have been made toward the synthesis of gibberellins. In addition to the exploration of C/D ring constructing methods (see section 3.1), many procedures have been accumulated on the synthesis of A/B ring skeleton of gibberellins, the results being shown in Table 4. Thus far, total syntheses of C_{19}-gibberellins (A_2, A_4, A_9, A_{10})[17] and C_{20}-gibberellin (A_{15})[18] have been performed after overcoming difficulties encountered during the sequence of multistage reactions. Works on formal total synthesis[19,20] from natural products and on chemical conversions between gibberellins[21,22] are also valuable from synthetic as well as biogenetic points of views.

3.3 Other Tetracyclic Diterpenes

In addition to thirteen toxic diterpenes of the andromedane skeleton and grayanotoxins[1], leucothol A, B, and D which have the novel carbon skeleton, have been elaborated from the poisonous shrub, *Leucothol grayana* Maximowicz[2]. Aphidicolin, having a new ring system of aphidi-

[10] *U. R. Ghatak, B. Sanyal*, Chem. Commun. 876 (1974).
[11] *U. R. Ghatak, B. Sanyal, S. Ghosh*, J. Am. Chem. Soc. *98*, 3721 (1976).
[12] *E. J. Corey, R. L. Danheiser*, Tetrahedron Lett. 4477 (1973).
[13] *E. J. Corey, T. M. Brennan, R. L. Carney*, J. Am. Chem. Soc. *93*, 7316 (1971).
[14] *L. J. Dolby, C. N. Skold*, J. Am. Chem. Soc. *96*, 3276 (1974).
[15] *Y. Yamada, K. Hosaka, H. Nagaoka, K. Iguchi*, Chem. Commun. 519 (1974).
[16] *A. J. Baker, A. C. Goudie*, Chem. Commun. 951 (1972).
[17] *K. Mori, M. Shiozaki, N. Itaya, M. Matsui, Y. Sumiki*, Tetrahedron *25*, 1293 (1969).
[18] *W. Nagata, T. Watanabe, Y. Hayase, M. Narisada, S. Kamata*, J. Am. Chem. Soc. *92*, 3202 (1970).
[19] *B. E. Cross, I. L. Gatfield*, Chem. Commun. 33 (1970).
[20] *M. Node, H. Nori, E. Fujita*, Chem. Commun. 898 (1975).
[21] *D. H. Bowen, D. M. Harrison, J. MacMillan*, Chem. Commun. 808 (1972).
[22] *J. R. Bearder, J. MacMillan*, Chem. Commun. 421 (1976).

[1] *H. Hikino, S. Koriyama, T. Ohta, T. Takemoto*, Chem. Pharm. Bull. *20*, 422 (1972).
[2] *H. Hikino, S. Koriyama, T. Takemoto*, Tetrahedron Lett. 3831 (1972).

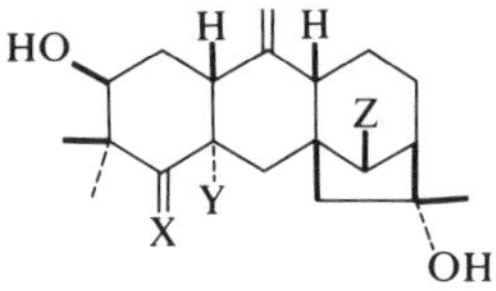

Leucothol A
X=O, Y=H, Z=H
Leucothol B
X=OH, H; Y=OH, Z=OH
Leucothol D
X=O, Y=OH, Z=OH
(*Leucothol grayana*)[2]

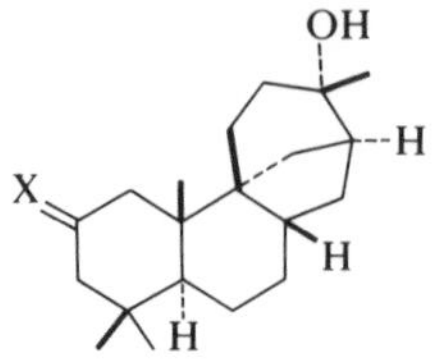

Stemodin
X=H, OH
Stemodinone
X=O
(*Stemodia maritima* L.)[4]

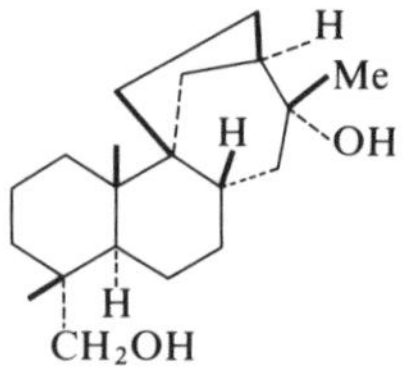

Stemarin
(*Stemodia maritima* L.)[5]

Biosynthetic pathway of aphidicolin

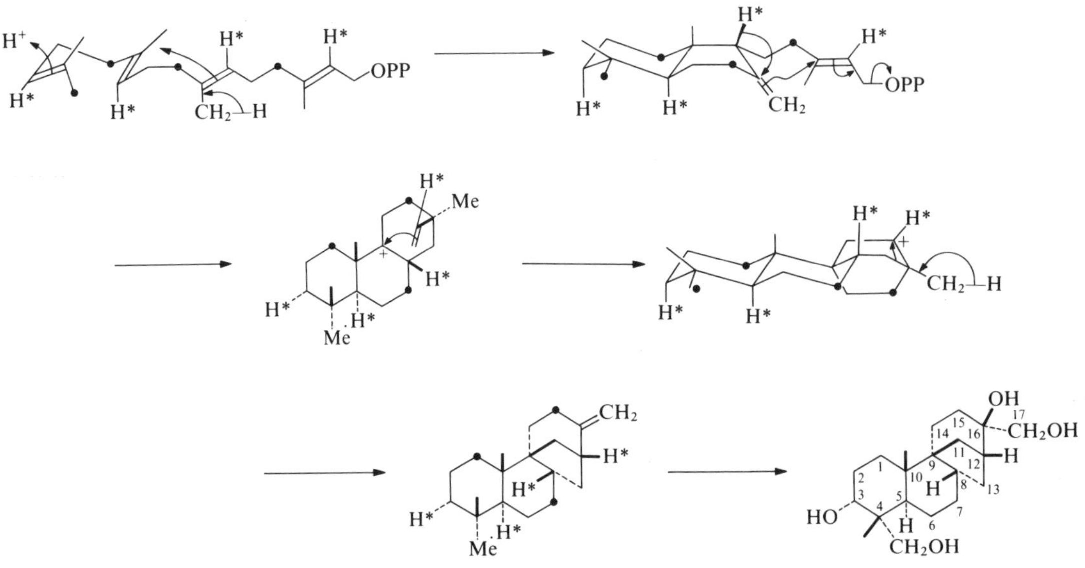

Aphidicolene

Aphidicolin
(*Cephalosporium aphidicola*)[3]

• from C_2 of mevalonate; H* from 4R-H of mevalonate

colane, is an antiviral diterpene, produced by the fungi *Cephalosporium aphidicola.* It is active against a range of DNA-containing virus, *e.g. Herpex simples*, and acts by inhibiting the synthesis of virus DNA[3]. Examination of the leaf constituents of the rare littoral *Stemodia maritima* L. (Scrophulariaceae) has brought to light three new diterpenes, stemodine, stemodinone, and stemarin[4,5]. The structure of stemodin and stemodinone bears a close resembrance to that of the aphidicolin excepting the stereochemistry at C-9, 12 and 16 (aphidicolane numbering). Incorporation experiments of 1,2-$^{13}C_2$-acetate and 2-^{14}C-4R-^{3}H-mevalonate precursors have disclosed the biosynthesis of aphidicolin, in which spectroscopic and degradative studies show that

[3] *K. M. Brundret, W. Dalziel, B. Hesp. J. A. J. Jarvis, S. Nėidle*, Chem. Commun. 1027 (1972).
[4] *P. S. Manchand, J. D. White, H. Wright, J. Clardy*, J. Am. Chem. Soc. *95*, 2705 (1973).
[5] *P. S. Manchand, J. F. Blount*, Chem. Commun. 894 (1975).
[6] *M. R. Adams, J. D. Bu'Lock*, Chem. Commun. 389 (1975).
[7] *S. W. Pelletier, W. H. De Camp, S. D. Lajsic, Z. Djarmati, A. H. Kapadi*, J. Am. Chem. Soc. *96*, 7815 (1974).

aphidicolin arises from initial chair boat cyclization of the diterpene chain followed by successive H-shift, cyclization, and rearrangement steps as shown in the scheme[6]. A variant of the scheme, with the alternative α-vinyl stereochemistry at C-13 of the pimarane cation intermediate, accounts for the biogenesis of stemodins.

X-ray analysis has revealed the structure of novel diterpene alkaloid, delphisine[7], from which the correct structure of neoline and other relevant alkaloids was elaborated.

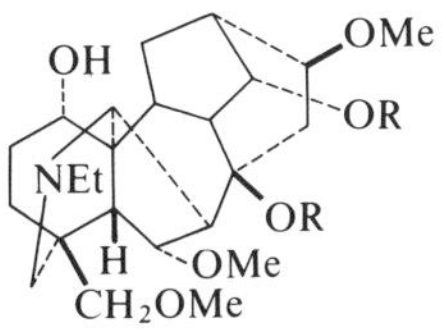

Delphisine
R = Ac
Neolin
R = H

4 Miscellaneous Diterpenes

Although geranyl geraniol (G·GOH) is considered as the intermediate precursor of cyclic diterpenoids, the isolation from natural sources had not been reported until the characterization from *Cedrela toona*[1]. A marine sponge (*Halichondria* sp.) produces linear geranyl-linalyl isonitrile (G·L-isonitrile) and its corresponding formamide and isothiocyanate[2]. Oxidogeranyl geranyl-methyl-1,4-hydroquinone was elaborated from a marine alga (*Sargassum tortile*) and was revealed to have a property to induce the settling of swimming larvae of *Coryne uchidai*[3]. The epoxide is considered to be a biogenetic precursor of taondiol isolated from a species of seaweed, *Taonia atmaria*[4]. Actually, chemical transformation of the former to the latter has been carried out by the action of Lewis acids[5].

Acyclic diterpenoids characterized from tobacco were interrelated with a constituent of tobacco, duvatriene-1,3-diol, by conversion of the latter to the former upon treatment with *p*-TsOH[6]. The duvatriene diol isolated from immature tobacco leaves has a plant growth inhibitory activity and controls the growth of lateral shoots of tobacco[7].

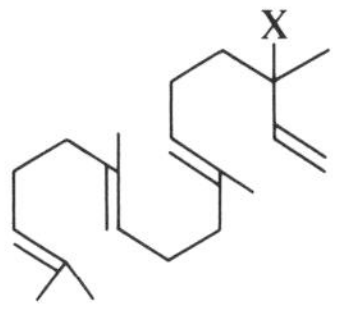

X = NC G. L-isonitrile
X = NHCHO G. L-formamide
X = NCS G. L-isothiocyanate

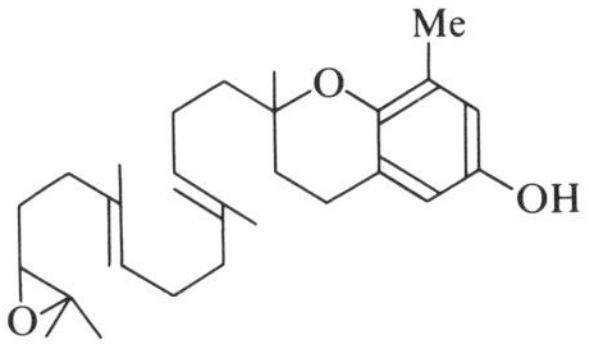

Oxido G. G-Me-*p*-hydroquinone

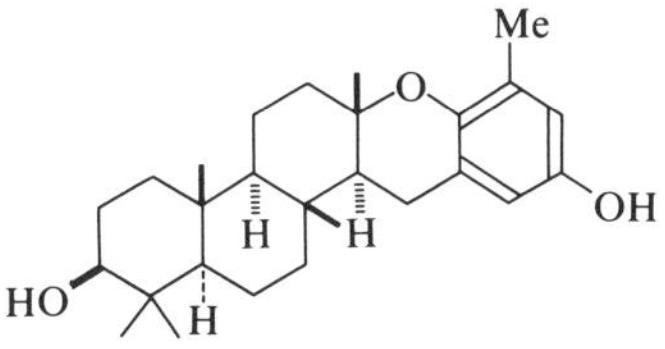

Taondiol

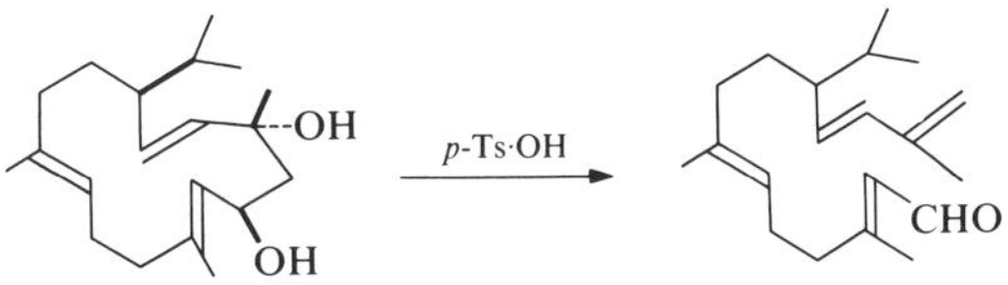

4,8,13-Duvatriene-1,3-diol

Acyclic diterpene from tobacco

It is of interest to note that novel cembrane type compounds such as nephthenol and sarcophine have been elaborated from several species of soft corals.[8-11,12]. Sarcophine, characterized from a soft coral *Sarcophtum glaucum*, is believed to be one of the repellents protecting the coral against

[1] *B. A. Nagasampagi, L. Yankov, Sukh Dev*, Tetrahedron Lett. 189 (1967).
[2] *B. J. Burreson, P. J. Scheuer*, Chem. Commun. 1035 (1974).
[3] *T. Kato, A. S. Kumanireng, I. Ichinose, Y. Kitahara, Y. Kakinuma, M. Nishihira, M. Kato*, Experientia *31*, 433 (1975).
[4] *A. G. Gonzalez, J. Darias, J. D. Marin*, Tetrahedron Lett. 2729 (1971).
[5a] *A. S. Kumanireng, T. Kato, Y. Kitahara*, Chemistry Lett. 1045 (1973).
[5b] *A. G. Gonzalez, J. D. Martin, M. L. Rodriguez*, Tetrahedron Lett. 3657 (1973).
[6a] *J. L. Courtney, S. McDonald*, Tetrahedron Lett. 459 (1967).
[6b] *A. J. Aesen, N. Junker, C. R. Enzell, J. E. Berg, A. M. Pilotti*, Tetrahedron Lett. 2607 (1975).
[7] *J. P. Springer, J. Clardy, R. H. Cox, H. G. Cutler, R. J. Cole*, Tetrahedron Lett. 2737 (1975).
[8] *F. J. Schmitz, D. J. Vanderah, L. S. Ciereszko*, Chem. Commun. 407 (1974).

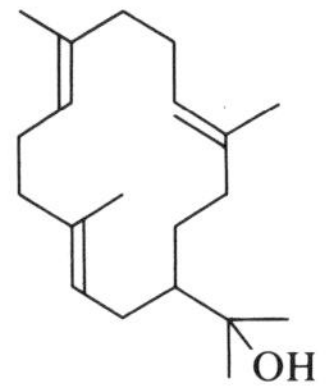

Nephthenol (soft coral, *Nephthea* sp.)[8]

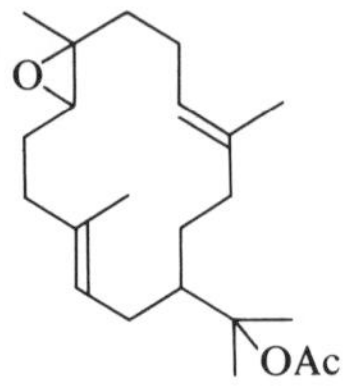

Epoxynephthenol acetate (soft coral, *Nephthea* sp.)[8]

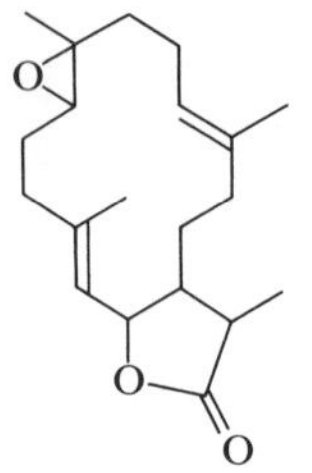

Sarcophine (soft coral, *Sarcophytum glaucum*)[9]

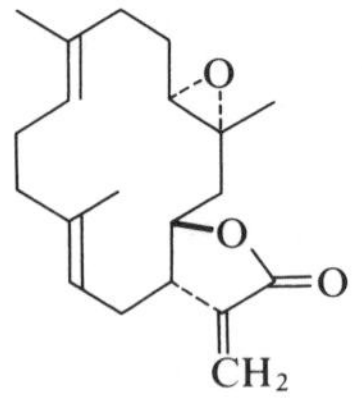

Lobophytolide (soft coral, *Lobophytum cristagalli*)[10]

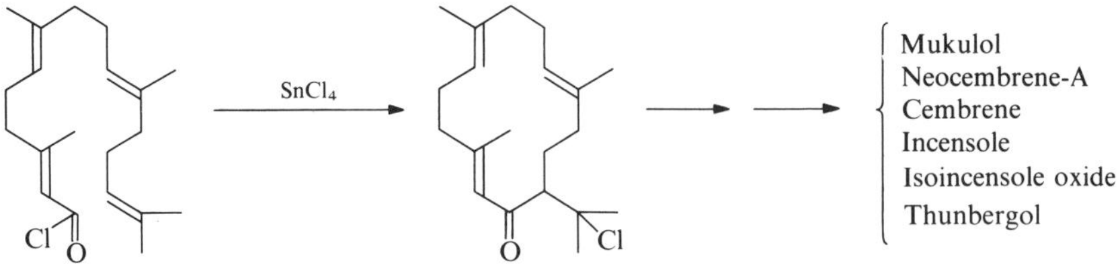

$SnCl_4$ → Chloro ketone → → Mukulol, Neocembrene-A, Cembrene, Incensole, Isoincensole oxide, Thunbergol

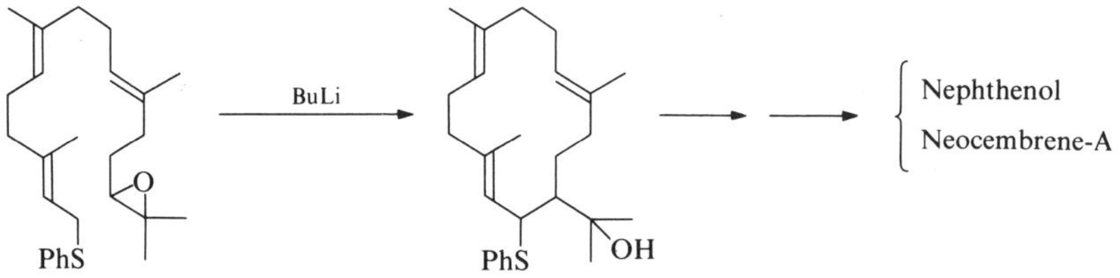

BuLi → Phenyl thioether → → Nephthenol, Neocembrene-A

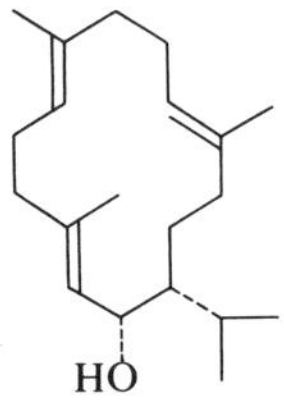

Mukulol (*Commiphora mukul*)[18]

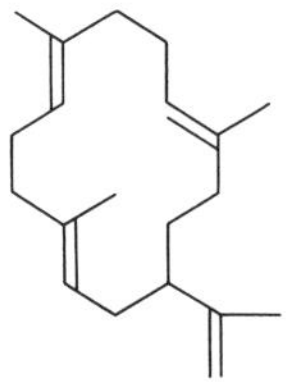

Neocembrene-A (*Nasutitermes exitiosus*)[17]
Cembrene-A (*Commiphora mukul*)[18]

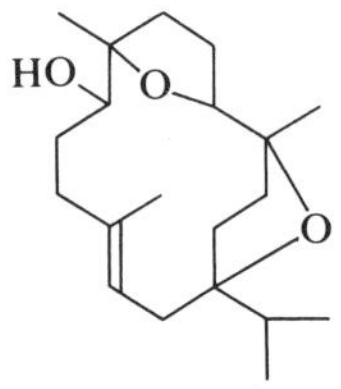

Isoincensole oxide (Frankincense, *Boswellia earteri*)[19]

predators[9]. Based on biogenetic consideration, cembrane skeleton was effectively synthesized from geranyl geranyl derivatives by two different routes. One is a cationic cyclization of geranyl geranic acid chloride to give chloroketone (71%)[13] and the other is an anionic cleavage of epoxide derivative, affording phenyl thioether[14]. These intermediates were successfully converted to natural cembrenoids[15,16] including neocembrene-A[17], a termite trail pheromone. Syntheses of cembrene[20] and casbene[21] via multi step reactions have also been carried out.

---

[9] *J. Bernstein, U. Shmeuli, E. Zadock, Y. Kashman, I. Neeman*, Tetrahedron *30*, 2817 (1974).

[10] *B. Tursch, J. C. Braekman, D. Daloze, M. Herin, R. Karlsson*, Tetrahedron Lett. 3769 (1974).

[11] *Y. Kashman, E. Zadock, I. Neeman*, Tetrahedron 3615 (1974).

[12] *P. J. Scheuer*, Chemistry of Marine Natural Products, Academic Press, New York, 1973.

[13] *T. Kato, T. Kobayashi, Y. Kitahara*, Tetrahedron Lett. 3299 (1975).

[14] *M. Kodama, Y. Matsuki, S. Ito*, Tetrahedron Lett. 3065 (1975).

Biogeneses of new classes of diterpenoids, pachydictyol A, dollabelladiene and dolastane elaborated from marine origins, are illustrated by the initial bond formation of C-1 of geranyl geranyl pyrophosphate with C-10 (path a) and C_{11}(path b), respectively. Path b is accompanied with bond formation between C_{10} and C_{14} (dotted line). Further cyclization of the former type ten membered intermediate leads to bicyclic pachydictyol A[22] and dictyols[23,24] while the latter pathway constitutes dollabelladiene[25], from which dolastane skeleton is formed by transannular connection of C-3 with C_8. Dolatriol has been elaborated from a poisonous kind of sea hares and is quite cytotoxic[26]. A novel tricyclic diterpene, decipidiene triol from *Eremophila decipiens*[27], possesses a rare cyclobutane ring in its decipiane skeleton, which might be biosynthesized from the common ten membered intermediate via path a.

Cyathins, antibiotic substances produced by the birds nest fungus *Cyathus helenae* Brodie[28], belong to another type of new diterpenes. Its carbon skeleton may be derived by cyclization at C_1–C_7 and C_6–C_{10} of G-G pyrophosphate followed by rearrangement of the resulting bicyclic cation. The biogenesis of crotofolin A[29] of a new carbon skeleton, for which the name crotofolane has been proposed, probably occurs via a transannular closure of a bicyclic precursor. The bicyclic compounds embodying the salient skeletal feature have been isolated from the *Euphorbiaceae* families[30].

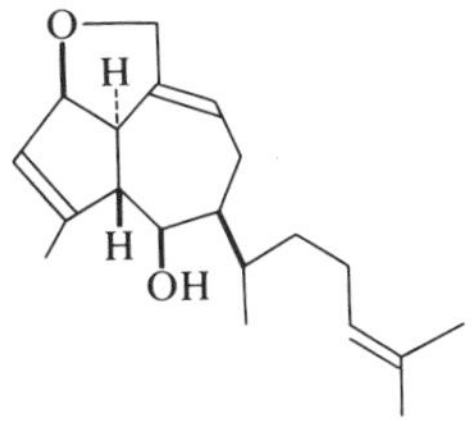

Dictyol A
(Brown alga, *Dictyota dichotoma*)[23,24]

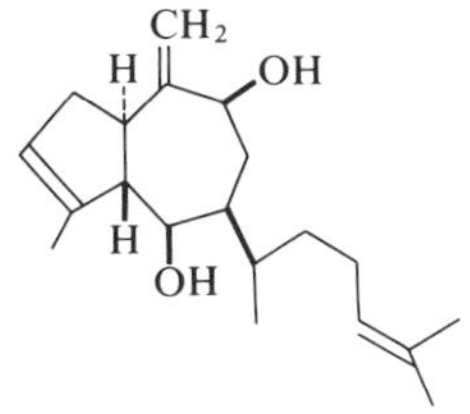

Dictyol B
(*Dictyota dichotoma*)[23,24]

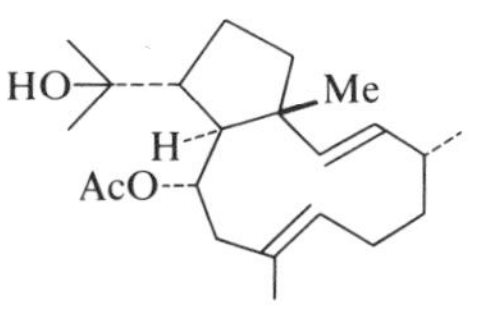

10-Acetoxy-18-hydroxy-2,7-dollabelladiene
(Sea hare, *Dollabella californica* Sterns)[25]

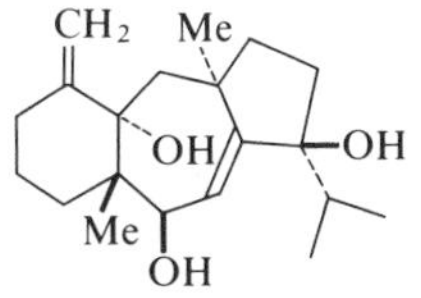

Dolatriol
(Sea hare, *Dolabella auricularia*)[26]

[15] *Y. Kitahara, T. Kato, T. Kobayashi, B. P. Moore*, Chemistry Lett. 219 (1976).

[16a] *T. Kato, Chin Chung Yen, T. Uyehara, Y. Kitahara*, Chemistry Lett. 565 (1977).

[16b] *T. Kato, M. Suzuki, M. Takahashi, Y. Kitahara*, Chemistry Lett. 465 (1977).

[17] *A. J. Birch, W. V. Brown, J. E. T. Corrie, B. P. Moore*, J. Chem. Soc., Perkin I 2653 (1972).

[18] *V. D. Patil, U. R. Mayak, Sukh Dev*, Tetrahedron *29*, 341 (1973).

[19] *M. L. Forcellese, R. Nicoletti, C. Santarelli*, Tetrahedron Lett. 3783 (1973).

[20] *W. G. Dauben, G. H. Beasley, M. D. Broadhurst, B. Muller, D. J. Peppard, P. Pesnelle, C. Suter*, J. Am. Chem. Soc. *97*, 4973 (1975).

[21] *L. Crombie, G. Kneen, G. Pattenden*, Chem. Commun. 66 (1976).

[22] *D. R. Hirschfeld, W. Fenical, G. H. Y. Lin, R. M. Wing, P. Radlick, J. J. Sims*, J. Am. Chem. Soc. *95*, 4049 (1973).

[23] *E. Fattorusso, S. Magno, L. Mayol. C. Santacrose, D. Sica, V. Amico, G. Oriente, M. Piattelli, C. Tringali*, Chem. Commun. 575 (1976).

[24] *L. Minale, R. Riccio*, Tetrahedron Lett. 2711 (1976).

[25] *C. Ireland, D. J. Faulkner, J. Finer, J. Clardy*, J. Am. Chem. Soc. *98*, 4664 (1976).

[26] *G. R. Pettit, R. H. Ode, C. L. Herald, R. B. von Dreele, C. Michel*, J. Am. Chem. Soc. *98*, 4677 (1976).

[27] *E. L. Ghisalberti, P. R. Jefferies, P. Sheppard*, Tetrahedron Lett. 1775 (1975).

[28] *W. A. Ayer, H. Taube*, Tetrahedron Lett. 1917 (1972).

[29] *W. R. Chan, E. C. Price, P. S. Manchand, J. P. Springer, J. Clardy*, J. Am. Chem. Soc. *97*, 4437 (1975).

[30a] *P. Narayanne, M. Rohl, K. Zechmeister, D. W. Engel, W. Hoppe, E. Hecker, W. Adolf*, Tetrahedron Lett. 1325 (1971).

[30b] *M. Hergenhahn, S. Kusumoto, E. Hecker*, Experientia *30*, 1438 (1974).

[31] *W. Fenical, J. Finer, J. Clardy*, Tetrahedron Lett. 731 (1976).

[32] *W. Fenical, B. Howard, K. B. Gifkins, J. Clardy*, Tetrahedron Lett. 3983 (1975).

[33] *S. M. Kupchan, C. W. Sigal, M. J. Matz, C. J. Gilmore, R. F. Bryan*, J. Am. Chem. Soc. *98*, 2295 (1976).

[34] *M. C. Wani, H. L. Taylor, M. E. Wall, P. Coggen, A. T. McPhail*, J. Am. Chem. Soc. *93*, 2325 (1971).

[35] *S. M. Kupchan, J. G. Sweeny, R. L. Baxter, T. Murae, V. A. Zimmerly, B. R. Sickles*, J. Am. Chem. Soc. *97*, 672 (1975).

[36] *S. M. Kupchan, Y. Shizuri, T. Murae, J. G. Sweeny, H. R. Haynes, M-Shing Shen, J. C. Barrick, R. F. Bryan, D. van der Helm, K. K. Wu*, J. Am. Chem. Soc. *98*, 5719 (1976).

[37] *M. Maruyama, A. Terahara, Y. Itagaki, K. Nakanishi*, Tetrahedron Lett. 299 (1967).

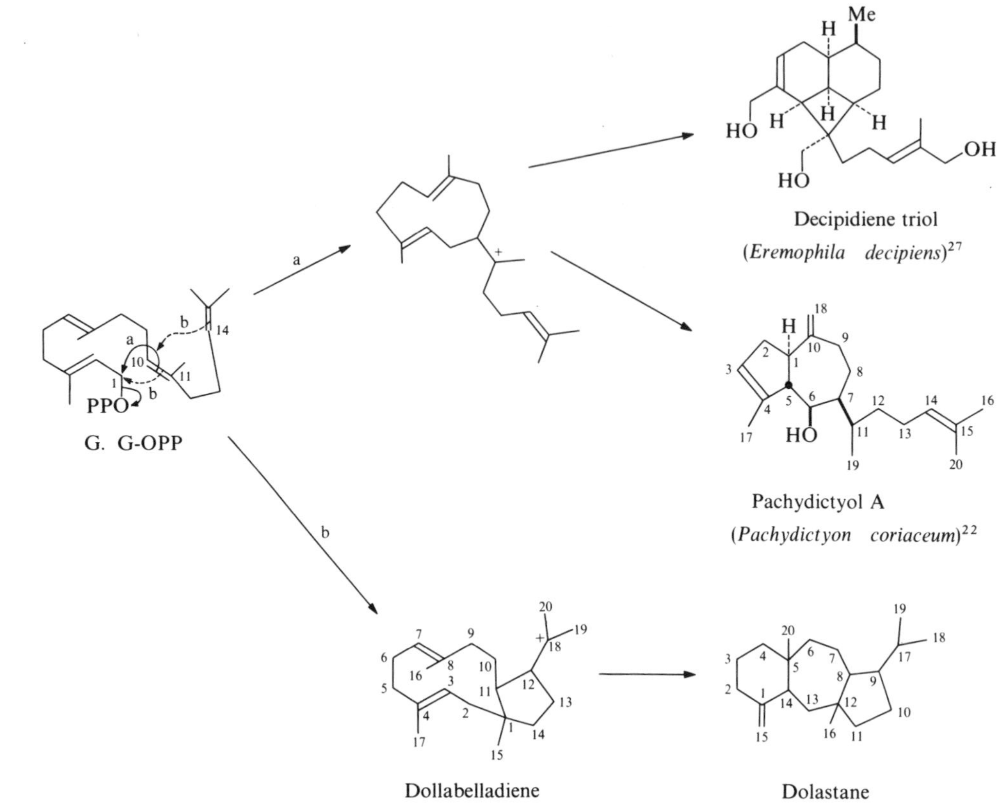

G. G-OPP

Decipidiene triol
(*Eremophila decipiens*)[27]

Pachydictyol A
(*Pachydictyon coriaceum*)[22]

Dollabelladiene

Dolastane

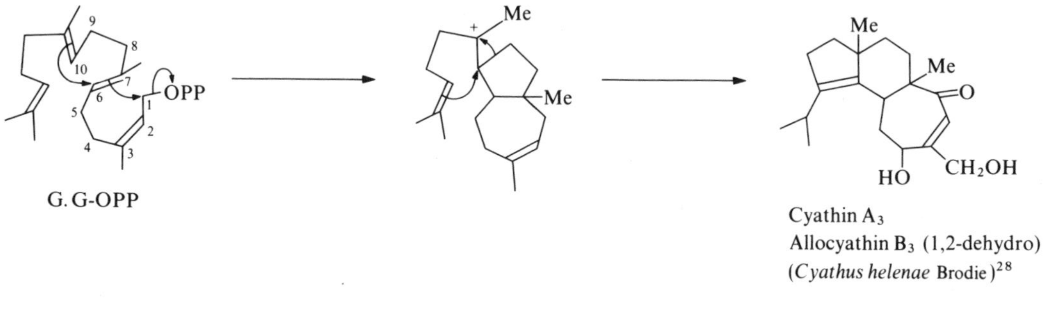

G. G-OPP

Cyathin A_3
Allocyathin B_3 (1,2-dehydro)
(*Cyathus helenae* Brodie)[28]

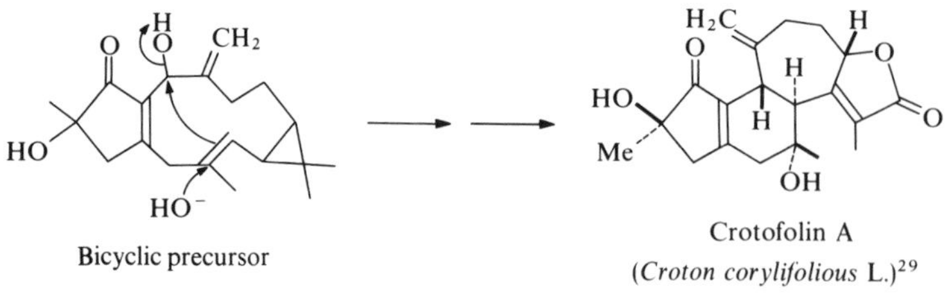

Bicyclic precursor

Crotofolin A
(*Croton corylifolious* L.)[29]

Biogenesis of bromine containing diterpene, sphaerococcenol A[31], isolated from a species of red seaweed, seems rather complicated. Sphaerococcenol A cannot be disected into isoprene subunits. However, the bromomethyl carbon (C-16) bonded to C-7 is presumed to have its terpenoid origin at C-6. This migration could conceivably occur from a C-6, C-7 cyclopropane intermediate which could react with Br^+ to yield the bromomethyl group. Irieols are dibrominated diterpenes possessing unusual carbon skeleton[32].

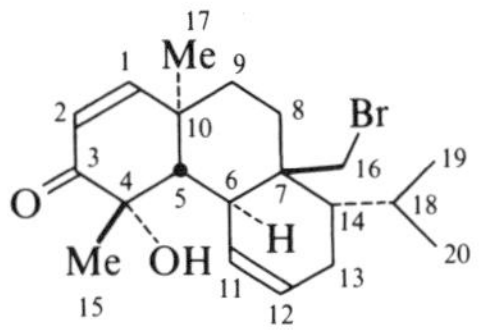

Sphaerococcenol A
(Red alga, *Sphaerococcus coronopifolius*)[31]

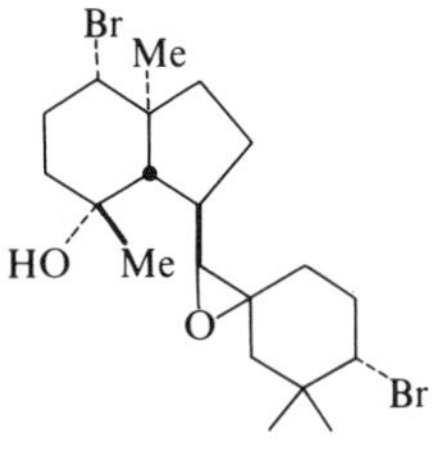

Irieol A
(Red alga, *Laurencia* sp.)[32]

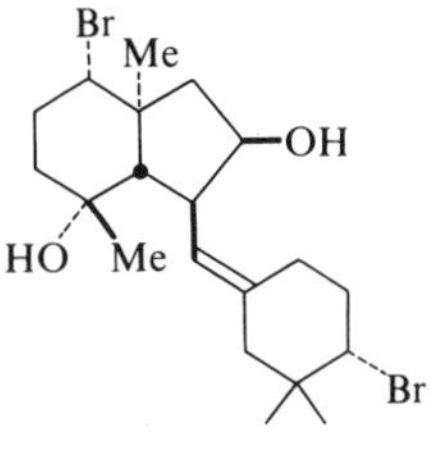

Iriediol
(*Laurencia* sp.)[32]

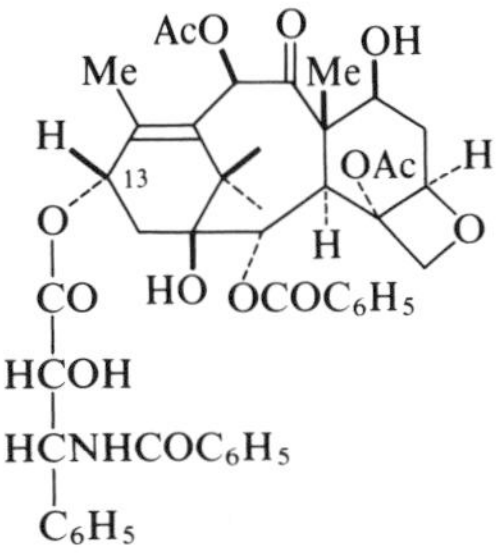

Taxol
(*Taxus brevifolia*)[34]

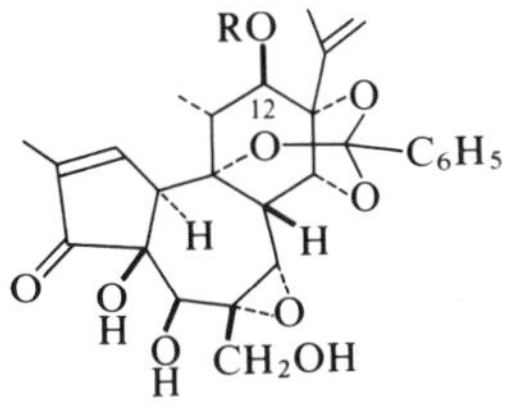

Gnididin
$R = COCH \overset{t}{=} CHCH \overset{t}{=} CH(CH_2)_4Me$
Gniditrin
$R = COCH \overset{t}{=} CH(CH \overset{t}{=} CH)_2(CH_2)_2Me$
Gnidicin
$R = COCH{=}CH{\cdot}C_6H_5$
(*Gnidia lamprantha*)[35]

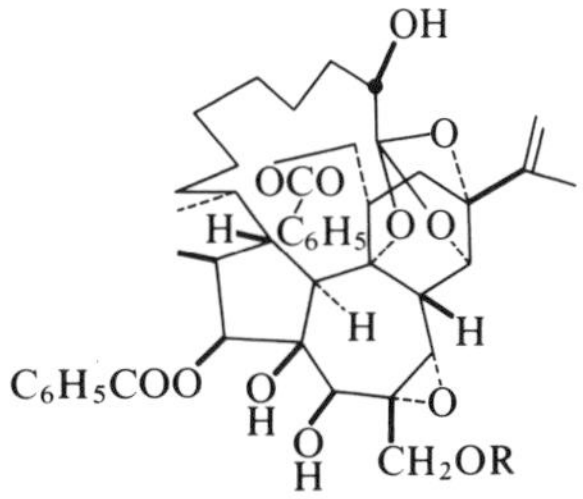

Gnidimacrin
$R = H$
Gnidimacrin 20-Palmitate
$R = COC_{15}H_{31}$
(*Gnidia subcordata*)[36]

Several compounds possessing an antileukemic and antitumor activities have been elucidated after being seeked out from natural sources. The representative diterpenes are jatrophone[33], taxol[34], gnididins[35], and gnidimacrins[36]. Jatrophone was found to react with thiol group on proteins such as bovine serum albumin and RNA polymerases in *E. coli*, giving a monoadduct product. The significance of the reactions of jatrophone and other electrophilic tumor inhibitors with thiols on protein has been speculated to support the hypothesis that these agents may act by selective alkylation of growth regulatory biological molecules.
Ester groups at C-13 of taxol and at C-12 of gnididins play an important role for indicating the antitumor activity since the activity disappearred almost completely when these ester groups were removed by hydrolysis. Gnidimacrins exhibit the most potent antileukemic activity of any members of the daphnetoxin class and are the first such compounds found to contain the novel macrocyclic orthoester structural features.

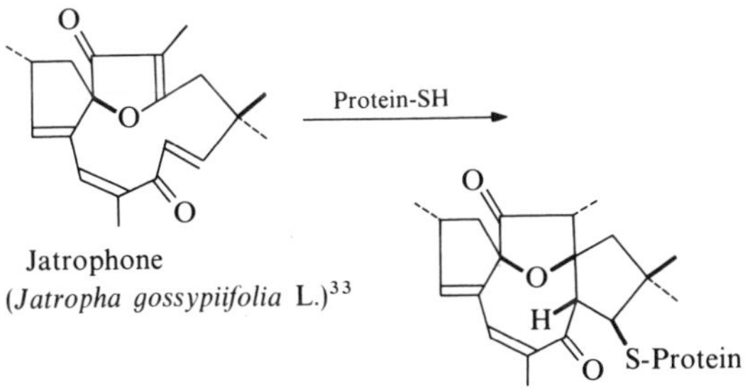

Jatrophone
(*Jatropha gossypiifolia* L.)[33]

Protein-SH

Protein monoadduct

The striking demonstration of the utility of the nuclear Overhauser effect is provided in the studies of stereochemistry and conformation of ginkgolides and taxine derivatives[37,38].
Several kinds of diterpenes which belong to this section are listed in Table 5.**

Recommended Literature for Further Reading

D. J. Faulkner, Tetrahedron *33*, 1421 (1977).

** Names in parentheses written below each compounds indicate the origin of the compound.

[38] *M. C. Woods, H. C. Chiang, Y. Nakadaira, K. Nakanishi*, J. Am. Chem. Soc. *90*, 522 (1968).

Table 5. Miscellaneous diterpenes

Crinitol
(Brown alga, *Cystoseira crinita*)[39]

Caulerpol
(Alga, *Caulerpa brownii*)[40]

(*Sarcophytum glaucum*)[11]

(*Sarcophytum glaucum*)[11]

(*Sarcophytum glaucum*)[11]

Asperdiol
(*Eunicea asperula*)[41]

Sinulariolide
(*Sinularia flexibilis*)[42]

Sinularin
(*Sinularia flexibilis*)[42]

Stylatulide
(sea pen, *Stylatula* sp.)[43]

Ingol
$R_1 = R_2 = R_4 =$ Acetate
$R_3 =$ Nicotinate
(*Euphorbia ingens*)[44]

Huratoxin
$R_1 = CH{=}CHCH{=}CH(CH_2)_8Me$
(*Hura crepitans*)[45]

Prostratin
(*Pimelea prostrata*)[46]

Baccatin IV
$R_1 = OH, R_2 = Ac$
1-Dehydroxybaccatin IV
$R_1 = H, R_2 = Ac$
Baccatin VI
$R_1 = OH, R_2 = Bz$
Baccatin VII
$R_1 = OH, R_2 = COC_5H_{11}$
(*Taxus baccata* L.)[47]

[39] *E. Fattorusso, S. Magno, L. Mayol. C. Santacrose, D. Sica, V. Amico, G. Oriente, M. Piattelli, C. Tringali*, Tetrahedron Lett. 937 (1976).

[40] *A. J. Blackman, R. J. Wells*, Tetrahedron Lett. 2729 (1976).

[41] *A. J. Weinheimer, J. A. Matson, D. van der Helm, M. Poling*, Tetrahedron Lett. 1295 (1977).

[42] *A. J. Weinheimer, J. A. Matson, M. B. Hossain, D. van der Helm*, Tetrahedron Lett. 2923 (1977).

[43] *S. J. Wratten, D. J. Faulkner, K. Hirotsu, J. Clardy*, J. Am. Chem. Soc. 99, 2825 (1977).

[44] *H. J. Opferkuch, E. Hecker*, Tetrahedron Lett. 3611 (1973).

[45] *K. Sakata, K. Kawazu, T. Mitsui, N. Masaki*, Tetrahedron Lett. 1141 (1971).

[46] *A. R. Cashmore, R. N. Seelye, B. F. Cain, H. Mack, R. Schmidt, H. Hecker*, Tetrahedron Lett. 1737 (1976).

[47] *D. P. D. C. de Marcano, T. G. Halsall*, Chem. Commun. 365 (1975).

3 Saponins

Contribution by

T. Kawasaki, Fukuoka

3 Saponins

Toshio Kawasaki

Faculty of Pharmaceutical Sciences, Kyushu University, Higashi-ku, Fukuoka, 812 Japan

1 Introduction

The saponins have been defined as a group of plant constituents possessing characteristic properties such as hemolytic and surface activities, ability to form molecular complexes with cholesterol and toxic effects on fishes and amphibians. They are widely distributed in the plant kingdom*[1] and have been regarded for a long time as the significant principles of many medicinal herbs. Some of them, for instance, "sarsasaponin (parillin)," "digitonin" and "gypsophila saponin," have been known for more than a century. However, because of the difficulties of isolation in a pure state and in structural investigations, they were only known to be glycosides with complex aglycones and the chemistry of saponins had been an undeveloped field until about fifteen years ago.

In the earlier days they were conventionally classified into neutral and acidic saponins, but later the aglycones (sapogenins) of the former were found to be either steroid or triterpenoid and those of the latter, triterpenoid compounds and the classification, steroid and triterpenoid saponins, has since been generally accepted. Meanwhile, studies on the glycosidal alkaloids of the plants of the genera *Solanum, Lycopersicum, Cestrum* and *Veratrum* led to the findings that their aglycones (alkamines) are the nitrogen analogs of steroid sapogenins and they have been called solanum alkaloids and regarded as the third group of saponins, basic steroid saponins. The steroid saponins and the solanum alkaloids differ from other natural steroid glycosides, phytosterol-, cardiac- and digitanol glycosides and some of the veratrum-, holarrhena- and buxus alkaloids in the structural features of their aglycones, which have the spirostane skeleton and its nitrogen analogs.

Recent progress in the chemistry of natural products has disclosed the existence of a number of compounds which belong structurally to one of the three groups mentioned above but fail to exhibit all or a part of the properties of the so-called saponins. In this chapter the saponins are discussed according to chemical classification, disregarding their properties. Since a number of excellent papers have appeared on the saponins[1] and, in particular, the comprehensive reviews on the triterpenoid saponins by Basu and Rastogi[2] and Woitke, Kayser and Hiller[3], on the solanum alkaloids by Schreiber[4] and on biochemistry of steroid saponins and solanum alkaloids by Heftmann[5] have been published*[2], the present author intends to refer mainly to the structural chemistry of steroid saponins and to some critical areas in general.

2 Chemical Structures

2.1 Steroid Saponins

The chemical studies on the aglycones of steroid saponins have been conducted extensively for more than thirty years and a great number of sapogenins have been isolated and their structures established to show that they are almost exclusively the spirostane (**1**) derivatives. On the other hand, as to their parent glycosides, the isolation in pure form, the structure elucidation and the subsequent investigation on the properties had not been successfully accomplished until 1954–1963*[3], when two diosgenin triglycosides, dioscin (**2**) and gracillin, and a digitogenin pentaglycoside, digitonin (**3**), were obtained as pure crystals and their complete structures were

*[1] A few have been isolated from animals (*e.g.*, holothurin from sea cucumber, asterosaponin from starfish).

*[2] Unless otherwise indicated the references for all the compounds and subjects in this chapter appear in these reviews.

*[3] Except trillin, diosgenin monoglucoside.

[1] a) *G. Vogel,* Planta Medica *11*, 362 (1963); b) *R. Tschesche, G. Wulff,* Planta Medica *12*, 272 (1964); c) *N. K. Kochetkov, A. J. Khorlin,* Arzneimittel Forsch. *16*, 101 (1966); d) *T. Kawasaki,* Sôgô Rinsyô *16*, 1053 (1967); e) *G. Wulff,* Deut. Apoth.-Ztg. *108*, 797 (1968); f) *Y. Birk, in* Toxic Constituents of Plant Foodstuffs, *I. E. Liener* (Ed.), p. 169, Academic Press, New York 1969; g) *R. Tschesche,* Kagaku-no Ryôiki *25*, 571 (1971); *Pharma. Int.* 17 (1971).

[2] *N. Basu, R. P. Rastogi,* Phytochemistry *6*, 1249 (1967).

[3] *H. D. Woitke, J. P. Kayser, K. Hiller,* Pharmazie *25*, 133, 213 (1970).

[4] *K. Schreiber,* Alkaloids *10*, 1 (1968).

[5] *E. Heftmann,* Lloydia *30*, 209 (1967); Lloydia *31*, 293 (1968); Steroid Biochemistry, Academic Press, New York 1970.

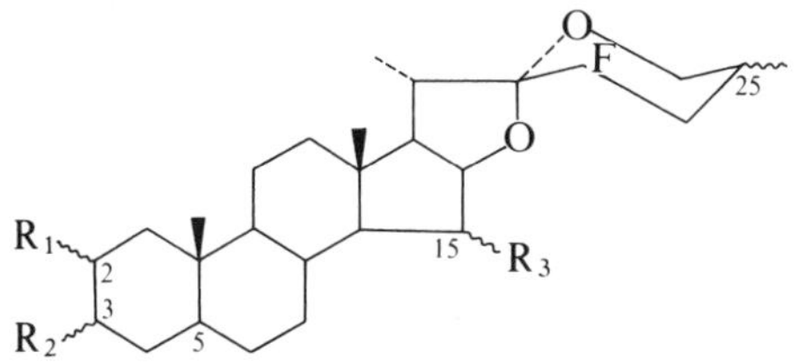

Spirostane Skeleton (**1**): 25R (D) or S (L); $R_1 = R_2 = R_3 = H$

Dioscin (**2**): Δ^5; 25R; $R_1 = R_3 = H$
$R_2 = \beta$-*O*-β-D-glc·pyr $\genfrac{}{}{0pt}{}{\overset{2}{<}\alpha\text{-L-rha·pyr}}{\underset{4}{\smallsetminus}\alpha\text{-L-rha·pyr}}$
(Chacotriose)

Digitonin (**3**): 5α H; 25R; $R_1 = \alpha$-OH, $R_3 = \beta$-OH

$R_2 = \beta$-*O*-β-D-gal·pyr $\overset{4}{-}$ β-D-glc·pyr $\genfrac{}{}{0pt}{}{\overset{3}{|}\ \beta\text{-D-glc·pyr};\ \overset{2}{<}\beta\text{-D-gal·pyr}}{\underset{3}{\smallsetminus}\beta\text{-D-xyl·pyr}}$
(Digitopentaose)

Yononin (**4**): 5β-H; 25R; $R_2 = \alpha$-OH, $R_3 = H$
$R_1 = \beta$-*O*-α-L-ara·pyr

Table 1. Typical steroid saponins

Steroid saponin	Aglycone	Sugar moiety
Dioscin (2)	diosgenin	rha $\overset{2}{\ }$ rha $\overset{4}{\ }$ > glc–(Chacotriose)
Prosapogenin A of 2	diosg.	rha $\overset{2}{-}$ glc–
Prosapogenin B of 2	diosg.	rha $\overset{4}{-}$ glc–
Trillin	diosg.	glc–
Gracillin (G)	diosg.	rha $\overset{2}{\ }$ glc $\overset{3}{\ }$ > glc–
Prosapogenin A of G	diosg.	glc $\overset{3}{-}$ glc–
Aspidistrin	diosg.	glc $\overset{2}{\ }$ xyl $\overset{3}{\ }$ > glc $\overset{4}{-}$ gal–(Lycotetraose)
Saponin P-d	disog.	rha $\overset{2}{\ }$ rha $\underset{4}{-}$ rha $\overset{4}{\ }$ > glc–
Digitonin (3)	digitogenin	glc $\overset{3}{-}$ gal $\overset{2}{\ }$ xyl $\overset{3}{\ }$ > glc $\overset{4}{-}$ gal–(Digitopentaose)
Deglucodigitonin	digitog.	(gal $\overset{2}{\ }$ xyl $\overset{3}{\ }$ > glc $\overset{4}{-}$ gal–)
Gitonin	gitogenin	gal $\overset{2}{\ }$ xyl $\overset{3}{\ }$ > glc $\overset{4}{-}$ gal–
F-Gitonin	gitog.	Lycotetraose
Tigonin	tigogenin	(Digitopentaose)
Lanatigonin-I	tigog.	glc $\overset{3}{-}$ gal $\overset{2}{\ }$ xyl $\overset{3}{\ }$ > glc $\overset{3}{-}$ gal–
Degalactotigonin	tigog.	Lycotetraose
Digalonin	digalogenin	(Digitopentaose)
Lanadigalonin	digalog.	glc $\overset{3}{-}$ gal $\overset{2}{\ }$ xyl $\overset{3}{\ }$ > glc $\overset{3}{-}$ gal–
Pennogenin diglycoside	pennogenin	rha $\overset{2}{-}$ glc–
Pennog. triglycoside	pennog.	Chacotriose
Pennog. tetraglycoside	pennog.	rha $\overset{2}{\ }$ rha $\underset{4}{-}$ rha $\overset{4}{\ }$ > glc–
Parillin	sarsasapogenin	glc $\overset{2}{\ }$ rha $\overset{4}{\ }$ glc $\overset{6}{\ }$ > glc–

glc–: β-D-glucopyranosyl, gal–: β-D-galactopyranosyl
rha–: α-L-rhamnopyranosyl, xyl–: β-D-xylopyranosyl

presented by our research group and by Tschesche and his collaborators, respectively.

Up to 1970 approximately eighty steroid saponins were isolated, among which about fifty have been assigned complete structures; some twenty regarded as pure and their components have been reported*4. In the early stage of investigation all of them seemed to have structural features not only in the aglycones but also in the sugar moieties which are different somewhat from those of the triterpenoid saponins and entirely from other steroid glycosides*5. Namely, with the exception of some glycosides which might be considered as prosapogenins, the sugar moieties are oligosaccharides*6 which consist of two or three kinds of sugar such as D-glucose, D-galactose, L-rhamnose and D-xylose and fork into branches having rhamnose and/or xylose residues at terminals. The site of sugar linkage with aglycone was regarded, even in the polyhydroxy spirostane derivatives, to be only the hydroxyl group at C-3 by analogy with the fact that the glycosides of 3β-monohydroxy steroids occur more frequently in plants. In F-gitonin[6] this was proved for the first time actually to be the case. The so-called "typical" steroid saponins which have the above features are listed in Table 1 (including their prosapogenins).

Recently a number of spirostanol glycosides which do not show these structural features with regard to the sugar moiety have been isolated and the glycosides of which aglycones are not spirostanol but its modification have also been reported. These are tentatively named "atypical" steroid saponins in this chapter.

Yononin (**4**)[7] represents the first example*7. Its aglycone, yonogenin, possesses, as shown by Takeda and his coworkers[8], a 3α-hydroxyl group and had been noted as an unusual dihydroxy sapogenin. The parent glycoside, yononin, has been found to have one mole of arabinose attached to the hydroxyl group at C-2 and not at C-3. Later, Kimura and his coworkers[9] reported that convallasaponin B has arabinose combined with a tertiary hydroxyl at C-5, although its aglycone bears the usual 3β-hydroxyl group. According to Ripperger and Schreiber[10], both paniculonins A and B have a disaccharide containing D-quinovose linked only with the hydroxyl group at C-6 from among those at positions C-3 (β), 6 and 23. Convallasaponins E, D and glucoconvallasaponin B isolated by the Kimura group[9] are also "atypical" because the first one is a linear 3-*O*-triarabinoside and the other two have two sugar parts attached to the hydroxyl groups at C-3 and 1, and C-3 and 5, respectively. Parillin had been considered to have a linear trisaccharide, but the structure has been revised by the Tschesche group to be a tetraglycoside (**9**) which forks into three branches representing another "typical" one.

In 1947 Marker and Lopez[11] proposed that the spirostane derivatives are formed from the corresponding furostane (**5**) glycosides in the presence of enzyme or acid. The idea had once seemed to be refuted by the data presented by Wall and his coworkers[12], but the isolations of jurubine (**6**) by the Schreiber group in 1966 and sarsaparilloside (**7**)[13] by Tschesche and his collaborators in 1967, which were respectively assigned the structures 25S-3β-amino-5α-furostane-22α,26-diol 26-*O*-β-D-glucopyranoside and 26-*O*-β-D-glucopyranosyl-25S, 5β-furostane-3β,22α,26-triol 3-*O*-tetraglycoside and were

*4 Some of them might be identical with known saponins.

*5 Reichstein has proposed a classification of the glycosides according to the sugar moiety: A-type (saponins) and B-type (cardiac glycosides) (*T. Reichstein*, Angew. Chem. *74*, 887 (1962)).

*6 Kochetkov has proposed to designate such glycosides as oligosides (Ref. [1c]).

*7 Hispidin, a parent saponin of hispidogenin (12β-hydroxy-5α,25R-spirostan-3-one) has been reported to be a dirhamnoside and, if the aglycone is not an artifact, the biose should be combined with the hydroxyl group at C-12 (*P. C. Maiti, et al.*, Chem. Ind. (London) 1653, (1965)).

[6] *T. Kawasaki, I. Nishioka, T. Komori, T. Yamauchi, K. Miyahara*, Tetrahedron *21*, 299 (1965).

[7] *T. Kawasaki, K. Miyahara*, Tetrahedron *21*, 3633 (1965).

[8] *K. Takeda, T. Okanishi, A. Shimaoka*, Chem. Pharm. Bull. (Tokyo) *6*, 532 (1958).

[9] *M. Kimura, M. Tohma, I. Yoshizawa, A. Fujino*, Chem. Pharm. Bull. (Tokyo) *16*, 2191 (1968) and references cited therein.

[10] *H. Ripperger, K. Schreiber*, Chem. Ber. *101*, 2450 (1968).

[11] *R. E. Marker, J. Lopez*, J. Am. Chem. Soc. *69*, 2389 (1947).

[12] *M. M. Krider, M. E. Wall*, J. Am. Chem. Soc. *74*, 3201 (1952); J. Am. Chem. Soc. *76*, 2938 (1954); *E. S. Rothman, M. E. Wall, C. R. Eddy*, J. Am. Chem. Soc. *74*, 4013 (1952); *M. E. Wall, S. Serota, L. P. Witnauer*, J. Am. Chem. Soc. *77*, 3086 (1955).

[13] *R. Tschesche, G. Lüdke, G. Wulff*, Tetrahedron Lett. 2785 (1967); Chem. Ber. *102*, 1253 (1969).

Furostane Skeleton (**5**): 25R or S; $R_1 = R_2 = R_3 = H$

Jurubine (**6**): 5α-H, 25R
$R_1 = NH_2$, $R_2 = OH$, $R_3 =$ -*O*-β-D-glc·pyr

Sarsaparilloside (**7**): 5β- H ; 25S

$R_1 =$ -*O*- β-D-glc·pyr (2)→β-D-glc·pyr; (4)→α-L-rha·pyr; (6)→β-D-glc·pyr

$R_2 = OH$, $R_3 =$ -*O*-β-D-glc·pyr

Jurubidine (**8**): 25R

Parillin (**9**): 25S

found to be converted *in vitro* with enzyme as well as acid to the corresponding spirostane derivatives, jurubidine (**8**) and parillin (**9**), have presented positive evidence supporting and reviving Marker's suggestion. In succession, convallamaroside (26-*O*-β-D-glucopyranosyl-5β-furost-25(27)-ene-1β,3β,22α,26-tetrol 3-*O*-α-L-rhamnopyranosyl-(1-3)-[β-D-glucopyranosyl-(1-4)]-α-L-rhamnopyranoside) was reported by the Tschesche group[1e,14] and the furostane 3,26-bisglycosides (**10**) corresponding to dioscin (**2**), gracillin, pennogenin di- and tetraglycosides and prosapogenin A of dioscin were isolated in our laboratory[16,37,38,*8]. The predominance of this type of compounds together with the related spirostane glycosides in the fresh materials has been demonstrated by means of thin-layer chromatography[37], and the conversion of the furostane bisglycoside (**10**a) to dioscin (**2**) by a *Dioscorea floribunda* leaf homogenate has been proved by the Heftmann group[15]. Therefore the furostane bisglycosides are thought probably to be the genetic precursors of the spirostane glycosides and they might be called the prototype compounds of the latter group[*9]. The pennogenin di-, tri- and tetraglycosides and three corresponding diosgenin glycosides were obtained[16] along with two each (**10**a, c, d, e) of their respective prototype compounds and this fact is

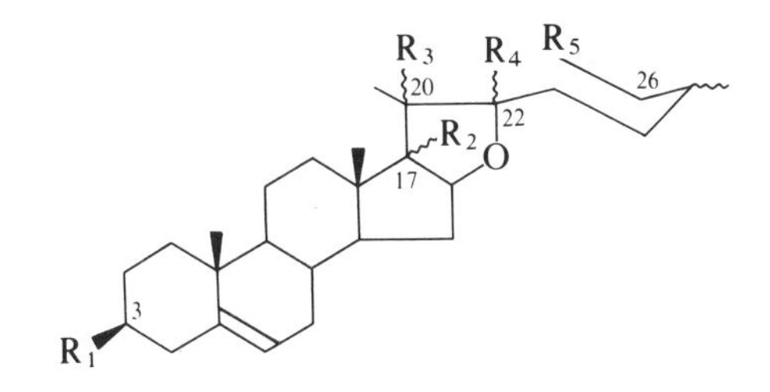

3,26-Bisglycosides (Prototype Compounds) corresponding to Pennogenin- and Diosgenin Glycosides (**10**):

$R_3 = β$-H, $R_4 = OH$, $R_5 =$ -*O*-β-D-glc·pyr

(a) $R_1 =$ -*O*-β-D-glc·pyr (2)→α-L-rha·pyr; (4)→α-L-rha·pyr; $R_2 = H$

(b) $R_1 =$ -*O*-β-D-glc·pyr (2)→α-L-rha·pyr; (3)→β-D-glc·pyr; $R_2 = H$

(c) $R_1 =$ -*O*-β-D-glc·pyr (2)—α-L-rha·pyr $R_2 = H$

(d) $R_1 =$ -*O*-β-D- glc·pyr (2)—α-L-rha·pyr $R_2 = α$-OH

(e) $R_1 =$ -*O*-β-D-glc·pyr (2)→α-L-rha·pyr; (4)→α-L-rha·pyr (4)—α-L-rha·pyr; $R_2 = α$-OH

Nologenin (**11**): $R_1 = R_2 = R_3 = R_5 = OH$, $R_4 = H$

*8 Smilaxsaponin B (*T. Kawasaki, et al.*, Yakugaku Zasshi *86*, 673 (1966)) and kikubasaponin (*T. Kawasaki, et al.*, Chem. Pharm. Bull. (Tokyo) *10*, 703 (1963)) have been identified (Ref. [38]) with the prototype compounds (**10**a, b, $R_4 = OCH_3$) of dioscin and gracillin, respectively.

14 *R. Tschesche, B. T. Tjoa, G. Wulff, R. V. Noronha*, Tetrahedron Lett. 5141 (1968).

15 *R. A. Joly, J. Bonner, R. D. Bennett, E. Heftmann*, Phytochemistry *8*, 857, 1445 (1969).

Table 2. Atypical steroid saponins

Steroid saponin	Aglycone	Sugar moiety
Yononin (**4**)	yonogenin	ara–(2β-OH)
Tokoronin	tokorogenin	ara–(1β-OH)
Tokorogenin Glucoside	tokorog.	glc–(1β-OH)
Neotokoronin	neotokorogenin	ara–(1β-OH)
Timosaponin A-III	sarsasapogenin	glc$\overset{2}{-}$gal–(Timobiose) (3β-OH)
Timosap. A-I	sarsasapog.	gal–(3β-OH)
Convallasaponin A	convallagenin A	ara–(3β-OH)
Glucoconvallasap. A	convallag. A	glc$\overset{2}{-}$ara–(3β-OH)
Convallasap. B	convallag. B	ara–(5β-OH)
Glucoconvallasap. B*[a]	convallag. B	glc–(3β-OH), ara–(5β-OH)
Convallasap. C	isorhodeasapogenin	rha$\overset{3}{-}$rha$\overset{2}{-}$ara–(3β-OH)
Convallasap. D*[a]	rhodeasapogenin	glc–(1β-OH), rha$\overset{2}{-}$xyl$\overset{3}{-}$rha–(3β-OH)
Convallasap. E	diosgenin	ara$\overset{2}{-}$ara$\overset{2}{-}$ara–(3β-OH)
Paniculonin A	paniculogenin	xyl$\overset{3}{-}$qui–(6α-OH)
Paniculonin B	paniculog.	rha$\overset{3}{-}$qui–(6α-OH)
Avenacoside A*[b]	nuatigenin	glc–(26-OH), $\begin{matrix}\text{glc}\\ \text{rha}\end{matrix}\overset{2}{\underset{4}{>}}$glc–(3β-OH)
Kryptogenin Glucoside*[c]	kryptogenin	glc–(3β-OH)
Dehydrokryptogenin Diglycoside*[c]	17(20)-dehydro-kryptogenin	rha$\overset{4}{-}$glc–(3β-OH)
Dehydrokryptog. Tetraglycoside*[c]	17(20)-dehydro-kryptog.	$\begin{matrix}\text{rha}\\ \text{rha}\underset{4}{-}\text{rha}\end{matrix}\overset{2}{\underset{4}{>}}$glc–(3β-OH)
Saponin P-a	diosgenin	$\begin{matrix}\text{rha}\\ \text{ara·fur}\end{matrix}\overset{2}{\underset{4}{>}}$glc–(3β-OH)
Prosapogenin of "Nogiran" Saponins	2-*O*-Ac-epimetagenin	tri-*O*-Ac-ara–(11α-OH)

ara–: α-L-arabinopyranosyl, ara·fur–: α-L-arabinofuranosyl
qui–: β-D-quinovopyranosyl, glc–, gal–, rha–, xyl–: cf. Table 1
*[a]: bisdesmosidic but not prototype compound
*[b]: F-ring being five-membered
*[c]: E,F-rings being opened

*[9] Tschesche has proposed (Ref. [1e,g,3]) to call the bis- and monoglycosides respectively bisdesmosidic (the name being suffixed with -side) and monodesmosidic (suffixed with -in).

[16] *T. Nohara, K. Miyahara, T. Kawasaki*, Chem. Pharm. Bull. (Tokyo) *23*, 872 (1975); *T. Nohara, Y. Ogata, M. Aritome, K. Miyahara, T. Kawasaki*, Chem. Pharm. Bull. (Tokyo) *23*, 925 (1975).

against the Marker's claim[17] that pennogenin is an artifact yielded secondarily from a nologenin glycoside "nolonin" on treatment with hydrochloric acid in ethanol, and his structure of nologenin, furost-5-ene-3,17,*20*,26-tetrol (**11**)[18], seems to be revised as 3β,17α,*22*,26-tetrol corresponding to pennogenin. It is of interest from a view point of the biogenesis of diosgenin glycosides.

It is also noted that avenacoside A reported by Tschesche and his coworkers[19] is a nuatigenin 3,26-bisglycoside and the aglycone, which is not a furostane derivative but has the five-membered F-ring, is converted with acid into a spirostan-25-ol, isonuatigenin.

All the component monosaccharides of steroid saponins so far reported are of the pyranose type with only one exception, the L-arabinofuranose residue found in a saponin (P-a)[20] from *Paris polyphylla*.The most recent finding by Yosioka and his collaborators[21] that a prosapogenin of "nogiran" saponins has the fully acetylated L-arabinose unit linked with the hydroxyl group at C-11 of 2-*O*-acetyl-epimetagenin attracts special attention.

"Atypical" steroid saponins (excluding prototype compounds) are as given in Table 2.

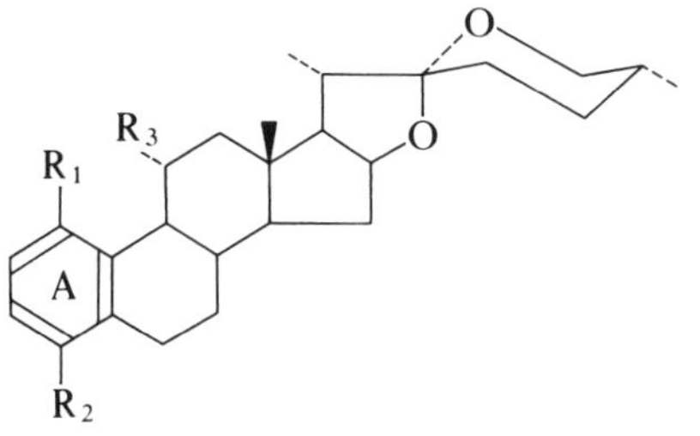

R3 R1 A R2 O O

Luvigenin (**12**): $R_1 = R_3 = H$, $R_2 = CH_3$

Meteogenin (**13**): $R_1 = CH_3$, $R_2 = H, R_3 = OH$

[17] *R. E. Marker, R. B. Wagner, P. R. Ulshafer, E. L. Wittbecker, D. P. J. Goldsmith, C. H. Ruof,* J. Am. Chem. Soc. *65*, 1199 (1943); *R. E. Marker, J. Lopez,* J. Am. Chem. Soc. *69*, 2386 (1947).

[8] *R. E. Marker, R. B. Wagner, P. R. Ulshafer, E. L. Wittbecker, D. P. J. Goldsmith, C. H. Ruof,* J. Am. Chem. Soc. *69*, 2167 (1947); *R. E. Marker,* J. Am. Chem. Soc. 2395 (1947).

[19] *R. Tschesche, M. Tauscher, H.-W. Fehlhaber, G. Wulff,* Chem. Ber. *102*, 2072 (1969).

[20] *T. Nohara, H. Yabuta, M. Suenobu, R. Hida, K. Miyahara, T. Kawasaki,* Chem. Pharm. Bull. (Tokyo) *21*, 1240 (1973).

[21] *I. Yosioka, K. Imai, I. Kitagawa,* Tetrahedron Lett. 1177 (1971); Tetrahedron *30*, 2283 (1974).

There are many structurally interesting sapogenins of which glycosides still remain to be isolated and characterized (*e.g.*, isoplexigenins[22], igagenin,[23] eduligenin and lowegenin[24]) and, in particular, elucidation of the structures of the parent saponins of luvigenin (**12**) and meteogenin (**13**), both of which have been found by the Takeda group[25] to have an aromatic A-ring and which have been suggested to be artifacts derived from polyhydroxy spirostane derivatives by dienol-benzene rearrangement, is expected.

2.2 Triterpenoid Saponins

With a few exceptions, for instance glycyrrhizic acid*[10] described by Lythgoe and Trippet (1950) and Marsh and Levy (1956) and asiaticoside described by Polonsky and her collaborators (1959–1961), it was not until 1962, when aralosides A and B were reported by Kochetkov and his collaborators, that a number of triterpenoid saponins were isolated in pure form and their structures investigated in detail. The number of compounds of this group for which structures have been fully established up to now (1970) or whose genins, component sugars and organic acids are known to some extent is close to two hundreds. The reviews[2,3] cover all these compounds reported until 1968 and even those for which the components are not thoroughly known. To the author's knowledge, about one hundred compounds of this group appeared in 1969–1970 and almost half of them were fully or partially characterized.

In view of the structures known so far, the features common to many triterpenoid saponins seem to be as follows:

a. The aglycones have the oleanane skeleton (**14**).

b. Except those which are regarded possibly as prosapogenins, they contain three to ten

*[10] Structure of the sugar moiety has been confirmed (*M. Okigawa, et al.,* Tetrahedron Lett. 2935 (1970)).

[22] *R. Freire, A. G. González, E. Suárez,* Tetrahedron *26*, 3233 (1970).

[23] *F. Yasuda, Y. Nakagawa, A. Akahori, T. Okanishi,* Tetrahedron *24*, 6535 (1968).

[24] *R. F. Barreira, A. G. González, J. A. S. Rocio, E. S. Lopez,* Phytochemistry *9*, 1641 (1970).

[25] *K. Takeda, T. Okanishi, K. Igarashi, A. Shimaoka,* Tetrahedron *15*, 183 (1961); *K. Igarashi,* Chem. Pharm. Bull. (Tokyo) *9*, 722, 729 (1961).

monosaccharide units which combine in the form of one or two oligosaccharide chains*[6] with the hydroxyl group at C-3 of the aglycones and sometimes with the 28-carboxyl group as well*[9]. The sugar moiety is linear or forks into branches and the monosaccharides involved are D-glucose, D-galactose, D-fructose, D-glucuronic and D-galacturonic acids, D-xylose, L-arabinose (sometimes as the furanose), D-fucose, L-rhamnose and D-quinovose, almost exclusively in pyranose form.

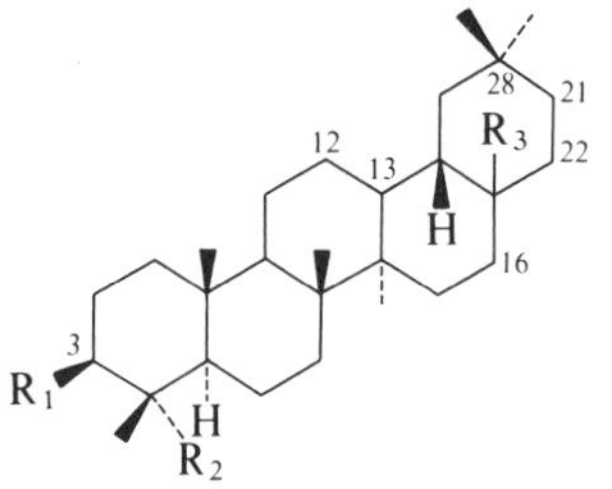

Oleanane Skeleton (**14**): $R_1 = H, R_2 = R_3 = CH_3$

Gypsoside A (**15**): Δ^{12}; $R_2 = CHO$

R_1 = -*O*-D-glc·pyr·uronic acid (3 → L-ara·pyr; 4 → D-glc·pyr $\frac{}{4}$ β-D-gal·pyr)

R_3 = -COO-L-rha·pyr (2 → D-xyl·pyr $\frac{3}{}$ D-xyl·pyr; 4 → D-fuc·pyr $\frac{}{3}$ D-xyl·pyr)

The saponins which have organic acids, such as acetic, butyric, isovaleric, tiglic, angelic and benzoic acids, combined with the hydroxyl groups at C-16, 21 and 22 are also known (*e.g.* escins, teasaponins).

Some examples of the typical ones are glycyrrhizic acid, gypsoside A (**15**), aralosides, kalopanax saponins, primula saponins and cyclamin. As is the case in the steroid saponins, there are also many triterpenoid saponins which do not have the above features and they will increase in number. Among them are: asiaticoside (a ursane derivative having a linear trisaccharide linked only to the 28-carboxyl group), ginsenoside Rg_1, trametoside[26], holothurin (all being tetracyclic triterpene derivatives, the first one having the glucose residues attached to the hydroxyl groups both at C-6 and 20, the second being the ester glucoside at the 21-carboxyl group and the last representing one of the animal saponins), saponins from *Myrsine australis* and from *Sophora japonica*, patrinoside C_1[27] and barringtonin[28] (containing in the sugar moieties, glucuronolactone, D-glucofuranose and L-sorbose, respectively).

Secondary alterations of the aglycones during the hydrolytic cleavage of the sugar linkage are often encountered in the triterpenoid saponins (*e.g.*, escins, ginsenosides) and the structures of genuine sapogenins have been a matter of particular concern.

2.3 Basic Steroid Saponins (Solanum Alkaloids)

The recent aspects of this field have been presented in full detail in a comprehensive treatise[4]. Though it was published in 1968, as far as the glycosides are concerned, only several new compounds, for instance sevcorine[29], solasurine[30], solakhasianine[31], edpetilinine[32] and veralosine[33] have appeared in the mean time. Successful chemical studies on the compounds of this group, as glycosides, were initiated by Kuhn and his coworkers, by whom the sugar moieties of solanine, solamargine, tomatine and demissine were assigned in 1955–1957; the branched-chain oligosaccharide structures were later found to be a feature not only of the solanum alkaloids but of many steroid- and triterpenoid saponins.

Among about fifty compounds so far reported nearly half of them have been completely characterized and their common features found to be the striking similarity of both aglycones (alkamines) and sugar parts to those of the steroid saponins. The skeletons of alkamines are usually spirosolane (22R, 25R- =solasodane (**16**), 22S,25S- =tomatidane (**18**)) and solanidane (**19**), which are the nitrogen analogs of spirostane (**1**) and its relatives,*[11] and the component oligosac-

[26] *H. Inouye, K. Tokura, T. Hayashi*, Tetrahedron Lett. 2811 (1970).

*[11] The 22-epimeric spirostane derivatives corresponding to those of tomatidane have not yet been found in nature.

[27] *V. G. Bukharov, V. V. Karlin, V. A. Talan*, Khim. Prir. Soedin *5*, 89 (1969); Chem. Abst. *71*, 70878 (1969).

[28] *K. S. M. Pillai*, Bull. Res. Inst. Univ. Kerala Ser. A6 15 (1959); Chem. Abst. *54*, 16746 (1960).

[29] *R. N. Nuriddinov, S. Yu. Yunosov*, Khim. Prir. Soedin *4*, 60 (1968); Chem. Abst. *69*, 27562 (1968).

[30] *D. K. Seth, R. Chatterjee*, J. Inst. Chem. Calcutta *41*, 194 (1968); Chem. Abst. *72*, 79405 (1970).

[31] *P. C. Maiti, S. Mookherjea*, Indian J. Chem. *6*, 547 (1968); Chem. Abst. *70*, 29197 (1969).

[32] *R. Shakirov, R. N. Nuriddinov, S. Yu. Yunusov*, Khim. Prir. Soedin *5*, 605 (1969); Chem. Abst. *73*, 25813 (1970).

[33] *A. M. Khashimov, R. Shakirov, S. Yu. Yunusov*, Khim. Prir. Soedin *6*, 343 (1970); Chem. Abst. *73*, 99170 (1970).

charides are similar or identical to those in some steroid saponins. Thus, for example, α-sola margine (**17**) corresponds to dioscin (**2**); some other representatives are α-solanine, α-chaconine, solasonine, α-solamarine, tomatine (5α-tomatidan-3β-ol 3-*O*-lycotetraoside) and demissine.

The first isolation and characterization by the Schreiber group of a novel type of solanum alkaloid, jurubine (**6**), is worthy of special attention in connection with Marker's suggestion as mentioned earlier (see 2–1). Its structure, bearing nitrogen atom at C-3 instead of in the side chain at C-17, is also noted. The precursors of spirosolane and solanidane derivatives, namely those corresponding to jurubine (**6**) as a prototype compound of jurubidine (**8**), have not yet been found. Kuhn and Löw have isolated leptine-1 which is the 23-*O*-acetate of a leptinidine (a solanidane derivative) triglycoside.

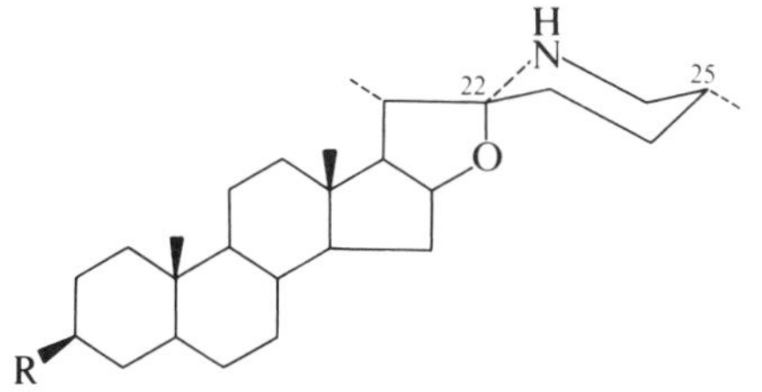

Solasodane (22R, 25R-Spirosolane) Skeleton(**16**): R = H

α-Solamargine(**17**): Δ^5; R =-*O*-β-D-glc·pyr (2 → α-L-rha·pyr; 4 → α-L-rha·pyr)

(Chacotriose)

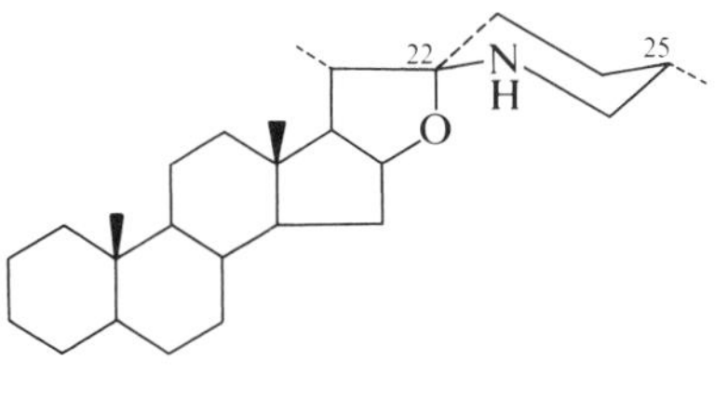

Tomatidane (22S, 25S-Spirosolane) Skeleton (**18**)

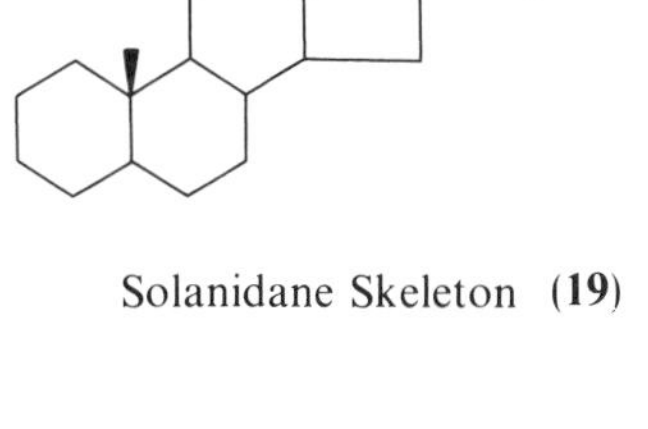

Solanidane Skeleton (**19**)

Some new types of alkamines[4,34] have been reported which have skeletons, for example 22,26-epiminocholestane, different from but related to the usual ones. Almost all of them have been known to exist as their glycosides in plants, but, except for sevcorine[29], edpetilinine[32] and veralosine[33], the parent glycosides still remain to be isolated.

3. Isolation and Structure Determination

The remarkable advances of saponin chemistry in the last fifteen years are greatly indebted to the development of modern methods of isolation, identification and structure determination. Above all, the development of chromatography has provided valuable means for separation of saponins and their derivatives and has opened a new horizon. Isolation in a homogeneous state of these rather large and complex compounds, which coexist with many analogs and homologs with relatively slight structural differences, could not have been achieved without application of modern chromatographic techniques. It has also played a most important role in the detection and identification and in the separation of prosapogenins in partial hydrolysates to determine the structures of sugar moieties. The methods are reviewed in the published papers[1b,c,f,2,3].

With respect to the early stage of preparation of crude saponins, the procedure employed by the Wall group[35], extraction of the aqueous suspension of the plant extractives with butanol saturated with water, is conveniently applied in combination with the conventional ones and precipitation of saponins as molecular complexes with cholesterol in alcoholic solutions is also useful when applicable. Recovery of saponins from complexes is achieved in the usual manner and more conveniently, especially when small amounts of saponins are available, by dissolving the complexes in a chloroform-methanol mixture followed by chromatography on alumina[36].

In the structure elucidation of aglycones, special consideration should be given to the question whether the sapogenin split off a saponin by

[34] *Y. Sato, in* Chemistry of the Alkaloids, *S. W. Pelletier* (Ed.), p. 591, Van Nostrand, Reinhold 1970.

[35] *E. S. Rothman, M. E. Wall, H. A. Walens*, J. Am. Chem. Soc. *74*, 5791 (1952).

[36] *T. Kawasaki, I. Nishioka,* Kyushu Branch Meeting of Pharm. Soc. Japan, Fukuoka 1965.

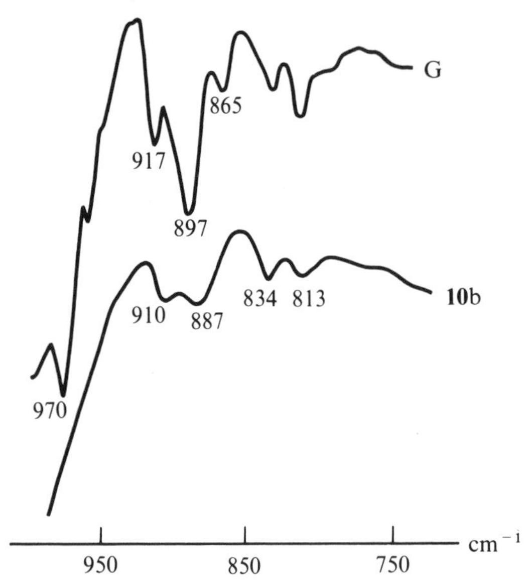

Fig. 1. Infrared Spectra (KBr disk) of Gracillin (G) and its Prototype Compound (**10**b).

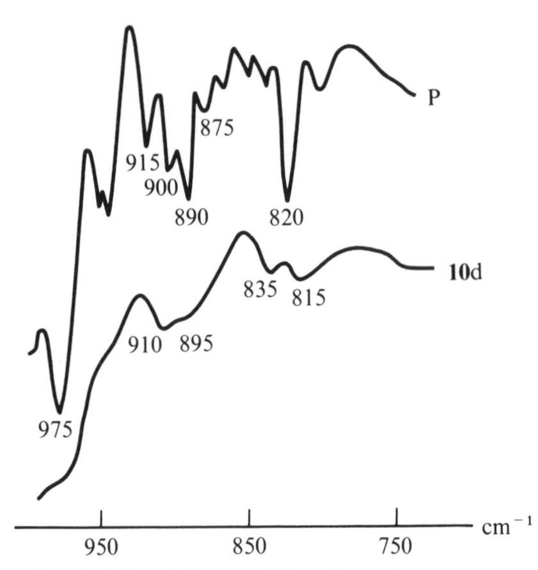

Fig. 2. Infrared Spectra (KBr disk) of Pennogenin Diglycoside (P) and its Prototype Compound (**10**d).

hydrolysis is an artifact or not. The particular methods for cleavage of the glycosidic linkage to give the genuine aglycone, such as Smith degradation and enzymatic hydrolysis, are also described in the reviews[3]. It is to be noted that, when the prototype compound of a spirostanol glycoside, for example, 3-*O*-oligoglycosyl-furostane-3,26-diol 26-*O*-glucoside, is hydrolyzed with enzyme, the glucosidic linkage at C-26 is preferentially cleaved and subsequently ring closure takes place to give 3-*O*-oligoglycosyl-spirostan-3-ol. The furostane bisglycoside structure can be assumed on thin-layer chromatograms, with some exceptions*[12], by positive Ehrlich reaction[37] and confirmed by the infrared spectra[38] which do not show the characteristic absorptions of spirostane derivatives[39] (Figs. 1,2). Unequivocal evidence is obtained by examination of the products of Marker's degradation or Baeyer-Villiger oxidation followed by hydrolysis.

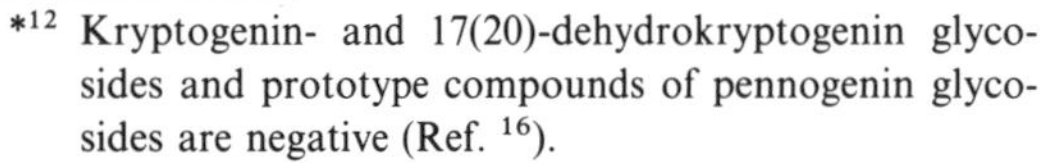

*[12] Kryptogenin- and 17(20)-dehydrokryptogenin glycosides and prototype compounds of pennogenin glycosides are negative (Ref. [16]).

[37] *S. Kiyosawa, M. Hutoh, T. Komori, T. Nohara, I. Hosokawa, T. Kawasaki*, Chem. Pharm. Bull. (Tokyo) *16*, 1162 (1968).

[38] *T. Kawasaki, T. Komori, K. Miyahara, T. Nohara, I. Hosokawa, K. Mihashi*, Chem. Pharm. Bull. (Tokyo) *22*, 2164 (1974).

[39] *M. E. Wall, C. R. Eddy, M. L. McClennan, M. E. Klumpp*, Anal. Chem. *24*, 1337 (1952); *C. R. Eddy, M. E. Wall, M. K. Scott*, Anal. Chem. *25*, 266 (1953).

In order to determine the chemical composition of a saponin and to characterize the sugar moiety, qualitative and quantitative (separate and total) determinations of the component monosaccharides are carried out with an acid hydrolysate of the saponin according to the usual procedure, but for this purpose it is essential that the hydrolysis be complete and cause no appreciable loss of released monosaccharides. Therefore, qualitative examination of not only monosaccharides but oligosaccharides, prosapogenins and original saponin in the hydrolysates obtained under various conditions and quantitative determination of known amounts of the sugars treated in parallel are prerequisite. When the acid hydrolysis is carried out in an alcoholic solution the liberated sugars lose their reducing power considerably and, if a saponin is hardly soluble in water, a dioxane-water mixture is recommended as a solvent and in some cases a stepwise hydrolysis is needed[40]. The sequence and the sites of linkage with each other as well as the ring sizes of the component monosaccharides are determined by partial hydrolysis, acetolysis, periodate oxidation and permethylation followed by methanolysis and subsequent examination and identification of their products. Importance of the chromatography of saponins for checking partial hydrolysis and for

[40] *T. Tsukamoto, T. Kawasaki, T. Yamauchi*, Pharm. Bull. (Tokyo) *4*, 35 (1956).

separating the resulting prosapogenins was mentioned earlier and other procedures are described in the reviews on sugar chemistry. It should be emphasized, however, that the structures of saponins could not be discussed until the Kuhn method had been presented as an effective means for the exhaustive methylation of saponins. The Hakomori method proposed later is more conveniently applied in many cases and should be noted as well.

It is due, for one thing, to the difficulty in molecular weight measurement that the molecular formulae of saponins have been determined by analyzing the aglycones and the sugars in the complete hydrolysates in conjunction with the elemental analysis data, but recently the method using vapor pressure osmometry has overcome the difficulty to some extent and has been employed sometimes satisfactorily. Nuclear magnetic resonance spectrometry of peracetylated saponin is of use, on occasion, to estimate the number of acetyl groups and, consequently, of monosaccharide units in combination with the qualitative data on the components. In a moderate sized saponin (mol. wt. of peracetate, 600–1600), the mass spectra of its peracetate obtained with low ionizing (50–22 eV) and accelerating (6.4–4.5 kV) voltages give significant information on molecular weight of the saponin and kinds of terminal mono- and disaccharides[41]. The application of field ionization mass spectrometry to the structure elucidation of some cardiac glycosides has been reported by Brown and his coworkers[42] and might be applicable also to some saponins.

In some special cases the structure of a saponin can be determined by conversion into a known saponin or its derivative. For example, aspidistrin has been assigned[43] the structure, diosgenin 3-*O*-lycotetraoside, on the basis of the fact that it was hydrogenated to afford degalactotigonin which gave on Marker's degradation a 3β-hydroxy-5α-pregn-16-en-20-one 3-*O*-glycoside identical with that derived from tomatine and a pennogenin glycoside has been proved[16] to be 3-*O*-α-L-rhamnopyranosyl-(1-2)-β-D-glucopyranoside by correlation with the hexamethyl ether of prosapogenin A of dioscin.

Finally a few problems among those which deserve further investigation in connection with the structural chemistry of saponins are suggested. They concern the more convenient and definite methods for assignment of the anomeric configurations and conformations of the individual component monosaccharides and for comparison of a saponin with another.

The determination of the mode of sugar linkage has been based on either the behavior during enzymatic hydrolysis with α- and β-glycosidases, or the molecular rotation difference (the Klyne rule). The former method is not employed in general for such a complex glycoside as saponin and the use of the latter has also certain difficulties; namely, it is not always applicable to an oligoglycoside and furthermore it requires several specific partial hydrolyses of the sugar moiety and isolation of all the prosapogenins in pure form and in large enough amounts for optical rotation measurements. In some limited cases, for instance when the nuclear magnetic resonance spectrum of a saponin peracetate or permethylate shows an anomeric proton signal as a doublet with a coupling constant (J value) of 6–9 Hz which is ascribable to that of a D-glucopyranose (or L-arabinopyranose) unit, the sugar could be regarded to have β-configuration in C1 (normal) conformation (α and C1 in L-arabinose), because the J value suggests the trans diaxial orientation of the protons at C-1 and 2 of a pyranose residue which is only conceivable in the structure mentioned above[44]. In many cases, however, the assignments of several anomeric proton signals on a spectrum to the individual monosaccharide units are not easily and definitely made and, moreover, if the J value is small, 1–3 Hz, suggesting the axial-equatorial or equatorial-equatorial orientation of 1,2 protons, the stereochemistry can not be discussed on this basis alone.

The most reliable method so far for the determination of identity of a saponin with another is the comparison*[13] of various physical and

[41] *T. Komori, Y. Ida, Y. Mutô (née Inatsu), K. Miyahara, T. Nohara, T. Kawasaki*, Biomed. Mass Spectromet. 65 (1975).

[42] *P. Brown, F. R. Bruschweiler, G. R. Pettit, T. Reichstein*, J. Am. Chem. Soc. *92*, 4470 (1970).

[43] *K. Miyahara, Y. Ida, T. Kawasaki*, Chem. Pharm. Bull. (Tokyo) *20*, 2506 (1972); *Y. Môri, T. Kawasaki*, Chem. Pharm. Bull. (Tokyo) *21*, 224 (1973).

*[13] Comparison of X-ray powder patterns as an aid to identify ginsenoside Rg_1 acetate with panaxoside acetate have been reported (*Y. Iida, et al.*, Tetrahedron Lett. 5449 (1968)).

[44] cf. *K. Miyahara, T. Kawasaki*, Chem. Pharm. Bull. (Tokyo) *17*, 1369, 1735 (1969).

chemical data of the two samples taken under the same conditions, but it is rather tedious and sometimes still incomplete.

Further investigations have also looked for applications of enzymatic reactions not only for the determination of anomeric configuration and of structures of genuine aglycones but to the preparation on a larger scale of a specified saponin by degradation (cf., preparation of F-gitonin[45]) or partial synthesis as well.

4 Properties

Since the probable significance of saponins in the therapeutic effects of many crude drugs of plant origin has been a matter of special interest, physical, biological and pharmacological investigations have been conducted for a long time, but those in the earlier days dealt almost exclusively with the impure saponins and the recent studies on pure compounds with known structures have led to a lot of new findings and observations. An example is the antimicrobial activities of saponins, in particular, of the "typical" steroid saponins and the solanum alkaloids which have been studied by the Tschesche group, Wolters, and Imai and his coworkers[46]. The properties of saponins have been covered in several excellent reviews[1-5].

To our present knowledge, however, various properties are not always significantly correlated with each other and with the structures and they remain to be further investigated. When the saponins are defined on the basis of their chemical structures, not all of the saponins necessarily have the properties which had been regarded as inherent to saponins. Furthermore, one of the main characteristics of the so-called saponins, for example, hemolytic activity, is not always accompanied by another, foaming ability, and *vice versa*, and these porperties can also be exhibited by some compounds other than saponins. The formation of a molecular complex with 3β-hydroxy steroid which seemed to be peculiar to saponins is not observed either in a number of compounds excluding those such as "typical" steroid saponins and solanum alkaloids (digitonin, F-gitonin and tomatine being reported to have the highest potency). The mechanisms and the potency-structure relationships of hemolysis and complex formation have not been fully clarified.

The conventional methods for qualitative and quantitative determinations of saponins in plants are based on their "saponic" properties, such as hemolytic and foaming abilities and toxicity to fishes. The diversities in structures and properties of saponins, all of which are not always correlated, suggest that the determination should be carried out carefully by using several criteria, preferentially in combination with the physical and chemical methods.

5 Concluding Remarks

Space does not permit reference to the important and interesting subjects, biogenesis and physiological significance of saponins in plants, but the author wishes only to mention the facts that, while the steroid saponins are widely distributed in the *Dioscorea* plants, none of them has been found in *D. bulbifera forma spontanea* and *D. hispida* and related species, and, instead, a series of furanoid-norditerpenoids (*e.g.*, diosbulbin B)[47] and the tropane derivatives (*e.g.*, dioscorine)[48] have been isolated, respectively. The solanum alkaloids have not been found in the *Dioscorea* plants, whereas they are often accompanied by the steroid saponins in those of the genera *Solanum*, *Lycopersicum*, *Cestrum* and *Veratrum*. It is interesting from the view points of chemotaxonomy and biogenesis.

The science of saponins was an undeveloped field fifteen years ago and is a somewhat confusing area today. The confusion is partly due to the term "saponin" which originated from the specified, "soap-like", properties which are not always shared by all saponins defined structurally. The chemistry of saponins will undoubtedly continue to make great advances and it is also expected that their biology and pharmacology will be studied more extensively in parallel and

[45] *T. Kawasaki, I. Nishioka, T. Yamauchi, K. Miyahara, M. Enbutsu*, Chem. Pharm. Bull. (Tokyo) *13*, 435 (1965).

[46] *S. Imai, T. Murata, S. Fujioka, E. Murata, M. Goto*, Ann. Rep. Takeda Research Lab. *26*, 66 (1967); *S. Imai, S. Fujioka, E. Murata, M. Goto, T. Kawasaki, T. Yamauchi*, Ann. Rep. Takeda Research Lab. *26*, 76 (1967).

*[14] According to Sandberg they were classified into hemolytic and non-hemolytic saponins (*F. Sandberg*, Pakistan J. Sci. Ind. Res. *4*, 258 (1961)). Wulff has discussed the problem in detail (Ref. [1e]).

[47] *T. Kawasaki, T. Komori, S. Setoguchi*, Chem. Pharm. Bull. (Tokyo) *16*, 2430 (1968); *T. Komori, S. Setoguchi, T. Kawasaki, Chem. Ber. 101*, 3096 (1968).

[48] *R. Hegnauer*, Chemotaxonomie der Pflanzen, Bd. II, p. 144, Birkhäuser Verlag 1963.

the relationships between structures and properties will be fully elucidated in the near future. In that event, the saponins could be defined*[14] exactly and their utilization in medicine and also in other fields will be developed and established.

6 Addendum

Introduction

The following review articles are recommended for further reading.

(1) *J. Elks, in* Rodd's Chemistry of Carbon Compounds, *S. Coffery* (Ed.), 2nd ed., Vol. II_E, p. 1, Elsevier, Amsterdam 1971 (steroid saponins).

(2) *M. Pinkas, F. Trotin, L. Bezanger-Beauquesne*, Prod. Probl. Pharm. *27*, 187 (1972) (saponins in general).

(3) *K. Takeda, in* Progress in Phytochemistry, *L. Reinhold, Y. Liwschitz,* (Eds.) Vol. 3, p. 287, Interscience, New York 1972 (steroid sapogenins and saponins of Dioscoreaceae).

(4) *G. B. Elyakov, Yu. S. Ovodov*, Khim. Prir. Soedin *8*, 697 (1972) (glycosides of Araliaceae).

(5) *K. Hiller, H. D. Woitke, in* Pharm. Appl. Thin-Layer Paper Chromatography, *K. Macek*, (Ed.) p. 393, Elsevier, Amsterdam 1972.

(6) *R. Tschesche, G. Wulff, in* Fortschritte der Chemie organischer Naturstoffe, *W. Herz, H. Grisebach, G. W. Kirby* (Eds.), Vol. 30, p. 461, Springer-Verlag, Berlin 1973 (chemistry and biology of saponins in general).

(7) *J. G. Roddick*, Phytochemistry *13*, 9 (1974) (basic steroid saponin (α-tomatine)).

(8) *G. Wulff, G. Roehle*, Angew. Chem. *86*, 173 (1974) (glycoside syntheses).

(9) *S. K. Agarwal, R. P. Rastogi*, Phytochemistry *13*, 2623 (1974) (triterpenoid saponins).

Steroid Saponins

A number of steroid saponins, typical (A) and atypical (including prototype compounds) (B), have been reported.

(A): review–(6); *V. P. Konyukhov, et al.*, Khim. Prir. Soedin 9, 741 (1973) (*Chem. Abst. 82*, 28555 (1975)); *M. P. Tomova, et al.*, Planta Medica *25*, 231 (1974); *V. A. Pareshnichanko, et al.*, Prikl. Biokhim. Mikrobiol. *11*, 94 (1975) (*Chem. Abst. 82*, 125560 (1975)); *I. Khanna, et al.*, Indian J. Chem. *13*, 781 (1975) (*Chem. Abst. 83*, 164507 (1975)); *I. P. Dragalin, et al.*, Khim. Prir. Soedin *11*, 747 (1975) (*Chem. Abst. 84*, 150899 (1976)).

(B): review–(6); *H. Sato, et al.*, Agr. Biol. Chem. *37*, 225 (1973); *A. Tada, et al.*, Chem. Pharm. Bull. (Tokyo) *21*, 308 (1973); *A. Akahori, et al.*, Chem. Pharm. Bull. (Tokyo) *21*, 1799 (1973); *I. Yosioka, et al.*, Chem. Pharm. Bull. (Tokyo) *21*, 2092 (1973); *Y. Tomita, et al.*, Phytochemistry *13*, 729 (1974); *R. Tschesche, et al.*, Chem. Ber. *107*, 53, 2828 (1974); Chem. Ber. *108*, 265 (1975); *S. Kiyosawa, et al., Yokugaku Zasshi 95*, 102 (1975); *I. P. Dragalin, et al.*, Khim. Prir. Soedin *11*, 805, 806 (1975) (*Chem. Abst. 84*, 132606, 135992 (1976)); Phytochemistry *14*, 1817 (1975); *K. Kawano, et al.*, Agr. Biol. Chem. *39*, 1999 (1975); *G. V. Lazurevskii, et al.*, Dokl. Akad. Nauk SSSR *221*, 744 (Biochem.) (1975) (*Chem. Abst. 83*, 43679 (1975)); *B. Wilkomirski, et al.*, Phytochemistry *14*, 2657 (1975).

26-*O*-Methyl-polypodosaponin which is not spirostanol glycoside shows the saponic properties and is regarded as a novel type of steroid saponin (review–(6)). Starfish saponins are hydrolysed to give pregnane, cholestane and ergostane derivatives (*S. Ikegami, et al.*, Tetrahedron Lett. 1601, 3725 (1972); *Y. M. Sheikh, et al.*, J. Am. Chem. Soc. *94*, 3278 (1972); Tetrahedron Lett. 372 (1972); *Y. Shimizu*, J. Am. Chem. Soc. *94*, 4051 (1972); *Y. Kamiya, et al.*, Tetrahedron Lett. 655 (1974); *I. Kitagawa, et al.*, Tetrahedron Lett. 967 (1975); *J. A. Croft, et al.*, Comp. Biochem. Physiol. B *48(4B)*, 535 (1974) (*Chem. Abst. 82*, 28794 (1975)); *W. J. Fleming, et al.*, Comp. Biochem. Physiol. B *55(3B)*, 267 (1976) (*Chem. Abst. 85*, 2744 (1976)).

Trillenoside, the first 18-norspirostanol glycoside, is noteworthy also in that the aglycone has the 21-methyl group hydroxylated and that apiofuranose is a component unit of the sugar moiety. The structure of aglycone was determined by X-ray analysis of its tetraacetyl monobrosylate (*T. Nohara, et al.*, Tetrahedron Lett. 4381 (1975)).

A saponin consisting of kammogenin and five units of 2-deoxyribose was reported (*R. C. Backer, et al.*, J. Pharm. Sci. *61*, 1665 (1972).

Triterpenoid Saponins

The triterpenoid saponins for which structures are established have much increased in number (review–(6), (9)). Several novel type saponins have been reported: for example, nor-dammarane glycosides (*D. K. Kulshrestha, et al.*, Indian J. Chem. *13*, 309 (1975) (*Chem. Abst. 83*, 79408 (1975)), those which has three sugar moieties at C-2, 3 and 28 of the aglycone (review–(6)) and which contains a pentasaccharide moiety forking into four branches [review–(6); *N. K. Abubakirov, et al.*, Khim. Prir. Soedin *11*, 658 (1975) (*Chem. Abst. 84*, 74571 (1976))].

Ribose [*V. Ya. Chirva, et al.*, Dokl. Akad. Nauk SSSR *217*, 969 (Biochem.) (1974) (*Chem. Abst. 81*, 136449 (1974))] and apiofuranose (*I. Kitagawa, et al.*, Chem. Pharm. Bull. (Tokyo) *23*, 2268 (1975); *A. Tada, et al.*, Chem. Pharm. Bull. (Tokyo) *23*, 2965 (1975)) have been found as a component monosaccharide of the sugar part.

Holotoxin A and B from the sea cucumber have been assigned the full structures, lanostane type triterpenoid oligoglycosides (*I. Kitagawa, et al.*, Chem. Pharm. Bull. (Tokyo) *24*, 275 (1976)).

Basic Steroid Saponins

See also: *P. Bite, et al.*, Hagy. Kem. Foly. *78*, 221 (1972) (*Chem. Abst. 77*, 31536 (1972)); *E. N. Novruzov, et al.*, Khim. Prir. Soedin *11*, 434 (1975) (*Chem. Abst. 84*, 40726 (1976)); *S. Singh, et al.*, Phytochemistry *13*, 2020 (1974) (new glycosides).

Isolation and Structure Determination

Some of the methods recently developed are as follows.

(1) Droplet countercurrent chromatography for separation and semimicro determination: *T. Tanimura, et al.*, Kagaku-no Ryôiki *29*, 895 (1975); *I. Kitagawa, et al.*, Chem. Pharm. Bull. (Tokyo) *24*, 266 (1976); *Y. Ogihara, et al.*, J. Chromatog., in press (1976).

(2) Selective cleavage of glucuronide linkage in saponin by photolysis, lead tetraacetate degradation and acetic anhydride-pyridine degradation: *I. Kitagawa, et al.*, Chem. Pharm. Bull. (Tokyo) *22*, 1339 (1974); Tetrahdedron Lett. 549 (1976); Abstract Papers of the 8th International Symposium on Carbohydrate Chemistry, 115 (1976).

(3) NMR spectrometry of free saponins: *K. Miyahara, et al.*, Chem. Pharm. Bull. (Tokyo) *22*, 1407 (1974).

(4) Mass spectrometry of triterpenoid saponins: *M. J.-M. Colard, et al.*, Biomed. Mass Spectromet. *2*, 156 (1975); *N. K. Kochetkov, et al.*, Bull. Acad. Sci. USSR Div. Chem. Sci. *22*, 2647 (1973); *T. Komori, et al.*, Org. Mass Spectromet. *9*, 744 (1974); *R. Higuchi, et al.*, Chem. Pharm. Bull. (Tokyo) *24*, 2610 (1976). Field desorption mass spectrometry: *H. D. Beckey, et al.*, Angew. Chem. *87*, 425 (1975).

(5) ^{13}C NMR spectrometry of steroid and triterpenoid sapogenins: *D. M. Doddrell, et al., Tetrahedron Lett. 2381 (1974); H. Eggert, et al.*, Tetrahedron Lett. 3635 (1975).
^{13}C NMR spectrometry has been applied to the structure elucidation and identification of dammarane type triterpenoid saponins (*R. Kasai, et al.*, Abstract Papers of the 20th Symposium on the Chemistry of Natural Products, 280 (1976)).

4 Alkaloids

Contribution by

Y. Hirata, Nagoya
M. Ihara, Sendai
T. Kametani, Sendai

4 Alkaloids

Yoshimasa Hirata (4.1 ~ 4.2)
Department of Chemistry, Faculty of Science, Nagoya University, Chikusa-ku, Nagoya, 464 Japan

Tetsuji Kametani and Masataka Ihara (4.3 ~ 4.5)
Department of Pharmaceutical Sciences, Tohoku University, Aobayama, Sendai, 980 Japan

1 Classification

The classification of alkaloids, a general term for compounds containing (a) basic nitrogen atom(s) in the molecule, can be based either on a structural point of view or on the plant in which the alkaloid is found. Usually both ways are used in combination.

1) Pyrrolidine and pyrrolizidine alkaloids: *e.g.*, hygrine, monocrotaline (also classified as senecio alkaloid)
2) Piperidine, pyridine, and quinolizidine alkaloids: anabasine, lupinine, sparteine, matrine
3) Isoquinoline alkaloids
   a) Simple isoquinolines: cotarine
   b) Benzylisoquinolines and pavine type: laudanosine, argemonine
   c) Bisbenzylisoquinolines: pycnamine
   d) Cularine alkaloids: cularine (dihydrooxepine series)
   e) Phthalide isoquinolines: narcotine
   f) Aporphine alkaloids: nuciferine
   g) Proaporphine alkaloids: pronuciferine
   h) Dibenzopyrocoline alkaloids: cryptoaustoline
   i) Protoberberine alkaloids: berberine
   j) Protopine alkaloids: protopine
   k) Benzophenanthridine alkaloids: chelidonine
   l) Spirobenzyl isoquinolines: ochotensimine
   m) Ipecacuanha alkaloids:
4) Morphine alkaloids (modified benzylisoquinoline type): morphine
   4′) Husbanan type: husbanonine
5) Quinoline alkaloids
   a) Cinchona alkaloids: quinine
   b) Furoquinolines: lunacrine
6) Erythrina alkaloids: erysodine, α-erythroidine
7) Amaryllidaceae alkaloids: lycorine, lycorenine, galanthamine, crinine, tazettine, montamine, belladine
8) Indole alkaloids
   a) Indole amines: serotonine
   b) Eserine type: eserine
   c) Harman type: harmaline
   d) Yohimbine type and corynantheine type: yohimbine, corynantheine
   e) Ajmaline-sarpagine type: ajmaline, sarpagine
   f) Ibogamine type: ibogamine
   g) 2-Acylindoles: vobasine
   h) Eburnamine type: eburnamine
   i) Aspidospermine type: aspidospermine
   j) Strychnine type: strychnine
   k) Oxindoles: mitraphylline
   l) Dimeric indole alkaloids: vincaleukoblastine (or vinblastine)
   m) Ergot alkaloids: ergotamine
9) Colchicine alkaloids: colchicine
10) Tropane alkaloids: tropine
11) Lycopodium alkaloids: annotinine, lycopodine, lyconnotine, serratinine, lycodine, selagine, cernuine
12) Diterpene alkaloids: veatchine type, atisine type, lycoctonine type (aconitum type), heteratisine type
13) Steroid alkaloids: conessine, jervine
14) Miscellaneous alkaloids: quinolone alkaloids, imidazole alkaloids, purine alkaloids

2 Structure Elucidation

The structures of organic compounds, especially of natural products, are determined either by a combination of chemical and spectroscopic techniques or by X-ray crystallographic analysis. Though the latter has recently become very popular, the chemical technique is still employed when the compound or its derivatives are not crystalline, or else when a certain quantity of the sample is available and offers us the hope of finding reactions of interest.

2.1 Structural Determination of Alkaloids by Means of X-Ray Crystallographic Analysis

Since the structural elucidation of small amounts of complex alkaloids has lately come into frequent question, X-ray crystallographic analysis, which provides more accurate results within a shorter time on a smaller amount of sample than the combination of chemical and spectroscopic methods, is being used extensively. In the special case of alkaloids, we can take advantage of their capability of salt formation to introduce heavy atoms, halogens (Br, I) into the molecules. Their tendency to produce good crystals also enhances the usefulness of this technique.

2.1.1 Examples of Structural Determination with Introduction of Heavy Atoms

aconitine[1], annotinine[2], calycanthine[3], caracurine[4], echitamine[5], gelsemine[6], batrachotoxin[7], camptothecin[8], daphniphylline[9], yuzurimine[10], vincristine (leurocristine)[11], neothiobinupharidine[12], tuberostemonine[13], tomatidine[14], dichotine[15] and acutimine[16]

2.1.2 Examples of Structural Determination without Introduction of Heavy Atoms

villalstonine[17], reserpine[18], jamine[19]

2.2 Structural Determination by the Combination of Chemical and Spectroscopic Methods

In comparison with prewar procedures that largely depended on chemical techniques, less sample and less investigation time are now required. To give some examples of these techniques, the following three instances of the investigation of alkaloids, all of which show characteristic structural features and reactions, will be described.

2.2.1 Dendrobine[20]

Its crystals had been obtained from *Dendrobium nobile L.* for a fairly long time before the structure was established in 1964. Though the molecule is relatively small, its skeleton and stereochemistry offer some interesting questions. A few alkaloids of the same series have been found in this and analogous plants.

$C_{16}H_{25}NO_2$, mp 134°
IR($CHCl_3$) 1763 cm^{-1} (due to γ-lactone)
NMR ($CDCl_3$), in δ 0.98 (6H, s, J=6Hz, $(CH_3)_2CH$–),

1.47 (3H, s, CH_3–C⟨),
2.52 (3H, s, CH_3–N⟨),
2.70 (1H, d, J=3Hz, ⟩N–CH⟨),
4.87 (1H, q, J=3.5 Hz, –CHOCO–)

a) Dendrobine **1** has a tertiary nitrogen (inert to nitrous acid) and a γ-lactone indicated by the infrared spectrum. Since it is inert to catalytic hydrogenation, it is believed to have four rings (including the γ-lactone). The NMR measurements indicate the existence of two secondary methyl groups (methyl groups attached to a secondary carbon, which are assigned to an isopropyl group by the formation of isobutyric acid on oxidation and by mass spectrometry), one tertiary methyl and one *N*-methyl group.

b) Since the NMR spectrum of **1** shows that the signal of the ⟩N–CH– appeared as a doublet, a partial structure ⟩N–CH–CH– is deduced. On the other hand, the formation of oxodendrobine **5** and its infrared spectrum (1670 cm^{-1}) suggest the presence of –N–CH_2–, the nitrogen atom being a part of a five-membered ring.

[1] *Przybylska, L. Marion*, Can. J. Chem. *37*, 1116, 1846 (1959).
[2] *Przybylska, L. Marion*, Can. J. Chem. *35*, 1075 (1957).
[3] *T. A. Hamor, J. M. Robertson*, J. Chem. Soc. 194 (1962).
[4] *A. T. McPhail, G. A. Sim*, Proc. Chem. Soc. 416 (1961).
[5] *S. A. Hamilton, T. A. Hamor, J. M. Robertson, G. A. Sim*, Proc. Chem. Soc. 63 (1961).
[6] *F. M. Lovell, R. Pepinsky, A. J. C. Wilson*, Tetrahedron Lett. 1 (1959).
[7] *T. Tokuyama, J. Daly, B. Witkop, I. L. Karle, J. Karle*, J. Am. Chem. Soc. *90*, 1917 (1968).
[8] *M. E. Wall, M. C. Wani, C. E. Cook, K. H. Palmer, A. T. McPhail, G. A. Sim*, J. Am. Chem. Soc. *88*, 3888 (1966).
[9] *N. Sakabe, Y. Hirata*, Tetrahedron Lett. 965 (1966).
[10] *M. Sakurai, N. Sakabe, Y. Hirata*, Tetrahedron Lett. 6309 (1966).
[11] *J. W. Moncrief, W. N. Lipscomb*, J. Am. Chem. Soc. *87*, 4963 (1965).
[12] *G. I. Birnbaum*, Tetrahedron Lett. 4149 (1965).
[13] *H. Harada, M. Irie, N. Masaki, K. Osaki, S. Uyeo*, Chem. Commun. 460 (1967).
[14] *O. Kennard, L. R. Sanseverino, J. S. Rollett*, J. Chem. Soc., 956 (1967); *E. Höhne, H. Ripperger, K. Schreiber*, Tetrahedron *23*, 3705 (1967).
[15] *C. Djerassi*, J. Am. Chem. Soc. *92*, 222 (1970).
[16] *M. Nishikawa, K. Kamiya, M. Tomita, Y. Okamoto, T. Kikuchi, K. Osaki, Y. Tomiie, I. Nitta, K. Goto*, J. Chem. Soc. B, 652 (1968).
[17] *C. E. Nordman, S. K. Kumra*, J. Am. Chem. Soc. *87*, 2059 (1965).
[18] *I. L. Karle, J. Karle*, Acta Crystallogr. *24B*, 81 (1968).
[19] *I. L. Karle, J. Karle*, Tetrahedron Lett. 2065 (1963).

[20] *S. Yamamura, Y. Hirata*, Tetrahedron Lett. 79 (1964); *S. Yamamura, Y. Hirata*, Nippon Kagaku Zasshi *85*, 377 (1964); *T. Onaka, S. Kamata, T. Maeda, Y. Kawazoe, M. Natsume, T. Okamoto, F. Uchimaru, M. Shimizu*, Chem. Pharm. Bull. *12*, 506 (1964); *Y. Inubushi, Y. Sasaki, Y. Tsuda, B. Yasui, T. Konita, J. Matsumoto, E. Katarao, J. Nakano*, Tetrahedron *20*, 2007 (1964); *Y. Inubushi, E. Katarao, Y. Tsuda, B. Yasui*, Chem. Ind. 1689 (1964); *Y. Inubushi, Y. Tsuda, E. Katarao*, Chem. Pharm. Bull. *13*, 1482 (1965); *Y. Inubushi, Y. Sasaki, Y. Tsuda, J. Nakano*, Tetrahedron Lett. 1519 (1965).

In addition, an ABX type signal due to $>N-CH_2-\overset{|}{C}H-$ is observed in the NMR spectrum of cyano-nordendrobine **2**. Hence **Ia** or **Ib** is given for the partial structure in the vicinity of the nitrogen atom.

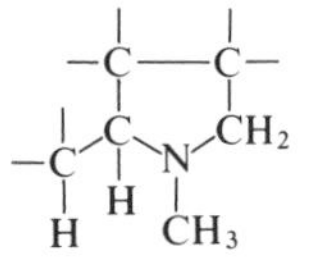

Ia 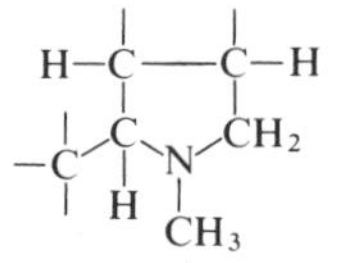 **Ib**

c) The ethereal oxygen of the γ-lactone of dendrobine **1** is supposedly attached to a secondary carbon. The multiplicity of the proton on the carbon bearing the γ-lactone ethereal oxygen in the NMR spectrum suggests the presence of two hydrogen atoms on the adjacent carbon atom(s) (α, α position or α, α' position). The secondary nature of the γ-lactone ethereal oxygen is supported also by the sequence of reactions **5**→**7**→**8**→**10**.

d) The NMR spectrum of dendrobinediol diacetate **4** shows an ABX signal (due to $AcOCH_2-CH\langle$) in the region between 4.4 and 3.8 ppm. Combined with the observations at c), the partial structure around the γ-lactone moiety is any of **IIa**–**IIc**.

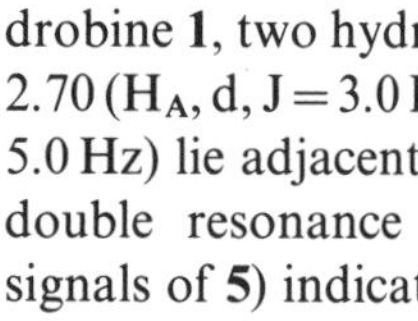

IIc

IIa $R^1=R^2=H$, $R^3=R^4=C$
IIb $R^1=R^2=C$, $R^3=R^4=H$
IIc $R^1=R^3=C$, $R^2=R^4=H$

e) According to the NMR spectrum of dendrobine **1**, two hydrogen atoms having signals at 2.70 (H_A, d, J = 3.0 Hz) and at 4.84 (H_B, q, J = 3.5, 5.0 Hz) lie adjacent to each other (confirmed by double resonance technique on H_A and H_B signals of **5**) indicating the relationship of the γ-lactone part and the pyrrolidine part as **III**.
In addition, the following facts support the partial structure **III**.

(1) The Hofmann degradation of the methiodide of dendrobinediol **3** yields a methine base **6**, which showed the band at 1704 cm^{-1} due to a ketone.

(2) The signal ($\rangle N-CH\langle$) which was observed as a doublet in the NMR spectrum of dendrobine **1** becomes a singlet in that of the keto ester **10**. This also indicates that C_q of **III** is a quaternary carbon.

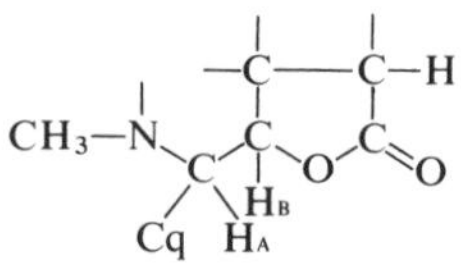

III

These findings exclude **IIa** and **IIb**, leaving only **IIc**. By combining the partial structures **I**, **II**, and **III**, the partial structure of dendrobine can be represented by either **IVa** or **IVb**.

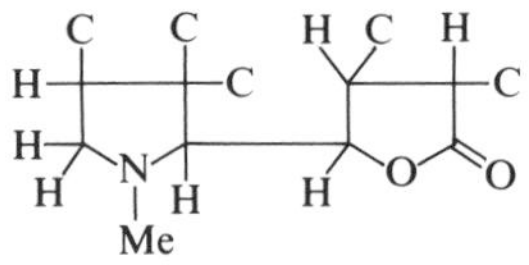

IVa

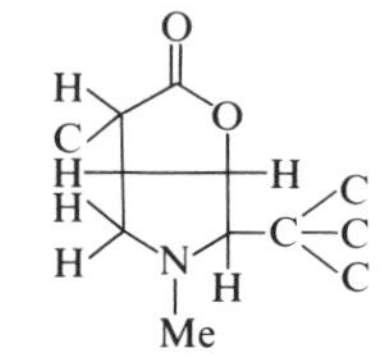

IVb

IVa is preferred since a selenium dehydrogenation reaction of dendrobine yielded 4-isopropyl-2-pyridone.

f) The reduction of the ketonic group of the methine base **6** with $NaBH_4$, followed by acetylation, oxidation with monoperphthalic acid to give an *N*-oxide and then by thermolysis, results in the formation of an *N*-free olefin **9**. The oxidation of **9** with OsO_4, and then with $Pb(OAc)_4$ yields a norketone **11**. The NMR spectrum of the bromoketone **12**, obtained on bromination of **11**, shows a new ABX type signal, whose X part is due to the hydrogen on the carbon atom carrying the bromine. The multiplicity of this signal suggests the partial structure $-\overset{O}{\overset{\|}{C}}-\underset{Br}{\underset{|}{C}}H-CH_2-$. As is expected from the NMR information, the dehydrobromination of **12** gives an α, β-unsaturated five-membered ring ketone **13**, the double bond of which was shown to be di-substituted by the NMR spectrum. [6.17 (1H, q, J = 6.0, 2.4 Hz), 7.60 (1H, q, J = 6.0, 1.8 Hz)]
This series of experiments expands the partial structure **IV** to **V**.

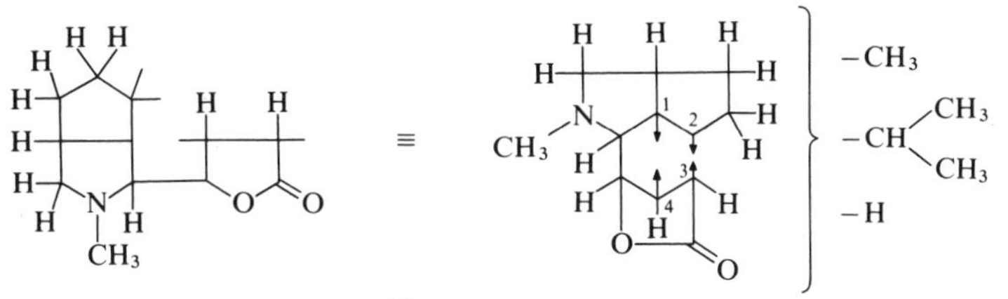

V

Addition of the remaining three fragments (tertiary CH_3, $-CH\langle{CH_3 \atop CH_3}$, H) to **V** together with the formation of a carbocyclic ring should give the whole structure. Taking into account:

(1) that C_1 (in V) is quaternary,

(2) that the carbonyl band of the remaining carbocyclic ketone in **6** and **10** appeared at $1704\,cm^{-1}$ and

(3) the multiplicity of the olefin proton of **13**, the remaining carbocyclic ring is considered to be a six-membered one formed by linking C_2 and C_3, with the uncharacterized hydrogen at C_2 of **V**, thus giving **VI**.

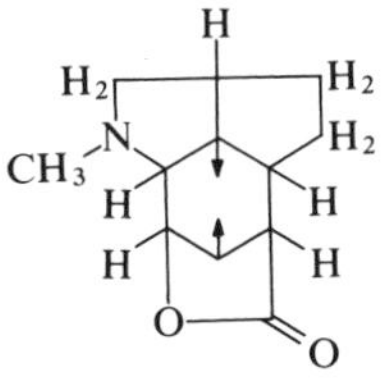

VI

The position of the isopropyl group is determined by the formation of 4-isopropyl-2-pyridone on Se dehydrogenation. Thus, the structure **1** is given for dendrobine.

g) Mass spectrometry played an important role in deducing the structure of dendrobine. It shows notable peaks at m/e = 263 (molecular ion peak, intensity 50), 235 (10), 220 (81), 210 (metastable peak), 184 (metastable peak), 178 (14), 136 (18), 109 (17), 108 (17), 96 (100), 58 (43).

h) The stereochemistry of dendrobine **1**:

(1) The steric reason requires the γ-lactone attached *cis* to the cyclohexane ring at 3- and 5-positions.

(2) A comparison of the pKa′ (in H_2O) of dendrobine **1** (7.80) and its derivatives **3** (9.12) and **4** (7.52) shows that the derivative possessing the hydroxyl group has a greater pKa′, indicating the presence of hydrogen bonding between N and OH. The NMR coupling constant (3–4 Hz) for C_2-H and C_3-H denotes the *cis* relationship of these two hydrogen atoms.

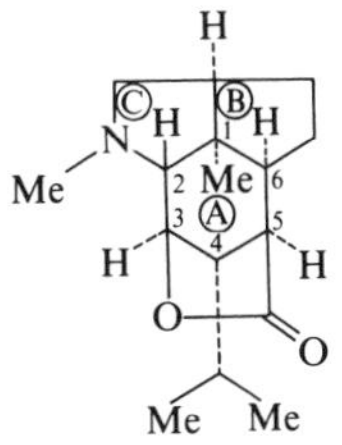

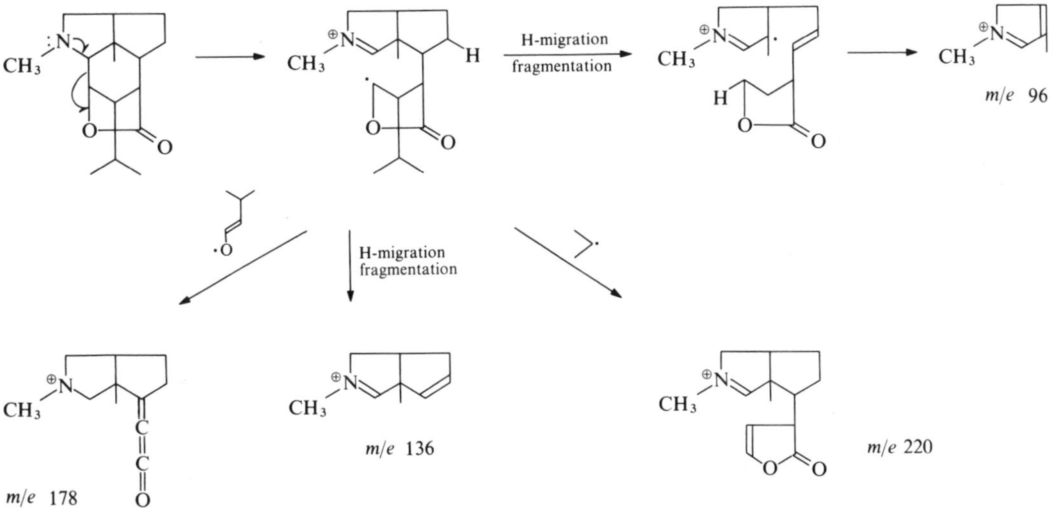

H-migration fragmentation

m/e 96

H-migration fragmentation

m/e 178

m/e 136

m/e 220

(3) The ring juncture of the B and C rings to the A ring is either *cis-cis* or *cis-trans*.

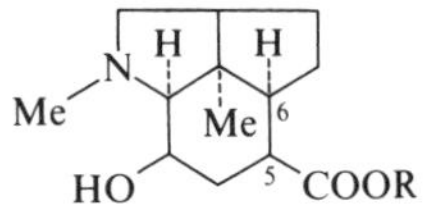

a

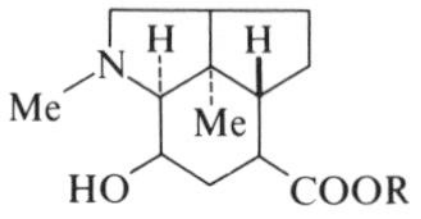

b

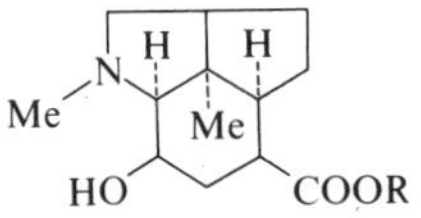

c

Configuration b is impossible because the coupling constant of C_5-H and C_6H is as large as 3.0 Hz (whether the A ring assumes the boat form or the chair form). Configuration c is also excluded by the coupling constant between C_5-H and C_6-H (J=3.0 Hz) as well as that between C_5-H and C_4-H (J=9.0 Hz), while the configuration a satisfactorily explains these J-values. Thus the configuration a is concluded to be the correct one.

When **3** is strongly heated in Ac_2O-pyridine, a monoacetoxy acetate is formed, which does not give the hydrochloride of **3** on treatment with an alkali followed by acidification with hydrochloric acid. Various spectral data give the structure d for this monoacetoxy acetate in which the nitrogen atom and the hydroxymethyl group of **3** are linked together. This finding again strongly supports the configuration a for the juncture of the A, B, and C rings.

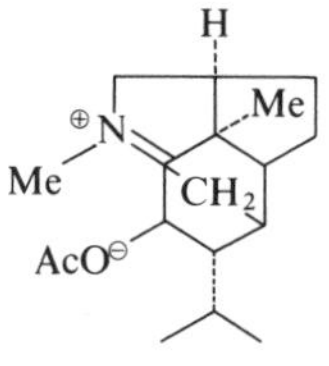

d

(4) The coupling constant (J=6.0 Hz) of C_3-H and C_4-H requires the *trans* relationship between the isopropyl and C_3-O. Thus, **1** is given for the stereochemistry of dendrobine.

i) The absolute configuration of dendrobine is determined by applying the lactone rule.

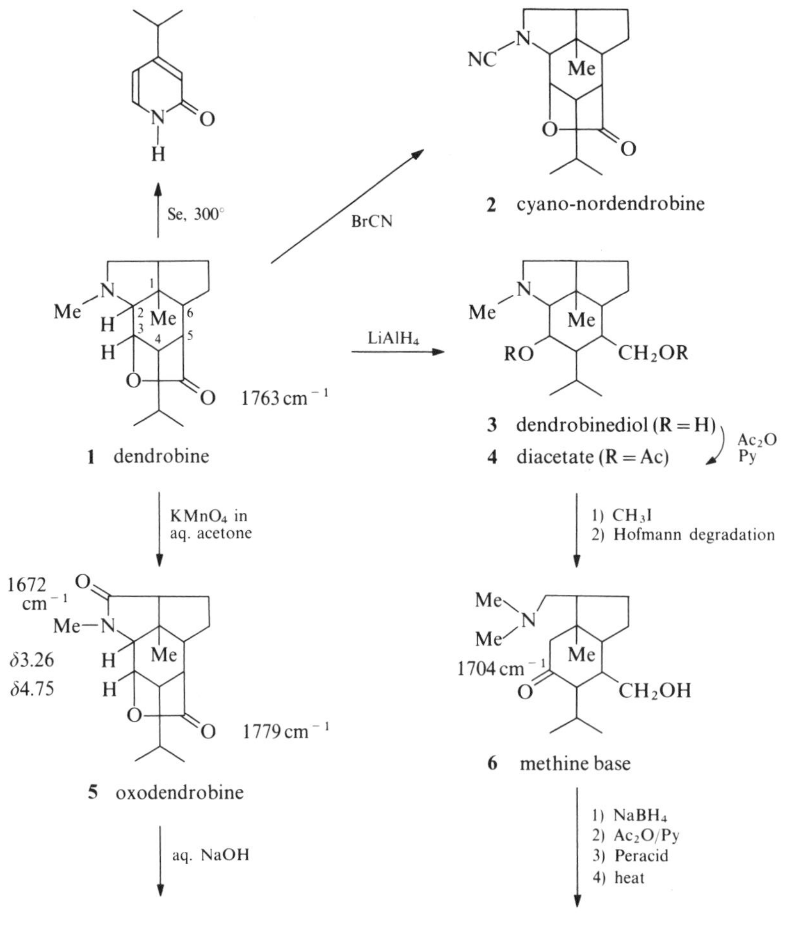
Se, 300°
BrCN
LiAlH₄
1763 cm⁻¹
2 cyano-nordendrobine
1 dendrobine
3 dendrobinediol (R = H)
4 diacetate (R = Ac)
Ac₂O
Py
KMnO₄ in
aq. acetone
1) CH₃I
2) Hofmann degradation
1672 cm⁻¹
δ3.26
δ4.75
1779 cm⁻¹
1704 cm⁻¹
5 oxodendrobine
6 methine base
aq. NaOH
1) NaBH₄
2) Ac₂O/Py
3) Peracid
4) heat

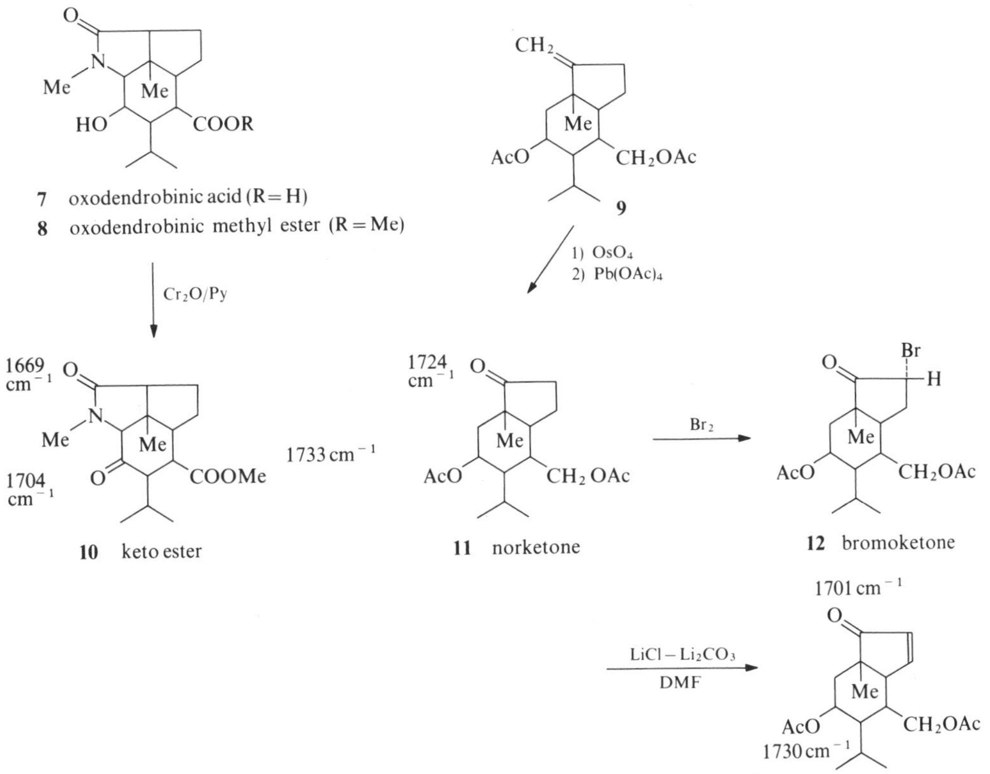

7 oxodendrobinic acid (R = H)
8 oxodendrobinic methyl ester (R = Me)

Cr_2O/Py

9

1) OsO_4
2) $Pb(OAc)_4$

10 keto ester

11 norketone

Br_2

12 bromoketone

$LiCl-Li_2CO_3$ / DMF

13 α,β-unsaturated ketone

2.2.2 Securinine[21]

This is an alkaloid obtained from *Securinega flueggoides Muell. Arg. (Euphorbiaceae)*

$C_{13}H_{15}NO_2$, mp 142–143° $[\alpha]_D^{20} = -1042°$ (EtOH).

The following functional groups are absent:

active hydrogen, N–Me, –O–Me, OH, NH, C–Me.

a) Spectral data: the presence of an α, β and γ, δ-unsaturated γ-lactone moiety.

Catalytic reduction of securinine **1** leads to a dihydro derivative **2**, which on further hydrogenation gives a tetrahydro derivative **3**.

	UV nm (log ε)
1	330 (3.24) 256 (4.27)
2	215 (4.33)
3	— —

	IR (CCl_4) cm^{-1}
1	1840 (w), 1760 (s), 1640 (s)
2	1815 (w), 1770 (s), 1652 (s)
3	1790 (s)

The presence of a γ-lactone is evidenced by the infrared spectrum of **3**, while the infrared spectrum of **2** showed the presence of an α, β-unsaturated γ-lactone, which is also revealed by the UV spectrum. The C=O band is split, indicating the presence of a hydrogen atom attached to the double bond of the α, β-unsaturated lactone. On comparison of the UV spectra of **1** and **2**, we find **1** has an α, β, γ, δ-unsaturated γ-lactone. The substitution pattern of the conjugated lactone system is given by the information of the vinyl proton region NMR of **1**, **2**, and **3** (60 MHz, $CDCl_3$). From the multiplicity of each signal of three vinyl protons (H_A, H_B, and H_C), I has been given for the substitution pattern of **1**.

I

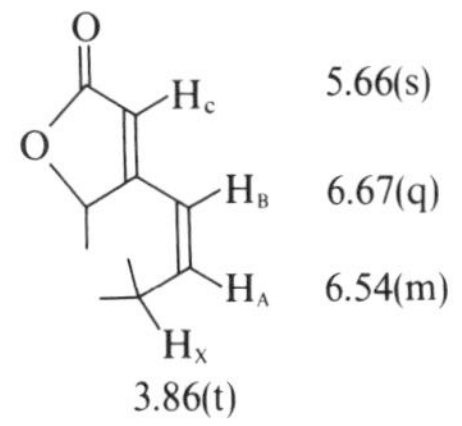

H_C 5.66(s)
H_B 6.67(q)
H_A 6.54(m)
H_X 3.86(t)

[21] *S. Saito, K. Kodera, N. Sugimoto, Z. Horii, Y. Tamura,* Chem. Ind. 1652 (1962); *S. Saito, K. Kodera, N. Shigematsu, A. Ide, Z. Horii, Y. Hamura,* Chem. Ind. 689 (1963); *T. Nakano, T. H. Yang, S. Terao,* Chem. Ind. 1651 (1962); *S. Saito, K. Kotera, N. Shigematsu, A. Ide, N. Sugimoto, Z. Horii, M. Hanaoka, Y. Yamawaki, Y. Tamura,* Tetrahedron *19*, 2085 (1963); *Z. Horii, M. Ikeda, Y. Yamawaki, Y. Tamura, S. Saito, K. Kodera,* Tetrahedron *19*, 2101 (1963).

In addition, **1** shows an ABX signal indicating the presence of H_X on the carbon atom adjacent to H_A.

b) Information on the skeleton:

(1) Tetrahydrosecurinine **3** yields phthalic acid upon oxidation with $KMnO_4$ in the presence of oxalic acid.

(2) The treatment of securinine **1** with KOH, followed by reduction with Raney Ni leads to the lactam alcohol A **9**, which on dehydrogenation with Pd-C gives 2(2′-tolyl) pyridine.

(3) The destructive distillation of securinine **1** over zinc dust yields phenol, *o*-cresol, *p*-toluidine and ammonia.

The observations of (1) and (2) allow the deduction of **II** for the skeleton, which, upon taking (3) into account, gives **III**.

II **III**

c) The planar structure of securinine:

Combining **I** and **III**, structure **1** is presented for securinine.

d) The stereochemistry of securinine:

If, when considering the stereochemistry securinine is regarded as the substituted pyrrolidine, then the two bonds C_{10b}–C_{3a} and C_{5a}–C_5 must be *cis* to each other in order for a molecule to be constructed. The remaining question is to establish the stereochemistry of C_{10a}–H and C_{10b}–O. The two lactam alcohols **9** and **10** shown in the chart are deduced to be stereoisomers as to C_{7a}, because both give 2 (2′-tolyl) pyridine on dehydrogenation. This stereoisomerism with reference to C_{7a} resulted from the catalytic reduction of the double bond of the α, β-unsaturated γ-lactone system of securinine **1**. This stereochemical problem was solved by the infrared spectral information with quinolizidine-A **11** and quinolizidine **12** obtained by the $LiAlH_4$ reduction of **9** and **10**, respectively. Theoretically, eight stereoisomers of this quinolizidine derivative are possible, which can be distinguished by checking for the existence of:

(1) *trans* quinolizidine band (Bohlmann band) and

(2) intramolecular hydrogen bond between the hydroxyl and the lone pair electrons on the nitrogen atom by infrared spectroscopy.

		trans band
Quinolizidine	A	+ (2809, 2760, 2682 cm^{-1})
	B	–
		intramolecular H-bond
Quinolizidine	A	– (3620 cm^{-1})
	B	+ (3505 cm^{-1})

Conformational analysis based on this fact gives **11** for quinolizidine A and **12** for quinolizidine B. In both compounds, **11** and **12**, C_{11a}–OH and C_{11b}–H are *cis* to each other and consequently C_{10a}–H and C_{10b}–O of securinine **1** are decided to be *trans* to each other. (The stereochemistry of quinolizidine A and B has been confirmed by stereospecific syntheses.)

e) Absolute configuration:

The C_{11b}–H bond of benzoquinolizidine (A) derived from securinine **1** corresponds to the C_{10a}–H bond of **1**:

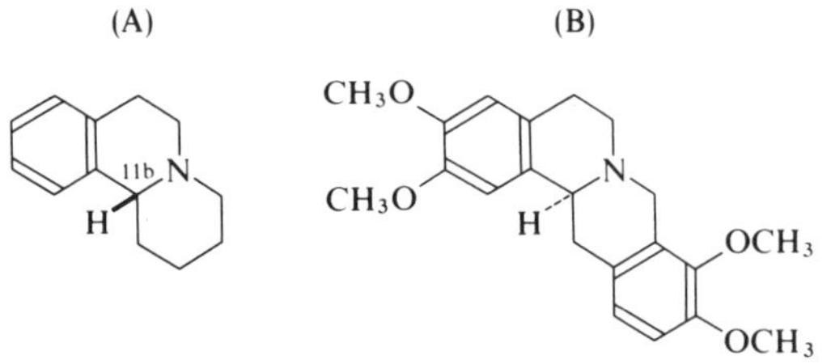

(A) (B)

All ORD curves of (A) and its hydrochloride and perchlorate exhibit the positive Cotton effect, while those of S(–)-tetrahydropalmatine (B), the absolute configuration of which is known, exhibit the negative effect. Consequently, C_{10a}–H of securinine must assume the R-configuration.

2.2.3 Meloscine[22]

This is an alkaloid from *Melodinus scandens Forst. (Apocynaceae).*

$C_{19}H_{20}N_2O$	mp 178–180°
UV (EtOH)	253 nm (log ε = 4.04), 280 (3.4) 289 (3.2)
IR (KBr)	3196 (NH) 1674 (–CONH–), 758 cm^{-1} (four adjacent hydrogens on an aromatic ring)

a) The reactions of meloscine and its partial structure:

(1) The partial structure **I** (p. 112) is deduced from the UV and IR spectra of meloscine **1**.

(2) Treatment of meloscine with MeOD/MeONa leads to a monodeutero derivative.

[22] *K. Bernauer, G. Englert, W. Vetter, E. K. Weiss*, Helv. Chim. Acta *52*, 1886 (1969).

H_2/PtO_2

$KMnO_4$

COOH

COOH

2

3

H_2/PtO_2-EtOH
(or Pd-C/AcOH)

Securinine **1**

BrCN

H_2/Pd-C
AcONa
EtOH

4

5

Al-Hg

Zn

8% HCl
reflux

6

7

1) KOH
2) H_2/Raney Ni
10%KOH

H_2/Pd-C

H_2/Pd-C

OH

OH

CH_3

CH_3

NH_2

8

H_2/PtO_2

H_2/Raney Ni
10% KOH

9 lactam alcohol A
(mp 184–185°)

10 lactam alcohol B
(mp 223–229°)

Pd-C
295°

Pd-C
295°

$LiAlH_4$

CH_3

11 quinolizidine A

2(2′-tolyl)pyridine

12 quinolizidine B

Since the lactam carbonyl is the sole activating group of this alkaloid, the deuterium must have been introduced in the α-position of the lactam, hence giving the partial structure **II**.
(3) Since the action of CH_3I on meloscine yields the methiodide **5**, the other nitrogen (Nb) of **1** is considered to be tertiary.
(4) The formation of a tetrahydro derivative **2** on catalytic hydrogenation suggests the presence of two double bonds (confirmed by NMR). Accordingly four rings are present besides the aromatic ring.
(5) The oxidation of the tetrahydro derivative **2** leads to two dilactams **3** and **4**.
(6) The catalytic reduction of meloscine methochloride gives three kinds of Emde base **6**, **7**, and **8**.
(7) The $LiAlH_4$ reduction of meloscine **1** leads to an *N*-substituted aniline type compound 9 (UV 248, 307 nm).

b) The NMR and mass spectra of meloscine and its derivatives:
The structure of this compound was established by skillful utilization of the NMR and mass spectra.

NMR Spectra: (100 MHz, δ)

Proton(s) at	Meloscine ($CDCl_3$)	Meloscine-Nb-methochloride (D_2O)
C-aromatic	6.8–7.4	6.85–7.55
N- 1	9.69	—
C- 3		2.77
C- 4		1.86, 2.11
C- 6	5.75	
C- 7	6.05	*ca.* 6.15
C- 8	*ca.* 3.2	*ca.* 3.95
C-10		
C-11		
C-19	3.53	4.5
C-20*	5.55 (Hc)	5.68
C-21*	4.79 (H_A), 4.91 (H_B)	5.01, 5.05
Nb–CH_3		3.44

* $J_{AB} = \sim 1$ $J_{AC} = 10$ $J_{BC} = 17$
Arrows refer to decoupling experiments.

(1) The signal pattern of four protons appearing at 6.8–7.4 ppm exhibits the presence of a 1.2-disubstituted benzene ring in **1**.
The signal at 9.69 (due to –CONH–) disappears on addition of D_2O, confirming the aforementioned partial structure **I**.
(2) Compound **1** shows an ABXY type signal ($-CH_A=CH_B-CH_xH_y-$) at 6.05 (C_7-H_A), 5.75 (C_6-H_B), and *ca.* 3.2 ($C_8-H_xH_y$; the position can be detected by the double resonance technique).
On irradiation at around 3.2 ppm, the complex pattern of the AB-part was changed to an AB quartet. (J = *ca.* 10 Hz, which indicates that the double bond is not involved in the five-membered ring.)
On the other hand, simultaneous irradiation at 5.75 (H_A) and 6.05 (H_B) (triple resonance) changes the XY-part signal to a doublet.
Comparing the XY-parts (C-8) of the ABXY-type signals in meloscine **1** and its methochloride **5**, the signals in the latter appear at the lower field by 0.75 ppm, thus suggesting the partial structure of $-CH_xH_y-Nb\langle$.
(3) The signal ($C_{19}-H$, singlet) appeared at 3.53 ppm in **1** was observed at the lower field by *ca.* 1 ppm in **5**. This finding suggests the partial structure $-\overset{|}{C}H-Nb\langle$.
From the results of (2) and (3) the partial structure **III** is deduced.
(4) An ABC type signal (cf., C-20, C-21) is found in the olefinic proton region of **1**, suggesting the partial structure **IV**.
(5) The methochloride **5** shows an ABX signal with the AB part at *ca.* 2 ppm ($-C_4-H_AH_B$) and the X-part at 2.77 ppm (C_3-H_z, quartet), which was confirmed by double resonance. (This ABX type signal is observed more clearly in the spectrum of **4**).
This ABX signal is overlapping with the other signals and so is not clear in the spectrum of **1**. In the methochloride of the previously mentioned monodeuterated meloscine, this part appears as an AB type quartet, indicating that it is the methine hydrogen of the partial structure **V** that can be deuterated. Thus the partial structure **II** can be correlated with **V**, giving the combined partial structure, **VI**.
Mass spectra:
The Emde bases **6** and **7** are isomers formed by reduction of each of the two double bonds (C_6-C_7 and $C_{20}-C_{21}$) originally present in meloscine **1** in addition to hydrogenolysis of the quarternary nitrogen ($Nb-C_8$), while the Emde base **8** underwent reduction of both double bonds and hydrogenolysis of the quarternary nitrogen.
The mass spectra of **6** and **7** show a very intense peak at m/e = 214 ($C_{13}H_{14}N_2O$, $M^+-C_7H_{12}$). The spectrum of **8** also shows the

same peak ($M^+-C_7H_{14}$). Accordingly, the C_7 fragment involves the carbon atoms associated with the two double bonds (C-6, 7, 8, 20, 21, cf. partial structures **III** and **IV**). Of the two remaining carbons of the C_7 fragment, one is undoubtedly the quaternary carbon (C-5) of the partial structure **IV** and the other is a methylene carbon.

Thus the fragmentation is rationalized as follows:

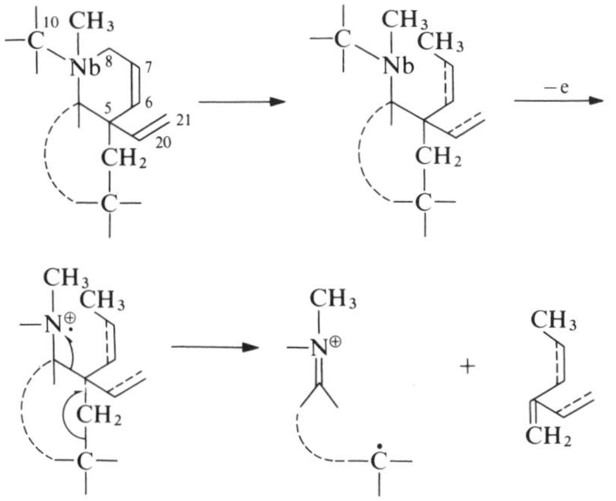

Therefore the partial structure **VII** can be presented.

c) The structure of meloscine:

Three deuterium atoms are incorporated in each of the dilactams **3** and **4** on treatment with CH_3OD-CH_3ONa. Since one of the deuteriums enters the C-3 position (cf. partial structures **II** and **VI**), both **3** and **4** are shown to have the moiety $-N_b-C(=O)-CH_2-$.

(1) The signal at 2.74 ppm (C_{11}-methylene, singlet; this also appears as a singlet in benzene-d_6 and in CF_3COOH) in the NMR spectrum of **4** (100 MHz, 220 MHz, $CDCl_3$) disappears on trideuteration. Hence $-N_b-C(=O)-CH_2-C$ (quaternary) is given as a partial structure.

(2) The signal at 2.23 ppm (C_7-methylene, multiplet) in the spectrum of **3** disappears on trideuteration, thus giving the partial structure $-N_b-C(=O)-CH_2-CH_2-$.

From the results of (1) and (2), the partial structure **VII** is expanded to **VIII**. The partial structures **I** and **III** together include all the carbons of meloscine. Taking **VI** into consideration, **I** and **VIII** can be combined together to give **IX**. Since C-12 of **IX** must be quaternary, **1** is the sole possible structure for meloscine, which is also supported by the mass spectrum.

(3) The mass spectrum of meloscine

The peak at m/e = 134 ($C_9H_{12}N$ as determined by high resolution mass spectrometry) is intense. This intense peak at 134 is still observed after treatment of **1** with MeOD, but its tetrahydro derivative shows the corresponding peak at m/e 138. Accordingly we find that the peak at 134 in **1** is associated with Nb, C-6, C-7, C-20, and C-21.

The tetrahydro derivative **2** shows a base peak at m/e 268 ($M-C_2H_5$), which is attributed to the elimination of an ethyl radical from **X**.

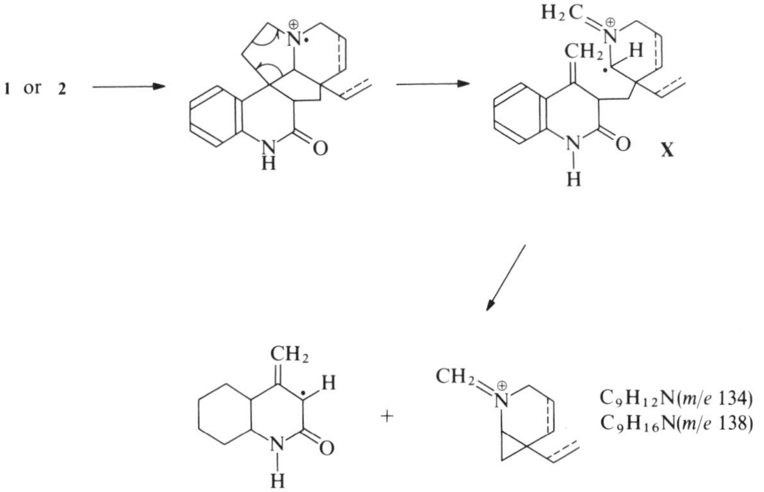

1 or 2 → → **X**

$C_9H_{12}N$ (*m/e* 134)
$C_9H_{16}N$ (*m/e* 138)

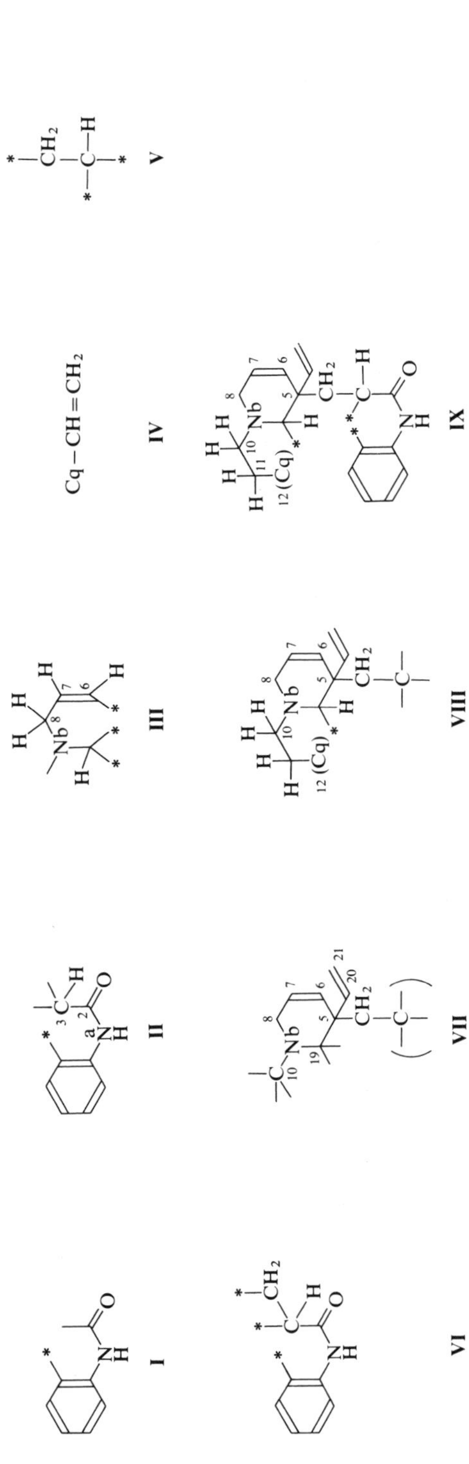

Fig. 1

I II III IV V

VI VII VIII IX

$Cq-CH=CH_2$ (IV)

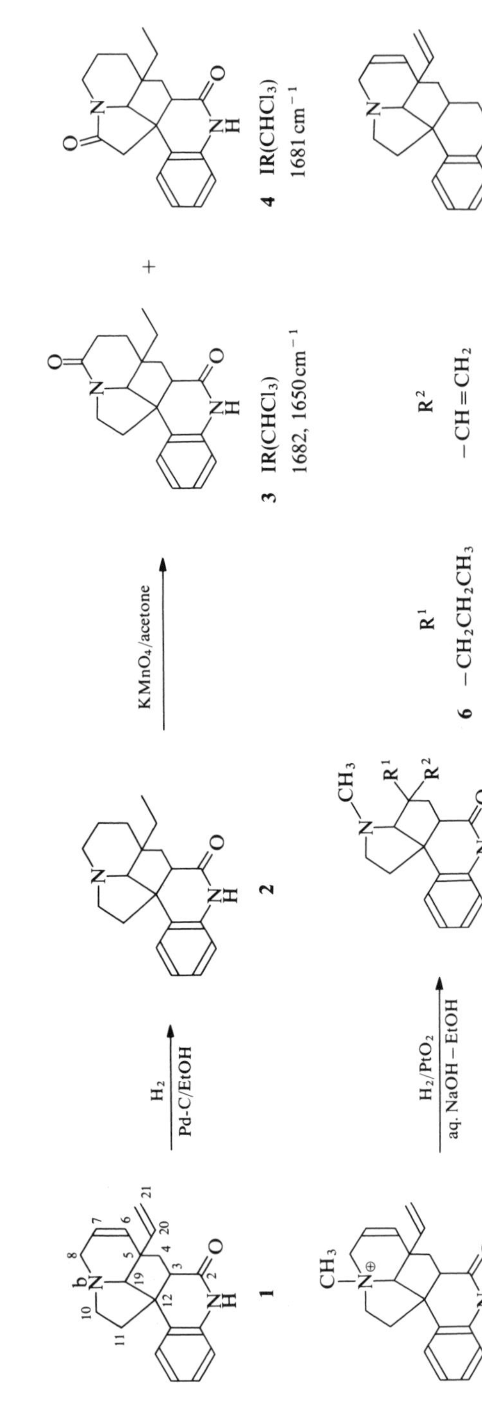

Fig. 2

1 $\xrightarrow[\text{Pd-C/EtOH}]{H_2}$ 2 $\xrightarrow{\text{KMnO}_4\text{/acetone}}$ 3 + 4

3 IR($CHCl_3$) 1682, 1650 cm^{-1}

4 IR($CHCl_3$) 1681 cm^{-1}

5 $\xrightarrow[\text{aq. NaOH}-\text{EtOH}]{H_2/PtO_2}$ 6, 7, 8

9

	R^1	R^2
6	$-CH_2CH_2CH_3$	$-CH=CH_2$
7	$-CH=CH-CH_3$	$-CH_2CH_3$
8	$-CH_2CH_2CH_3$	$-CH_2CH_3$

3 Total Synthesis

As regards the total syntheses of alkaloids, we are concerned with the following: a final confirmation of structure assigned by chemical and physical evidence, and an appreciation of the pharmacological activities of synthesized materials. The chemical structures of the alkaloids are deduced by the classical methods of organic chemistry, namely, by the degradation of the bases to simpler fragments and by the application of the modern tools of ultraviolet, infrared, nuclear magnetic resonance, mass spectrometry and X-ray crystallography, greatly facilitated by modern high speed computing techniques. The final proof of any structure deduced from complex data, however, remains the totally unambiguous synthesis of the material of that structure and comparison of this product with material from natural sources. Herein, we describe such a proof for the structure of capaurimine.

The structure of capaurimine, isolated from the *Corydalis* species, had been assigned to **3** by Manske[1] on the basis of degradative evidence. Recently, however, Kametani[2] found that the characters of compound **3** synthesized through the Mannich reaction of **1** were different from those of the natural product. The structure of capaurimine was then reinvestigated and deduced to be **8**[3]. This deduction was confirmed by synthesis[4] and further X-ray analysis[5] of the *p*-bromobenzoyl derivative of the natural compound. Condensation of the phenolic tetrahydroisoquinoline **4** with formalin under physiological conditions gave preferentially the required base **5** rather than the isomer **6**. Methylation of **5**, followed by debenzylation of the resulting base **7**, afforded (±)-capaurimine **8**. In the course of our work, it was also revealed that capaurimine derivatives take the *cis*-quinolizidine conformation because of a non-bonded interaction between oxygen at the C-1 position and hydrogens at the C-13 position.

The total synthesis of natural products can be accomplished only after consideration of the whole of chemical knowledge and theory. Of course, to achieve the synthesis of a complicated compound, ingenuity is further needed.

Morphine **18**, for example, which is pharmacologically important as an analgesic and possesses five rings and five asymmetric centers, presents a challenge to the organic chemist concerned with natural products. The synthesis of morphine was completed first by Gates and

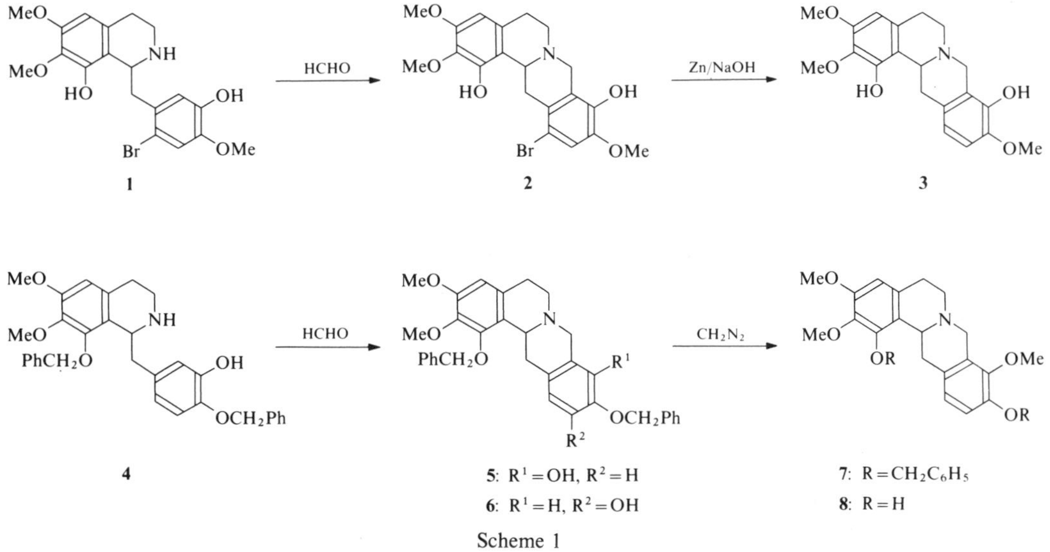

1 —HCHO→ 2 —Zn/NaOH→ 3

4 —HCHO→ 5, 6 —CH_2N_2→ 7, 8

5: $R^1 = OH$, $R^2 = H$

6: $R^1 = H$, $R^2 = OH$

7: $R = CH_2C_6H_5$

8: $R = H$

Scheme 1

[1] *R. Manske*, J. Am. Chem. Soc. *68*, 1800 (1947); *R. Manske*, Can. J. Res. *18B*, 80 (1940); *R. Manske*, Can. J. Res. *20B*, 49 (1942); *T. Kametani, M. Ihara, T. Honda*, Chem. Commun. 1301 (1969); *T. Kametani, M. Ihara, T. Honda*, J. Chem. Soc. C 1060 (1970).

[2] *T. Kametani, K. Fukumoto, H. Yagi, H. Iida, T. Kikuchi*, J. Chem. Soc. C 1178 (1968).

[3] *T. Kametani, M. Ihara, T. Honda*, J. Chem. Soc. C 2342 (1970).

[4] *T. Kametani, T. Honda, M. Ihara*, J. Chem. Soc. C 2396 (1971).

[5] *T. Kametani, M. Ihara, C. Kabuto, Y. Kitahara, H. Shimanouchi, Y. Sasada*, Chem. Commun. 1241 (1970).

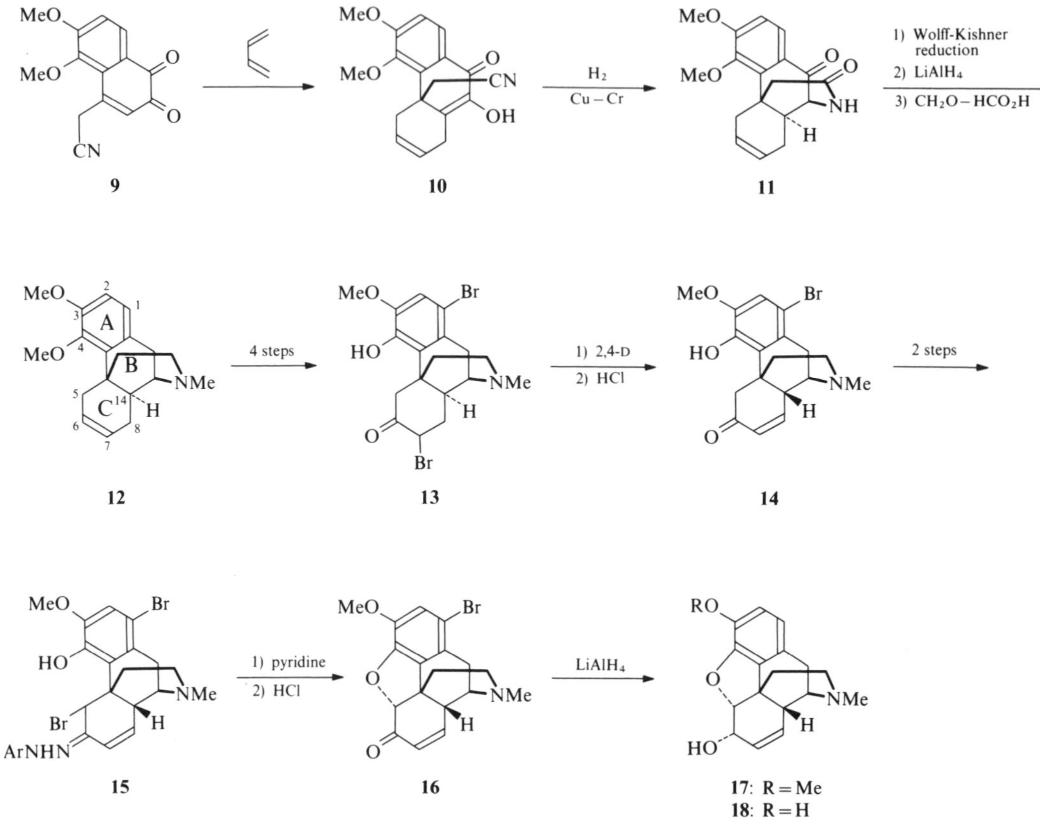

Scheme 2

Tschudi[6] in 1952. The key step in this synthesis was a Diels-Alder addition of butadiene to 4-cyanomethyl-1,2-naphthoquinone **9**. On catalytic hydrogenation, the adduct **10** underwent ring closure to keto-lactam **11**, the carbonyl group of which was removed by Wolff-Kishner reaction. Further reduction of the amide carbonyl group with lithium aluminium hydride, followed by *N*-methylation, gave the base **12**. At this stage the optical resolution was effected but the product was proved to be epimeric at C-14 (rings B/C *trans*) with the natural series. However, this steric problem was resolved, as shown later. The double bond of **12** was hydrated and the ether group at the C-4 position was selectively demethylated. Oxidation, followed by bromination, gave the 6-ketone **13**. In the course of debromination of **13** with 2,4-dinitrophenylhydrazine to produce the α,β-unsaturated ketone structure of codeinone, inversion occurred at the center adjacent to the conjugated system to make the natural orientation at the C-14 position. After splitting with hydrochloric acid, the enone **14** was reduced to the ketone, dibromination of which, followed by treatment with 2,4-dinitrophenylhydrazine, afforded the hydrazone **15**. Then heating with pyridine and splitting with acid gave the bromo-compound **16**. Reductive removal of the bromine atom at the C-1 position gave codeine **17**, which was converted into morphine **18** by cleavage of the ether group.

The approach by Elad and Ginsburg[7] to dihydrocodeinone **22** constituted the second synthesis of morphine. This synthesis involved the building of the ethanamine chain by way of an unusual reaction. Acylation of the aminodiketophenanthrene **19** with acetylglycollyl chloride, followed by closure of the 9,13-bridge, gave the product **21**, which had the desired carbon skeleton of morphine with the correct stereochemistry at C-14 position. Dihydrocodeinone **22**, prepared from **21** in seven steps, was resolved via (+)-tartaric acid salt.

[6] *M. Gates, G. Tshudi*, J. Am. Chem. Soc. *74*, 1109 (1952); *M. Gates, G. Tshudi*, J. Am. Chem. Soc. *78*, 1380 (1956).

[7] *D. Elad, D. Ginsburg*, J. Am. Chem. Soc. *76*, 312 (1954); *D. Elad, D. Ginsburg*, J. Chem. Soc. 3052 (1954).

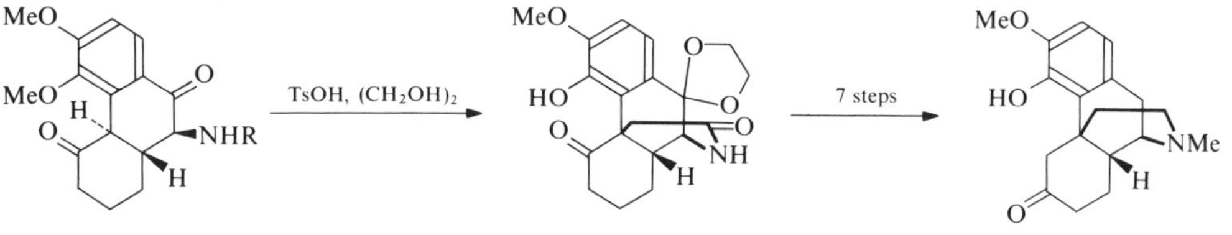

MeO MeO H O O NHR H TsOH, $(CH_2OH)_2$ MeO HO O O O O NH H 7 steps MeO HO NMe H O

19: R=H
20: $R=COCH_2OAc$

21

22

Scheme 3

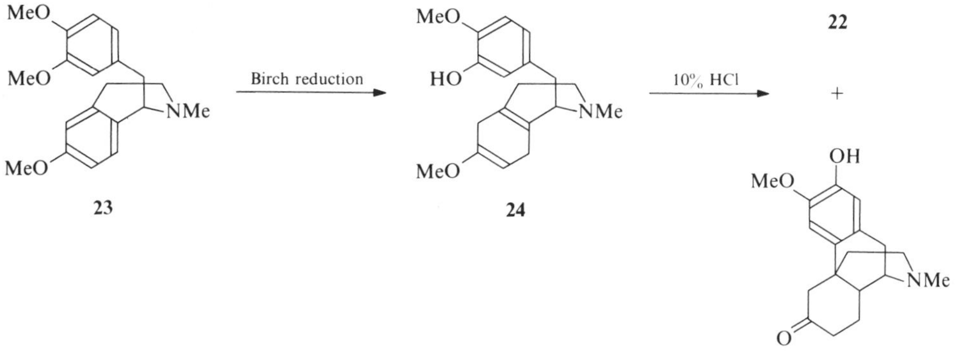

MeO MeO NMe MeO Birch reduction MeO HO NMe MeO 10% HCl 22 + OH MeO NMe O

23

24

25

Scheme 4

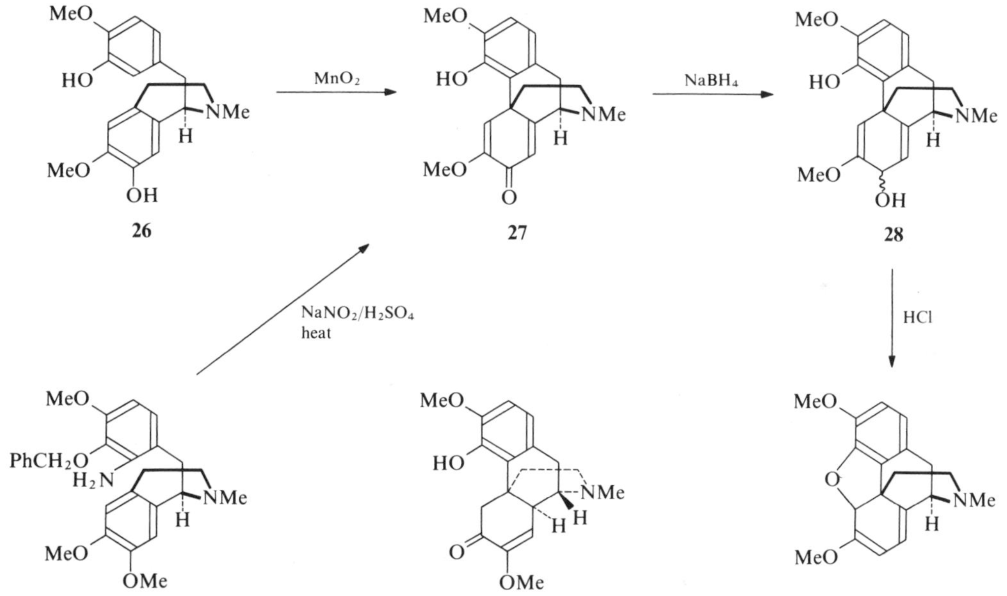

MeO HO NMe H MeO OH MnO_2 MeO HO NMe H MeO O $NaBH_4$ MeO HO NMe H MeO OH

26

27

28

$NaNO_2/H_2SO_4$ heat

HCl

MeO $PhCH_2O$ H_2N NMe H MeO OMe MeO HO NMe H H O OMe MeO O NMe H MeO

30

31

29

Scheme 5

The first synthesis of the skeleton of the morphine alkaloids was achieved by Grewe[8], whose method was in fact a version of the biogenetic approach and involved an acid catalysed cyclization of a benzylhexahydroisoquinoline to a morphine derivative. Although this method attracted the attention of many workers, the final goal of total synthesis along the lines of the opium alkaloid was achieved by Morrison[9] as follows. Birch reduction of the tetrahydroisoquinoline **23** gave the phenolic hexahydroisoquinoline **24**, the treatment of which with hydrochloric acid afforded dihydrocodeinone **22** together with the isomer **25**.

[8] *R. Grewe, A. Mondon, E. Nolte*, Ann. *564*, 161 (1949).

[9] *G. Morrison, R. Waite, J. Shavel*, Tetrahedron Lett. 4055 (1967).

Morphine alkaloids are biosynthesized through phenolic oxidative coupling of reticuline **26** as discussed in Chapter 4. Barton[10] synthesized radioactive salutaridine **27** biogenetically by phenol oxidation of radioactive reticuline **26**. The dienone **27** was reduced with sodium borohydride to allylic alcohols **28**, which were converted under acidic treatment into thebaine **29**. Since thebaine had been chemically correlated with morphine via codeinone and codeine, the total synthesis of morphine was accomplished. Using a modified Pschorr reaction, Kametani[11] recently synthesized morphine through salutaridine in better yield than by way of oxidation. The diazotization of (−)-aminoisoquinoline **30**, prepared by a resolution of the racemate, followed by the thermal decomposition of the resulting diazonium salt, gave salutaridine **27**, which was converted into thebaine **29** according to Barton's method. The antipode of thebaine was also synthesized from the (+)-aminoisoquinoline. Since thebaine had already been converted into the antipode of sinomenine **31**, this synthesis also constituted the synthesis of sinomenine.

Thus, the syntheses of natural products along the biogenetic route are very attractive both from the point of view of the development of the synthetic method and of the potential in the elucidation of the biogenetic route.

[10] *D. Barton, G. Kirby, W. Steglich, G. Thomas*, Proc. Chem. Soc. 203 (1963); *D. Barton, D. Bhakuni, R. James, G. Kirby*, J. Chem. Soc. C 128 (1967).

[11] *T. Kametani, M. Ihara, K. Fukumoto, H. Yagi*, J. Chem. Soc. C 2030 (1969).

Recommended Books and Reviews

R. Manske, H. Holmes, The Alkaloids, Vol. I–XII, Academic Press, New York 1950ff.

H. Boit, Ergebnisse der Alkaloids-Chemie bis 1960 unter besonderer Berücksichtung der Fortschritte seit 1950, Academie-Verlag, Berlin 1961.

E. Rodd, Chemistry of Carbon Compounds, Vol. IVC, Elsevier Publ. Co., Amsterdam 1960.

S. Pelletier, Chemistry of the Alkaloids, Van Norstrand Reinhold Co., New York 1970.

T. Kametani, The Chemistry of Isoquinoline Alkaloids, Hirokawa Publ. Co., Tokyo and Elsevier Publ. Co., Amsterdam 1968.

K. Bently, The Isoquinoline Alkaloids, Pergamon Press, London 1965.

D. Ginsburg, The Opium Alkaloids, Interscience Publ. Co., New York 1962.

S. Pfeifer, L. Kuhn, Proaporphine, Pharmazie *20*, 659 (1965).

K. Stuart, M. Cava, The Proaporphine Alkaloids, Chem. Rev. *68*, 321 (1968).

W. Taylor, Indole Alkaloids, Pergamon Press, London 1966.

U. von Euler, Tobacco Alkaloids and Related Compounds, Pergamon Press, London 1965.

S. Pelletier, The Chemistry of the C-20-Diterpene Alkaloids, Quart. Rev. *21*, 525 (1967).

N. Anand, J. Bindra, S. Ranganathan, Art in Organic Synthesis, Holden-Day, Inc., New York 1970.

R. Manske, The Alkaloids, Academic Press, New York, Vol. 13, 1971 and Vol. 14, 1973.

J. Saxton, The Alkaloids, The Chemical Society, London, Vol. 1, 1971, Vol. 2, 1972, Vol. 3, 1973, Vol. 4, 1974, Vol. 5, 1975; Vol. 6, 1976.

D. Hey, K. Wiesner, MTP International Review of Science, Vol. 9, Alkaloids, Butterworths, London 1973.

M. Shamma, The Isoquinoline Alkaloids, Academic Press, New York 1972.

T. Kametani, The Chemistry of Isoquinoline Alkaloids, Vol. 2, The Sendai Institute of Heterocyclic Chemistry, Sendai 1974.

J. Glasby, Encyclopedia of the Alkaloids, Plenum Press, New York 1975.

K. Stuart, Chem. Rev. *71*, 47 (1971).

T. Kametani, K. Fukumoto, Synthesis 657 (1972).

A. Schultz, Chem. Rev. *73*, 385 (1973).

T. Kametani, K. Fukumoto, F. Satoh, Bioorg. Chem. *3*, 430 (1974).

S. Tobinaga, Bioorg. Chem. *4*, 110 (1975).

T. Kametani, Heterocycles *3*, 775, 901, 1199 (1975); *4*, 675, 909, 1065, 1211, 1319, 1477 (1976).

M. Hesse, Indolalkaloide in Tabellen, Springer-Verlag, Berlin 1964 and 1968.

W. Döpke, Ergebnisse der Alkaloid-Chemie 1960–1968, Akademie-Verlag, Berlin 1976.

S. Coffey, Rodd's Chemistry of Carbon Compounds, Vol. IV D, Elsevier Publ. Co., Amsterdam 1977.

4 Biogenetic Route

The various methods of synthesizing alkaloids by way of biogenesis have been established mainly by the extensive work of British groups. Battersby found that tyrosine-2-^{14}C **1** is incorporated into papaverine-1,3-^{14}C **6** in *Papaver somniferum*[1]. This suggests that the skeleton of benzylisoquinoline is formed by the condensation of an amine **3** and an aldehyde **4** derived from aromatic amino acids as C_6–C_2–N and C_6–C_2 units.

Reticuline **7**, biosynthesized by the methylation of norlaudanosoline **5**[2], is an important precursor of berberine, aporphine and morphine type alkaloids. A large number of very widely distributed alkaloids have the benzylisoqui-

[1] *A. Battersby, B. Harper*, J. Chem. Soc. 3526 (1962).

[2] *A. Battersby, G. Evans*, Tetrahedron Lett. 1275 (1965).

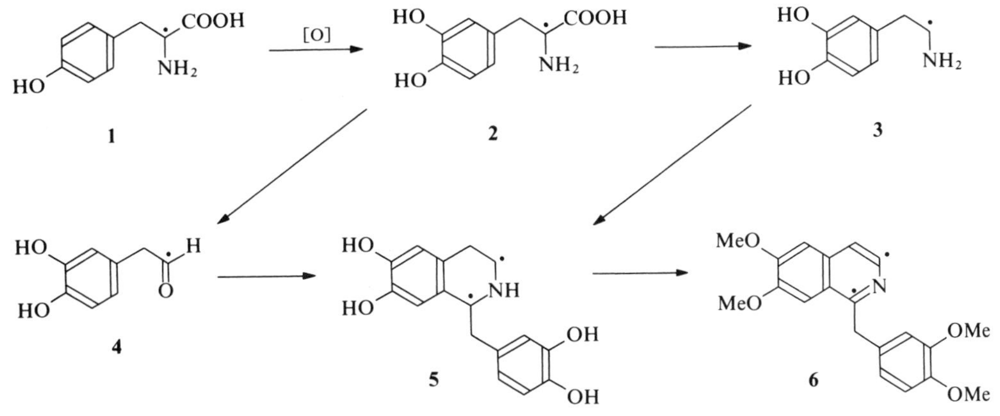

Scheme 1

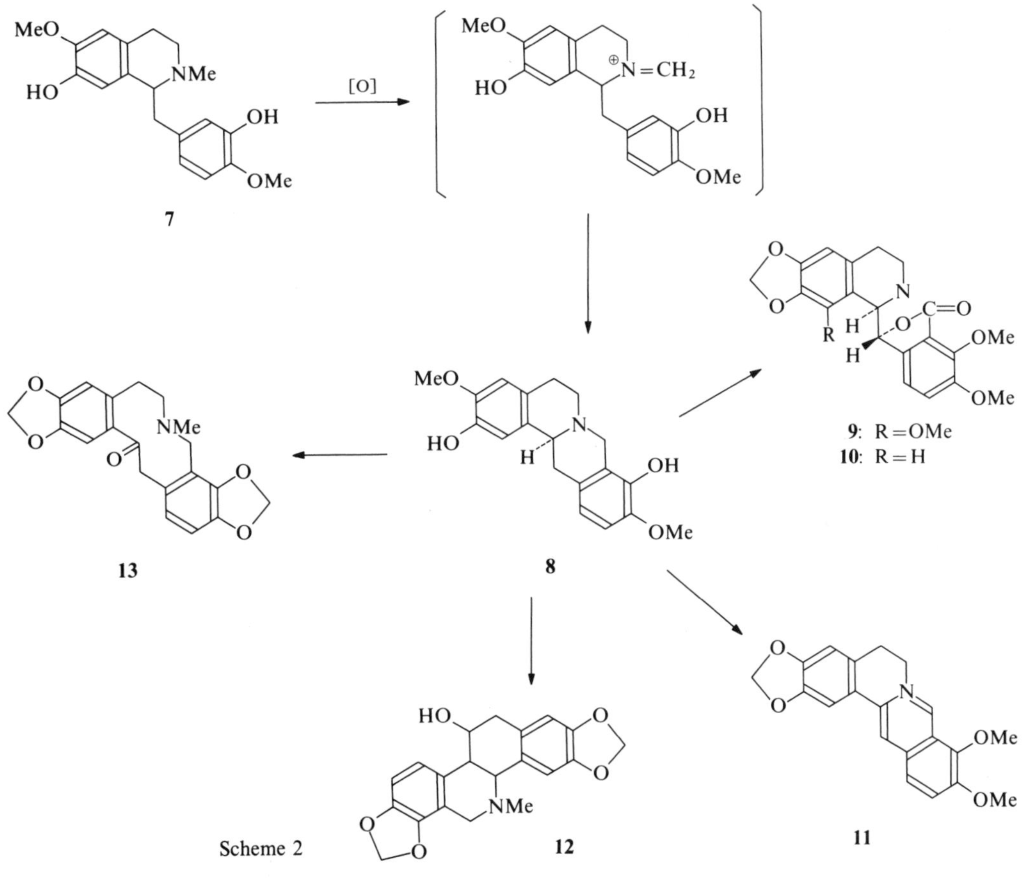

9: R=OMe
10: R=H

Scheme 2

noline structure with one additional carbon atom. Four skeletons of this type, *i.e.*, the protoberberine, the protopine, the phthalide-isoquinoline and the benzophenanthridine families, are known. It was proved by tracer works[3] that reticuline **7** is a common precursor of these alkaloids. When tyrosine-2-^{14}C was fed to *Papaver somniferum*[4], *Hydrastis canadensis*[5], and *Chelidonium majus*[6], two ^{14}C were incorporated into the expected positions: (−)-narcotine **9**, (−)-hydrastine **10**, berberine **11** and chelidonine

[3] *D. Barton, R. Hesse, G. Kirby*, Proc. Chem. Soc. 267 (1963); *D. Barton, R. Hesse, G. Kirby*, J. Chem. Soc. 6379 (1965); *A. Battersby, M. Hirst*, Tetrahedron Lett. 669 (1965); *A. Battersby, R. Francis, E. Ruveda, J. Stauton*, Chem. Commun. 89 (1965).

[4] *A. Battersby, D. McCaldin*, Proc. Chem. Soc. 365 (1962).

[5] *J. Gear, I. Spenser*, Can. J. Chem. *41*, 783 (1963).

[6] *E. Leete*, J. Am. Chem. Soc. *85*, 473 (1965).

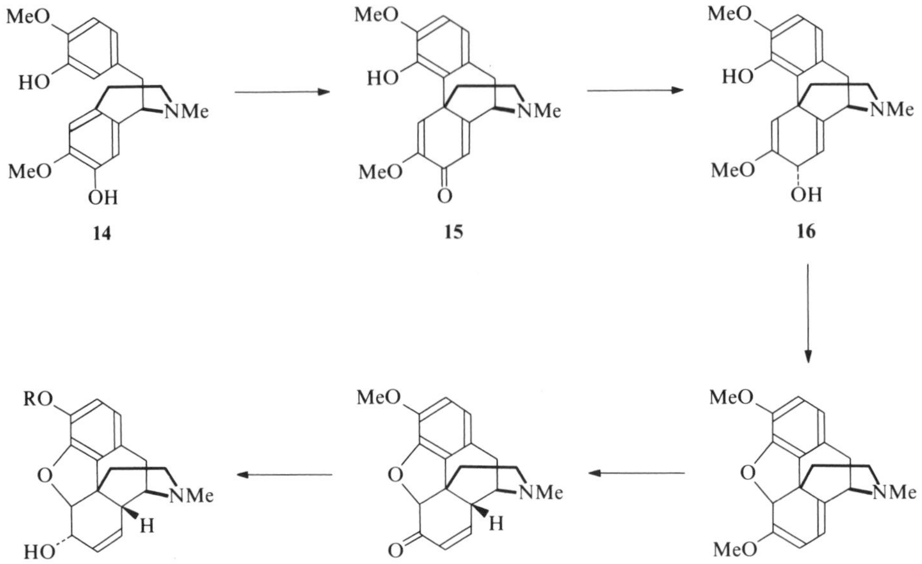

14 15 16

19: R=Me
20: R=H

18 17

Scheme 3

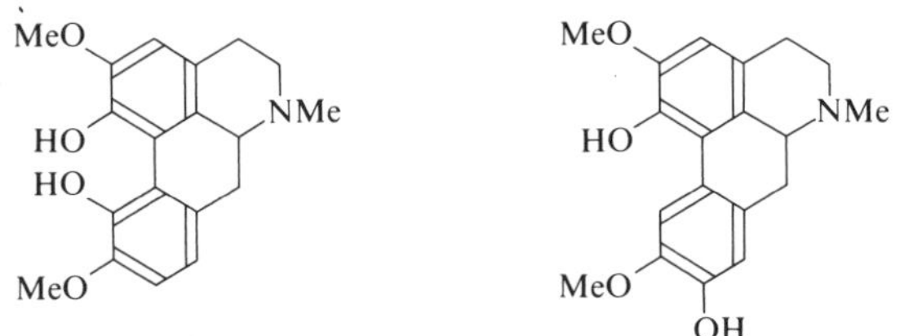

21 22

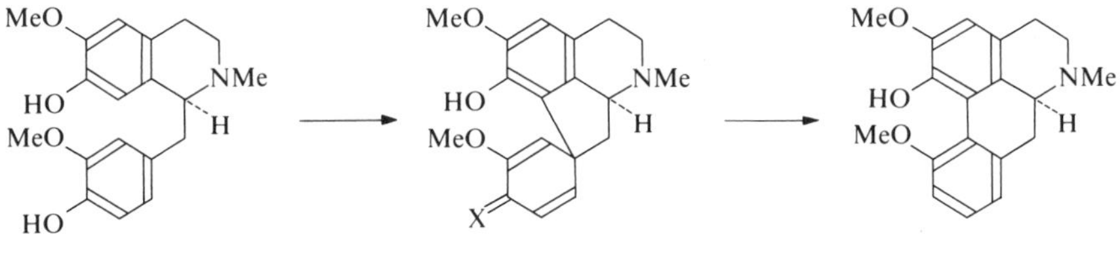

23

24: X=O
25: X=H, OH

26

Scheme 4

12. It was further shown that (−)-scoulerine **8** is an intermediate from reticuline **7** to protopine **13**, (−)-narcotine **9** and chelidonine **12**[7].

(−)-Reticuline **14** was also incorporated into thebaine **17** and other opium alkaloids in *P. somniferum*[8]. The direction of the pathway [thebaine **17**→codeine **19**→morphine **20**] was proved on the basis of $^{14}CO_2$ feeding experiments. The CO_2 label appeared consecutively in these compounds[9]. Recently, the existence of codeinone **18** as an intermediate between thebaine **17** and codeine **19** was proved[10]. Barton and his co-workers[11] further established that salutaridine

[7] *A. Battersby, R. Francis, M. Hirst, R. Southgate, J. Staunton*, Chem. Commun. 602 (1967); *A. Battersby, M. Hirst, D. McCaldin, R. Southgate, J. Staunton*, J. Chem. Soc. C 2163 (1968).

[8] *A. Battersby, D. Foulkes, R. Binks*, J. Chem. Soc. 3323 (1965).

[9] *H. Rapoport, F. Stermitz, D. Baker*, J. Am. Chem. Soc. *82*, 2765 (1960).

[10] *A. Battersby, E. Brochmann-Hanssen, J. Martin*, Chem. Commun. 483 (1967); *G. Blaschke, H. Parker, H. Rapoport*, J. Am. Chem. Soc. *89*, 1540 (1967).

[11] *D. Barton, G. Kirby, W. Steglich, G. Thomas, A. Battersby, T. Dobson, H. Ramuz*, J. Chem. Soc. 2423 (1965).

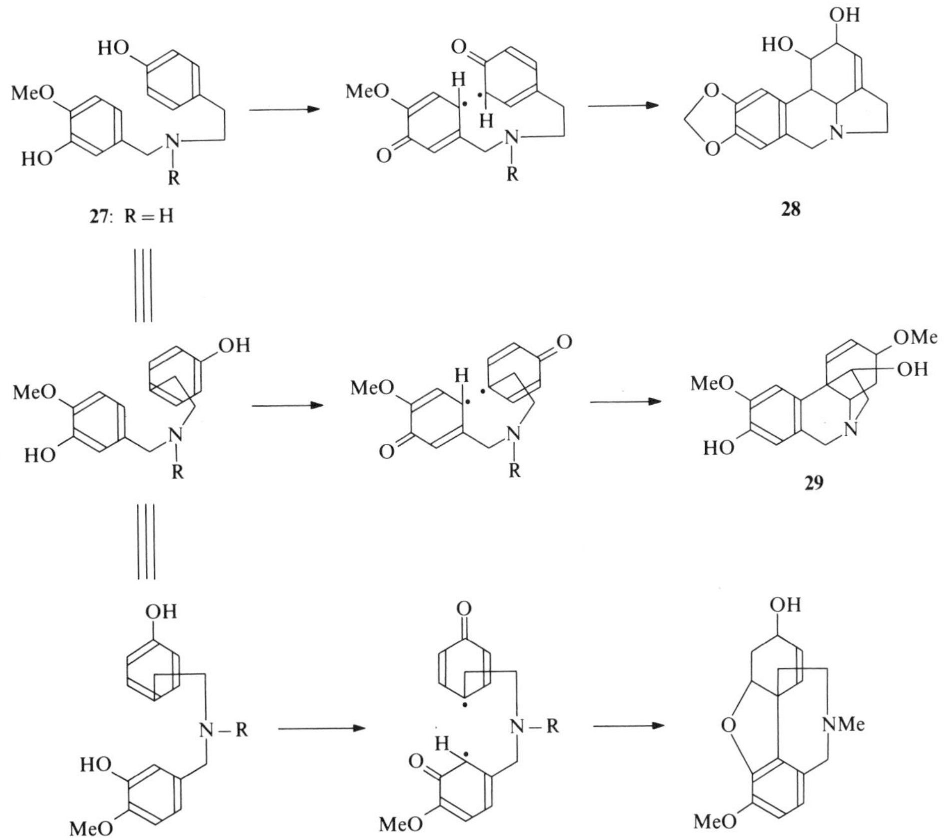

Scheme 5

15 is formed by phenolic oxidative coupling of (−)-reticuline **14** also found in opium and that salutaridinol–I **16** is a precursor of thebaine **17**. The biogenetic pathway to morphine from reticuline was confirmed as below.

1,2,10,11- and 1,2,9,10-tetrasubstituted aporphine alkaloids, for example corytuberine **21** and isoboldine **22**, were formed by direct oxidative coupling of reticuline[12]. On the other hand, the 1,2,11-trisubstituted aporphine, isothebaine **26**, was biosynthesized by oxidative coupling of (+)-orientaline **23**, followed by reduction of the resulting proaporphine **24** and successive dienol-benzene rearrangement of **25**[13].

Barton[14] proposed that phenol oxidation of *O*-methylnorbelladine **27** formed from C(6)–C(2) and C(6)–C(1) units derivable from aromatic amino acids gives three types of *Amaryllidaceae* alkaloids; the lycorine **28**, the haemanthamine **29** and the galanthamine **30** families. This biogenesis was proved by many tracer experiments using *O*-methylnorbelladine **27** or its biogenetic equivalents[15].

Several hypotheses for the biogenesis of colchicine **33**, which has interesting physiological effects, have been proposed. It was proved that phenylalanine and tyrosine take part in the formation of the skeleton. Leete[16] and Battersby[17] showed by many tracer studies that ring A and the carbons at C-5,6 and 7 positions in colchicine are formed from phenylalanine **31** via cinnamic acid **32** but that tyrosine does not

[12] *R. Robinson*, The Structure Relations of Natural Products, Clarendon Press, Oxford 1955.

[13] *A. Battersby, R. Brown, J. Clements, G. Iverach*, Chem. Commun. 2030 (1965); *A. Battersby, T. Brown*, Chem. Commun. 171 (1966); *A. Battersby, T. Brockson, R. Ramage*, Chem. Commun. 464 (1969).

[14] *D. Barton, T. Cohen*, Festschrift Arthur Stoll, p. 117, Brikhauser, Basel 1957.

[15] *D. Barton, G. Kirby*, Proc. Chem. Soc. 392 (1960); *D. Barton, G. Kirby*, J. Chem. Soc. 806 (1962); *D. Barton, G. Kirby, J. Taylor, G. Thomas*, Proc. Chem. Soc. 254 (1961); *D. Barton*, Proc. Chem. Soc. 293 (1963); *A. Battersby*, Proc. Chem. Soc. 189 (1963).

[16] *E. Leete, P. Nemeth*, J. Am. Chem. Soc. *82*, 6055 (1960); *E. Leete*, J. Am. Chem. Soc. *85*, 3666 (1963).

[17] *A. Battersby, R. Binks, J. Reynolds, D. Yeowell*, J. Chem. Soc. 4257 (1964).

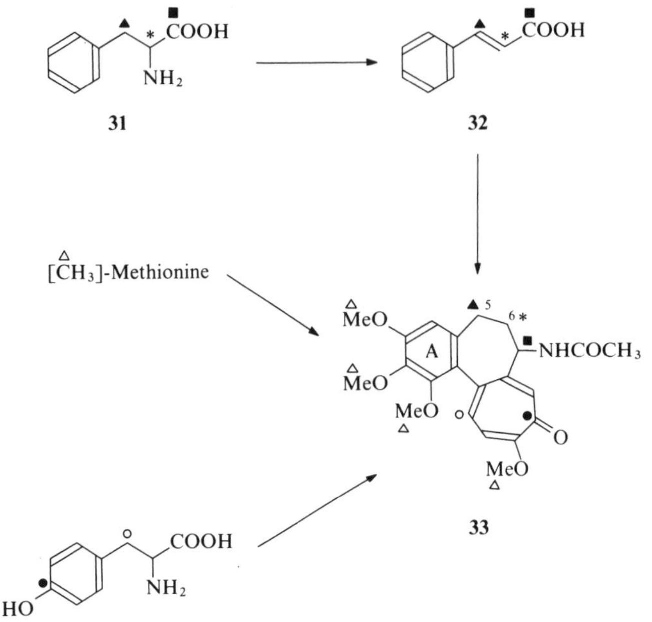

31 → 32

[CH₃]-Methionine

33

Scheme 6

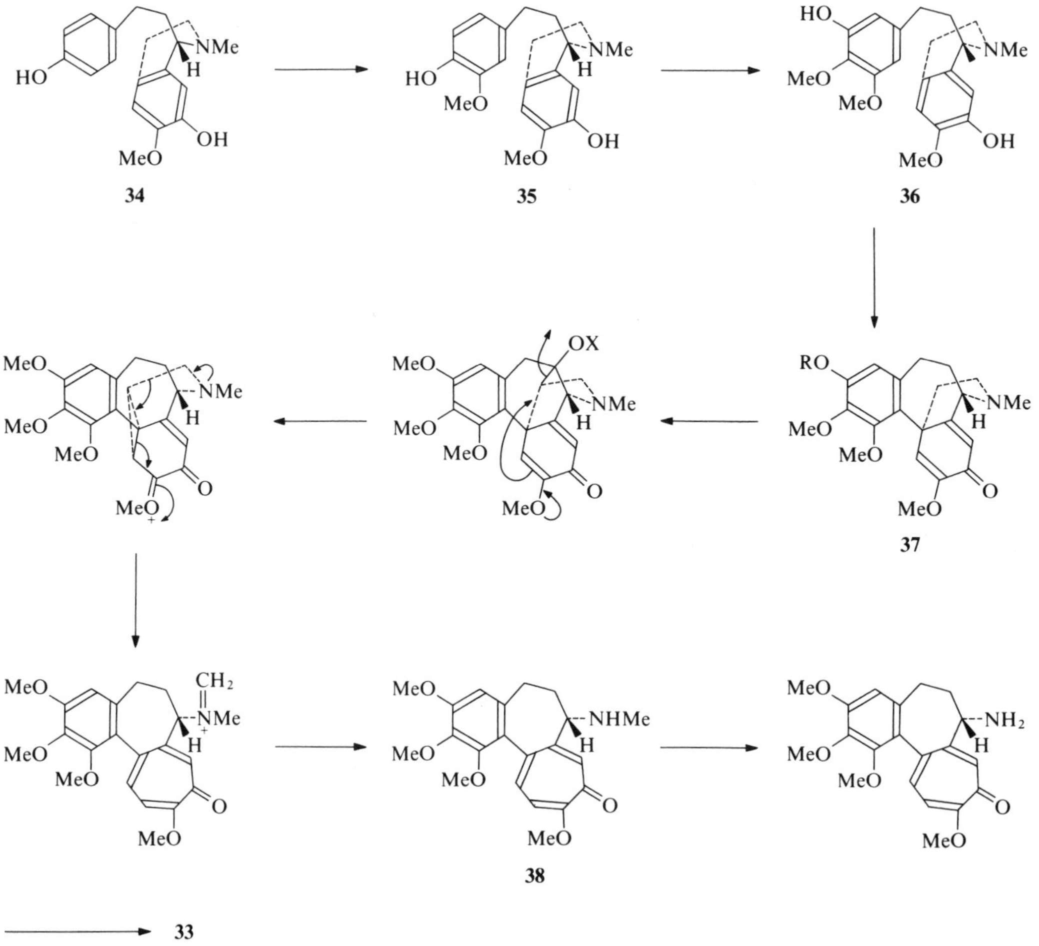

34 → 35 → 36 → 37

38

→ 33

Scheme 7

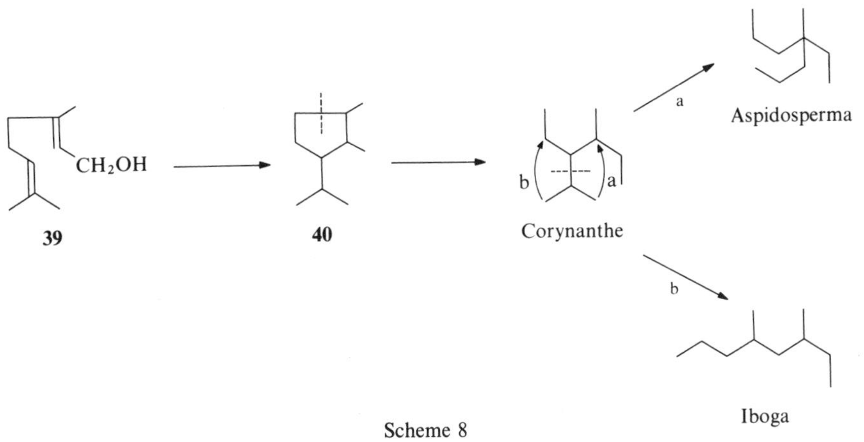

39 40 Corynanthe

a Aspidosperma

b Iboga

CH_2OH

Scheme 8

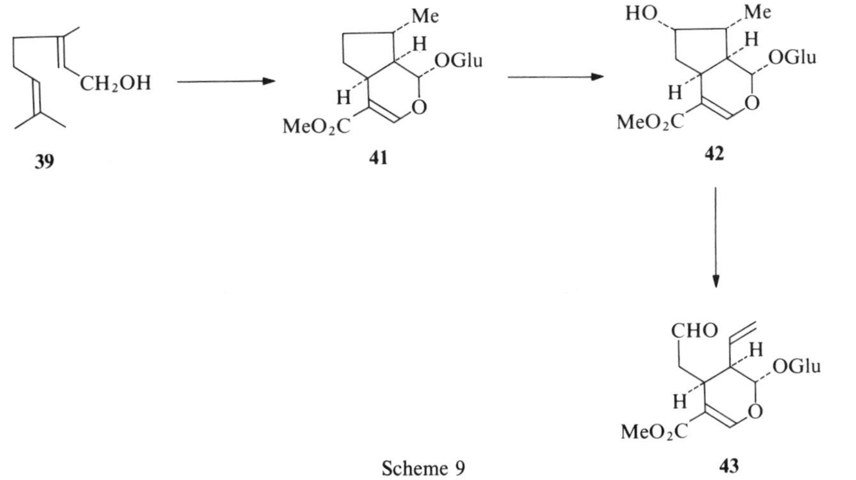

39 41 42 43

CH_2OH Me H OGlu MeO_2C HO CHO

Scheme 9

participate in this formation. Moreover, it was proved that the tropolone part of colchicine is derived from tyrosine[18].

Battersby proposed the biosynthetic pathway to colchicine as below. It was verified on the basis of tracer experiments[19] that the 1-phenethyl-tetrahydroisoquinolines, **34**–**36**, *O*-methyl-androcymbine (**37**, R = Me), and demecolcine **38**, which is also present in the *Colchicum* plant, are precursors of colchicine. But evidence for the mechanism of ring-expansion (**37**→**38**) has not yet been reported.

An aminoethyl residue attached to the β-position of the indole ring in most of these alkaloids suggests that tryptophane or tryptamine are precursors. This hypothesis was proved by many tracer evidence in the case of ajmaline[20], serpentine[20], chimonanthine[21], ibogaine[22], vindoline[22,23], lysergic acid[24], evodiamine[25] and rutecarpine[25]. Many indole alkaloids consist of C_9–C_{10} units in addition to the tryptamine unit. They are classified into three types: the Corynanthe, the Aspidosperma and the Iboga families. Thomas[26] and Wenkert[27] suggested a terpenoid origin for this unit.

This hypothesis was supported by many tracer experiments and recently the terpenoid cor-

[18] *A. Battersby, R. Herbert,* Proc. Chem. Soc. 260 (1964); *E. Leete,* Tetrahedron Lett. 333 (1965).

[19] *A. Battersby, R. Herber, E. McDonald, R. Ramage, J. Clements,* Chem. Commun. 603 (1966); *A. Barker, A. Battersby, E. McDonald, R. Ramage, J. Clements,* Chem. Commun. 390 (1967).

[20] *E. Leete,* Tetrahedron *14,* 35 (1961); *E. Leete,* J. Am. Chem. Soc. *82,* 6338 (1960).

[21] *E. Schutte, B. Maier,* Arch. Pharm. *298,* 459 (1965).

[22] *E. Leete, M. Yamazaki,* Tetrahedron Lett. 1499 (1964).

[23] *E. Leete, A. Amend, I. Kompis,* J. Am. Chem. Soc. *87,* 4168 (1965).

[24] *K. Mothes, F. Weygand, D. Groger, H. Grisebach,* Z. Naturforschung *14b,* 41 (1958).

[25] *M. Yamazaki, A. Ikuta,* Tetrahedron Lett. 3221 (1966).

[26] *R. Thomas,* Tetrahedron Lett. 544 (1961).

[27] *E. Wenkert,* J. Am. Chem. Soc. *84,* 98 (1962).

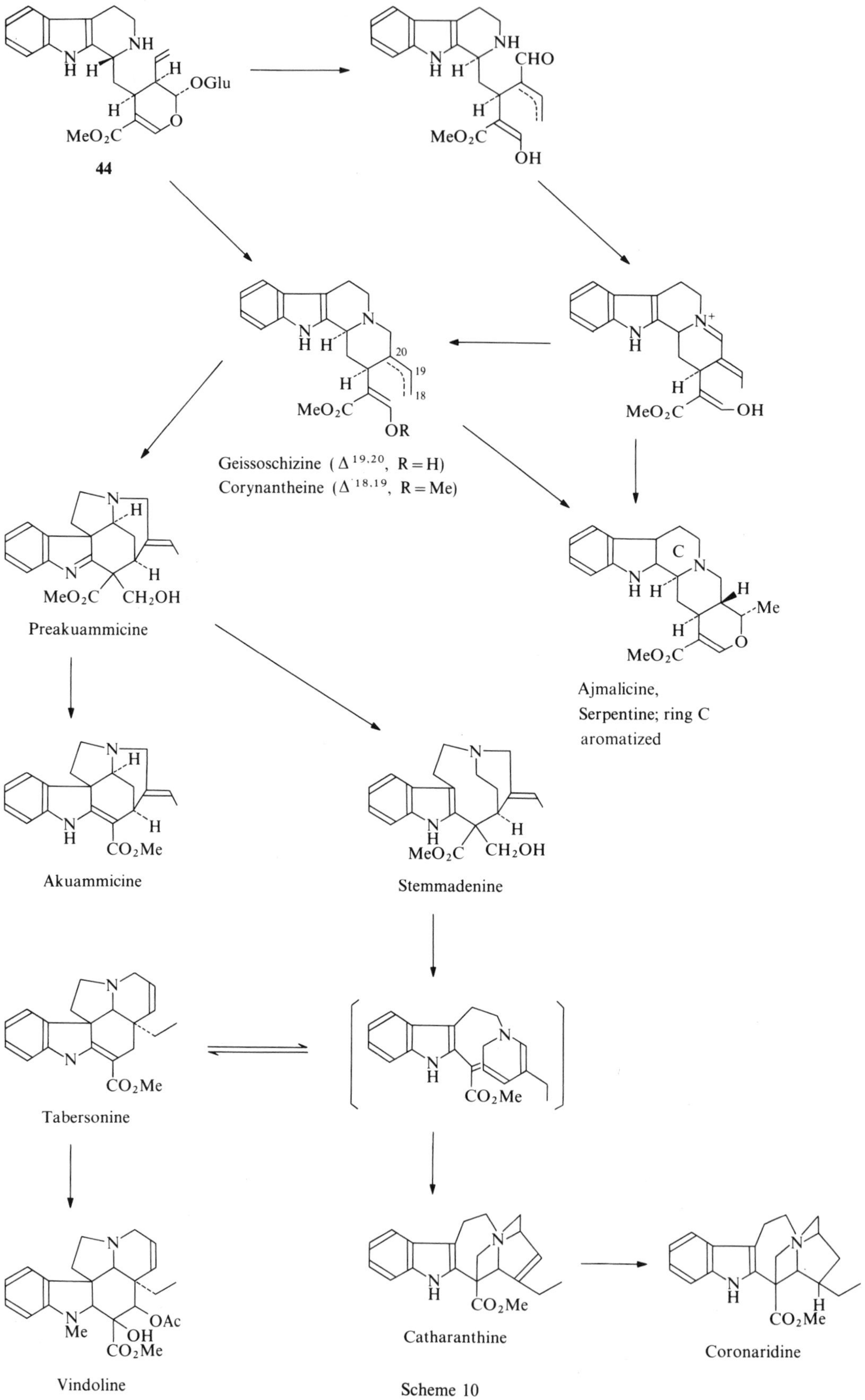

44

Geissoschizine ($\Delta^{19,20}$, R = H)
Corynantheine ($\Delta^{18,19}$, R = Me)

Preakuammicine

Ajmalicine,
Serpentine; ring C
aromatized

Akuammicine

Stemmadenine

Tabersonine

Vindoline

Catharanthine

Coronaridine

Scheme 10

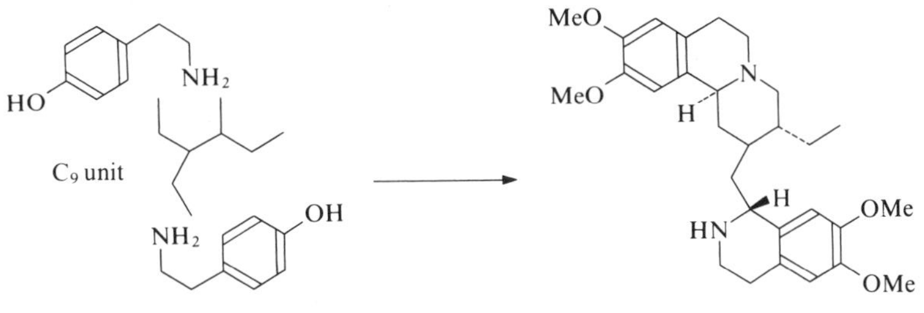

45

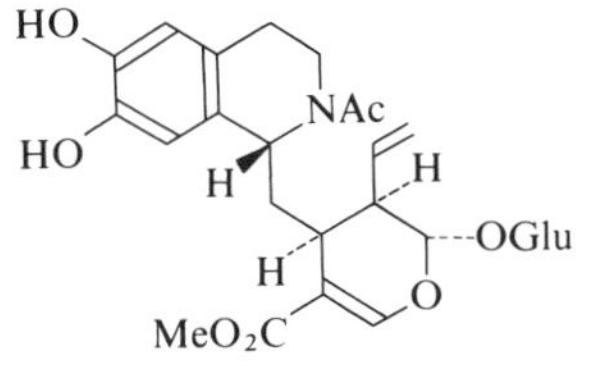

46

Scheme 11

responding to **40** and the opened compound were isolated from nature as loganine **42**[28] and secologanine **43**[29]. It was established that loganine **42**, derived from geraniol **39** via deoxyloganine **41**[30], is a key precursor of secologanine **43**. This aldehyde **43** was converted *in vivo* into all three major types of indole alkaloids by way of vincoside **44**, isolated together with isovincoside and *N*-acetylvincoside[31].

Scott[32] suggested the order [Corynanthe → Aspidosperma → Iboga] for the biosynthetic pathway of three major families. The studies of the sequential appearance of various alkaloid types in germinating *Vinca rosea* and tracer experiments in the same plant revealed the biological conversion among the several alkaloids as follows[33].

Many biogenetic theories of *Ipecauanha* alkaloids, represented by emetine **45**, have been proposed. It was recently established [34] that two aromatic units are derived from tyrosine and that the C_9 unit of the middle part is derived from the same terpenoids as the indole alkaloids, *i.e.*, geraniol and loganine. The compound corresponding to vincoside of the indole alkaloids group, ipecoside **46**, was also found in the *Ipecacuanha* plant[30, 35].

5 Pharmacological Activities

Many alkaloids have important and peculiar pharmacological activities and are utilized as medicines. Medicinal applications of representative alkaloids are shown in Table 1.

---

28 *A. Battersby, R. Kapil, J. Martin, L. Mo*, Chem. Commun. 133 (1968); *A. Battersby, R. Brown, R. Kapil, J. Martin, A. Plunkelt*, Chem. Commun. 890 (1966); *P. Loew, D. Arigoni*, Chem. Commun. 137 (1968).

29 *A. Battersby*, Pure Appl. Chem. *14*, 117 (1967); *A. Battersby, A. Burnett, P. Parsons*, J. Chem. Soc. 1187 (1969).

30 *A. Battersby, A. Burnett, P. Parsons*, Chem. Commun. 826 (1970).

31 *A. Battersby, A. Burnet, P. Parsons*, Chem. Commun. 1282 (1968); *A. Battersby, A. Burnett, E. Hall, P. Parsons*, Chem. Commun. 1582 (1968); *A. Battersby, A. Burnett, R. Parons*, J. Chem. Soc. C 1193 (1969).

32 *A. Qureshi, A. I. Scott*, Chem. Commun. 945, 947 (1968).

33 *A. Qureshi, A. I. Scott*, Chem. Commun. 948 (1968); *A. I. Scott, R. Cherry, A. Qureshi*, J. Am. Chem. Soc. *91*, 4932 (1969); *J. Kutney, W. Cretney, J. Hadifield, E. Hall, V. Nelson, D. Wigfield*, J. Am. Chem. Soc. *90*, 3566 (1968); *A. Battersby, E. Hall*, Chem. Commun. 793 (1969).

34 *A. Battersby, R. Binks, W. Lawrie, G. Parry, B. Webster*, J. Chem. Soc. 7459 (1965); *A. Battersby, B. Gregory*, Chem. Commun. 134 (1969).

35 *A. Battersby, B. Gregory, H. Spencer, J. Turner*, Chem. Commun. 219 (1967).

Recommended Books and Reviews

R. Robinson, The Structural Relations of Natural Products, Clarendon Press, Oxford 1955.

D. H. R. Barton, T. Cohen, Festschrift Arthur Stoll, Birkhauser, Basel 1957.

P. Bermfeld, Biosynthesis of Natural Compounds, Pergamon Press, London 1964.

T. Robinson, The Biochemistry of Alkaloids, Springer-Verlag, Berlin, Heidelberg, New York 1968.

A. Battersby, Alkaloid Biosynthesis, Quart. Rev. *15*, 259 (1961).

T. Geissman, Biosynthesis, The Chemical Society, London, Vol. 1, 1972, Vol. 2, 1973, Vol. 3, 1975.

Table 1.

Compound	Structure	Activity
Protoalkaloids		
Muscarine (**1**)	(**1**)	Parasympathomimetic
Ephedrine (**2**)	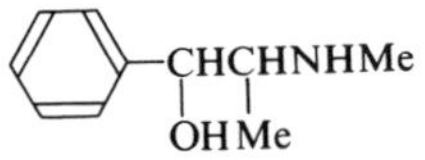 (**2**)	Sympathomimetic, vasoconstrictor, anti-allergic, mydriatic, antidote for morphine and barbiturate
Mescaline (**3**)	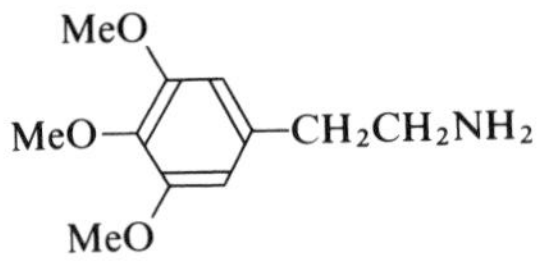 (**3**)	Psychotomimetic, hallucinogen
Narceine (**4**)	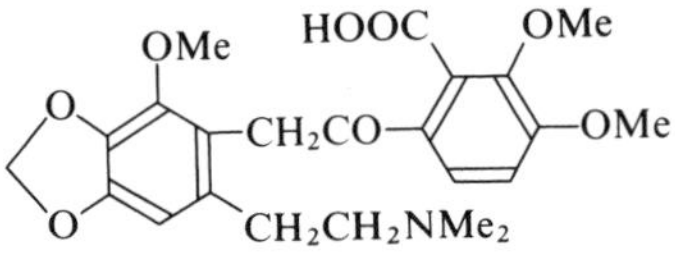 (**4**)	Narcotic, analgesic, antitussive
Colchicine group		
Colchicine (**5**) ($R^1 = Me$, $R^2 = Ac$)	(**5**)	Uricosuric, antipyretic, acute gout
Colchiceine (**5**) ($R^1 = H$, $R^2 = Ac$)		Uricosuric, antipyretic, acute gout
Demecolcine (**5**) ($R^1 = R^2 = Me$)		Antineoplastic agent, antipyretic
Tropan Alkaloids		
Cocaine (**6**) ($R^1 = COOMe$, $R^2 = Ph$)	(**6**)	Surface anesthetic, ointment
Tropacocaine (**6**) ($R^1 = H$, $R^2 = Ph$)		Local and spinal anesthetic
Atropine (Hyoscyamine) (**6**) ($R^1 = H$, $R^2 = CH(Ph)CH_2OH$)		Anticholinergic, mydriatic, respiratory stimulant
Apoatropine (**6**) ($R^1 = H$, $R^2 = C(Ph){=}CH_2$)		Antispasmodic
Scopolamine (**7**)	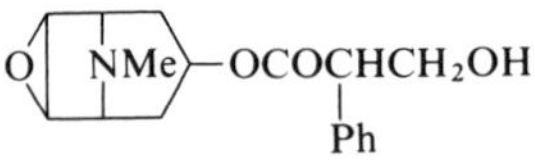 (**7**)	Anticholinergic, CNS depressant, mydriatic, antiparkinson agent, sedative
Dioscorine (**8**)	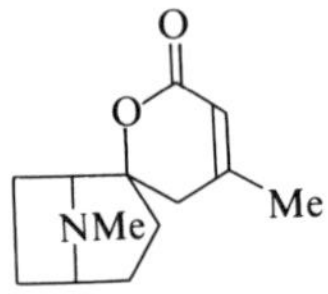 (**8**)	Analeptic

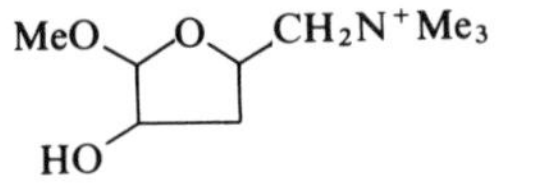

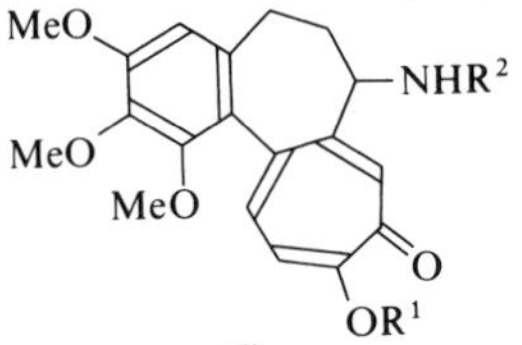

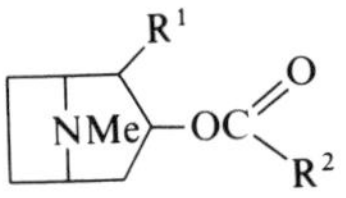

Table 1. (continued)

Alkaloid	Structure	Activity
Piperidine and Pyridine Alkaloids		
Isopelletierine (**9**) ($R^1=R^2=H$, $R^3=CH_2COCH_3$)		Anthelmintic, teniacide
Lobeline (**9**) ($R^1=Me$, $R^2=CH_2CH(OH)Ph$, $R^3=CH_2COPh$)	(**9**)	Respiratory stimulant, ruminatoric
Lobelanine (**9**) ($R^1=Me$, $R^2=R^3=CH_2COPh$)		Respiratory stimulant, ruminatoric
Lobelanidine (**9**) ($R^1=Me$, $R^2=R^3=CH_2CH(OH)Ph$)		Respiratory stimulant, ruminatoric
Arecoline (**10**) ($R=Me$)	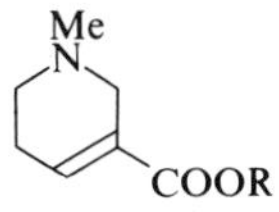 (**10**)	Parasympathomimetic, miotic, anthelmintic
Arecaidine (**10**) ($R=H$)		Parasympathomimetic, miotic, anthelmintic
Nicotine (**11**)	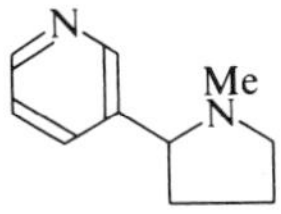 (**11**)	Ganglionic depressant, depolarizing drug, insecticide
Securinine (**12**)	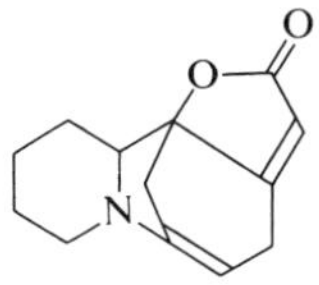 (**12**)	Neurasthenic states, cardiac insufficiency
Quinolizidine Alkaloids		
Sparteine (**13**)	(**13**)	Antiarrhythmic
Anagyrine (**14**)	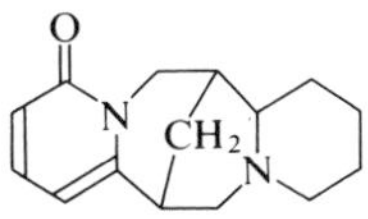 (**14**)	Edema
Purine Alkaloids		
Caffeine (**15**) ($R^1=R^2=R^3=Me$)	(**15**)	CNS, respiratory, cardiac stimulant
Theobromine (**15**) ($R^1=R^3=Me$, $R^2=H$)		Diuretic, myocardial stimulant, vasodilator, smooth muscle relaxant
Theophylline (**15**) ($R^1=R^2=Me$, $R^3=H$)		Diuretic, myocardial stimulant, smooth muscle relaxant
Imidazole Alkaloid		
Pilocarpine (**16**)	(**16**)	Parasympathomimetic, topical miotic

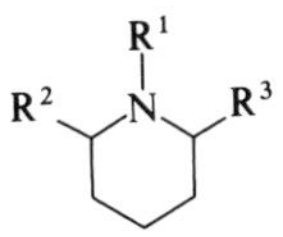

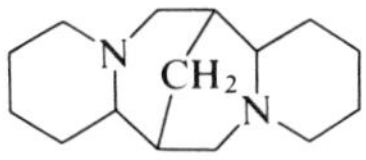

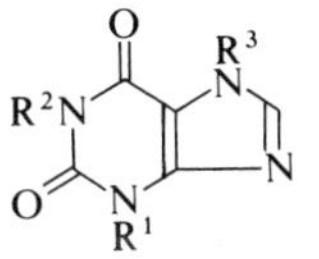

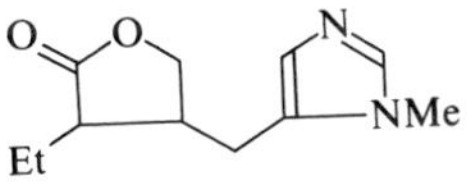

Table 1. (continued)

Compound	Structure	Activity
Isoquinoline Alkaloids		
Salsoline (**17**) ($R^1=R^3=R^5=H$, $R^2=R^4=Me$)	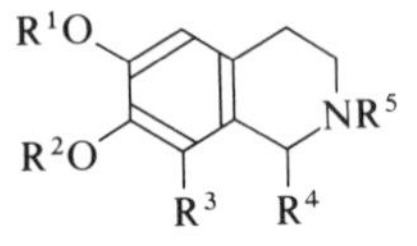 (**17**)	Antihypertensive
Cotarnine (**17**) ($R^1+R^2=CH_2$, $R^3=OMe$, $R^4=OH$, $R^5=Me$)		Hemostatic
Papaverine (**18**)	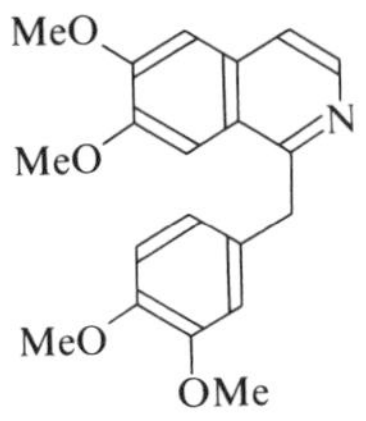 (**18**)	Smooth muscle relaxant, antispasmodic
Noscapine (1-Narcotine) (**19**)	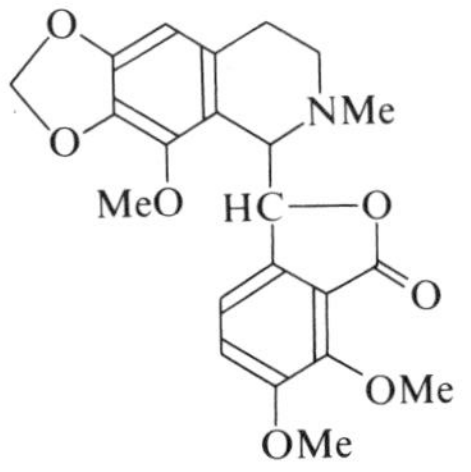 (**19**)	Antitussive
Protoberberine Alkaloids		
Berberine (**20**) ($R^1+R^2=CH_2$)	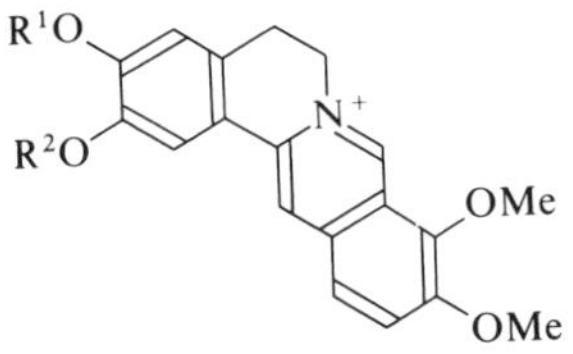 (**20**)	Antimalarial, febrifuge, carminative, dressing for indolent ulcers
Jatrorrhizine (**20**) ($R^1=H$, $R^2=Me$)		Bitter tonic
Benzophenanthridine Alkaloid		
Chelidonine (**21**)	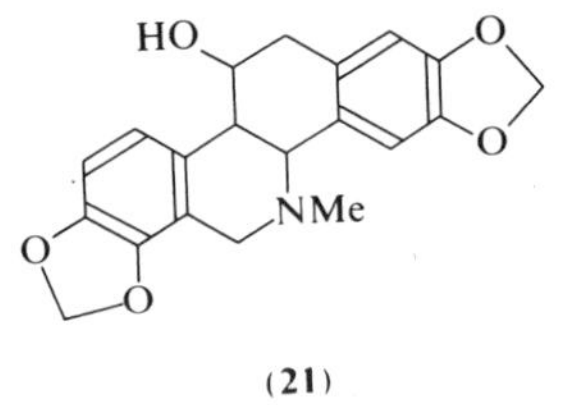 (**21**)	Smooth muscle depressant
Aporphine Alkaloids		
Bulbocapnine (**22**) ($R^1+R^2=CH_2$, $R^3=OH$, $R^4=H$)	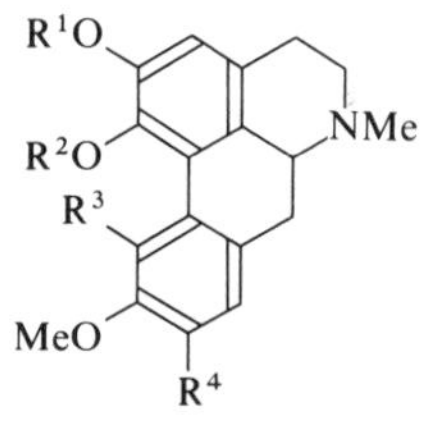 (**22**)	Muscular tremors, vestibular nystagmus
Boldine (**22**) ($R^1=R^3=H$, $R^2=Me$, $R^4=OH$)		Diuretic

Table 1. (continued)

Alkaloid	Structure	Activity
Morphine Alkaloids		
Morphine (**23**) ($R=H$)	(**23**)	Narcotic analgesic
Codeine (**23**) ($R=Me$)		Analgesic, antitussive
Bisbenzylisoquinoline Alkaloids		
Tubocurarine (**24**)	(**24**)	Skeletal muscle relaxant, diagnostic agent in myasthenia gravis
Tetrandrine (**25**)	(**25**)	Antipyretic, analgesic
Ipecacuanha Alkaloids		
Emetine (**26**) ($R=Me$)	(**26**)	Amebicide, emetic
Cephaeline (**26**) ($R=H$)		Amebicide, emetic
Erythrina Alkaloids		
Erysothiopine (**27**) (R^1 or $R^2=H$ or SO_2CH_2COOH)	(**27**)	Curare-like action
Erysothiovine (**27**) ($R^1=SO_2CH_2COOH$, $R^2=Me$)		Curare-like action
β-Erythroidine (**28**)	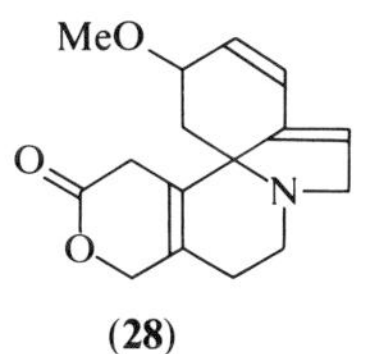 (**28**)	Curarimimetic

Table 1. (continued)

Compound	Structure	Activity
Amaryllidaceae Alkaloids Galanthamine (**29**)	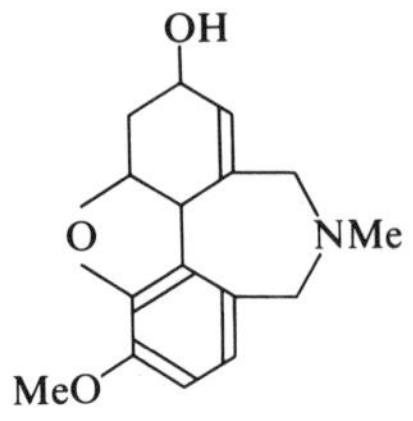 (**29**)	Cholinesterase inhibitor
Lycorine (**30**)	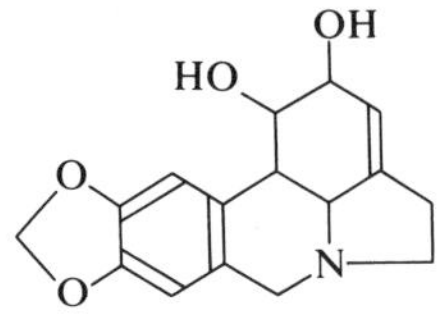 (**30**)	Amebicide
Tylophorine Group Tylocrebrine (**31**)	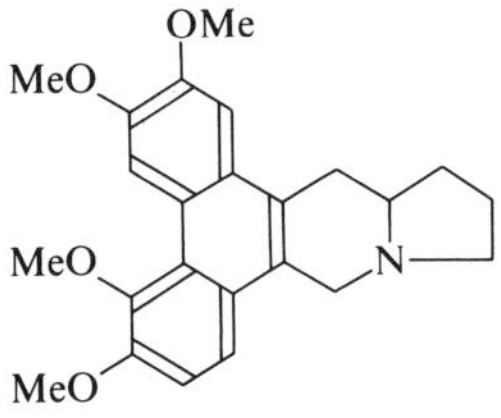 (**31**)	Antileukemic agent
Indole Alkaloids Harmaline (**32**) (R = Me)	(**32**)	Amine oxidase inhibitor, CNS stimulant
Harmalol (**32**) (R = H)		Anthelminic, narcotic
Harmine (**32**) (R = Me, ring C = aromatized)		CNS stimulant
Physostigmine (**33**)	(**33**)	Parasympathomimetic, anticholinesterase agent
Geneserine (**33**) ($N_1 \rightarrow O$)		Parasympathomimetic
Ajmalicine (**34**)	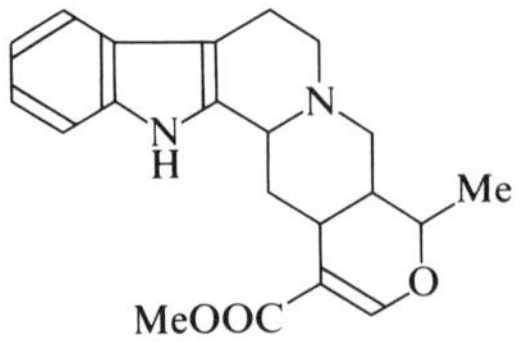 (**34**)	Hypotensive, tranquilizer, sedative

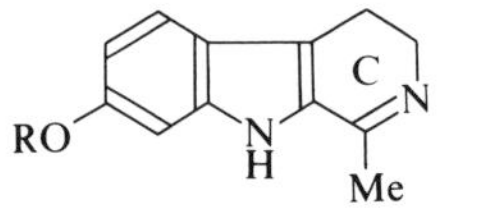

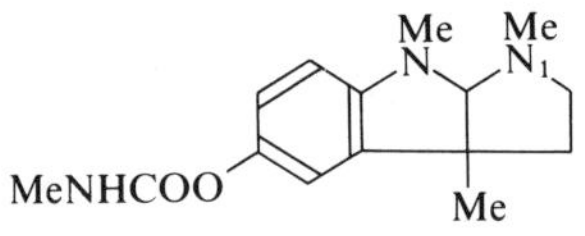

Table 1. (continued)

Compound	Structure	Activity
Yohimbine (**35**) ($R^1 = R^2 = R^3 = H$)		Adrenergic blocking agent, aphrodisiac, local anesthetic, mydriatic
Reserpine (**35**) ($R^1 = OMe$, $R^2 = Me$, 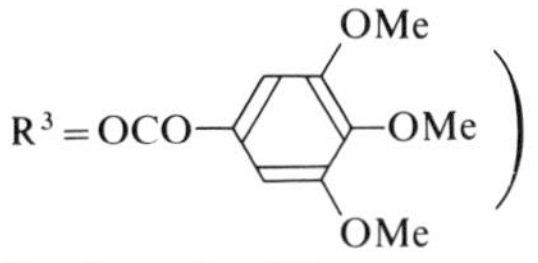 $R^3 = OCO$–$C_6H_2(OMe)_3$)		Hypotensive, tranquilizer, sedative
Rescinnamine (**35**) ($R^1 = OMe$, $R^2 = Me$, 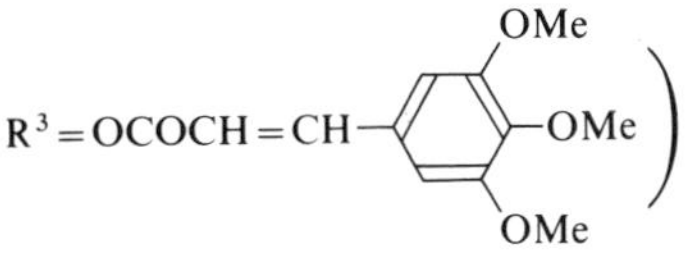 $R^3 = OCOCH{=}CH$–$C_6H_2(OMe)_3$)	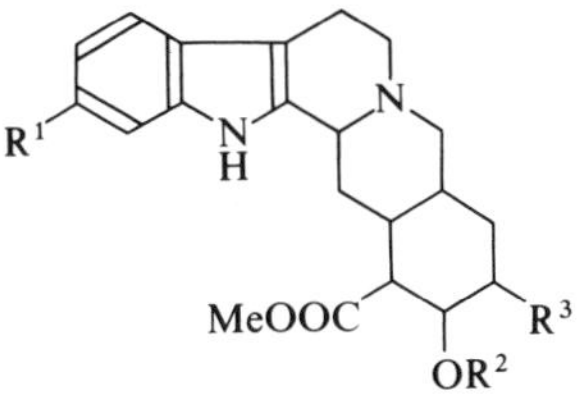 (**35**)	Tranquilizer, sedative, antihypertensive
Deserpidine (**35**) ($R^1 = H$, $R^2 = Me$, 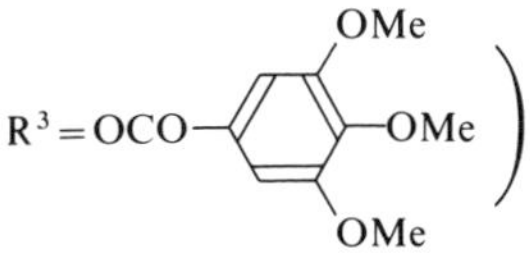 $R^3 = OCO$–$C_6H_2(OMe)_3$)		Hypotensive, tranquilizer
Perivine (**36**)	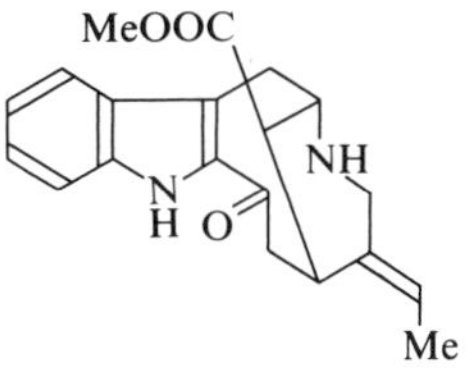 (**36**)	Antineoplastic agent
Echitamine (Ditaine) (**37**)	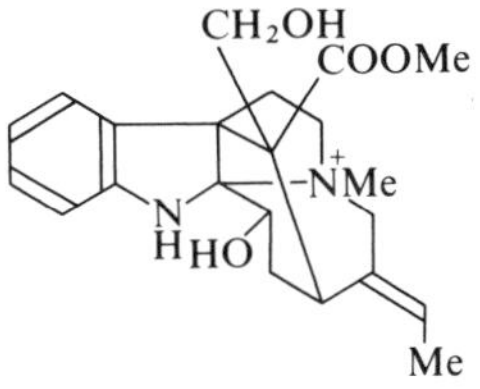 (**37**)	Antimalarial, curare-like action
Ajmaline (**38**)	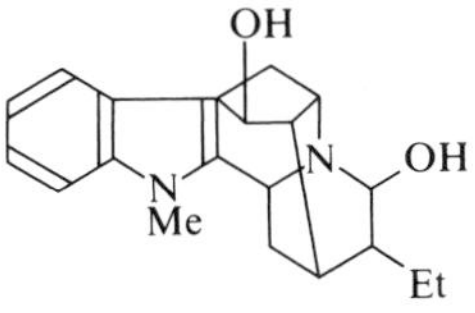 (**38**)	Hypotensive, sedative
Strychnine (**39**) ($R^1 = R^2 = H$)	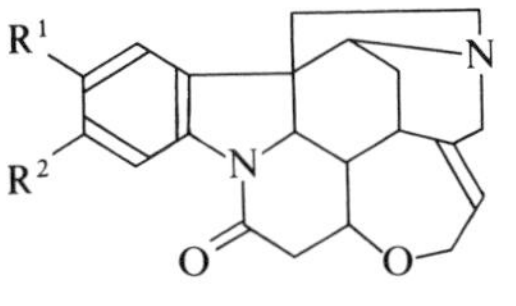 (**39**)	CNS stimulant, bitter tonic, circulatory stimulant
Burcine (**39**) ($R^1 = R^2 = OMe$)		CNS stimulant, bitter tonic, circulatory stimulant

Table 1. (continued)

Compound	Structure	Activity
Ibogaine (**40**) (R^1=OMe, R^2=H)	(**40**)	Antidepressant
Tabernanthine (**40**) (R^1=H, R^2=OMe)		Analgesic, serotonine antagonist
Vincamine (**41**)	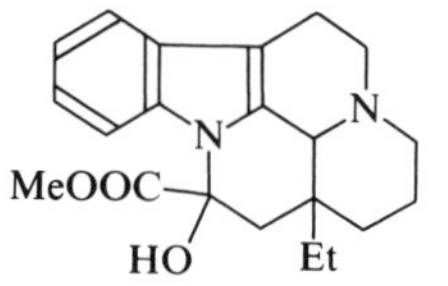 (**41**)	To improve intellectual capacity in patients with cerebrovascular disorders
Aspidospermine (**42**)	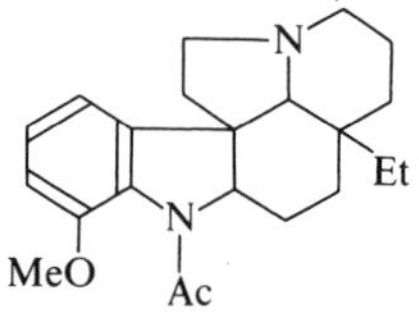 (**42**)	Respiratory stimulant
Gelsemine (**43**)	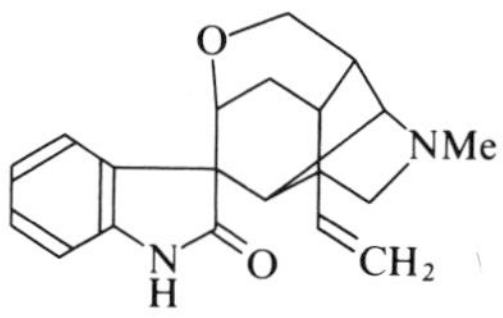 (**43**)	CNS stimulant
Vinblastin (**44**)	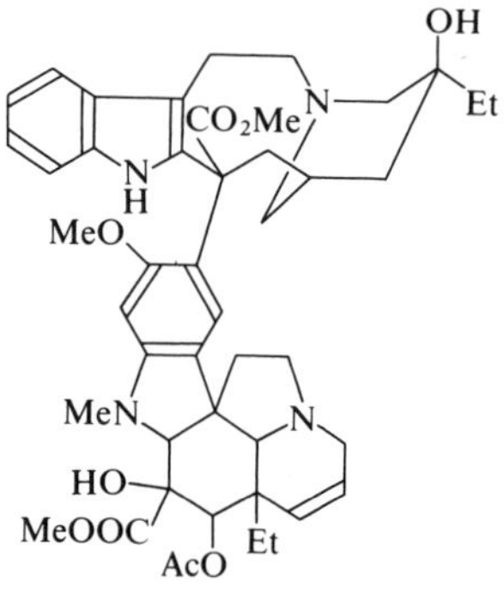 (**44**)	Antineoplastic
C-Toxiferine I (**45**)	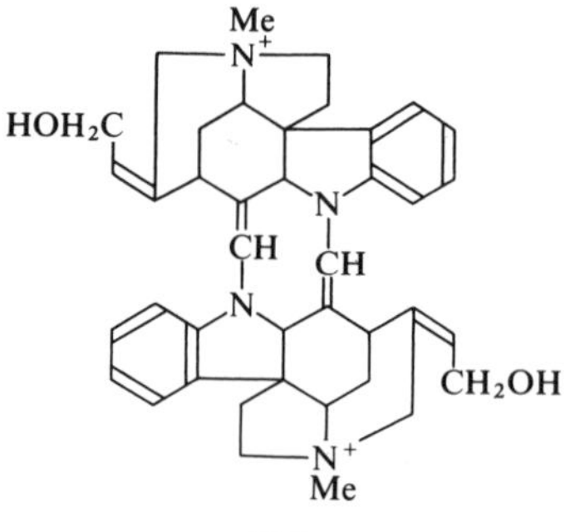 (**45**)	Curare-like action

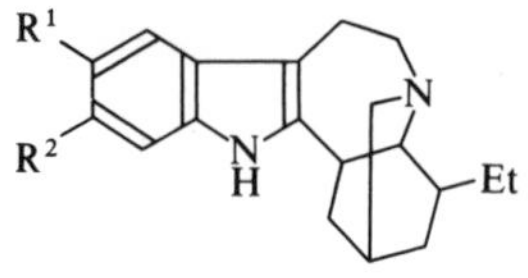

Table 1. (continued)

	Structure	Activity
C-Calebassine (**46**)	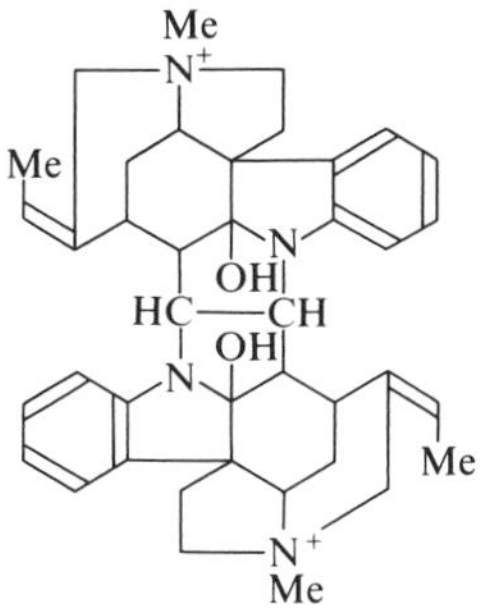 (**46**)	Curare-like action
Ergot Alkaloids		
Lysergic acid (**47**) (R = OH)	(**47**)	Hallucinogen, psychotomimetic
Ergonovine (**47**) (R = NHCH(Me)CH$_2$OH)		Oxytocic
Ergotamine (**48**) (R^1 = Me, R^2 = CH$_2$Ph)	(**48**)	Analgesic vasoconstrictor in vascular headache, oxytocic
Ergocornine (**48**) (R^1 = R^2 = CHMe$_2$)		Peripheral vascular disorders
Ergocristine (**48**) (R^1 = CHMe$_2$, R^2 = CH$_2$Ph)		Peripheral vascular disorders
Cinchona Alkaloids		
Cinchonidine (**49**) (R^1 = H, R^2 = CH=CH$_2$)	(**49**)	Antimalarial
Cinchonine (Stereoisomer of cinchonidine)		Antimalarial
Quinidine (**49**) (R^1 = OMe, R^2 = CH=CH$_2$)		Cardiac depressant (antiarrhythmic), antimalarial
Quinine (stereoisomer of quinidine)		Suppressive antimalarial, analgesic, antipyretic, oxytocic, sclerosing agent, local anesthetic, bitter and stomachic
Hydroquinine (**49**) (R^1 = OMe, R^2 = Et)		Antimalarial
Quinoline Alkaloids		
Galipine (**50**)	(**50**)	Aromatic bitter
Camptothecin (**51**)	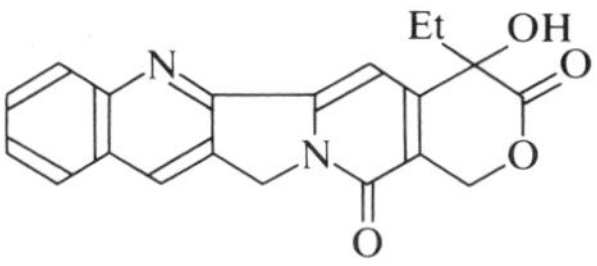 (**51**)	Antileukemic and antitumor activities

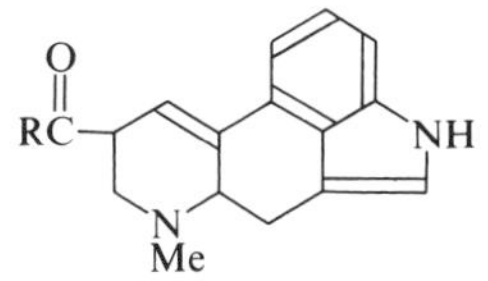

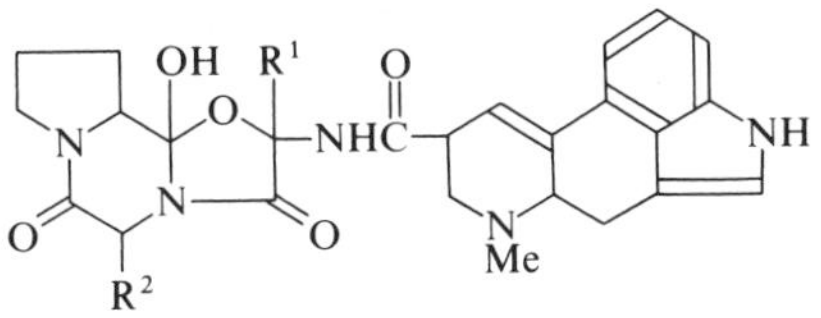

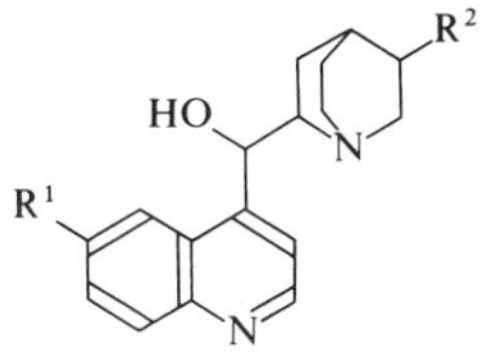

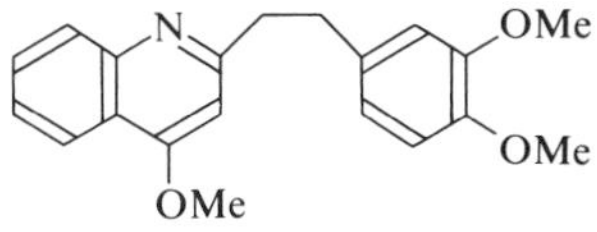

Table 1. (continued)

Compound	Structure	Activity
Diterpene Alkaloids		
Ajaconine (**52**)	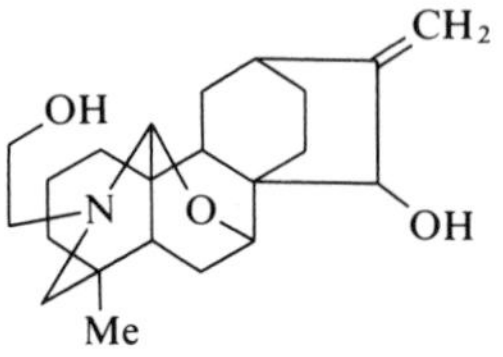 (**52**)	Pediculicide
Atisine (**53**)	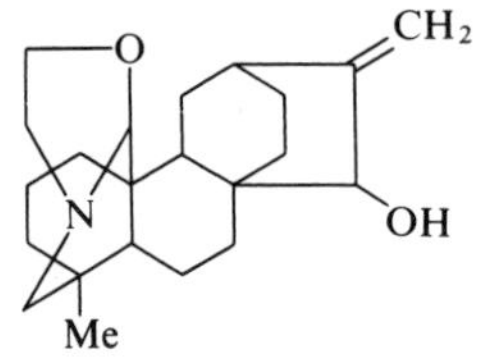 (**53**)	Febrifuge, gastric anesthetic
Aconine (**54**) ($R^1=R^2=H$)	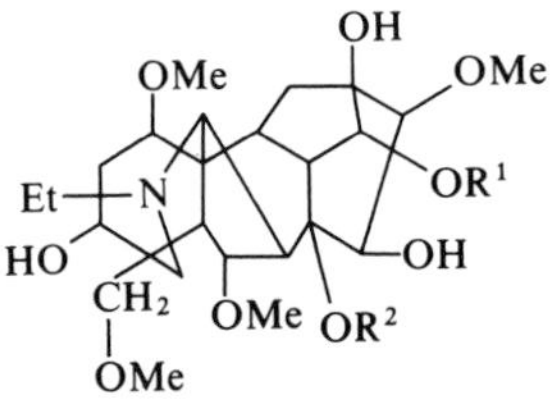 (**54**)	Febrifuge, gastric anesthetic
Aconitine (**54**) (R^1=COPh, R^2=Ac)		Febrifuge, gastric anesthetic
Veratrum Alkaloids		
Germine (**55**) ($R^1=R^2=R^3=R^6$=H, $R^4=R^5$=OH)	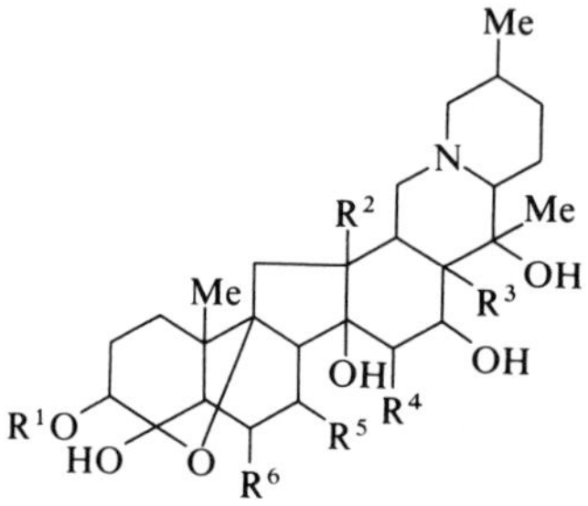 (**55**)	Antihypertensive
Zygadenine (**55**) ($R^1=R^2=R^3=R^5=R^6$= R^4=OH)		Antihypertensive
Protoveratrine (**55**) (R^1=COC(Me)(OH)CH$_2$CH$_3$, $R^2=R^3$= H, R^4=OCOCH(Me)CH$_2$CH$_3$, R^5 =R^6=OAc)		Antihypertensive
Cevadine (Veratrine) (**55**) (R^1=COC(Me)=CHCH$_3$, $R^2=R^3$= OH, $R^4=R^5=R^6$=H)		Insecticide
Veratramine (**56**)	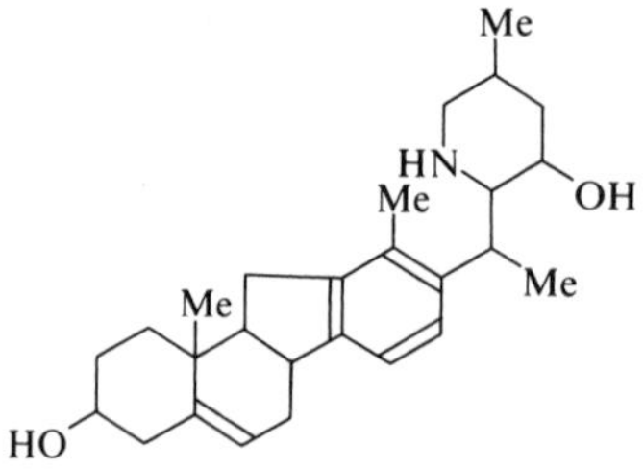 (**56**)	Antihypertensive

5 Natural Pigments

Contributions by

Ch. Baumann, Giessen
J. W. Buchler, Aachen
A. Gossauer, Braunschweig
T. Goto, Nagoya
E. R. H. Jones, Oxford
T. Kubota, Osaka
R. A. Nicolaus, Naples
V. Thaller, Oxford
R. H. Thomson, Old Aberdeen
T. Tokoroyama, Osaka
K. Tsukida, Kobe
S. Yamamura, Nagoya

5.1 Pyran Compounds

*Toshio Goto and Shosuke Yamamura**
Department of Agricultural Chemistry, Nagoya University, Chikusa-ku, Nagoya 464; *Faculty of Pharmacy, Meijo University, Tempaku-ku, Nagoya, 468 Japan

1 Introduction

The parent compounds, α-pyran and γ-pyran, are not aromatic compounds, but formal "aromaticity" can be expected in their fully conjugated carbonyl derivatives, α-pyrone and γ-pyrone, although the actual aromatic character of these pyrones is very much less than that of the furans, another important class of oxygen heterocycles. In this chapter, only pyrone derivatives are described[1].

2 Pyrones Derived from a Polyketide Involving Acetate as the Starter Unit

There are many natural products having the pyronone (3,4-dihydro-2H-pyran-2,4-dione) moiety which are derived biosynthetically via the acetate-malonate pathway, which is connected closely with the biosynthesis of polyphenols. For example, the $CH_3COCH_2COCH_2COCH_2COOH$ type intermediate, which is considered to be produced by condensation of acetyl CoA with 3 moles of malonyl CoA, could cyclize to afford either a pyrone (**4** or **5**) or a phenol (**1**, **2** or **3**)[2]. Indeed, tetraacetic lactone (**4**)[3] can be chemically converted to orsellinic acid (**1**) under very mild conditions[2]. Recently, biogenetic type transformations of pyrones to polyphenols have been actively investigated[4]. The polyketide chain is normally assembled by the condensation of acetyl CoA with chain-building units of malonyl CoA which undergo decarboxylation, but a starter unit other than the acetyl CoA may also be involved (see Section 3).

One of the simplest natural pyronones derived from a polyketide involving acetate as the starter unit is opuntiol (**6**)[5], a constituent of cactus, but a modified pyronone of this type, versicolin (**7**)[6], found as a metabolite of *Aspergillus versicolor*, is assumed to be derived from a sugar (see Sec. 4). The side chain at C_6 can be a long polyketide chain such as that in citreoviridin (**8**)[7], a toxic metabolite of *P. citreoviride*. Substituents may be introduced at the C_3 and/or C_5 positions of pyronone as shown in the case of radicinin (**9**)[8] and arenol (**10**)[9]. The biosynthetic pathway of radicinin (**9**), a phytotoxic metabolite of *Stremphylium radicinum*, has been studied using ^{14}C[10] and ^{13}C[11] as the tracer. Both results are consistent with the scheme (**9a**), which shows the presence of two acetate-malonate units. In the case of citromycetin (**11**)[12], a matabolite of *P. frequentans*, the presence of two chains has also been demonstrated.

Isocoumarins (**12**), which are distributed widely in nature, belong to the polyketide group. For example, the isocoumarin nucleus of canescin (**14**)[13], a metabolite of *P. canescens*, is produced via the acetate-malonate pathway, but the carbon atom indicated is introduced from me-

(iso: isolation; st: structure; syn: synthesis; bio: biosynthesis)

[1] (a) *W. Bors, M. Magalhaes, O. Gottlieb*, Fortschritte der chem. org. Naturstoffe *20*, 131 (1962); (b) *F. Dean*, Naturally Occurring Oxygen Ring Compounds, Butterworths, London 1963; (c) *D. Crout*, The Biosynthesis of Carbocyclic Compounds, *in* Topics in Carbocyclic Chemistry, *D. Lloyd* (Ed.), p. 63, Logos Press, London, 1969; (d) *T. Money*, Biogenetic Type Synthesis of Phenolic Compounds, Chem. Rev. *70*, 553 (1970).

[2] *R. Bentley, P. Zwitkowits*, J. Am. Chem. Soc. *89*, 676 (1967).

[3] (iso) *G. Marx, S. Tanenbaum*, J. Am. Chem. Soc. *90*, 5302 (1968); (syn) *H. Guilford, A. Scott, D. Kingle, M. Yalpani*, Chem. Commun. 1127 (1968); *S. Yamamura, K. Kato, Y. Hirata*, Chem. Commun. 1580 (1968), J. Chem. Soc. (C), 2461 (1969).

[4] See [1] (d); *P. Hedgecock, P. Praill, A. Whitear*, Chem. Ind. 1268 (1966); *A. Birch, D. Cameron, R. Rickards*, J. Chem. Soc. 4395 (1960); *T. Money, J. Douglas, A. Scott*, J. Am. Chem. Soc. *88*, 624 (1966); *T. Money, I. Qureshi, G. Webster, A. Scott*, J. Am. Chem. Soc. *87*, 3004 (1965).

[5] *A. Ganguly, I. Govindachari, P. Mohamed*, Tetrahedron *21*, 93 (1965).

[6] *A. Dhar, S. Bose*, Tetrahedron Lett. 4871 (1969).

[7] *N. Sakabe, T. Goto, Y. Hirata*, Tetrahedron Lett. 1825 (1964).

[8] (str) *J. Grove*, J. Chem. Soc. 3234 (1964); (syn) *K. Kato, Y. Hirata, S. Yamamura*, J. Chem. Soc. (C), 1997 (1969).

[9] *J. Vrkoc, L. Dolejs, P. Sedmera, S. Vasickova, F. Sorm*, Tetrahedron Lett. 247 (1971).

[10] *J. Grove*, J. Chem. Soc. (C), 1860 (1970).

[11] *M. Tanabe, H. Seto, L. Johnson*, J. Am. Chem. Soc. *92*, 2157 (1970).

[12] (str) *S. Gatenback, K. Mosbach*, Biochem. Biophys. Res. Commun. *11*, 166 (1963); (bio) *A. Birch, S. Hussain, R. Richards*, J. Chem. Soc. 3494 (1964); *J.* Grove, J. Chem. Soc. (C), 1860 (1970).

[13] *A. Birch, F. Gager, L. Mo, A. Pelter, J. Wright*, Australian J. Chem. *22*, 2429 (1969); Tetrahedron Lett. 1519 (1969).

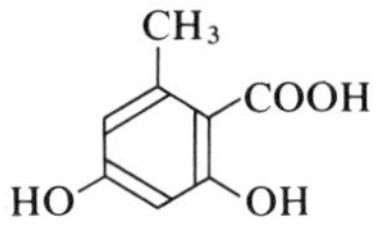

(1) 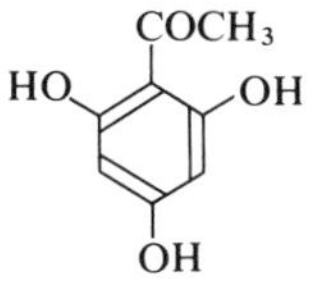 (2) 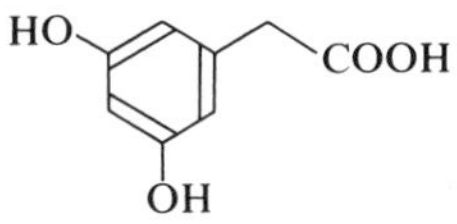 (3) 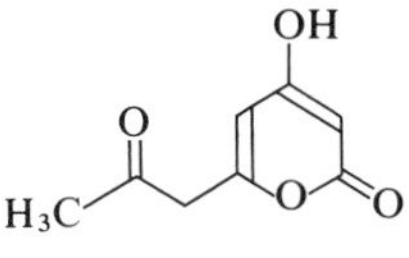 (4)

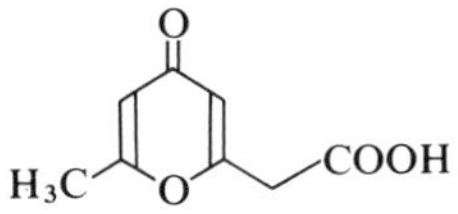

(5) 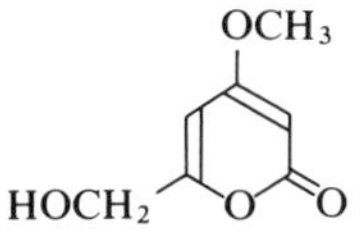 (6) 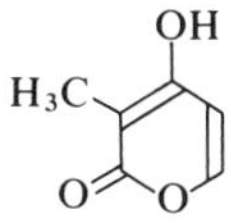 (7)

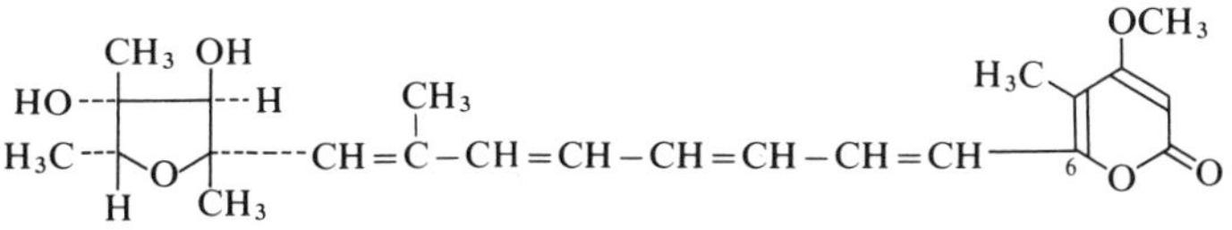

(8)

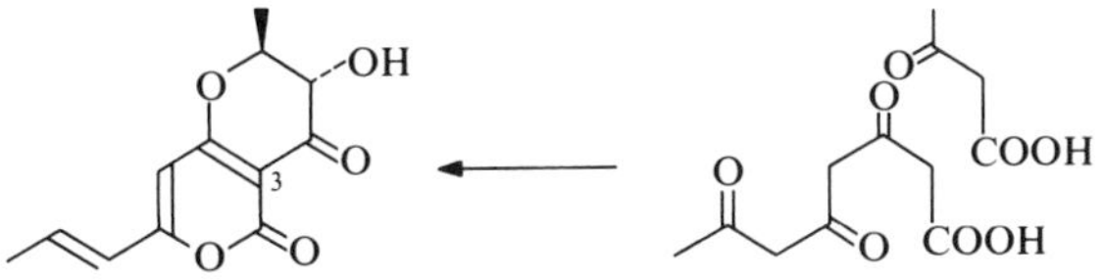

(9) (9a)

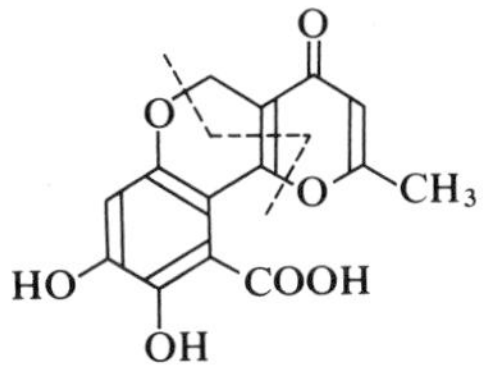

(11) 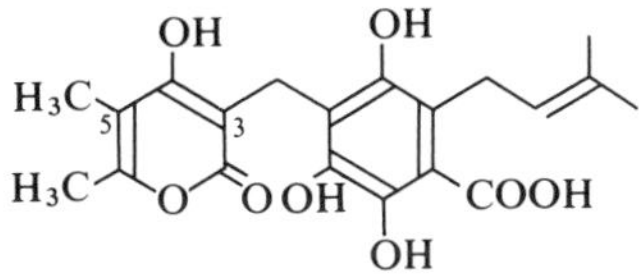 (10)

Fig. 1.

thionine. Glomellin (**15**) is not a natural product but an artifact produced by the action of a weak base on glomelliferic acid (**16**) from depside of *Parmelia glomelifera*[14]. The group of extended pyrones (**13**) exemplified by rotiorin (**17**)[15], a pigment of *P. sclerotiorum*, is structurally related to the isocoumarins (**12**). The two six-membered rings with the side chain in rotiorin (**17**) are derived from an acetate-malonate chain but the four-carbon β-oxolactone fragment is considered to be derived from acetoacetate without incorporation of malonate. Ascochitin (**18**)[16], a metabolite of *Ascochyta pisa*, is the first example of pironoquinonoids. Funicone (**19**)[17], a metabolite of *P. funiculosum*, is assumed to be produced from the extended pyrone (**19a**) by oxidation. The structure of sclerin (**20**)[18], a metabolite of *Sclerotina libertiana*, is related to the isocoumarins but it was proved to be derived from two acetate-malonate chains (**20a**). The monomeric unit of the dimeric metabolite of *P.*

[14] *Y. Asahina, H. Nogami*, Ber. *70*, 1498 (1937); *K. Minami*, J. Pharm. Soc. Jap. *64*, 315 (1944).

[15] *J. Holker, J. Stannton, W. Whalley*, J. Chem. Soc. 16, (1964); *R. Gray, W. Whalley*, Chem. Commun. 762 (1970).

[16] (str) *I. Iwai, H. Mishima*, Chem. Ind. 186 (1965); (syn) *M. Galbraith, W. Whalley*, Chem. Commun. 620 (1966).

[17] *L. Meilini, G. Nasini, A. Selva*, Tetrahedron *26*, 2739 (1970).

[18] *T. Kubota, T. Tokoroyama, S. Oi, Y. Satomura*, Tetrahedron Lett. 631 (1969).

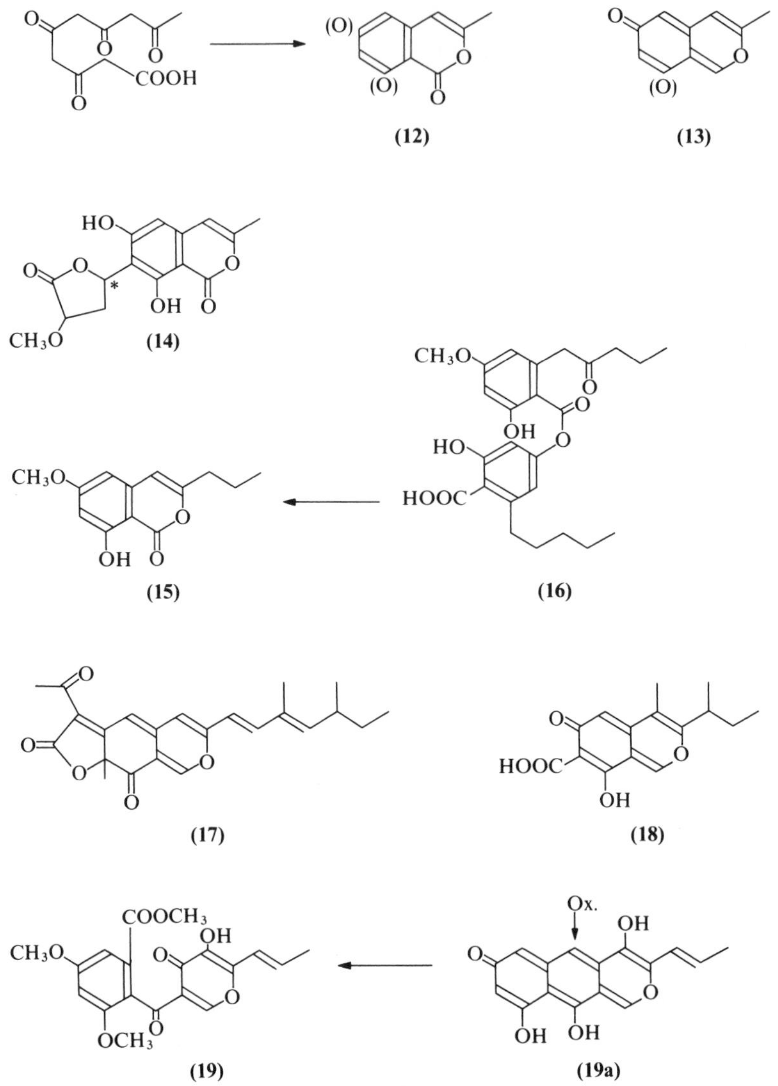

(12) (13)

(14)

(15) (16)

(17) (18)

(19) (19a)

Fig. 2.

duclauxi, duclauxin (**21**)[19], is produced via the biosynthetic pathway (**21a→21**) involving the loss of one carbon atom (*).

Chromones (**22**) with a methyl group at C_2 are widely distributed in nature. Barakol (**23**)[20], which was found recently in *Cassia siamea*, has a novel dioxaphenalene nucleus and is easily dehydrated to **23a** in a desiccator.

Aflatoxins (**24**: aflatoxin B_1)[21] are hepatotoxic and highly carcinogenic metabolites of *Aspergillus flavus*. Although aflatoxins are derivatives of coumarin (see Section 4), an early suggestion that phenylalanine and tyrosine might be precursors has been shown to be erroneous[22]. According to the suggested mechanism by Büchi *et al.*[23], aflatoxins are biosynthesized from a polyketide involving a rearrangement of an intermediate peroxide (**24b**). Total synthesis of racemic aflatoxin B_1 (**24**) has been carried out[24].

[19] *U. Sankawa, H. Taguchi, Y. Ogihara, S. Shibata*, Tetrahedron Lett. 2883 (1966).

[20] *B. Bycroft, A. Hassaniali-Walji, A. Johnson, T. King*, J. Chem. Soc. (C), 1686 (1970).

[21] *T. Asao, G. Büchi, M. Abdel-Kadler, S. Chang, E. Wick, G. Wagan*, J. Am. Chem. Soc. *85*, 1705 (1963); *87*, 882 (1965).

[22] *J. Donkersloot, D. Hsieh, R. Mateles*, J. Am. Chem. Soc. *90*, 5020 (1968).

[23] *M. Biollaz, G. Büchi, G. Milne*, J. Am. Chem. Soc. *92*, 1035 (1970).

[24] *G. Büchi, D. Foulkers, M. Kurono, G. Mitchell, R. Schneider*, J. Am. Chem. Soc. *89*, 6745 (1967).

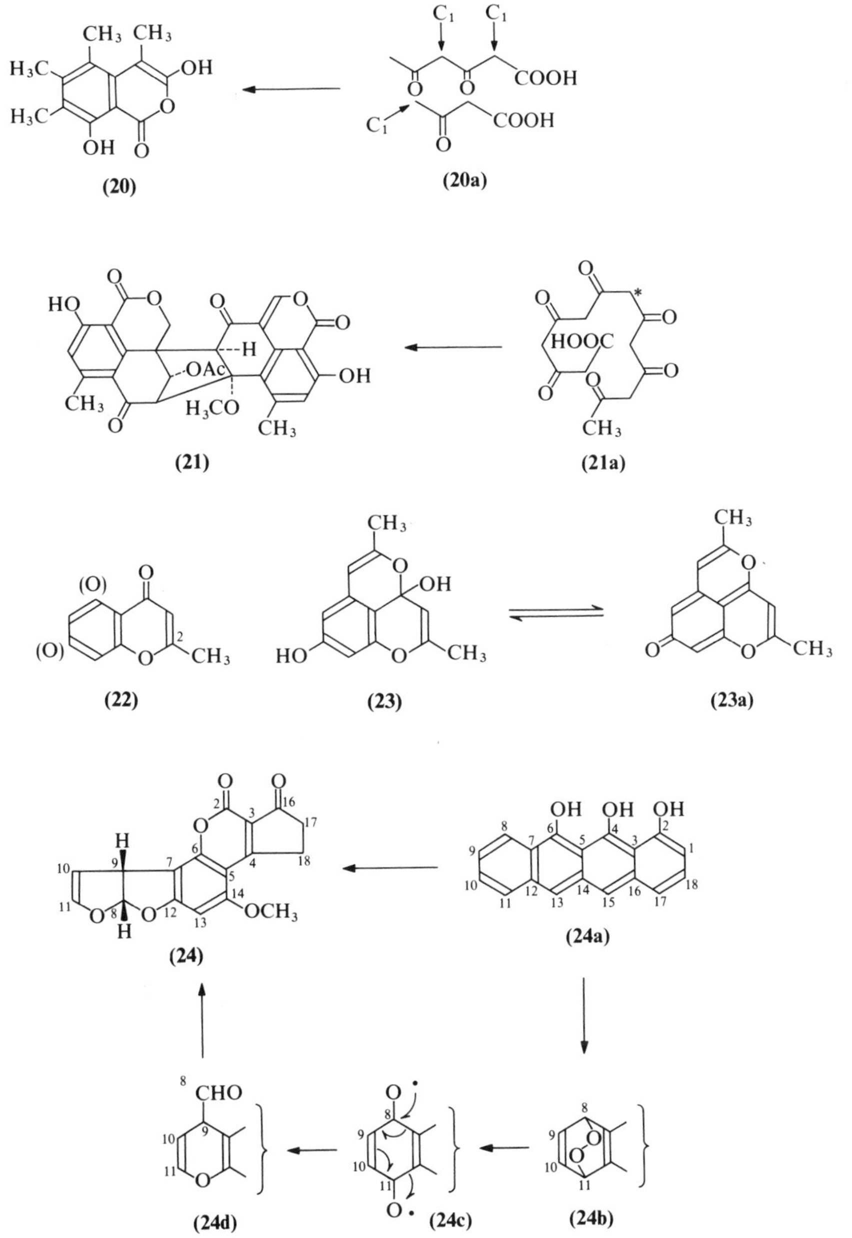

(20) (20a)

(21) (21a)

(22) (23) (23a)

(24) (24a)

(24d) (24c) (24b)

Fig. 3.

3 Pyrones Derived from a Polyketide Involving a Starter Other than Acetate

The acetate starter may be replaced by other carboxylic acids such as cinnamic acid. Typical monocyclic pyrones having such a starter are paracotoin (**25**) from koto bark, yangonin (**26**) from kawa root, anibin (**27**), a constituent of rosewood *Aniba rosaeodora*, and aureothin (**29**)[25], a toxic metabolite of *Streptomyces thioluteus*. A biogenetic type synthesis[26] of several of these compounds (**26–28**, etc.) has been carried out.

Flavonoids (**31**) are colored compounds widely distributed in plants and are biosynthesized from a polyketomethylene chain with a C_6–C_3 starter unit which is derived via the shikimate pathway through cinnamic acid derivatives[1c]. Isoflavonoid (**32**)[1c] biosynthesis from chalcone

[25] *Y. Hirata, H. Nakata, K. Yamada, K. Okuhara, T. Naito*, Tetrahedron *14*, 252 (1961); see also (luteoreticulin) *Y. Koyama, Y. Fukakusa, N. Okimura, S. Yamagishi, T. Arai*, Tetrahedron Lett. 355 (1969).

[26] *T. Harris, C. Harris*, J. Org. Chem. *31*, 1032 (1966); *T. Harris, C. Combs, Jr.*, J. Org. Chem. *33*, 2399 (1968).

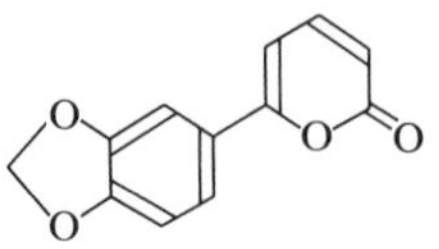

(25)

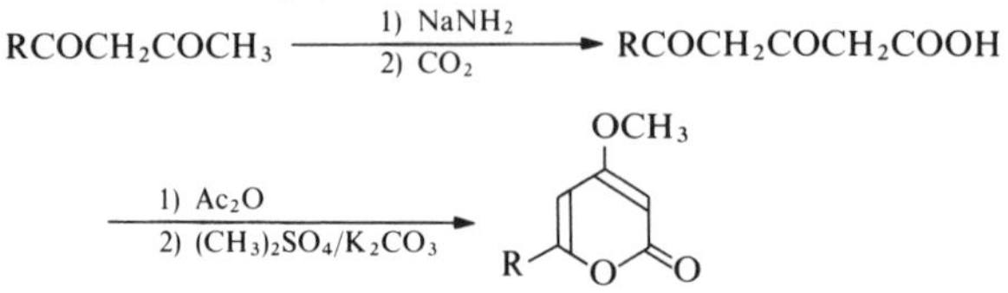

$$RCOCH_2COCH_3 \xrightarrow[2)\ CO_2]{1)\ NaNH_2} RCOCH_2COCH_2COOH$$

$$\xrightarrow[2)\ (CH_3)_2SO_4/K_2CO_3]{1)\ Ac_2O}$$

Starter (RCOOH)	Product	
p-Methoxycinnamic acid	Yangonin	**(26)**
Nicotinic acid	Anibin	**(27)**
3,4-Dihydroxycinnamic acid	Hispidin	**(28)**

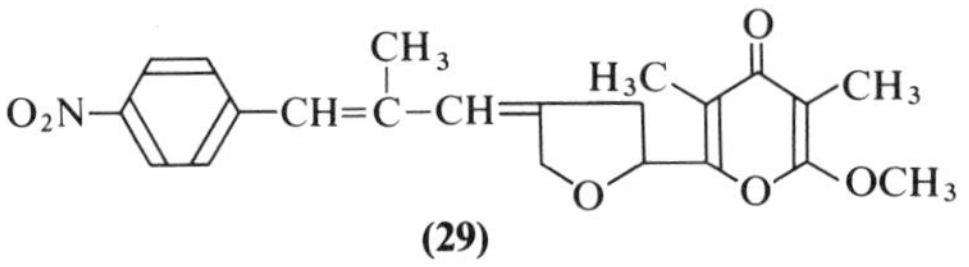

(29)

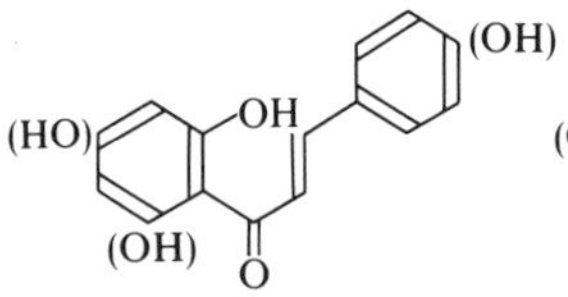

(30) 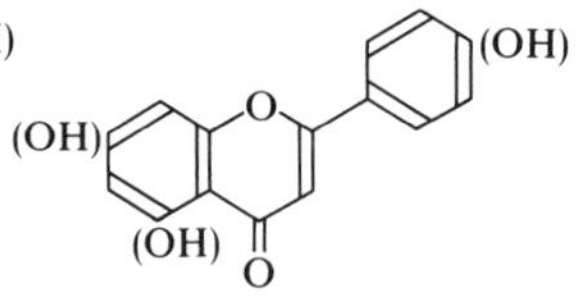 (31) (33)

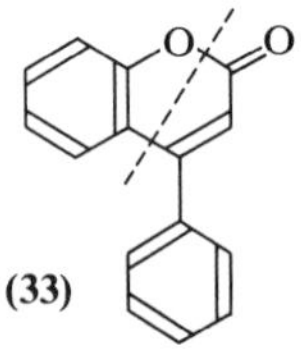

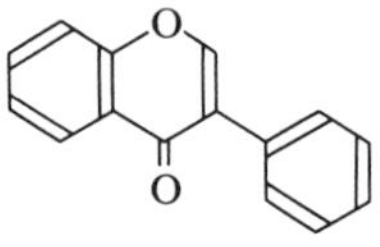

(32) 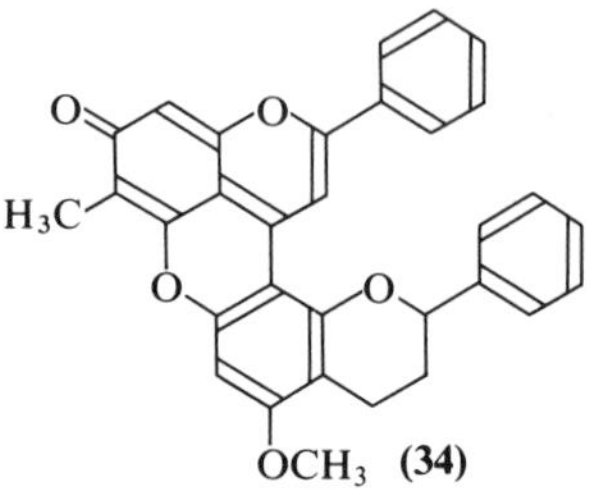 (34)

Fig. 4.

(**30**) precursors involves an oxidation and a 1,2-aryl shift. A biogenetic-type synthesis[27] of an isoflavonoid from a chalcone has been achieved, which is the first chemical analogy for this rearrangement. Dracorubin (**34**)[28], a dark red pigment from *Dracaena draco*, has the modified flavonoid structure. Structures of several bisflavonoids have been elucidated[29]. Neoflavones (**33**)[30] have been shown not to be produced from chalcones (**30**).

4 Metabolites Which are not Derived from Polyketides

Stizolobic (**35**) and stizolobinic acid (**36**), which

[27] *W. Ollis, K. Ormand, I. Sutherland*, Chem. Commun. 1237 (1968).

[28] *A. Robertson, W. Whalley, J. Yates*, J. Chem. Soc. 3117 (1950).

[29] *Y. Fukui, N. Kawano*, J. Am. Chem. Soc. *81*, 6331 (1959); *K. Nakazawa, M. Ito*, Chem. Pharm. Bull. (Japan) *11*, 283 (1963); *W. Baker, A. Finch, W. Ollis, K. Robinson*, J. Chem. Soc. 1477 (1963); *V. Murti, P. Raman, T. Seshadri*, Tetrahedron *23*, 397 (1967); *A. Pelter, R. Warren, J. Usmani, R. Rizvi, M. Ilyas, W. Rahman*, Experientia *25*, 351 (1969).

[30] *W. Ollis*, Experientia *22*, 777 (1966); *G. Kunesch, J. Polonsky*, Chem. Commun. 317 (1967).

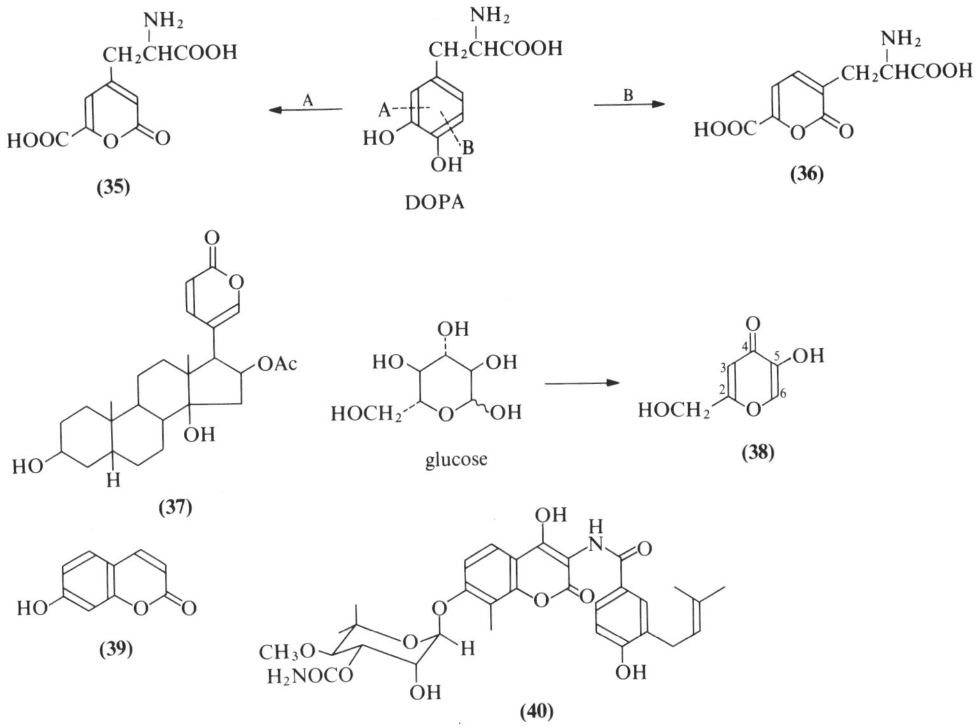

Fig. 5.

are amino acids produced by *Stizolobium hassjoo*, are assumed to be biosynthesized from DOPA by a metapyrocatechase type oxidative cleavage[31].

Bufodienolides include glycosides of plant origin and toad poison such as bufotalin **(37)**[32]. They are steroid derivatives with an α-pyrone moiety attached at C_{17} and have strong biological activities like cardienolides.

γ-Pyrones without an oxygen functionality at C_2 and C_6 are often encountered in nature. As exemplified by kojic acid **(38)**[1b], most of them have a γ-pyrone nucleus with $-CH_3$, $-CH_2OH$ or $-COOH$ at the 2 and/or 6 position and –OH at the 3 and/or 5 position. Biosynthetically, such γ-pyrones are sugar derived.

Coumarins[1c] are distributed widely in plants mostly as derivatives of umbelliferone **(39)**, which is derived via the shikimic acid pathway involving *p*-hydroxycinnamic acid as an intermediate. The coumarin nucleus of novobiocin **(40)**[33], an antibiotic produced by *Streptomyces niveus*, is produced directly from tyrosine without degradation.

5 Modification of Structures by Incorporation of Isoprenoids

Many natural coumarins, chromones, flavonoids, etc. have been isolated with a isoprenoid side chain. The chain is further modified into a furan or pyran ring as exemplified by umbelliferone derivatives **(41–43)**[34]. In general, the furan system arises by oxidative fission of the isopropyl group in isopropylfurano compounds such as **41**. A heartwood constituent, ptaeroxylin **(44)**[35], has an interesting oxepin ring in the molecule.

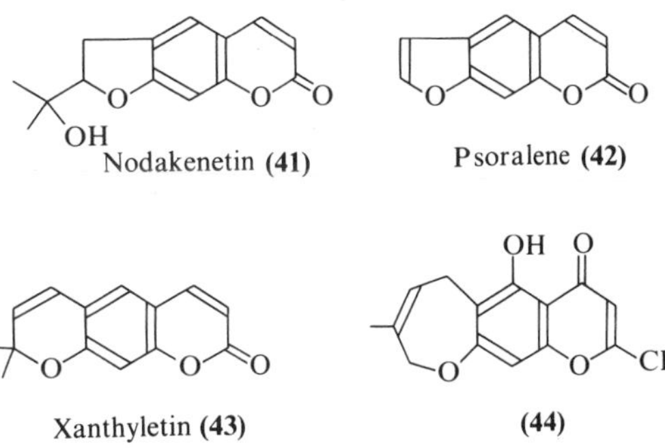

Nodakenetin **(41)** Psoralene **(42)** Xanthyletin **(43)** **(44)**

Fig. 6.

[31] *S. Senoh, S. Imamoto, Y. Maeno, T. Tokoroyama, T. Sakan*, Tetrahedron Lett. 3431 (1964).

[32] *L. Fieser, M. Fieser*, Steroids, p. 782, Reinhold Publ. Co., New York 1959.

[33] *K. Chambers, G. Kenner, M. Robinson, B. Webster*, Proc. Chem. Soc. 291 (1960); *C. Bunton, G. Kenner, M. Robinson, B. Webster*, Tetrahedron *19*, 1001 (1963).

Addendum

Recently, further notable investigations on many naturally occurring pyrones have been carried out. In particular, the successful application of FT ^{13}C NMR spectroscopy is invaluable not only in the structural elucidations of these substances but also in their biosyntheses. Several simple pyrones have been isolated: nectriapyrone (**45**)[36], mucidone (**46**)[37], aloenin (**47**)[38], rosellisin (**48**)[39], hallactone (**49**)[40] and LL-Z 1220 (**50**)[41]. Nectriapyrone (**45**) with antibacterial activity and hallactone A (**49**), an insect toxin from *Podocarpus hallii*, belong to terpenoids. A new antibiotic LL-Z 1220 (**50**) is a γ-pyrone with the unique cyclohexene diepoxide ring system. Austdiol (**51**)[42], a toxic metabolite from maize meal cultures, belongs to the extended γ-pyrones represented by votiorin (**17**). Citreoviridin (**8**) has been proved to be biosynthesized from a simple C_{18} polyketide followed by introduction of five methyl groups derived from methionine[43]. Aurovertin B (**52**) is also a toxic metabolite from the fungus *Calcarisporium arbuscula*, and its structure is quite similar to that of **8**[44]. Biosynthesis of aureothin (**29**) has also been elucidated by means of FT ^{13}C NMR method[45]. Büchi *et al.* suggested that aflatoxins are biosynthesized from a single C_{18} polyketide involving a rearrangement of an intermediate peroxide (**24b**). However, recent studies using FT ^{13}C NMR technique indicated the intermediacy of an anthraquinone containing a linear C_6-chain derived from a single C_{20} polyketide[46]. Many interesting and complex flavonoids and coumarins have been isolated as plant constituents or fungal metabolites, as exemplified by semiglobrinol (**53**)[47], albofungin (**54**)[48], kidamycin (**55**)[49] and stemonone (**56**)[50].

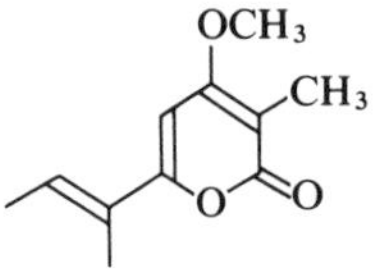

OCH_3 CH_3 O O

(**45**)

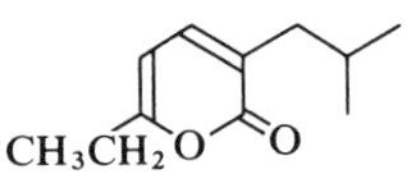

CH_3CH_2 O O

(**46**)

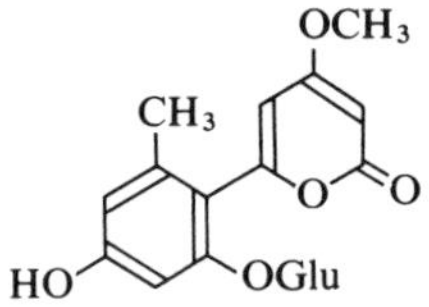

OCH_3 CH_3 O O HO OGlu

(**47**)

[34] *S. Brown*, Biosynthesis of Coumarins, p. 15, Pergamon Press, Oxford 1966.

[35] *F. Dean, D. Taylor*, J. Chem. Soc. (C), 114 (1966); *F. Dean, B. Parton, N. Somvichien, D. Taylor*, Tetrahedron Lett. 3459 (1967).

[36] (str) *M. S. F. Nair, S. T. Carey*, Tetrahedron Lett. 1655 (1975); (syn) *T. Reffstrup, P. M. Boll*, Tetrahedron Lett. 1903 (1975).

[37] *G. Sipma, B. Van der Wal, D. K. Ketternes*, Tetrahedron Lett. 4159 (1972).

[38] (str) *T. Suga, T. Hirata, K. Tori*, Chem. Lett. 715 (1974); (bio) *T. Suga, T. Hirata, F. Koyama, E. Murakami*, Chem. Lett. 873 (1974).

[39] *M. S. R. Nair, S. T. Carey*, Tetrahedron Lett. 3517 (1975).

[40] *G. B. Russell, P. G. Fenemore, P. Singh*, Chem. Commun. 166 (1973); see also (inumakilactone E) *T. Hayashi, K. Kakisawa, S. Ito, Y. P. Chen, H. Y. Hsu*, Tetrahedron Lett. 3385 (1972).

[41] *D. B. Borders, P. Shu, J. E. Lancaster*, J. Am. Chem. Soc. *94*, 2540 (1972).

[42] *R. Vleggar, P. S. Steyn, D. W. Nagel*, J. Chem. Soc. Perkin I, 45 (1974).

[43] *D. W. Nagel, P. S. Steyn, N. P. Ferreira*, Phytochemistry *11*, 3215 (1972).

[44] *L. J. Mulheirn, R. B. Beechey, D. P. Leworthy*, Chem. Commun. 874 (1974).

[45] *M. Yamazaki, F. Katoh, J. Ohishi, Y. Koyama*, Tetrahedron Lett. 2701 (1972); *R. Cardillo, C. Fuganti, D. Ghiringhello, D. Giangrasso, P. Grasselli*, Tetrahedron Lett. 4875 (1972).

[46] *R. Thomas*, personal communication to M. O. Moss, Aflatoxin and Related Mycotoxins, Phytochemical Ecology, *J. B. Harborne* (Ed.), p. 140, Academic Press, London 1972, *D. P. H. Hsieh, J. N. Seiber, C. A. Reece, D. L. Fitzell, S. L. Yang, J. I. Dalezios, G. N. LaMar, D. L. Budd, E. Motell*, Tetrahedron *31*, 661 (1975); *P. S. Steyn, R. Vleggaar, P. L. Wessels, D. B. Scott*, Chem. Commun. 193 (1975); *D. P. H. Hsieh, R. C. Yao, D. L. Fitzell, C. A. Reece*, J. Am. Chem. Soc. *98*, 1020 (1976) and references cited therein.

[47] *T. H. Smalberger, A. J. Van der Berg, R. Vleggar*, Tetrahedron *29*, 3099 (1973).

[48] *A. I. Gurevich, M. G. Karapetyan, M. N. Kolosov, V. N. Omelchenko, V. V. Onoprienko, G. I. Petrenko, S. A. Papravko*, Tetrahedron Lett. 1751 (1972).

[49] *M. Furukawa, Y. Iitaka*, Tetrahedron Lett. 3287 (1974).

[50] *D. Shiengthong, T. Donavanik, V. Uaprasert, S. Roengsumran, R. A. Massy-Westropp*, Tetrahedron Lett. 2015 (1974).

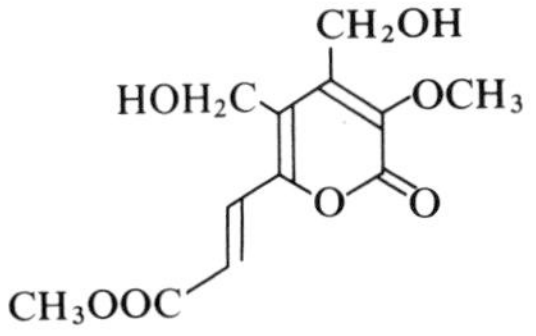

CH2OH HOH2C OCH3 O O CH3OOC

(48)

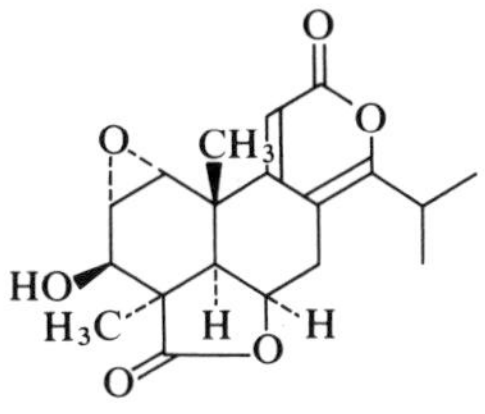

O CH3 O O HO H3C H H O O

(49)

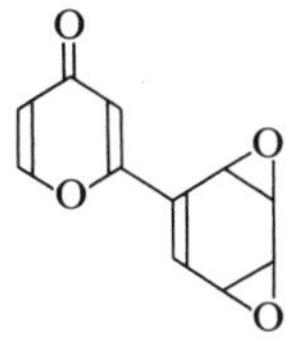

O O O O

(50)

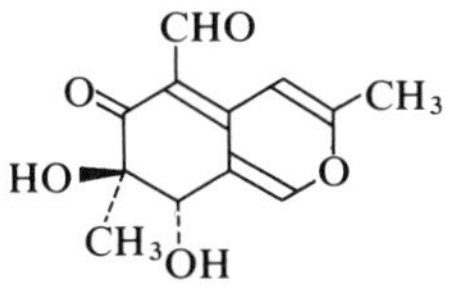

CHO O CH3 HO O CH3 OH

(51)

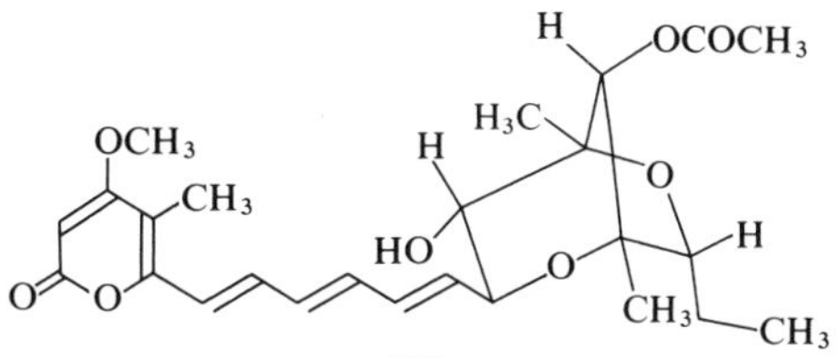

H OCOCH3 H3C OCH3 H CH3 O HO O H O O CH3 CH3

(52)

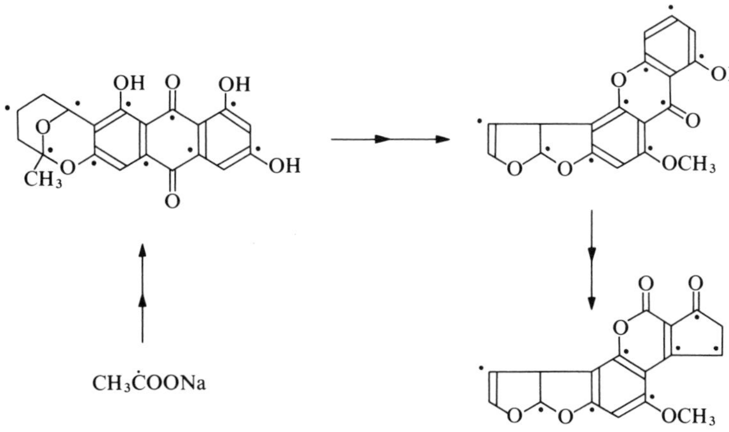

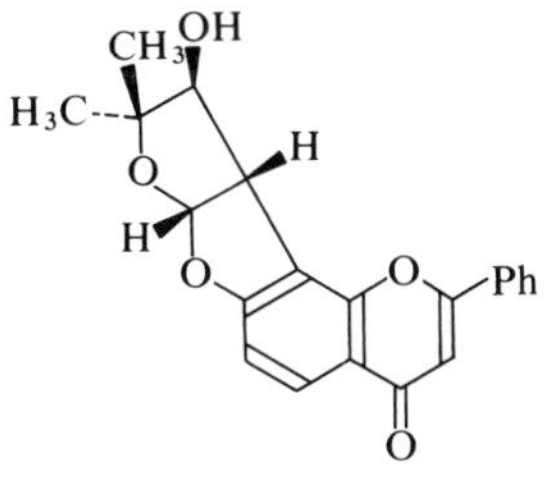

OH O OH OH CH3 O O O

OH O OCH3

O O OCH3

CH3ĊOONa

CH3 OH H3C H O H O O Ph O

(53)

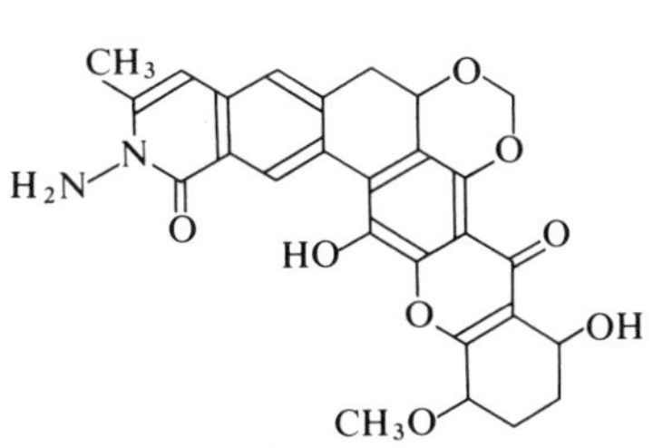

CH3 O H2N N O O O HO O O OH CH3O

(54)

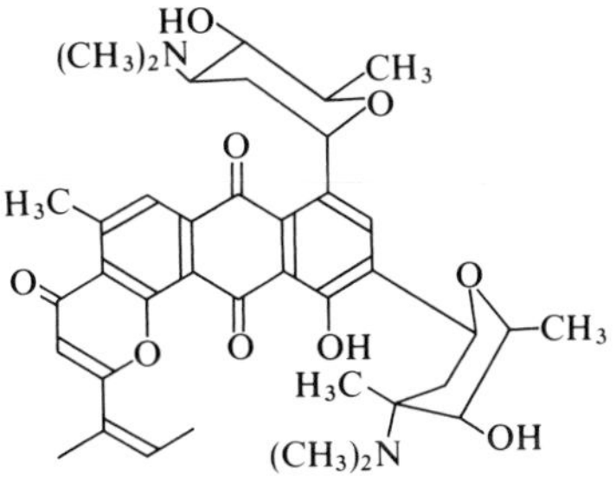

HO (CH3)2N CH3 O O H3C O O O OH CH3 H3C (CH3)2N OH

(55)

CH3O O O O OH O OCH3 OCH3

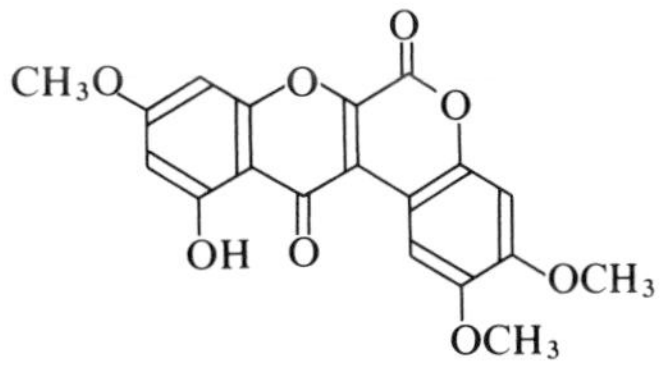

(56)

5.2 Pyrrol Compounds

5.2.1 The Chemistry of Naturally Occurring Bile Pigments

Abert Gossauer
Institut für Organische Chemie der Technischen Universität
Schleinitzstrasse
D-3300 Braunschweig, West Germany

1 Introduction

The best known bile pigments occur in nature as degradation products of haemoglobin and related chromoproteins (see Scheme 1). All known bile pigments related to protoporphyrin IX (the chromophore of haeme: **1**) belong to the IX α series. That means they are formed by oxidative cleavage of the porphyrin ring at the methine bridge which links together the rings A and B of the macrocycle, the missing meso C-atom being lost as carbon monoxide[1].

On the other hand, different types of bile pigment-like compounds have been isolated from plants (*e.g.* phycobilins and phytochrome) and some invertebrates, where they participate to important, and often vital, biological processes. Bile pigments are usually represented by linear formulae (*e.g.* **2** and **3**) which assume E-configurated exocyclic double bonds. Most likely, however, bile pigments whose rings II and III are linked by a methine bridge, thus forming a pyrromethene moiety, are Z-configurated at the middle double bond, and their molecules are helix-shaped (c.f. Ref.[2]). Recently the X-ray analysis of biliverdin dimethylester[3] and bilirubin[4] showed that they occur as all-Z and Z-Z-isomers respectively. On the other hand, the occurrence of stereoisomers at the exocyclic double bonds has been suggested to be responsible for the photochromy of phytochrome[5]. Actually, this kind of isomerism has been encountered in some 5(1*H*)-pyrromethenone[6] and 3,4-dihydro-5(1*H*)-pyrromethenone derivatives[7] as well as in Etiobiliverdin IVγ[7a].

Hitherto all bile pigments related to protoporphyrin IX with the exception of mesobilirhodins (cf. section 5) have been synthesized by condensation of two appropriate building blocks belonging to one or two of the three types (**8**–**10**) represented on Scheme 2. Starting materials are the 5(1*H*)-pyrromethenone carboxylates **8** (R=COOMe, COOEt or COOBu^t) which are readily accessible by base-catalyzed condensation[8] of a pyrrole aldehyde **7** with the appropriate 3-pyrrolin-2-one **6**. Catalytical hydrogenation of the exocyclic double bond of the obtained 5(1*H*)-pyrromethenone derivative **8** leads to the corresponding 5(2*H*)-dipyrrylmethanone **9** which in turn, can be reduced under more drastic conditions yielding the 3,4-dihydro derivative **10**. In order to synthesize the desired bile pigment, one component is transformed into the corresponding 5′-aldehyde (R=CHO) and then condensed with the other component, present as the 5-unsubstituted derivative (R=H) or as *tert*-butyl 5′-carboxylate (R=COOBu^t), in the presence of acid. Some practical examples of the reactions represented on Scheme 2 are given in the following sections.

The present article intends to give a comprehensive review on the chemistry of naturally occurring bile pigments only, with special emphasis on the recent approaches developed to their chemical synthesis. Supplementary information can be obtained from the classical works by H. Fischer[9] and R. Lemberg[10] as well as from later reviews on bile pigments[11–16]. The most

---

[1] *R. Schmid, A. F. McDonagh*, Ann. N. Y. Acad. Sci. *244*, 533 (1975) and references given therein.

[2] *A. Moscowitz, W. C. Krueger, I. T. Kay, G. Skewes, S. Bruckenstein*, Proc. Nat. Acad. Sci. U.S. *52*, 1190 (1964).

[3] *W. S. Sheldrick*, J. Chem. Soc. (Perkin II), 1457 (1976).

[4] *R. Bonnett, J. E. Davis, M. B. Horsthouse*, Nature *262*, 326 (1976).

[5] *M. J. Burke, D. C. Pratt, A. Moscowitz*, Biochemistry, *11*, 4025 (1972).

[6] *H. Falk, G. Grubmayr, U. Herzig, O. Hofer*, Tetrahedron Lett. 559 (1975).

[7] *A. Gossauer, M. Blacha, W. S. Sheldrick*, J.C.S. Chem. Commun. 764 (1976).

[7a] *H. Falk, K. Grubmayr*, Angew. Chem. *89*, 487 (1977); Angew. Chem. Internat. Ed. *16*, 470 (1977).

[8] *H. Plieninger, H. Lichtenwald*, Hoppe-Seyler's Z. Physiol. Chem. *273*, 206 (1942).

[9] *H. Fischer, H. Orth*, Die Chemie des Pyrrols., Vol. II-1, Akademische Verlagsgesellschaft m.b.H. Leipzig 1937, Johnson Reprint Corp., New York 1968.

[10] *R. Lemberg, J. W. Legge*, Haematin Compounds and Bile Pigments, Interscience Publ., New York 1949.

[11] *W. Siedel*, Fortschr. Chem. Org. Naturstoffe *3*, 81 (1939), Springer-Verlag, Wien.

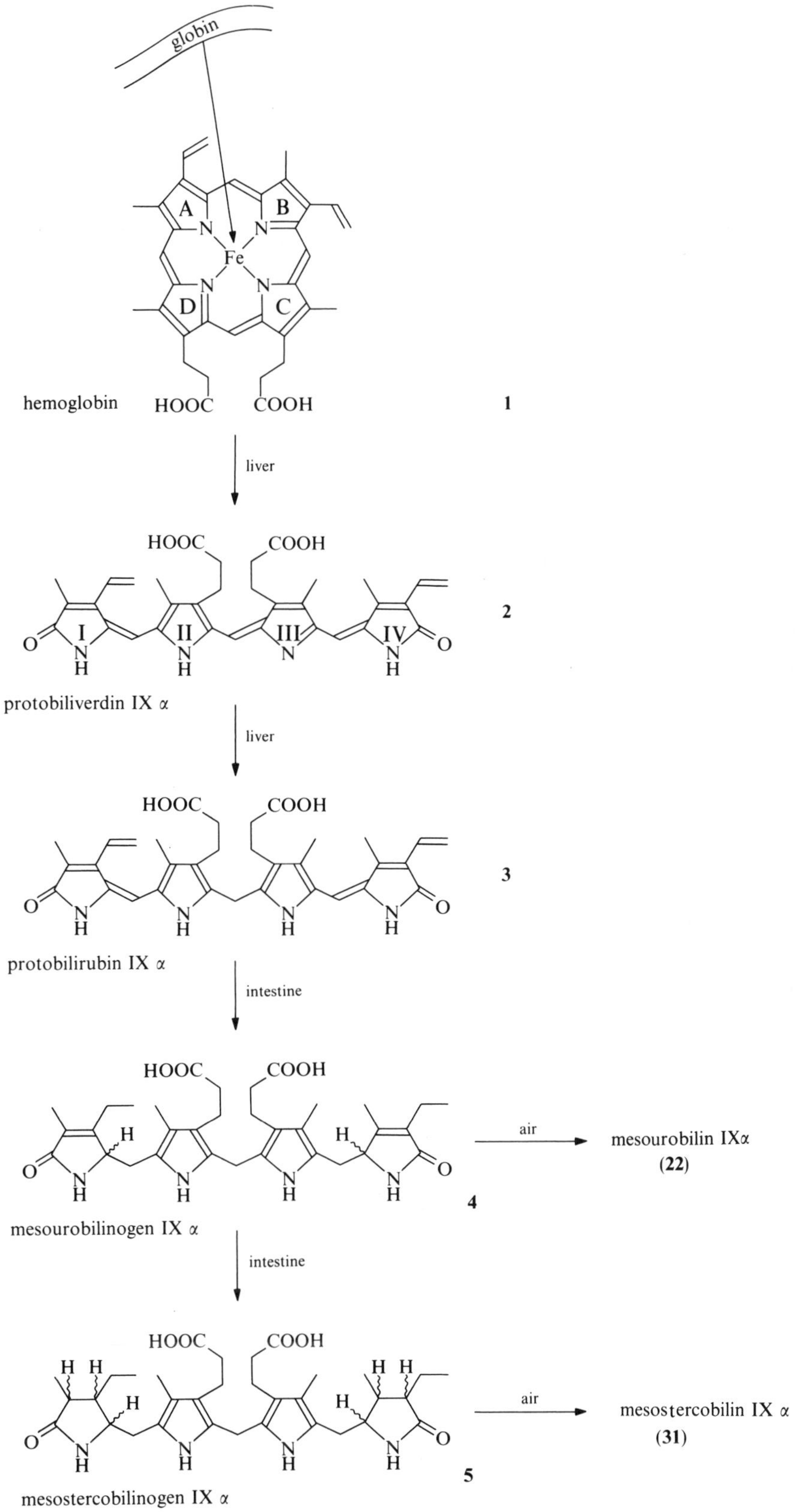

globin

A B D C Fe N N N N

hemoglobin HOOC COOH **1**

liver

HOOC COOH

I II III IV

O N H N H N N H O **2**

protobiliverdin IX α

liver

HOOC COOH

O N H N H N H N H O **3**

protobilirubin IX α

intestine

HOOC COOH

H H

O N H N H N H N H O

mesourobilinogen IX α **4**

air → mesourobilin IXα (**22**)

intestine

HOOC COOH

H H H H

H H

O N H N H N H N H O

mesostercobilinogen IX α **5**

air → mesostercobilin IX α (**31**)

Scheme 1. Biosynthesis of bile pigments.

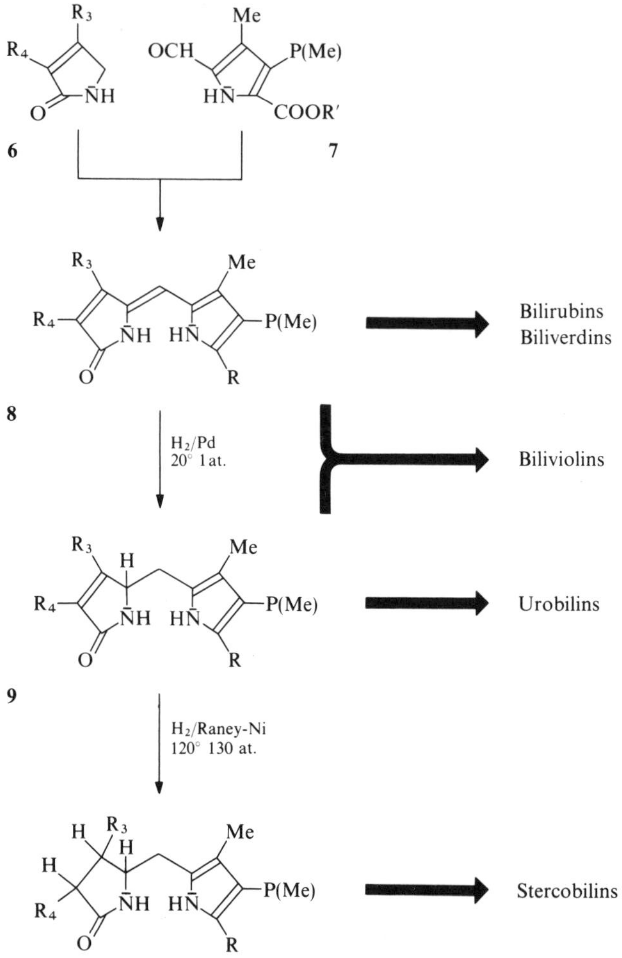

$P(Me) = CH_2-CH_2-COOCH_3$

Scheme 2. General procedure for the synthesis of bile pigments related to protoporphyrin IX.

important aspects of the biochemistry[17] and analytical methods[18] of bile pigments as well as of the chemistry of pyrroles and pyrrolones, which are intermediates for their synthesis[19] have been reviewed extensively elsewhere.

2 Protobiliverdin IXα (Biliverdin IXα, Uteroverdin, Oocyan)

This bile pigment, which is probably the biological precursor of bilirubin in the spleen, liver and other tissues[20,21] can be obtained by dehydrogenation of the latter with ferric chloride[22-25]. The obtained product is, however, a

[12] *C. H. Gray*, The Bile Pigments, Methuen, London 1953.

[13] *R. Lemberg*, Rev. Pure Appl. Chem. (Australia) *6*, 1 (1956).

[14] *W. Rüdiger*, Angew. Chem. *82*, 527 (1970); Angew. Chem. Internat. Ed. *9*, 473 (1970).

[15] *A. H. Jackson, K. M. Smith, in* The Total Synthesis of Natural Products, *J. W. ApSimon* (Ed.), Vol. 1, Wiley, New York 1973.

[16] *M. F. Hudson, K. M. Smith*, Chemical Soc. Revs. *4*, 363 (1975).

[17] *T. K. With*, Bile Pigments, Academic Press, New York 1968.

[18] *W. Rüdiger*, Fortschr. Chem. Org. Naturstoffe *29*, 209 (1971), Springer-Verlag, Wien.

[19] *A. Gossauer*, Die Chemie der Pyrrole, Springer-Verlag, Heidelberg 1974.

[20] *R. Lemberg*, Biochem. J. *29*, 1322 (1935).

[21] *R. Tenhunen, M. E. Ross, H. S. Marver, R. Schmid*, Biochemistry *9*, 298 (1970).

[22] *C. H. Gray, A. Lichtarowicz-Kulszycka, D. A. Nicholson, Z. J. Petryka*, J. Chem. Soc. 2264 (1961).

[23] *R. Lemberg*, Liebigs Ann. Chem. *499*, 25 (1932).

[24] *A. W. Nichol, D. B. Morell*, Biochim. Biophys. Acta *177*, 599 (1969).

[25] *P. Manitto, D. Monti*, Gazz. Chim. Ital. *104*, 513 (1974).

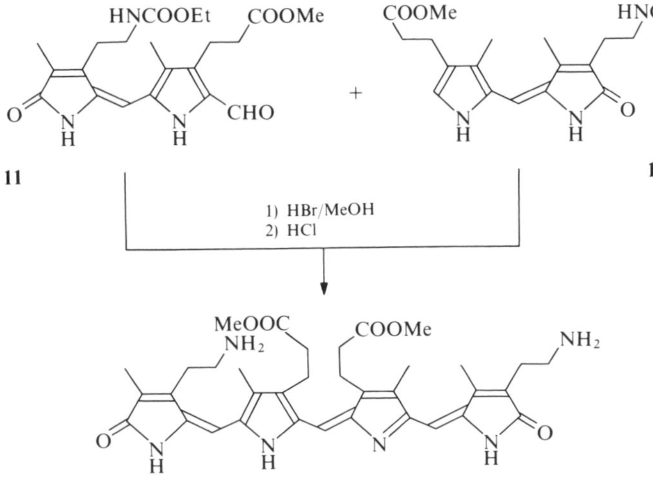

11 + 12

1) HBr/MeOH
2) HCl

13

mixture of isomers[26,27]. More convenient, therefore, is the use of dichloro-dicyano-*p*-benzoquinone as oxidizing agent[28] (cf. Ref.[29]). Biliverdin IXα occurs in different organs and tissues of vertebrates (*e.g.* dog placenta), in the root nodules of some plants, in the egg shells of some birds, in the bones, fins, and skin of certain fish, in the blue coral *Heliopora coerulea*, and in the integumental cells and haemolymphs of insects, such as the praying mantis, locust, and grasshopper (cf. Ref.[14,18]).

The chemical synthesis of biliverdin IXα was achieved by condensation of the 5(1*H*)-pyrromethenone derivatives **11** and **12** followed by cleavage of the urethane group of the obtained product, exhaustive methylation of the zinc chelate* of the resulting 1,19(21*H*, 24*H*)-bilindione **13** and, finally, Hofmann degradation of the 2-aminoethyl groups of the latter[30].

3 Protobiliverdin IXγ (Pterobilin)

Protobiliverdin IXγ (**14**) is the only natural bile pigment so far known which does not belong to the α-series. It occurs in the integumental cells of the caterpillar of the cabbage white butterfly[31,32]. Some neopterobilins (phorcabiline and sarpedobiline) which have been isolated from related species of lepidoptera are most probably formed from pterobilin during isolation of the pigment[33].

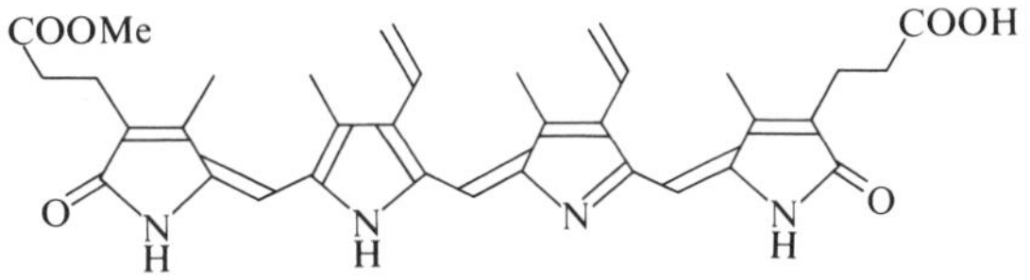

14

4 Protobilirubin IXα (Bilirubin IXα)

This bile pigment is the first isolable degradation product of the haeme. High concentrations of bilirubin in blood are found in jaundice patients. Bilirubin is only very slightly soluble in neutral aqueous media, in blood it occurs non-covalent bonded to protein (*e.g.* serum albumin[34]) as well as mono- and diglucuronic ester ("bilirubin conjugates")[35].

Commercial samples of bilirubin which can be easily obtained by extraction of ox bile stones[36] are usually not pure[37]. Synthetic bilirubin has been prepared by reduction of biliverdin IXα

* In order to avoid methylation at the lactam groups of **13**.

[26] *R. Bonnett, A. F. McDonagh*, J. Chem. Soc. (D), 238 (1970).

[27] *M. S. Stoll, C. H. Gray*, Biochem. J. *117*, 271 (1970).

[28] *H. Plieninger, F. El-Barkawi, K. Ehl, R. Kohler, A. F. McDonagh*, Liebigs Ann. Chem. *758*, 195 (1972).

[29] *H. Fischer, H. Reinecke*, Hoppe-Seyler's Z. Physiol. Chem. *265*, 9 (1940).

[30] *H. Fischer, H. Plieninger*, Hoppe-Seyler's Z. Physiol. Chem. *274*, 231 (1942).

[31] *W. Rüdiger, W. Klose, M. Vuillaume, M. Barbier*, Experientia *24*, 1000 (1968), Experientia *25*, 487 (1969).

[32] *M. Vuillaume, M. Barbier*, C. R. hebd. Séances Acad. Sci. *268*, 2286 (1969).

[33] *M. Choussy, M. Barbier*, Helv. Chim. Acta *58*, 2651 (1975).

[34] *J. G. Jacobsen*, FEBS Lett. *5*, 112 (1969).

[35] *C. C. Kuenzle*, Biochem. J. *119*, 387, 395, 411 (1970).

[36] Ref. [9], p. 634.

[37] *A. F. McDonagh, F. Assisi*, FEBS Lett. *18*, 315 (1971).

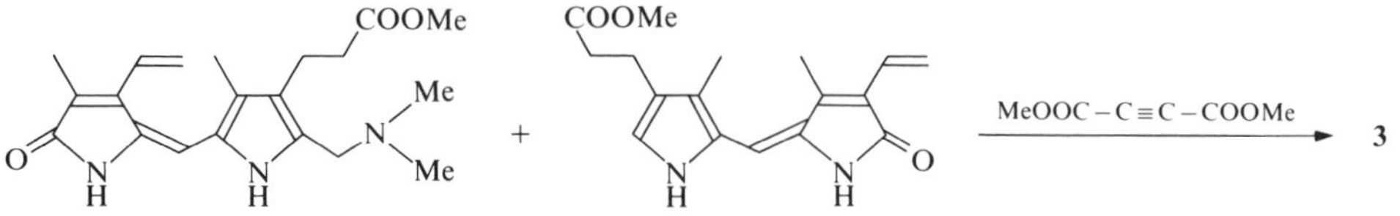

15 + 16 $\xrightarrow{MeOOC-C\equiv C-COOMe}$ 3

with hydrosulfite[30] as well as, directly, by condensation of the Mannich base **15** with the 5′-unsubstituted 5(1*H*)-pyrromethenone **16** in presence of dimethyl acetylenedicarboxylate[28]. The most convenient procedure to prepare dialkyl esters of bilirubin has been reported to be the reaction with 1-alkyl-3-*p*-tolyl triazene[38]. Esterification with diazomethane is usually accompanied by formation of a mono- and a bis-lactam ether of the type **17**[24,39,40]. The corresponding diethoxy derivative **18** can be prepared more conveniently by reaction of bilirubin with triethyloxonium tetrafluoroborate[41]. Compound **18** has been cyclized in the presence of nickel or cobalt ions (cf. Ref.[42]) yielding the tetradehydro corrin derivatives **19d** and **19b** respectively[41].

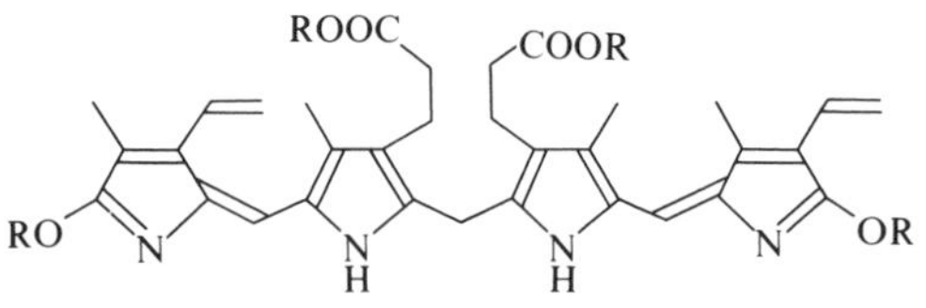

17 $R=CH_3$
18 $R=C_2H_5$

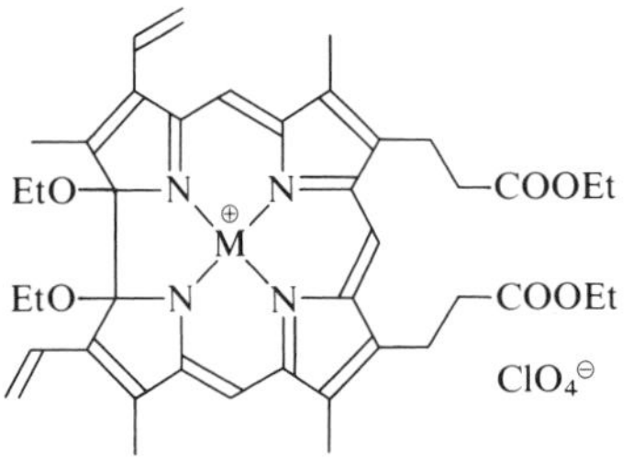

19a M = Ni(II).
19b M = Co (II).

Bilirubin is able to form metal complexes[43–47]. In contrast to the metal chelates formed by all bile pigments whose inner rings are joined by a methine bridge, however, the metal complexes of bilirubin are normally not isolable compounds which contain more than one metal atom per two ligand molecules.

Much work has been done to elucidate the structure of bilirubin in solution[24,48–53a]. It is now well established that the free propionic acid chains and the lactame and pyrrole rings are involved in intramolecular bonding giving rise to two energetically favorable conformations which are enantiomeric to each other and separated by a free-energy barrier of 17.9 ± 0.5 kcal mole^{-1}[53a]. In the solid state, the presence of intramolecular hydrogen bonds has been shown by X-ray diffraction analysis[3].

In strongly acidic media, bilirubin adds protic nucleophiles (*e.g.* alcohols or thiols) to the vinyl groups according to Markownikoff's rule[53b]. The same products are obtained when bilirubin, dissolved in chloroform, is irradiated with UV-light ($\lambda > 300$ nm) in the presence of alcohols[53c] or thiols[53d].

---

[38] *D. W. Hutchinson, B. Johnson, A. J. Kneil*, Biochem. J. *133*, 493 (1973).
[39] *H. Fischer, H. Plieninger, O. Weissbarth*, Hoppe-Seyler's Z. Physiol. Chem. *268*, 197 (1941).
[40] *C. C. Kuenzle, M. H. Weibel, R. R. Pelloni*, Biochem. J. *133*, 357 (1973).
[41] *H. H. Inhoffen, H. Maschler, A. Gossauer*, Liebigs Ann. Chem. 141 (1973).
[42] *D. Dolphin, R. L. N. Harris, J. L. Huppatz, A. W. Johnson, J. T. Kay*, J. Chem. Soc (C), 30 (1966).
[43] *P. O. Carra*, Nature (London) *195*, 899 (1962).
[44] *R. A. Velapoldi, O. Menis*, Clin. Chem. *17*, 1165 (1971).
[45] *C. C. Kuenzle, R. R. Pelloni, M. H. Weibel, P. Hemmerich*, Biochem. J. *130*, 1147 (1972).
[46] *D. W. Hutchinson, B. Johnson, A. J. Knell*, Biochem. J. *133*, 399 (1973).
[47] *J. D. VanNorman, R. Szentirmay*, Bioinorg. Chem. *4*, 37 (1974).
[48] *J. Fog, E. Jellum*, Nature *198*, 88 (1963).
[49] *R. Brodersen*, Acta Chem. Scand. *20*, 2895 (1966).
[50] *R. Brodersen, H. Flodgaard, J. Krogh Hansen*, Acta Chem. Scand. *21*, 2284 (1967).
[51] *C. C. Kuenzle, M. H. Weibel, R. R. Pelloni, P. Hemmerich*, Biochem. J. *133*, 364 (1973).
[52] *P. Manitto, G. Serverini Ricca, D. Monti*, Gazz. Chim. Ital. *104*, 633 (1974).
[53] *P. Manitto, D. Monti*, a) J.C.S. Chem. Commun. 122 (1976); b) Experientia *29*, 137 (1973); c) Experientia *27*, 1147 (1971); d) Experientia *28*, 379 (1972).
[54] *R. Bonnett, J. C. M. Stewart*, J. Chem. Soc. (Perkin I), 224 (1975).

Peculiar attention merits in the last years the study of the photochemistry of bilirubin[54,55], particularly in connection with the treatment of neonatal jaundice by phototherapy. The chemical work done in this area has been reviewed recently[56].

5 Urobilins

Bilirubin and its conjugates pass with the bile into the intestine, where they are hydrogenated by microorganisms of the fecal flora (mainly Clostridia[57]) under formation of urobilinogen (**4**) which upon mild re-oxidation by air yields urobilins. Urobilinogen which was prepared by H. Fischer by reduction of mesobilirubin IXα[58] (cf. Ref.[59]) as a colorless crystalline substance of melting point 197–202° was shown to be identical with the product isolated from several pathological urines[60]. Until now, three different urobilins have been claimed to be isolated from natural sources[61]: mesourobilin (formerly *i*-urobilin: racem. **22**) which occurs as a racemate[62], (+)-mesourobilin (*d*-urobilin of molecular weight 590:**22**)[63] and *d*-urobilin of molecular weight 588 which is optical active ($[\alpha]_D$ of the hydrochloride $= +4520°$)[64]. The high optical rotation of the urobilins has been attributed to the occurrence of helix-shaped molecules conformatively fixed by intramolecular hydrogen bonding thus forming inherent dissymetric chromophores[2]. The structures of both racemic[65] and optical active ($[\alpha]_D^{20}$ of the hydrochloride $= +4500°$)[66] mesourobilin as well as the absolute configuration of the latter[67] have been confirmed by chemical synthesis. Starting from the optical active 5(2*H*)-dipyrrylmethanone **20** (+)-mesourobilin IXα could be

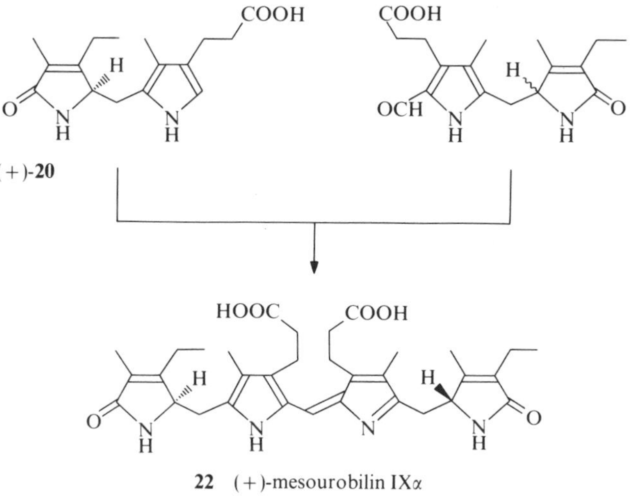

(+)-**20** **21**

22 (+)-mesourobilin IXα

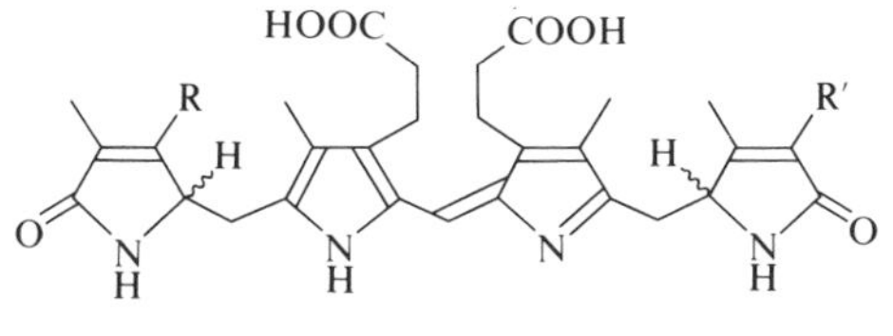

23 R = CH = CH$_2$, R′ = Et

24 R = Et, R′ = CH = CH$_2$

55 *C. S. Foote, T.-Y. Ching*, J. Am. Chem. Soc. *97*, 6209 (1975) and references given therein.

56 *A. F. McDonagh*, Ann. N.Y. Acad. Sci. *244*, 553 (1975).

57 *C. J. Watson*, Ann. Intern. Med. *70*, 839 (1969).

58 *H. Fischer*, Zeitschr. Biol. *65*, 163 (1915).

59 *Z. J. Petryka*, Ann. N.Y. Acad. Sci. *206*, 201 (1973).

60 *C. J. Watson*, J. Biol. Chem. *114*, 47 (1936).

61 *Z. J. Petryka, M. Weimer, D. A. Lightner, M. Chedekel, F. A. Bovey, A. Moscowitz, C. J. Watson*, Ann. N. Y. Acad. Sci. *244*, 521 (1975).

62 *H. Fischer, H. Halbach*, Hoppe-Seyler's Z. Physiol. Chem. *238*, 59 (1936).

63 *C. H. Gray, D. C. Nicholson*, J. Chem. Soc. 3085 (1958).

64 *M. Chedekel, F. A. Bovey, A. I.-R. Brewster, Z. J. Petryka, M. Weimer, C. J. Watson, A. Moscowitz, D. A. Lightner*, Proc. Nat. Acad. Sci. U.S. *71*, 1599 (1974).

65 *W. Siedel, E. Meier*, Hoppe-Seyler's Z. Chem. *242*, 101 (1936).

66 *H. Plieninger, K. Ehl, A. Tapia*, Liebigs Ann. Chem. *736*, 62 (1970).

67 *H. Brockmann, Jr., G. Knobloch, H. Plieninger, K. Ehl, J. Ruppert, A. Moscowitz, C. J. Watson*, Proc. Nat. Acad. Sci. U.S. *68*, 2141 (1971).

obtained after condensation with the racemic aldehyde **21** and successive separation of the resulting epimeric bile pigments[66]. In spite of the availability of both synthetic 2-[68] and 18-vinylurobilin[69] (**23** and **24** respectively), one of which has been suggested to be identical with d-urobilin of molecular weight 588[61,64], the structure of the latter has not yet been demonstrated convincingly.

On treatment with ferric chloride, urobilinogen yields a complex mixture of products whose main components are biliviolinoid compounds[70]. Mesobiliviolin **25** and isomesobiliviolin **26** have been prepared recently by unambiguous syntheses[71]. With base, mesourobilin (racem. **22**) rearranges to a mixture of mesobilirhodin and isomesobilirhodin stereoisomers (**27** and **28** respectively)[72,73] whose structures have been elucidated recently by stereospecific chemical syntheses of their dimethyl esters[74] (cf.Ref.[75]). Until now the occurrence of both mesobiliviolins and mesobilirhodins in nature has not been demonstrated.

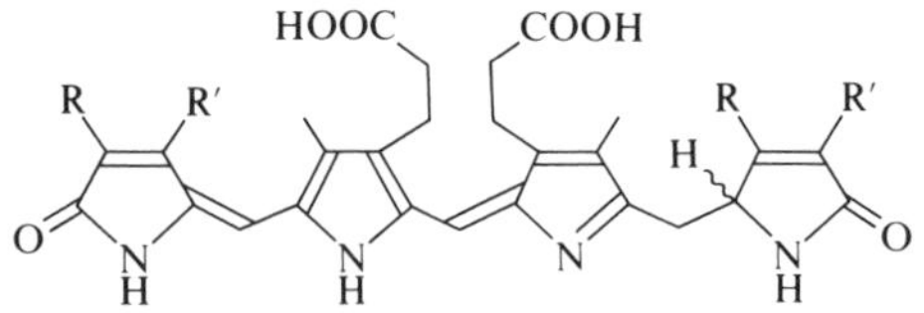

25 R = Me, R′ = Et

26 R = Et, R′ = Me

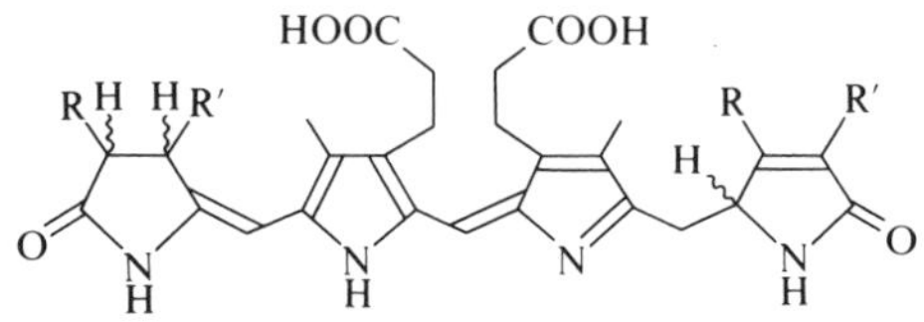

27 R = Me, R′ = Et

28 R = Et, R′ = Me

[68] *H. Plieninger, R. Steinsträsser*, Liebigs Ann. Chem. *723*, 149 (1969); see also: *H. Plieninger, K. H. Hentschel, R. D. Kohler*, Liebigs Ann. Chem. 1522 (1974).

[69] *A. Gossauer, J.-P. Weller*, Chem. Ber. *111*, 486 (1978).

[70] *M. S. Stoll, C. H. Gray*, Biochem. J. *117*, 271 (1970).

[71] *H. Plieninger, K. Ehl, K. Klinga*, Liebigs Ann. Chem. *743*, 112 (1971).

[72] *W. Rüdiger, H. P. Köst, H. Budzikiewicz, V. Kramer*, Liebigs Ann. Chem. *738*, 197 (1970).

[73] *P. O'Carra, S. D. Killilea*, Tetrahedron Lett. 4211 (1970).

[74] *A. Gossauer, G. Kühne*, Liebigs Ann. Chem. 664 (1977).

[75] *A. Gossauer, D. Miehe*, Liebigs Ann. Chem. 352 (1974).

6 Half Stercobilin

This bile pigment which probably consists of a mixture of two isomers **29** and **30** was discovered by Watson[76] in the feces of patients with idiopathic hemolytic anemia. As stercobilin (see below) half-stercobilin is levorotatory. Its synthesis has not yet been performed.

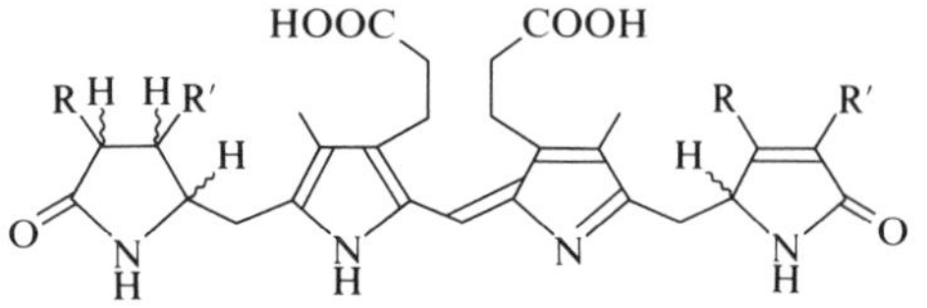

29 R = Me, R′ = Et

30 R = Et, R′ = Me

7 Mesostercobilin IXα (Stercobilin IXα)

Further hydrogenation of urobilinogen by the intestinal bacteria (cf. Section 5) leads to the formation of half-stercobilinogens (see above) and, finally, stercobilinogen (**5**). Unlike the readily crystallizable mesourobilinogen, the corresponding precursor of urobilin, stercobilinogen has never been prepared in crystalline form. Natural (−)-stercobilin (**31**) was first obtained crystalline by Watson from human feces[77,78]. It occurs also in pathological urines[79]. Stercobilin is an optical active compound ($[\alpha]_D^{20} = -4000°$) which shows a large amplitude of the 490 nm Cotton effect in nonpolar solvents[80]. The absolute configuration at all the six chiral centers of the molecule has been established[67] by chromic acid degradation to theyl methyl succinimides of already known absolute configuration as well as for the asymmetric atoms C-4 and C-16, on the basis of the relative configuration of an optical active 3,4-dihydroisoneobilirubinic acid derivative which was prepared as intermediate for the synthesis of a stereoisomer of natural stercobilin[81]. (−) Stercobilin yields a characteristic crystalline

[76] *C. J. Watson, A. Moscowitz, D. A. Lightner, Z. J. Petryka, E. Davis, M. Weimer*, Proc. Nat. Acad. Sci. U.S. *58*, 1957 (1967).

[77] *C. J. Watson*, Hoppe-Seyler's Z. Physiol. Chem. *204*, 57 (1932); Hoppe-Seyler's Z. Physiol. Chem. *208*, 101 (1932).

[78] *C. J. Watson*, J. Biol. Chem. *105*, 469 (1934).

[79] *C. J. Watson*, Hoppe-Seyler's Z. Physiol. Chem. *221*, 145 (1933).

[80] *A. Moscowitz*, Proc. Nat. Acad. Sci. U.S. *52*, 1190 (1964).

[81] *H. Plieninger, J. Ruppert*, Liebigs Ann. Chem. *736*, 43 (1970).

[82] *C. J. Watson*, Hoppe-Seyler's Z. Physiol. Chem. *233*, 39 (1935).

complex with ferric chloride, a reaction which has not been observed for other urobilinoids thus far examined[82,83].

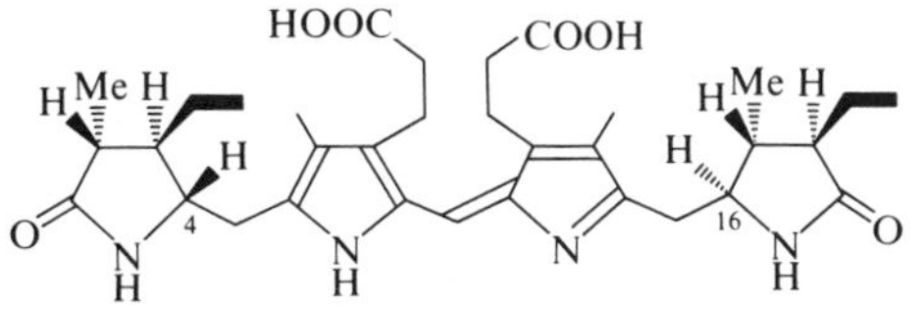

31 (−)-mesostercobilin

8 Phycobilins

Phycobilins are bile pigment-like compounds which occur as prosthetic groups of chromoproteins (so-called biliproteins) isolated from the red and blue-green algae. Such derivatives play an important role as photosynthetic pigments in the physiology of these systems. Because their absorption range comprises wavelengths where the chlorophyll absorption is already weak, the algae become able to absorb light which penetrates relatively deep into the sea water.

Photosynthetically active red and blue biliproteins, called phycoerythrins and phycocyanins, respectively, have been isolated from different algae (rhodophyta, cyanophyta, cryptophyta and one or two members of the chlorophyta). The occurence of bile pigments in the red algae (rhodophyta) and blue-green algae (cyanophyta) was pointed out some years ago by R. Lemberg[84−86]. C-phycocyanin, the main biliprotein in blue-green algae was first reported to contain a prosthetic group of the mesobiliviolin type, which was isolated by digestion of the chromoprotein in concentrated hydrochloric acid at 80°, and characterized as mesobiliviolin IXα (**25**).

Under somewhat milder conditions (conc. HCl at 25°) C.O'hEocha succeeded in the isolation of three pigments from C-phycocyanin, which he called phycobilin 630, phycobilin 655 and phycobilin 608 according to their main absorption maxima in acidic chloroform solution.

Because one of them (phycobilin 630) was converted to a mesobiliviolin-type pigment under the conditions employed by R. Lemberg for hydrolyzing the biliproteid, it was considered to be the prosthetic group (phycocyanobilin) of C-phycocyanin[87].

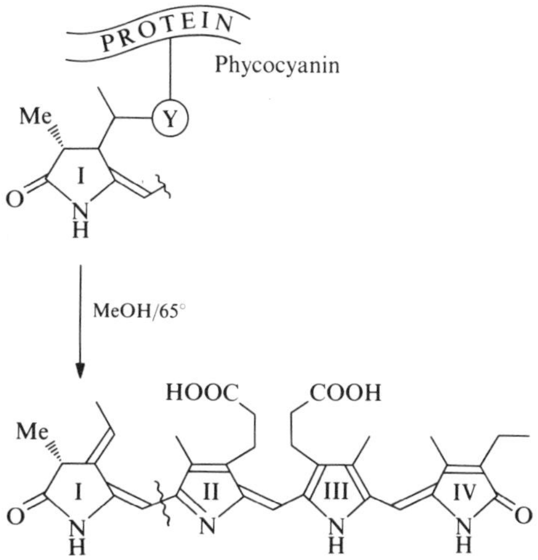

PROTEIN

Phycocyanin

MeOH/65°

32 phycocyanobilin

On the other hand, a so called "blue pigment" is obtained by refluxing C-phycocyanin in methanol[88−92]. Because approximately 20–30 residues of the chromophore are bound to C-phycocyanin, whose estimated molecular weight amounts to 300.000[93], a maximal yield of 3.6–5.6% (w/w) of phycocyanobilin can be obtained from the biliprotein. At the present time, it is well established that the chromophores of the phycoerythins and phycocyanins are covalently bonded to the protein[94,95] (cf. **32** and **39**). On treatment with boiling methanol an elimination reaction occurs and the corresponding phycobilins which are liberated—phycocyanobilin (=phycobiliverdin: **32**[14]) and phycoerythobilin (=phycobiliviolin: **39**[14]) respectively—are characterized by the presence of an ethylidene group at one of the lactam rings. It seems now to be evident, that the different pigments isolated

[83] *C. J. Watson, Z. J. Petryka*, Anal. Biochem. *30*, 159 (1969).
[84] *R. Lemberg*, Liebigs Ann. Chem. *461*, 46 (1928).
[85] *R. Lemberg*, Liebigs Ann. Chem. *477*, 195 (1930).
[86] *R. Lemberg, G. Bader*, Liebigs Ann. Chem. *505*, 151 (1933).
[87] *C. O'hEocha*, Biochemistry *2*, 375 (1963).
[88] *Y. Fujita, A. Hattori*, J. Biochem. (Tokyo) *51*, 89 (1962).
[89] *Y. Fujita, A. Hattori*, J. Gen. Appl. Microbiol. (Tokyo) *9*, 253 (1963).
[90] *P. O'Carra, C. O'hEocha*, Phytochemistry *5*, 993 (1966).
[91] *W. Siegelman, B. C. Turner, S. B. Hendricks*, Plant Physiol. *41*, 1289 (1966).
[92] *H. L. Crespi, U. Smith, J. J. Katz*, Biochemistry *7*, 2232 (1968).
[93] *O. Kao, D. S. Berns*, Biochem. Biophys. Res. Commun. *33*, 457 (1968).
[94] *H.-P. Köst, W. Rüdiger, D. J. Chapman*, Liebigs Ann. Chem. 1582 (1975).
[95] *E. Köst-Reyes, M.-P. Köst, W. Rüdiger*, Liebigs Ann. Chem. 1594 (1975).

earlier from the algae chromoproteins by treatment with concentrated hydrochloric acid at room temperature (see above) are in reality artefacts formed by a Michael-type addition of nucleophiles (HCl, water or methanol) to the ethylidene double bond of the primary formed phycobilins[96].

Hitherto the best characterized of the pigments which are isolated from the red and blue algae is the dimethyl ester of phycocyanobilin, whose structure has been elucidated by spectroscopic[98] and degradation methods[99], as well as by total synthesis[97,100]. The configuration at the double bond of the ethylidene group as well as at the asymmetric C-3 atom are also known[101].

The striking difference between the structure of phycocyanobilin and those of the bile pigments which arise from the degradation of hemoglobin is the substitution pattern at the ring I, whereas the part of the molecule, which comprises the rings II, III, and IV looks like mesobiliverdin IXα. However, because of the failure of the conventional methods to reduce the endocyclic double bond at the lactam ring of 5(1*H*)-pyrromethenones under preservation of the exocyclic one (cf. Scheme 2) none of the synthetic approaches to the synthesis of bile pigments related to protoporphyrin IX is suitable to obtain compounds of the phycobilin type. Thus, in the first reported synthesis of phycocyanobilin dimethylester Eschenmoser's sulfide contraction method was used to prepare the 3,4-dihydro-5(1*H*)-pyrromethenone derivative **35** starting from E-2-ethylidene-3-methyl-thiosuccinimide (**33**)[100]. A recent improvement of this synthesis consists of the condensation of the latter with the benzyl ester of a α-triphenylphosphoranyliden-α-(pyrrol-2-yl)-acetic acid derivative **34** to obtain **35** which is transformed into the corresponding 5′-aldehyde **36** and, after cleavage of the benzyl ester group, condensed with methyl isoneoxanthobilirubinate (**37**). After condensation, the free hydroxycarbonyl group at C-5 of **38** is cleaved by treatment with trifluoroacetic acid. The overall yield of obtained phycocyanobilin dimethyl ester amounts to 33%, referred to **33**[97].

Owing to the presence of a chirality center at C-2 phycocyanobilin is an optical active compound ($[\alpha]_D = +660°$ in chloroform). Noteworthy there is a marked difference between the optical rotatory dispersion curve of phycocyanobilin and that of C-phycocyanin. In fact the former is almost the mirror image of the protein curve. In virtue of the neighborhood of the lactam carbonyl group the hydrogen atom at C-2 of phycocyanobilin can be readily exchanged whereby racemization of the molecule occurs. Furthermore facile isomerization of phycocyanobilin dimethylester to a product identical with mesobiliverdin IXα takes place when the former is boiled under reflux in the presence of 1 *N* potassium hydroxide in methanol for 15 min[98]. Of particular interest is the fact, that in addition to the hydrogen atoms at C-2 and at a

[96] *R. J. Beuhler, R. C. Pierce, L. Friedman, H. W. Siegelman*, J. Biol. Chem. *251*, 2405 (1976).

[97] *A. Gossauer, R.-P. Hinze*, J. Org. Chem. *43*, 283 (1978).

[98] *W. J. Cole, D. J. Chapman, H. W. Siegelman*, Biochemistry *7*, 2929 (1968).

[99] *W. Rüdiger, P. O'Carra*, Eur. J. Biochem. *7*, 509 (1969).

[100] *A. Gossauer, W. Hirsch*, Liebigs Ann. Chem. 1496 (1974).

[101] *H. Brockmann, Jr., G. Knobloch*, Chem. Ber. *106*, 803 (1973).

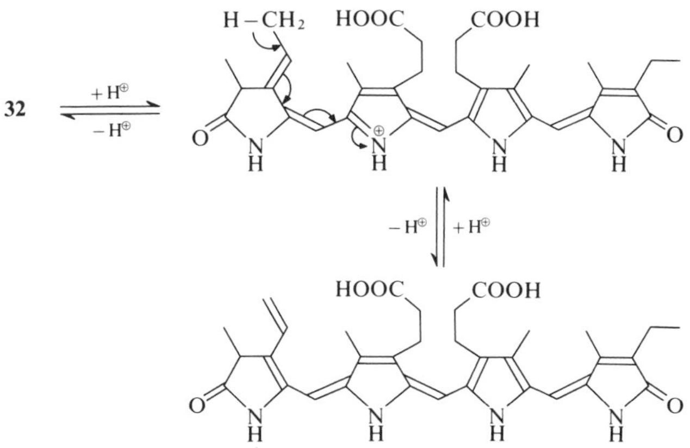

Scheme 3. Plausible mechanism for the exchange of H-atoms at the ethylidene group of phycocyanobilin.

methine bridge (probably C-15) also the hydrogen atoms of the ethylidene methyl group are exchanged by deuterium by refluxing phycocyanobilin dicarboxylic acid with CH_3OD. A tentative interpretation of this behavior has been given on the basis of a tautomeric form of **32** for the phycocyanobilin dicarboxylic acid[92]. Although the occurrence of such tautomeric forms cannot be excluded on principle, it must be pointed out, that exchange of the hydrogen atoms of the ethylidene methyl group can be straightforward explained on the basis of the usual formula **32** for phycocyanobilin according to the mechanism represented on Scheme 3.
The structure of **32** agress well with the NMR spectra of *both* phycocyanobilin dicarboxylic acid and its dimethyl ester respectively. Noteworthy the parent ion of phycocyanobilin dicarboxylic acid at *m/e* 588 corresponds to a molecular weight which is two mass units higher than expected[92,102].
When treated with zinc acetate, solutions of the phycocyanobilin form zinc complex salts which are red-fluorescent under ultraviolet irradiation. These zinc complexes, like those of other bile pigments, are unstable under acidic conditions, when zinc ions are displaced by protons.
Like phycocyanobilin, the chromophore released from C-phycocyanin, phycoerythrobilin (phycobiliviolin—purple pigment: **39**) can be isolated from some red chromoproteins (R-, B- and C-phycoerythins) by treatment with boiling methanol[87,88].

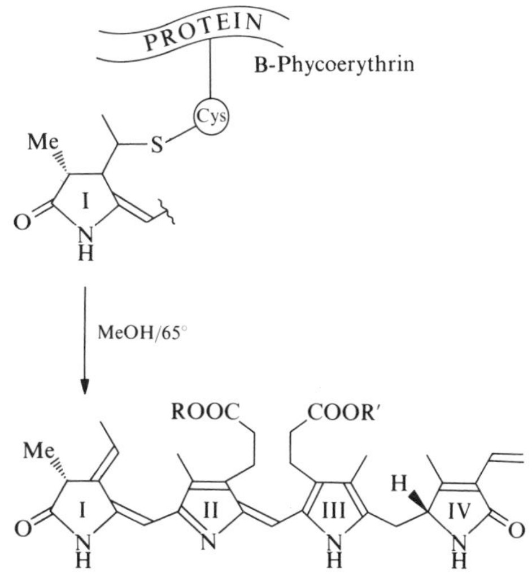

39 ($R=R'=H$) Phycoerythrobilin
40 ($R=H$, $R'=CH_3$) Aplysioviolin

The structure of phycoerythrobilin has been elucidated by spectroscopic[103] and degradation-methods[98,104] as well as, recently, by the chemical synthesis of its dimethyl ester[105] achieved by condensation of the 3,4-dihydro-5(1*H*)-pyrromethane derivative **35**, with methyl 5′-formylvinylisoneobilirubinate (**41**) under analogous conditions to those used to obtain phycocyanobilin dimethyl ester (see above). Phycoerythrobilin is thus isomeric with phycocyanobilin which is in fact obtained on treatment of the former with 12 *N* hydrochloric acid at room temperature[103]. On longer standing in concentrated hydrochloric

[102] *B. L. Schram, H. H. Kroes*, Eur. J. Biochem. *19*, 581 (1972).

[103] *D. J. Chapman, W. J. Cole, H. W. Siegelman*, J. Am. Chem. Soc. *89*, 5976 (1967).

[104] *W. Rüdiger, P. O'Carra, C. O'hEocha*, Nature *215*, 1477 (1967).

[105] *A. Gossauer, R.-P. Hinze, E. Klahr, J.-P. Weller*, to be published.

[106] *P. O'Carra, C. O'hEocha, D. M. Carroll*, Biochemistry *3*, 1343 (1964).

[107] *W. J. Cole, C. O'hEocha, A. Moscowitz, W. R. Krueger*, Eur. J. Biochem. *3*, 202 (1967).

acid solution at 15° phycoerythrobilin is converted slowly to an optically active urobilinoid pigment[85,106,107], while at 100°, in the same solvent, it forms, in addition, an isomeride seemingly identical with mesobiliviolin[106].

COOMe

35 + OCH H N H N H O

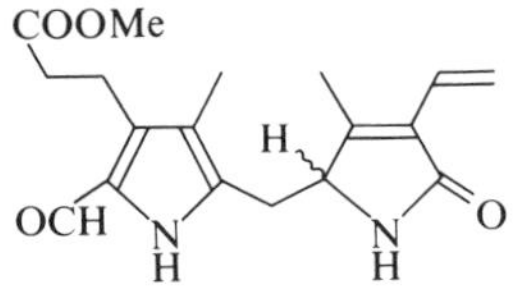

41

⟶ **39** (R = R′ = CH_3)

When oxidized with ferric chloride in methanol phycoerythrobilin yields a green pigment which is spectrally identical with mesobiliverdin[106]. When oxidized with chromic acid phycoerythrobilin[99,104], as phycocyanobilin, yields only E-ethylidene-methyl-succinimide among the degradation products. On the contrary a 2:7 mixture of the Z- ad E-isomers has been recently[108] isolated from B-phycoerythrin under the same conditions. This suggests that phycoerythrobilin is not present in the biliproteid but it is formed during its treatment with methanol. As phycocyanobilin, phycoerythrobilin boiled in 1 *N* KOH in methanol yields mesobiliverdin[91].
The hydrogen atom at C-2 of phycoerythrobilin which is alpha to a carbonyl group, as well as of a methine bridge (probably C-5?) and the methyl hydrogen atoms of the ethylidene group, have been found to be readily exchangeable by deuterium on refluxing the pigment in neutral methanol-d_1[109]. This behavior parallels that of phycocyanobilin.
Optical rotatory dispersion studies of the crystalline urobilins obtained from phycoerythrobilin pointed out that the absolute configuration of the asymmetric center at C-16 of phycoerythrobilin is the same as that in *d*-urobilin[107] (*i.e.* R-configurated[110]).
The absolute configuration of phycoerythrobilin at both C-2 and C-16 has been now unequivocally elucidated by total chemical synthesis of its optical active dimethyl ester (**39**, R = R′ = CH_3)[105].
Spectrophotometric titration of phycoerythrobilin indicated that it is a monobasic pigment having a pKa of about 6.4[106].

[108] *H.-P. Köst, W. Rüdiger*, Tetrahedron Lett. 3417 (1974).
[109] *H. L. Crespi, J. J. Katz*, Phytochemistry *8*, 759 (1969).
[110] *H. Plieninger, H. Böhm*, unpublished results.

When treated with zinc acetate, phycoerythrobilin forms a zinc complex salt which is orange fluorescent ($\lambda_{max}^{CHCl_3}$ = 592 nm). Phycoerythrobilin itself emitted a very stable orange fluorescence (λ_{max} = 623 nm) in 2,6-lutidine solution.
Like other bile pigment metal chelates, the zinc complex salt of phycoerythrobilin is unstable under acidic conditions, when zinc ions are displaced by protons.

9 Aplysioviolin

Closely related to phycoerythrobilin is aplysioviolin, the main violet pigment from the defensive excretion of the sea hare *Aplysia limacina*. Aplysioviolin which probably arises from the phycoerythrins contained in the food (red algae) of these snails is much more loose bound to the protein as phycoerythrobilin and can, therefore, be obtained easily in high yields by aceton extraction.
Based on analytical data W. Rüdiger[111] has proposed for aplysioviolin the structure **40**. Aplysioviolin is therefore the monomethylester of phycoerythrobilin. Particularly the position of the esterified propionic acid chain at C-12 is the only one compatible with the presence of 2,5-diformyl-opsopyrrolecarboxylic acid among the products obtained by oxidative degradation of the pigment, under conditions which leave ester groups unchanged[112]. When esterified with cold methanol-hydrochloric acid aplysioviolin yields a dimethyl ester which is identical with that of phycoerythrobilin, whereas in hot acidic methanol addition of one mol of the solvent—presumably at the ethylidene group—occurs in addition to dehydrogenation. With a pKa value of 6.0 aplysioviolin is slightly less basic as phycoerythrobilin (pKa ≈ 6.4[106]).

10 Phytochrome

A chromoprotein which regulates the periodic growth and development of the green plants, dependent on the conditions of illumination, is called phytochrome[113]. It was isolated for the first time by Siegelman and Firer from oat seedlings[114] (cf. Ref.[115]).

[111] *W. Rüdiger*, Hoppe-Seyler's Z. Physiol. Chem. *348*, 129, 1554 (1967).
[112] *W. Rüdiger*, private communication.
[113] *H. Smith*, Phytochrome and Photomorphogenesis, McGraw-Hill, New York 1975.
[114] *H. W. Siegelman, E. M. Firer*, Biochemistry *3*, 418 (1964).

The biological activity of phytochrome is probably associated with a photo-reversible equilibrium between a red ($\lambda_{max}=660$ nm) and a far-red ($\lambda_{max}=730$ nm) form of the pigment of which only the latter is physiologically active[116].

42

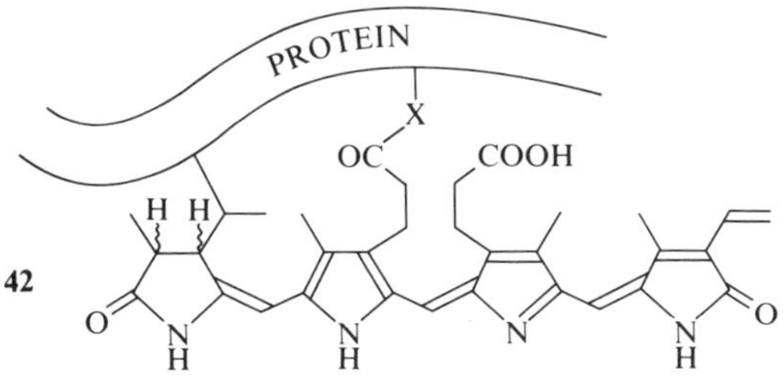

[115] *H. W. Siegelman, S. B. Hendricks*, Fed. Proc. *24*, 863 (1965).

[116] *H. Mohr*, Lectures on Morphogenesis, Springer Verlag, Heidelberg 1972 and references given therein.

The elucidation of the structure of phytochrome has been hampered by the difficulties involved in obtaining large enough quantities of the chromophore. Its relationship to the phycobilin seems, however, to be unquestionable and structure **42** has been suggested[117]. Among the different models proposed by several authors[92,118–120] to explain the photochromy of phytochrome, a change of configuration at the exocyclic double bonds induced by light appears the most suggestive one[5].

[117] *S. Grombein, W. Rüdiger, H. Zimmermann*, Hoppe-Seyler's Z. Physiol. Chem. *356*, 1709 (1975).

[118] *W. Rüdiger, D. L. Correll*, Liebigs Ann. Chem. *723*, 208 (1969).

[119] *H. Scheer, C. Krauss*, Photochem. Photobiol. *25*, 311 (1977).

[120] *S. Schoch, W. Rüdiger*, Liebigs Ann. Chem. 559 (1976).

5.2.2 Macrocyclic Tetrapyrrole Pigments

Johann W. Buchler
Institut für Anorganische Chemie der Technischen Hochschule Aachen, D-5100 Aachen, Templergraben 55, West Germany

1 The Role of Macrocyclic Tetrapyrrole Pigments in Nature

In macrocyclic tetrapyrrole pigments, four pyrrole nuclei form a ring system which is supplemented in corrin (**1**) by three (C-5, C-10, C-15) and in porphin (**2**) by four methine bridges (C_α, C_β, C_γ, C_δ). These compounds fulfil their biochemical functions only as metal chelates in which the central hydrogen atoms located on the nitrogen atoms are substituted by bi- or trivalent metal ions[1, 3, 4, 7, 8, 11, 12, 16, 34]. Some of the closed, rigid chelates are very stable to hydrolytic cleavage. The chromophorous system of vitamin B_{12} occurs as cobalt(III) chelate of **1**. Haems are iron chelates of **2** bearing different so-called "peripheric" substituents on C-1 to C-8; they form the prosthetic groups of haemoproteids (haemoglobins, myoglobins, cytochromes, catalases, and peroxidases). Chlorophylls (photosynthetic pigments in higher plants, algae, and bacteria) are derived from the magnesium chelates of the chlorin (7,8-dihydroporphin) **3** or of bacteriochlorin (3,4,7,8-tetrahydroporphin) **4**.

[1] *R. Willstätter, A. Stoll*, Untersuchungen über Chlorophyll, J. Springer Verlag, Berlin 1913.

[2] *H. Fischer*, Naturwissenschaften *17*, 611 (1929).

[3] *H. Fischer, H. Orth*, Die Chemie des Pyrrols, Akademische Verlagsgesellschaft m. b. H. Leipzig, I. Bd. 1934, II. Bd. I. Hälfte 1937, II. Hälfte 1940; Nachdruck Johnson Reprint Co., New York 1968.

[4] *A. Stoll, E. Wiedemann*, Fortschr. Chem. Forsch. *2*, 538 (1952).

[5] *H. Fischer, in* Org. Syntheses, Coll. Vol. *3*, 442; Siehe auch: *L. Gattermann, H. Wieland*, Die Praxis des organischen Chemikers, 37. Aufl., p. 355, de Gruyter Verlag, Berlin 1956.

[6] *R. B. Woodward*, Angew. Chem. *72*, 651 (1960); Pure Appl. Chem. *2*, 383 (1961).

[7] *J. E. Falk*, Porphyrins and Metalloporphyrins, Elsevier Publ. Co., Amsterdam, London, New York 1964; *K. M. Smith*, Porphyrins and Metalloporphyrins (new edition), Elsevier Publ. Co., Amsterdam, Oxford, New York 1975.

[8] *L. P. Vernon, G. R. Seely*, The Chlorophylls, Academic Press, New York, London 1966.

[9] *M. Weissbluth*, Struct. Bonding (Berlin) *2*, 1 (1967).

[10] *E. Bayer, P. Schretzmann*, Struct. Bonding (Berlin) *2*, 181 (1967).

[11] *T. W. Goodwin*, Porphyrins and Related Compounds, Academic Press, New York, London 1968.

[12] *H. H. Inhoffen, J. W. Buchler, P. Jäger*, Fortschr. Chem. org. Naturstoffe *XXVI*, 284 (1968); *H. H. Inhoffen, P. Jäger, R. Mählhop*, Justus Liebigs Ann. Chem. *749*, 109 (1971).

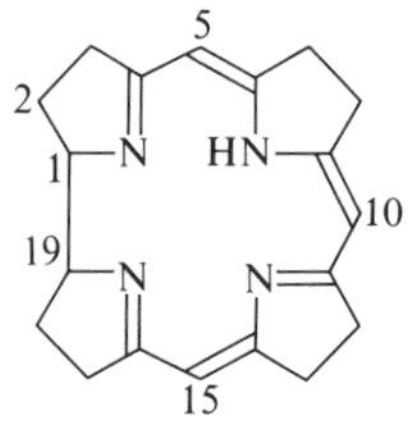

6	R
a	CH_3
b	CHO

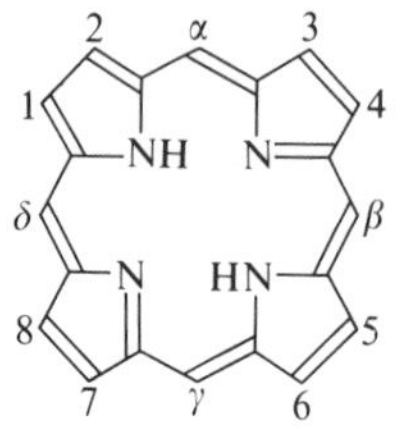

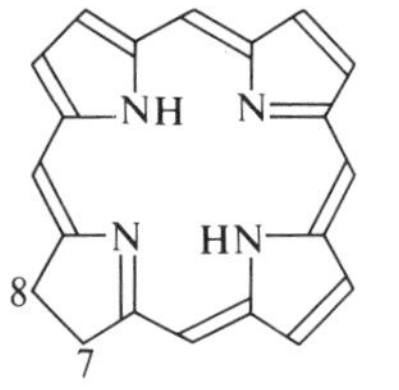

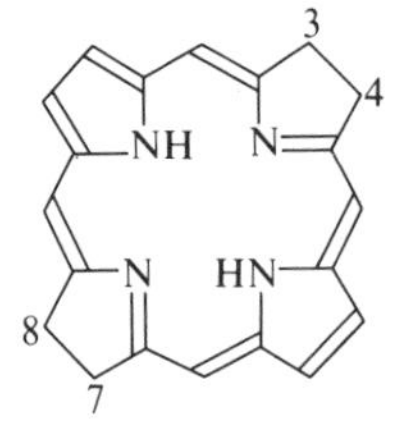

Protohaem **5** is attached to the globin via the axial ligands (proximal histidine) and via the peripheric side-chains. It is responsible for the oxygen transport in the blood (haemoglobin) and muscle tissue (myoglobin) by the reversible bonding of an oxygen molecule to the iron as axial ligand Y[10]. In the absence of globin, oxygen gives rise to conversion of the iron(II) ion in **5** to iron(III) in a reaction being reversible upon the action of reducing agents. This redox reaction determines the function of **5** in cytochrome b as well as in catalases and peroxidases[16].

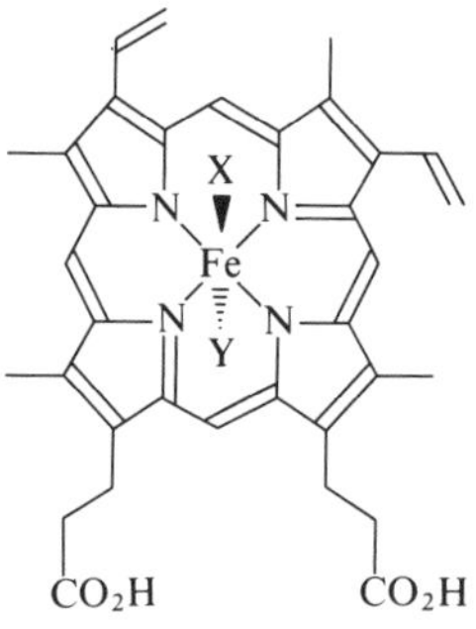

The chlorophylls a (**6a**) and b (**6b**) characterized by Willstätter[1] in higher plants in the constant ratio 3:1, induce the photosynthesis. Magnesium is particularly suitable as the central metal atom for the energy storage and transport, the photochemical electron supply, and the biosynthesis of **6**[8]. It is the lightest metal ion which is stable in aqueous medium in the porphin system. Its closed shell with only 10 electrons disturbs just slightly the fluorescence properties of the chromophore. The capability of bonding with axial ligands X, Y furthers the aggregation required for the energy transport. Porphinmagnesium complexes can be more readily oxidized to radical cations as compared with other porphin complexes[33]. This favors the electron emission as well as the closure of the isocyclic five-membered ring in **6**[17]. Also the transition from bridge-hydrogenated porphin precursors to chlorin complexes is facilitated by central ions with completely filled shells and lower stability[20].
The corrin ligand can stabilize Co(I), Co(II), and Co(III) as well as cobalt alkyls in aqueous medium; the latter is essential for the function of vitamin B_{12}.

2 Isolation

Haems are obtainable as chlorohaemins from defibrinated biological material by dissolving the latter in boiling glacial acetic acid and crystallizing in the presence of chloride ions; thus, from 1 *l* of beef blood can be obtained approx. 4 g of haemin (**5**, X=Cl, no Y, Fe(III) !)[1,3,5].

[13] *R. B. Woodward*, Pure Appl. Chem. *17*, 519 (1968).

[14] *H. Budzikiewicz*, Advan. Mass Spectrometry *4*, 313 (1968).

[15] *G. N. Schrauzer*, Accounts Chem. Res. *1*, 97 (1968).

[16] *G. S. Marks*, Heme and Chlorophyll, D. van Nostrand Co. Ltd., London 1969.

[17] *M. T. Cox, T. T. Howarth, A. H. Jackson, G. W. Kenner*, J. Am. Chem. Soc. *91*, 1232 (1969).

[18] *H. Wolf, I. Richter, H. H. Inhoffen*, Justus Liebigs Ann. Chem. *725*, 177 (1969).

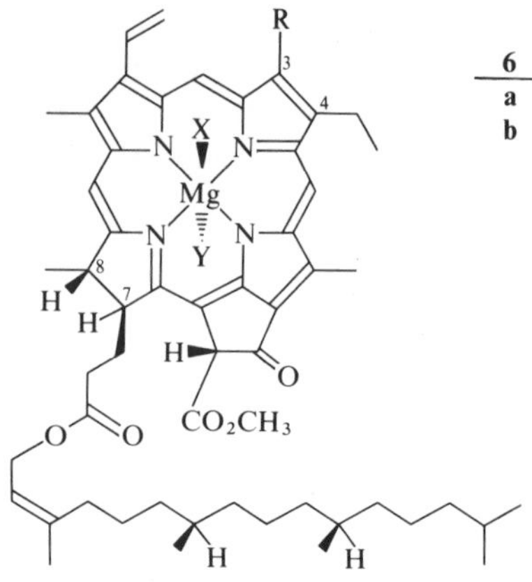

[19] *H. Brockmann, Jr., I. Kleber*, Angew. Chem. *81*, 626 (1969), Angew. Chem. Internat. Ed. Engl. *8*, 610 (1969); *H. Brockmann, Jr.*, Liebigs Ann. Chem. *754*, 139 (1971).

[20] *J. W. Buchler, H. H. Schneehage*, Angew. Chem. *81*, 912 (1969), Angew. Chem. Internat. Ed. Engl. *8*, 893 (1969); Tetrahedron Lett. 3803 (1972).

Chlorophylls **6a** and **6b** free from protein components[4] ('quantasomes', 'chloroplastin') can be extracted from dried leaves with *e.g.* 90% ethanol and separated chromatographically on saccharose with petroleum ether/propanol (0.5%) whereby carotinoids are also eliminated[4,8].

For all further operations, it is recommended to convert all carboxy groups into methoxycarbonyl groups by esterification or reesterification and to demetalate the pigment. In this way, the solubility is increased and the chromatographic behavior improved. The diterpene alcohol phytol (see Chapter 2) which is attached to the propanoic acid side chain at C-7 of **6** can be replaced by other alcohols, *e.g.* enzymatically by means of chlorophyllase[1,8]. Cleavage of the magnesium can be achieved with mineral acids. Hereby the chlorophylls are converted into pheophorbides which, in contrast to the former, are insensitive to light, air, acids, and alkali. Thus, artefact formation is avoided in the further experiments[4].

The removal of the iron(III) ion, which is considerably more strongly coordinated with the nitrogen atoms than the magnesium ion, has to be effected under mild conditions when labile side chains (*e.g.* formyl or vinyl groups) are present. Hydrochloric acid in glacial acetic acid is used for demetalation after reduction of the haemin iron to the more readily cleavable iron(II) by means of iron(II) acetate[7] (too strong reduction and too strongly acidic media have to be avoided as the porphin system is apt to undergo cleavage and uncontrolled resynthesis under these conditions). Thus, **5** yields, after esterification, protoporphyrin IX dimethyl ester **7a**.

Metal-free pigment-methyl esters can be readily separated on fine silica gel by means of column and preparative layer chromatography[18]. The combination of thin-layer chromatography with electron excitation spectroscopy is indispensable for purification tests and analyses of mixtures.

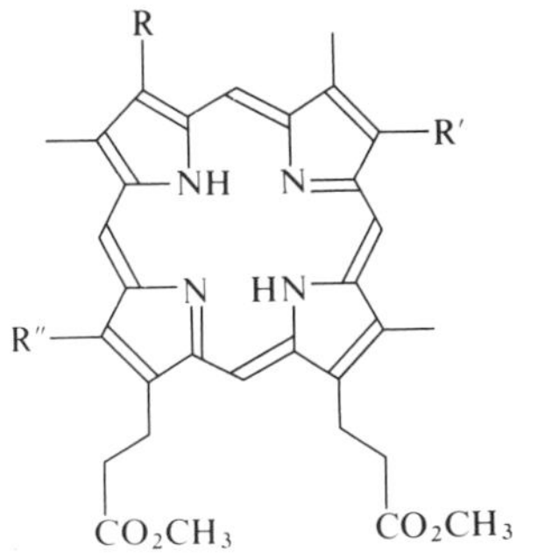

R, R′, R″, NH, N, N, HN, CO_2CH_3, CO_2CH_3

7

7	R	R′	R″
a	$CH=CH_2$	$CH=CH_2$	CH_3
b	H	H	CH_3
c	X[1)]	$CH=CH_2$	$CH=O$
d	H	H	H
e	C_2H_5	C_2H_5	CH_3
f	$CH=O$	$CH=CH_2$	CH_3
g	H	$CH=CH_2$	CH_3
h	Z[2)]	Z[2)]	CH_3

1) $X = CH(O{-}Y)-CH_2(CH_2CH=C(CH_3)CH_2)_3H$

2) $Z = CH(CH_3)-S-CH_2CH(NH_2)CO_2H$

3 Constitution and Structure Elucidation

The constitution and structure elucidation is today predominantly based on physical methods. When the molecule is stable and vaporizable in high vacuum below 300°, information on its elemental composition and constitution can be derived from the (if necessary, high resolution) mass spectrum. The structure of the molecule can then be determined by means of X-ray structural analysis using single crystals; however, only few complete structural data of this type have hitherto been reported on porphins with unsymmetric peripheric substitution[7,21]. Consequently, a combination of the usual spectroscopic methods[7,8,11,12,16,34] has to be applied which, in recent times, include laser-Raman spectroscopy[28], Mößbauer spectro-

---

21 *E. B. Fleischer*, Accounts Chem. Res. *3*, 105 (1970); *J. L. Hoard*, Science *174*, 1295 (1971).

22 *A. J. Bearden, W. R. Dunham*, Struct. Bonding (Berlin) *8*, 1 (1970).

23 *K. Wüthrich*, Struct. Bonding (Berlin) *8*, 53 (1970).

24 *A. Eschenmoser*, Quart. Rev. *24*, 366 (1970); Naturwissenschaften *61*, 513 (1974).

25 *A. W. Johnson*, Pure Appl. Chem. *23*, 375 (1970).

26 *P. Bamfield, R. Grigg, A. W. Johnson, R. W. Kenyon*, J. Chem. Soc. (C), 1259 (1968); *R. Grigg, A. W. Johnson, M. Roche*, J. Chem. Soc. (C), 1928 (1970).

27 *H. H. Inhoffen, J. Ullrich, H. A. Hoffmann, G. Klinzmann, R. Scheu*, Justus Liebigs Ann. Chem. *738*, 1 (1970).

28 *H. Bürger, K. Burczyk, J. W. Buchler, J.-H. Fuhrhop, F. Höfler, B. Schrader*, Inorg. Nucl. Chem. Lett. *6*, 171 (1970).

29 *M. W. Roomi, S. F. MacDonald*, Can. J. Chem. *48*, 139 (1970); Can. J. Chem. *49*, 3544 (1971).

30 *H. Brockmann, Jr., I. Kleber*, Tetrahedron Lett. 2195 (1970).

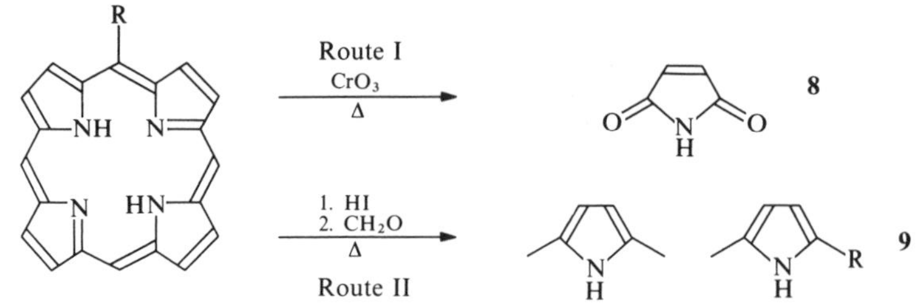

Scheme 1.

scopy[22], electron spin resonance, and magnetic measurements[9] as well as spectropolarimetric procedures[18].

Although paramagnetic ions generally disturb nuclear magnetic resonance spectroscopic investigations and should therefore be eliminated before the measurement, NMR spectra of iron porphins have also been recorded, particularly of haemoproteids[23]. The distinctive signals originating from the prosthetic groups are, due to the interaction with the unpaired electron spins of the iron ion, so strongly shifted as compared with the signals of the protein moiety that they can be distinguished from the latter.

When there is no comparable material available, degradation reactions have to be carried out to establish the peripheric or meso substituents. In resorcinol melts, labile substituents are readily substituted by hydrogen. Thus, deuteroporphyrin IX (**7b**) is formed from **7a** and cytodeuteroporphyrin **7d**[3,7,12,16] from porphyrin a which is the chromophoric system of cytochrome a having constitution **7c**[31] which is now known with the exception of the residue Y.

According to Scheme 1, the total degradation of the macrocycle can either be performed oxidatively or reductively.

Each of the four pyrrole rings is oxidized with chromium(VI) oxide in 50% sulfuric acid to give the corresponding maleic acid imide, *e.g.* **8** (Route I). Sensitive side-chains should be protected prior to degradation, *e.g.* through catalytic hydrogenation of **5** and demetalation to mesoporphyrin IX (**7e**), or degradation should be carried out in 2 *N* H_2SO_4 which requires, however, more time[11].

In the reductive degradation using hydrogen iodide[3,16], subsequent treatment with paraformaldehyde (Scheme 1, route II) is recommended[29]. In the latter process, all the thus formed free α-positions of the pyrrole rings are methylated and thus, the products (*e.g.* **9**) are stabilized and their number reduced. These products can be identified by gas chromatography; meso substituents remain attached to one of the adjacent pyrrole rings as substituted methyl groups, whereas they are cleaved upon applications of route I.

The absolute configurations of **6a** and **6b** and of bacteriochlorophyll could be determined by oxidative degradation and additional NMR and spectropolarimetric measurements[12,18,19,34]. Bacteriochlorophyll differs from chlorophyll a by the presence of two further hydrogen atoms at C-3 and C-4 as well as by an acetyl group in place of a vinyl group at C-2. Bacteriochlorophyll b which is closely related to chlorophyll a is presumably an epimer with respect to C-3 or C-4[30]. Both compounds contain the chromophoric system of **4**.

Recently, also chlorophyll derivatives with an unnatural configuration have been reported[40].

4 Total Syntheses

Numerous further syntheses[3,4,8,11,12,16,34] followed the final constitutional determination of **5** (H. Fischer[2], 1929), which was furnished by the synthesis of haemin, and the successful synthesis of the chlorophyll **6a** (R.B. Woodward[6], 1960). The following section describes some details of the synthesis of unsymmetrical porphyrins.

According to Scheme 2, four suitably substituted pyrrole systems are connected to one another stepwise. This process involves introduction of the meso-bridges, *e.g.* in the form of formyl or bromomethyl groups, which react with pyrroles unsubstituted in the α-position and generally

[31] *G. A. Smythe, W. S. Caughey*, Chem. Commun. 809 (1970).

[32] *A. D. Adler, F. R. Longo, F. Kampas, J. Kim*, J. Inorg. Nucl. Chem. *32*, 2443 (1970).

[33] *J.-H. Fuhrhop*, Struct. Bonding (Berlin) *18*, 1 (1974); Angew. Chem. *86*, 363 (1974), Angew. Chem. Internat. Ed. Engl. *13*, 321 (1974).

[34] *K. M. Smith*, Quart. Rev. *25*, 31 (1971).

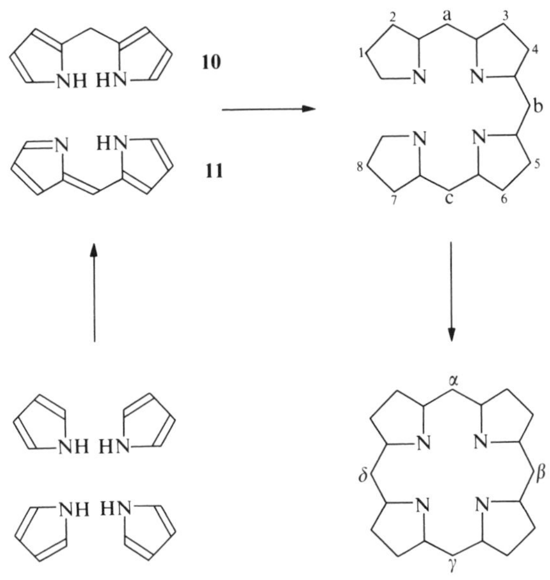

Scheme 2

generated in situ. Two pyrrole systems each are linked to the pyrromethanes **10** or pyrromethenes **11** which are in turn converted into the bilane derivatives **12** which are finally terminally bridged to give the porphin skeleton **13**. In most cases, **13** contains completely or partially hydrogenated methine bridges which are readily oxidized (usually in situ with air) to the porphin skeleton **2**. As, under the reaction conditions, the bilanes **12** occasionally isomerize owing to exchange and twisting of the pyrrole rings, the syntheses are advantageously carried out via less labile a- or b-oxobilanes, and b-bilenes[34,35,38] respectively, or via a,c-biladienes[25,26,34], the bridgehead positions of which are partially unsaturated. In this way, also spirographisporphyrin **7f**, the ligand system of the haem of polychete worms, and pemptoporphyrin **7g** (a fecal metabolite) could be synthesized.

In principle, the biosyntheses of **5** and **6** starting from porphobilinogens proceed similarly[11].

A series of porphyrins, *e.g.* porphyrin c (**7h**), which is contained in the prosthetical group of cytochrome c [7,16] as an iron chelate, may also be prepared by partial syntheses involving reconstruction of the side-chains of other porphins. In this context, the photoxidation of **7a** and conversion of the resultant product into **7f** as well as the 'functionalization' of the 3-methyl group of chlorophyll derivatives, which allows a transition from the a- into the b-series[12,34], should be mentioned.

The synthesis of vitamin B_{12} (Eschenmoser[24] and Woodward[13]) has stimulated development of surprising new methods of linking the necessary pyrrole units with one another (imino ester condensation, sulfide contraction, photochemical cyclization).

5 Preparation of Chemical Analogs

The difficulties arising in the preparation of large amounts of substances from natural material or by synthetic routes as well as the in vitro sensitivity of the products has led many groups to investigate the preparation of more easily accessible and manipulable model substances which are similar in function or structure or both to the natural tetrapyrrole pigments. Among the non-pyrrole-like substances should be mentioned the planar cobalt(II) chelate of N,N′-ethylene-bis-(salicylidenimine), which reversibly binds oxygen[10], or bis-(dimethylglyoximato)cobalt(I) anion, which, in many reactions, is similar to vitamin B_{12s}[15].

In the field of macrocyclic tetrapyrrole pigments

[35] *A. H. Jackson, G. W. Kenner, K. M. Smith*, J. Chem. Soc. (C), 502 (1971) and preceding articles.

[36] *J. W. Buchler, G. Eikelmann, L. Puppe, K. Rohbock, H. H. Schneehage, D. Weck*, Justus Liebigs Ann. Chem. *745*, 135 (1971).

[37] *H. H. Inhoffen, J. W. Buchler, L. Puppe, K. Rohbock*, Justus Liebigs Ann. Chem. *747*, 133 (1971).

[38] *P. J. Crook, A. H. Jackson, G. W. Kenner*, Justus Liebigs Ann. Chem. *748*, 26 (1971).

two methods are used. Firstly, the central metal atom is varied without changing the porphin ligands, *e.g.* with meso-tetraphenylporphin complexes **14** or octaethylporphin complexes **15**, whereby nearly all metals of the Periodic Table can be introduced[7,32,36]. These compounds are used to investigate the influence of the central metal atom, for example on the redox properties[33] or the hydrogenation of the porphin system[20]; these investigations are facilitated by the high symmetry of the ligand systems. The *in vivo* occurring, direct, constitutionally specific and stereospecific hydrogenation of substituted porphins (**2**) to chlorins (**3**) and bacteriochlorins (**4**) has hitherto not been achieved *in vitro*; in the total synthesis of chlorophyll[6], this hydrogenation has only been indirectly realized.

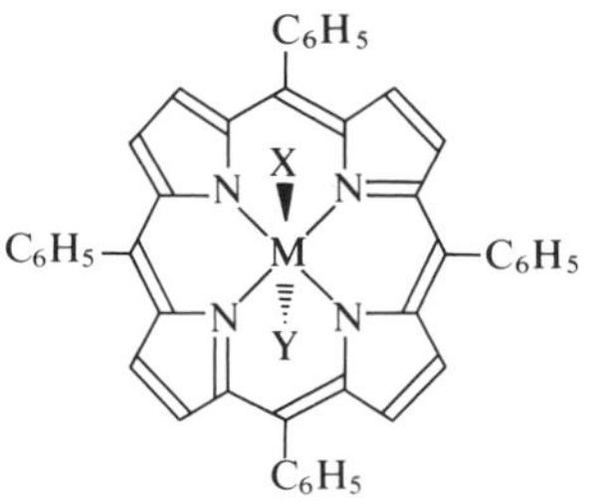

14

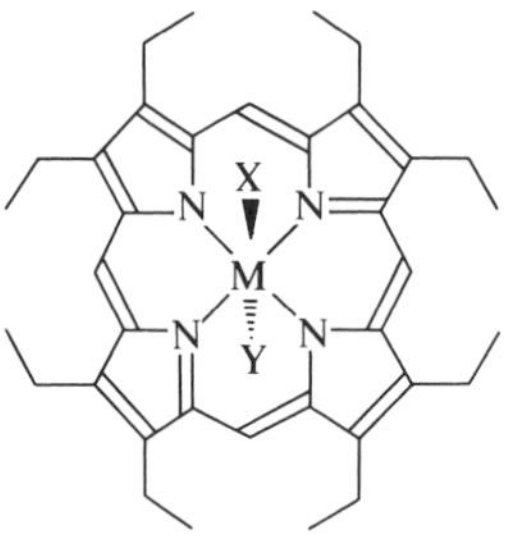

15

The second method is aimed at the preparation of novel tetrapyrrole ligand systems which are envisaged, with respect to their peripherical substitution, to approach the natural pigments stepwise. This procedure will be described by some typical examples.

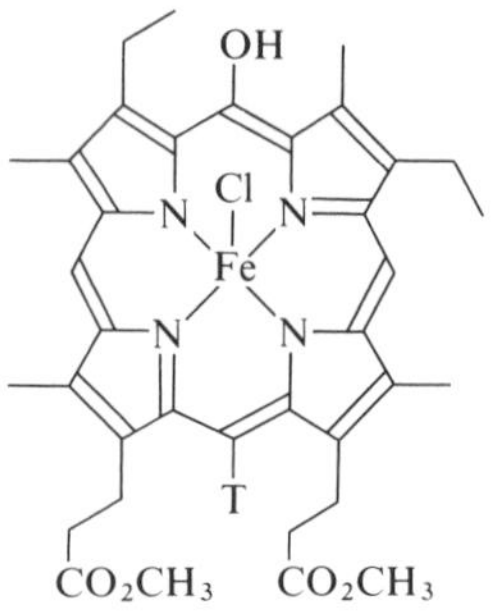

16 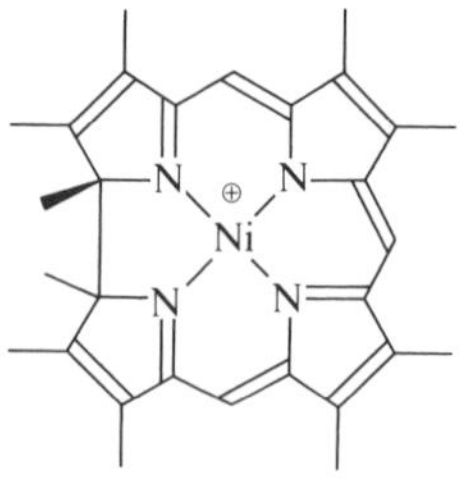 **17**

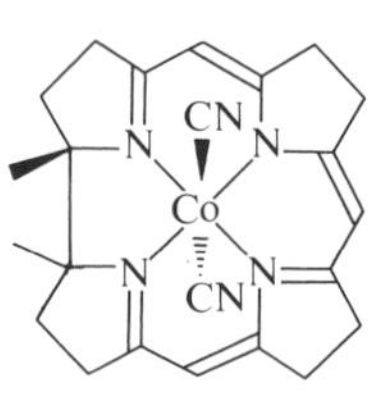

18 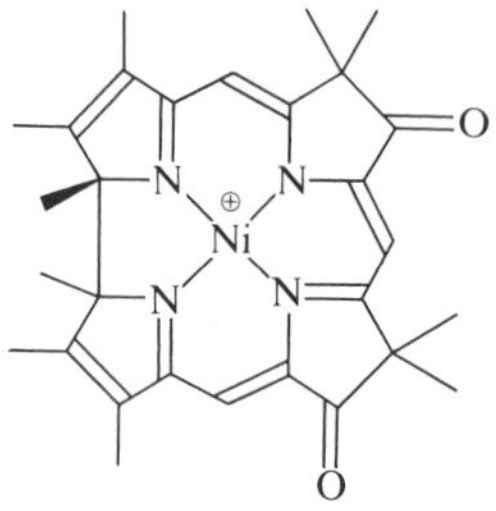 **19**

The formation of bile pigments has been investigated via the preparation of synthetic and labeled meso-hydroxyporphins such as **16**[38]. The synthesis of tetradehydrocorrin-complex ions such as **17** has been studied[25,39] making use of a metal-dependent cyclization of biladiene(a, c) precursors (the reversibility of which has been reported[37]) as well as the alkylation and hydrogenation of the peripheric double bonds of these systems to give corrins such as **18**.

Compound **17** and the metal-free compound **15** have been hydroxylated with osmium(VIII) oxide and the resultant diols rearranged to "Gemini ketones" using sulfuric acid[27]. Thus, **19** was, for example, obtained the oxo functions and peripheric double bonds of which may give incentive to further synthetic routes[27].

New results on the chemistry of synthetic iron porphyrins are particularly noteworthy. Thus, Baldwin[41] and Collman[42] have independently succeeded in synthesizing sterically hindered haems which are capable of reversibly binding molecular oxygen even at room temperature. Collman[43] and especially Holm and Ibers[44] have

[39] *A. W. Johnson, W. Overend*, Chem. Commun., 710 (1971); *R. Grigg, A. W. Johnson, G. Shelton*, J. Chem. Soc. (C) 2287 (1971) and preceding articles.

[40] *H. Wolf, H. Scheer*, Tetrahedron. Lett. 1111, 1115 (1972).

[41] *J. Almog, J. E. Baldwin, J. Huff*, J. Am. Chem. Soc. *97*, 227 (1975) and preceding articles.

[42] *J. P. Collman, R. R. Gagné, C. A. Reed, T. R. Halbert, G. Lang, W. T. Robinson*, J. Am. Chem. Soc. *97*, 1425 (1975) and preceding articles.

[43] *J. P. Collman, T. N. Sorrell, B. M. Hoffman*, J. Am. Chem. Soc. *97*, 913 (1975).

[44] *S. C. Tang, S. Koch, G. C. Papaefthymiou, S. Foner, R. B. Frankel, J. A. Ibers, R. H. Holm*, J. Am. Chem. Soc. *98*, 2414 (1976).

isolated iron(III) porphyrins, containing thiolate residues as axial ligands, which have hitherto been only little known and which have considerable importance as cytochrome models.

Literature Recommended for Further Reading

D. Dolphin, The Porphyrins, Vol. I–VII, Academic Press, New York, London, in press (1978).

5.3 Quinonoid Compounds

Ronald H. Thomson
Department of Chemistry, University of Aberdeen, Meston Walk, Old Aberdeen, Scotland.

1 Quinones

1.1 Introduction

About 500 quinones have been isolated from natural sources[1]. The great majority are benzoquinones, naphthoquinones, or anthraquinones with simple substituents (alkyl, hydroxyl, methoxyl) but more highly condensed systems also occur as quinones and there is considerable structural variation. This structural diversity results in a wide range of color from pale yellow through red, blue and green, to almost black. Despite this their contribution to natural colouring is relatively small. Many are present in roots or in wood and they are most evident in sea urchins, in greenfly (a common pest on roses) and in certain fungi and lichens. Quinones are fairly widely distributed in higher plants, fungi and bacteria, but in the animal kingdom[29] they have been found* only in arthropods, echinoderms and annelids.

Apart from the "bioquinones" (ubiquinones, plastoquinones and the vitamins K) very little is known of the function of natural quinones. The use of volatile benzoquinones by certain arthropods as a chemical defense against predators has been discussed elsewhere. Many plant quinones are weakly antibiotic and may serve as a protection against microbial attack[27], but direct evidence is very limited and as they often have no apparent function (*e.g.*, some microorganisms can thrive both with and without the production of quinones) it is frequently suggested that they are metabolic end-products. However further investigation may show that this is not the case for it has been found recently that the anthraquinones in *Cassia*[2] and in *Rumex*[3] are metabolized during fruit development, and in ergot (*Claviceps purpurea*)[4] and in *Penicillium frequentans*[5] it is established that anthraquinones are the precursors of other products which are formed via cleavage of the quinone ring.

Quinones are isolated by normal methods and no general procedure can be advocated as there is so much variation in structure and polarity. In preliminary tests they are most easily recognized by their redox properties; they are decolorized by mild reducing agents (and frequently reoxidized in air) which distinguishes quinones from nearly all other natural pigments. Sodium dithionite is particularly useful in this respect. Many quinones possess phenolic groups which give useful color reactions with alkalis and in non-hydroxylated compounds an unsubstituted quinonoid position can be detected by the Kesting-Craven color test. Final identification now relies mainly on spectroscopic data although degradative methods may still be required for novel structures. Details may be found elsewhere[1].

1.2 Benzoquinones

Simple quinones of this type[1] are represented by compounds (**1**) to (**4**). Toluquinone (**1**) is excreted by numerous arthropods[6], (**2**; R=H and OH) and (**4**) are mould metabolites, while primin (**3**) is the dermatitic agent found on the leaves of *Primula obconica*.

* Except the ubiquinones which occur more widely—see Part 2.

[1] *R. H. Thomson*, Naturally Occurring Quinones, 2nd ed., Academic Press, London 1971.

[2] *J. W. Fairbairn, A. B. Shrestha*, Phytochemistry *6*, 1203 (1967).

[3] *J. W. Fairbairn, F. J. Muhtadi*, Phytochemistry *11*, 215 (1972).

B. Franck, Angew. Chem. Internat. Ed. Engl. *8*, 251 (1969).

S. Gatenbeck, L. Malmström, Acta Chem. Scand. *23*, 3493 (1970); *R. F. Curtis, C. H. Hassall, D. R. Parry*, Chem. Commun. 1512 (1970).

[6] *J. Weatherston*, Quart. Rev. (London) *21*, 287 (1967); *H. Schildknecht*, Angew. Chem. Internat. Ed. Engl. *9*, 1 (1970).

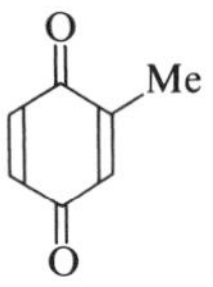

(1) 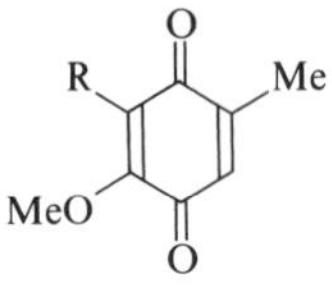 **(2)** 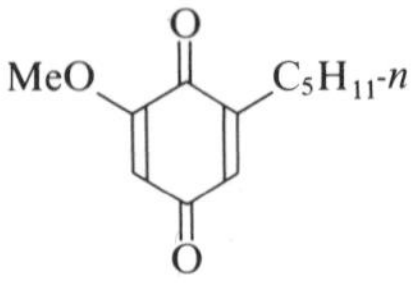 **(3)** 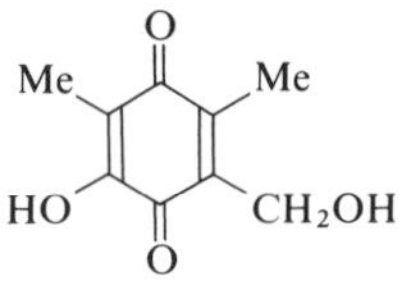 **(4)**

They show normal benzoquinone chemistry (see vol. V) and can be identified spectroscopically. Thus coprinin (**2**; R=H), $\lambda_{max}^{CHCl_3}$ 264, 365 nm (log ε 4.32, 2.88), $\nu_{max}^{CCl_4}$ 1681, 1653 cm^{-1}, shows NMR signals for methyl(*d*) and methoxyl(*s*) protons, coupling of the former to H-3 gives a quartet at τ 3.60, while the signal from H-6 is a singlet (τ 4.26). In primin (**3**), however, the signal from H-5, adjacent to the methoxyl group, is a doublet arising from coupling with H-3. The two substituents are therefore 2,6-orientated. The alkyl side chain is defined by the NMR and mass spectra, the base peak of the latter at *m/e* 153 (M − 55) arising by α-cleavage of the *n*-pentyl chain and rearrangement to give a dihydroxymethoxy-tropylium ion. In shanorellin (**4**) homo-allylic coupling results in a triplet (3H) at τ 8.07 and a quartet (2H) at 5.47 showing that one methyl group is adjacent to the hydroxymethyl group but the relative positions of the other two substituents could not be established by spectroscopic methods and the problem was solved by X-ray crystallographic analysis.

Primin (**3**) was synthesized by oxidation of 6-methoxy-2-*n*-pentylphenol with Fremy's salt while coprinin (**2**; R=H) can be obtained by addition of methanol to toluquinone (**1**), catalysed by zinc chloride or, more efficiently, by Thiele acetylation of (**1**) followed by oxidative hydrolysis and methylation.

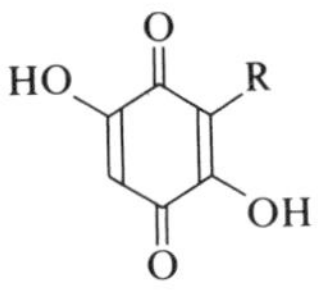

(5) 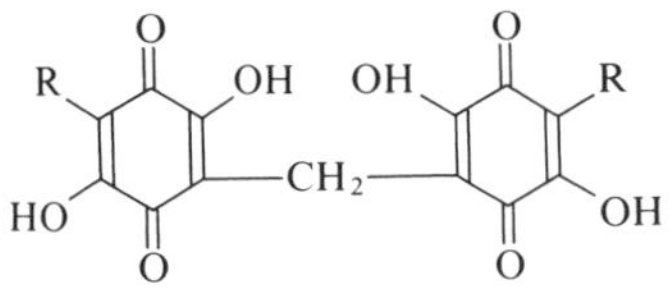 **(6)**

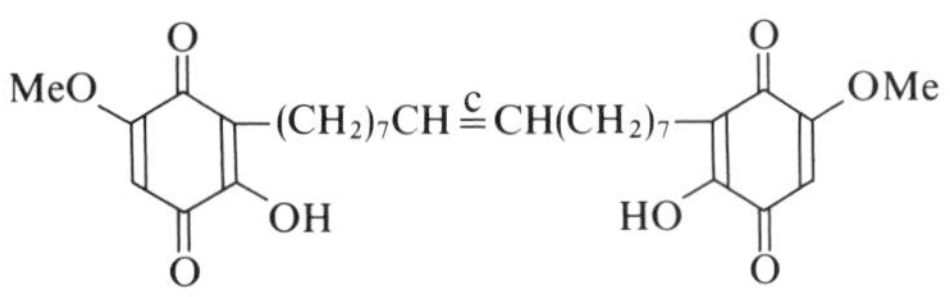

(7)

Long alkyl side chains are a feature of the quinones elaborated by Myrsinaceae; embelin (**5**; R=*n*-$C_{11}H_{23}$), rapanone (**5**; R=*n*-$C_{13}H_{27}$) and ardisiaquinone A (**7**) are present in *Ardisia sieboldiana*, and vilangin (**6**; R=*n*-$C_{11}H_{23}$)* [prepared from (**5**; R=*n*-$C_{11}H_{23}$) by reaction with formaldehyde] occurs with embelin in *Embelia ribes*. The side chains can be detached and identified by oxidation with alkaline hydrogen peroxide, *e.g.*, embelin yields lauric acid. Conversely, embelin has been synthesized by alkylation of 2,5-dihydroxybenzoquinone with lauroyl peroxide. Analogous fungal metabolites with polyprenyl side chains are bovinone [**5**; R=$(CH_2CH{=}C(Me)CH_2)_4H$] and amitenone [**6**;* R=$(CH_2CH{=}C(Me)CH_2)_4H$] found in *Boletus bovinus*.

Benzoquinones of a different type (**8**) are found in rosewoods (*Dalbergia* spp.) along with other neoflavanoids[7]. The UV absorption of these dalbergiones (**8**) is similar to that of (**2**; R=H) but as permanganate oxidation of (**8**; R=OMe) gives *p*-anisic acid there must be a benzenoid group attached to the alkyl side chain. The remaining unidentified fragment (C_3H_4) is unsaturated and must account for the optical activity. These requirements are met by the propene structure (**8**) which was deduced from the NMR spectra and confirmed synthetically by Claisen rearrangement of the cinnamyl ether (**9**) to (**10**), followed by oxidation with Fremy's salt. This gave the racemic dalbergione (**8**; R=H) and it is of interest that the R-form occurs in *D. nigra* and *D. latifolia* while the S-form is found in *D. miscolobium* and *D. baroni*. A quasiracemate of R-(**8**; R=H) and S-(**8**; R=OMe) has been isolated from *D. nigra*.

Another group of benzoquinones, found in fungi and lichens, are terphenyl derivatives, examples being polyporic acid (**11**), thelephoric acid (**12**) and phlebiarubrone (**13**) a rare *o*-benzoquinone. Typically polyporic acid gave *p*-terphenyl on zinc dust distillation and benzoic acid on oxidation of

* The quinones (**6**) may be artefacts.

[7] *W. D. Ollis*, Experientia *22*, 777 (1966).

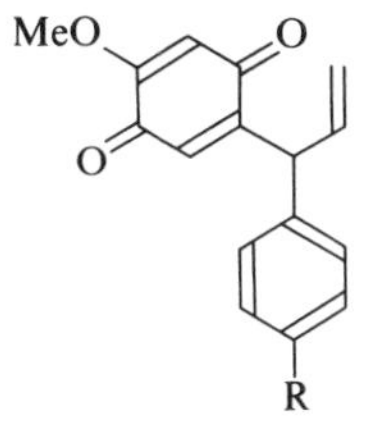

(8) 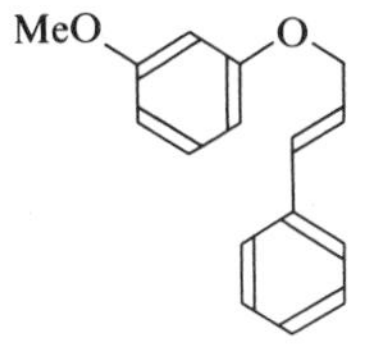 (9) 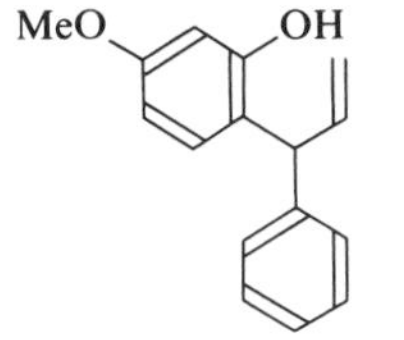 (10)

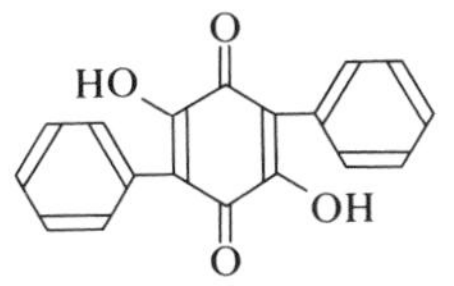

(11) 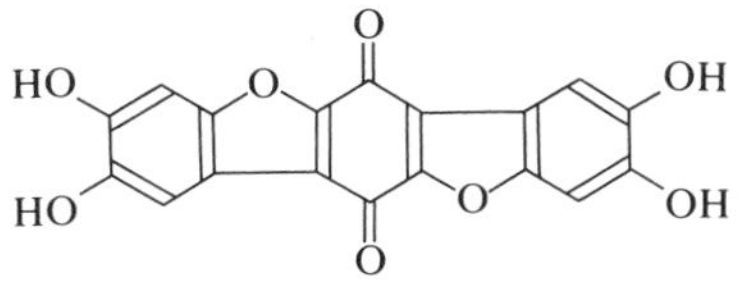 (12) 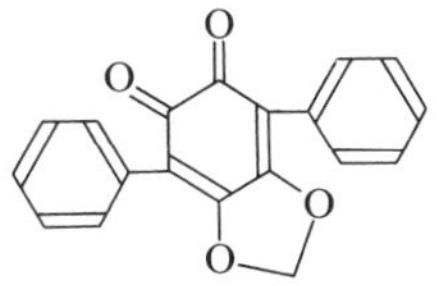 (13)

its diacetate with chromic acid. It can be synthesized from 2,5-diphenylbenzoquinone by bromination, followed by alkaline hydrolysis, or from 2,5-dihydroxybenzoquinone by Meerwein arylation. Polyporic acid occurs with pulvinic acid pigments in certain lichens and feeding experiments have shown that the former can be metabolized to the latter. Polyporic acid can also be oxidized *in vitro* to pulvinic anhydride.

The ubiquinones and plastoquinones are discussed in this Volume, Part 2, p. 117.

1.3 Naphthoquinones

Most naphthoquinones[1] are found in higher plants, chimaphilin (**14**), lawsone (**15**) (the colouring principle of henna) and juglone (**16**; R = H) being simple examples. All three probably exist *in vivo* in a colorless reduced form as a glucoside. Chimaphilin is present in *Pyrola media* along with 2-methyl-5-(γ,γ-dimethylallyl)-benzoquinone which can be cyclized to (**14**) *in vitro* by irradiation or Lewis acid catalysis. Similarly deoxylapachol (**17**; R = H) can be cyclized to 2-methylanthraquinone, a co-metabolite in teak wood (*Tectona grandis*), while lapachol (**17**; R = OH) can be isomerized by acid to either α(**18**)- or β(**19**)-lapachone which are co-metabolites in lapacho wood (*Tabebuia avellanedae*).

Dimeric naphthoquinones are not unusual, particularly in *Diospyros* spp. which elaborate *inter alia* 7-methyljuglone (**16**; R = Me), at least five dimers [*e.g.*, diospyrin (**20**)], a trimer and a tetramer. *o*-Naphthoquinones are relatively common, notably in the wood of *Mansonia altissima* where nine out of the ten mansonones, *e.g.*, (**21**) are *o*-quinonoid. These can usually be distinguished from *p*-isomers spectroscopically[1], and form quinoxalines.

Numerous natural pigments are derivatives of naphthazarin. The two *peri*-hydroxyl groups have a marked effect on the spectroscopic properties of these compounds which aids identification, and they also permit tautomerism. Thus

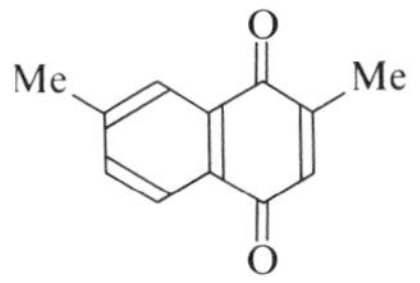

(14) 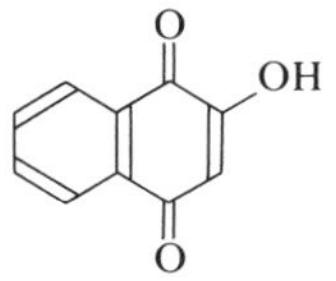 (15) 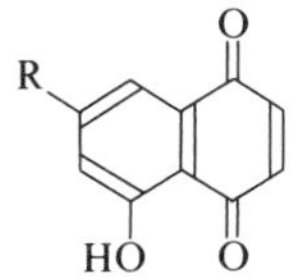 (16)

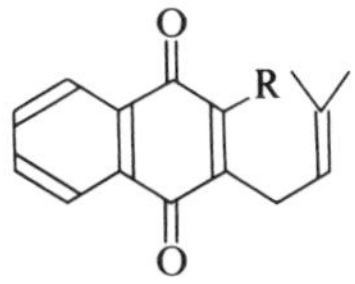

(17) 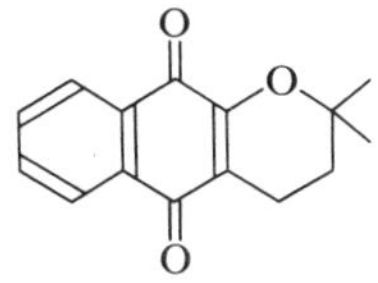 (18) 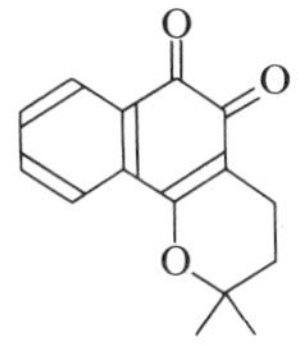 (19)

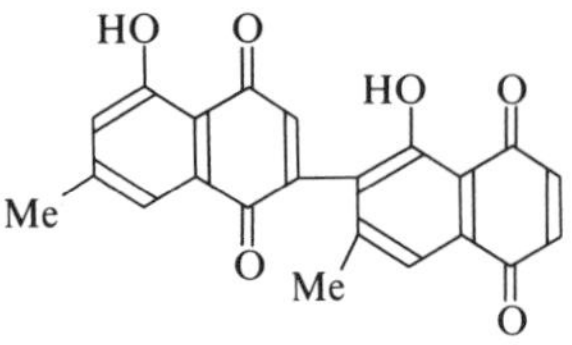

(20) 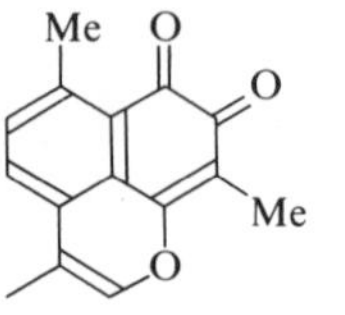 (21) 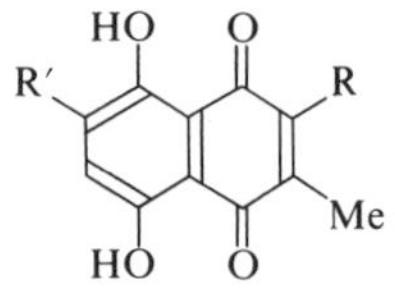 (22)

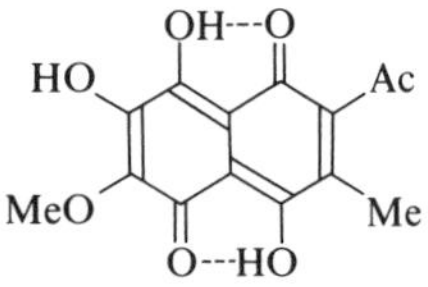

(23) 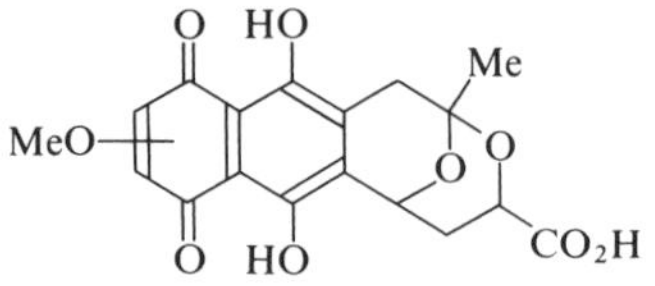 (24) 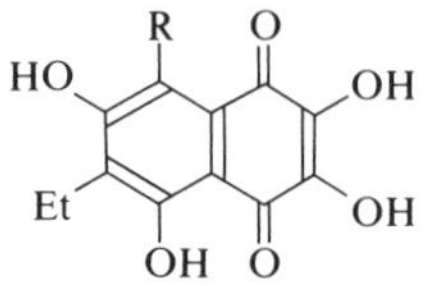 (25)

2-methylnaphthazarin (**22**; R=R′=H) can be prepared by condensing either quinol with methylmaleic anhydride, or methylquinol with maleic anhydride. 1,5-Quinone tautomers have also to be considered and cordeauxione (from the leaves of *Cordeauxia edulis*) exists as (**23**) in the solid state. *Fusaria* moulds are a good source of naphthazarins, *inter alia* (**22**; R′=MeO, R=$CH_2CH(OH)Me$ and CH_2COMe) and marticin (**24**). A group of over 20 spinochrome pigments, found in the calcareous parts of sea urchins and related animals, are derivatives either of naphthazarin or juglone. The quinones (**25**; R=H and OH) are typical.

Phylloquinone and the menaquinones (vitamins K) are discussed in this Volume, Part 2, p. 135.

1.4 Anthraquinones

This is the largest group of natural quinones[1]. 9,10-Anthraquinone itself has been found in the cuticular wax of perennial rye grass (*Lolium perenne*)[28] but characteristically these compounds are polyhydroxy(methoxy) derivatives with/without a carbon side chain (commonly methyl) in a β-position, frequently found as glycosides. Alizarin (**26**; R=R′=H) and purpurin (**26**; R=OH, R′=H), found in madder root (*Rubia tinctorum*), formerly an important natural dyestuff, are representative of a large group of quinones, particularly abundant in the Rubiaceae, which are substituted in one ring only. On the other hand emodin (**27**), having substituents in both benzenoid rings, typifies another large group found in fungi and higher plants. Dimeric quinones such as skyrin (**28**), from *Penicillium islandicum*, are also included in this group. They can be reduced to their monomers (*e.g.*, **28**→**27**) with alkaline dithionite, which is a reaction of considerable value for structural determination. It is of interest that chrysotalunin (**29**) has been found in soil, although its true origin is unknown. Julichrome $Q_{2\cdot 3}$ (**30**) is one of more than 20 closely related pigments elaborated by *Strep. shiodaensis*[8].

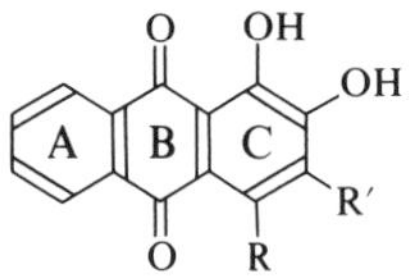

(26) 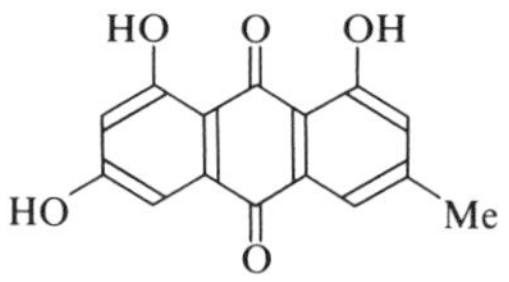 (27) 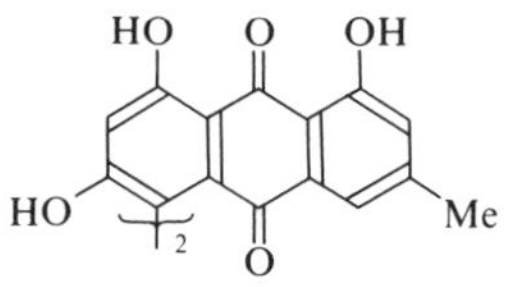 (28)

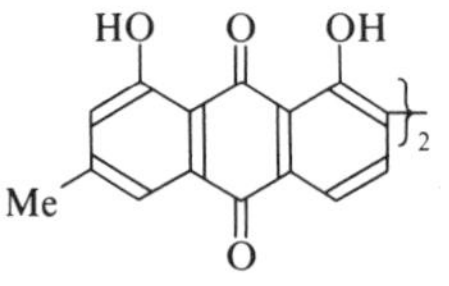

(29) 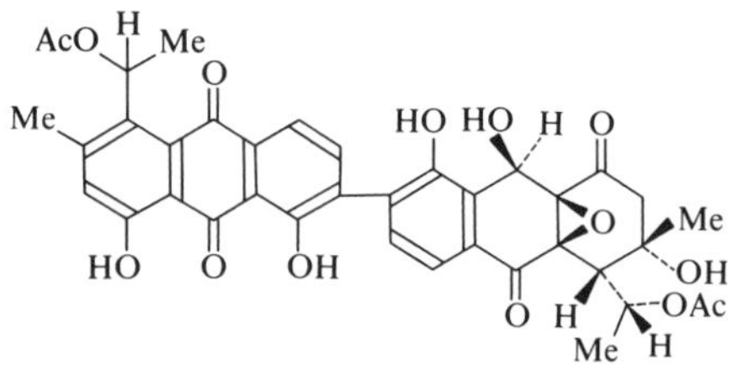 (30)

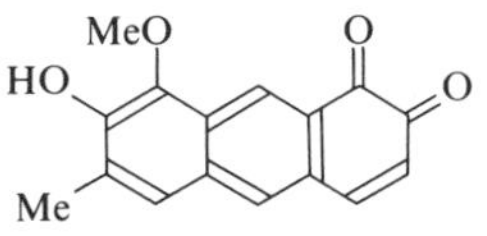

(31) 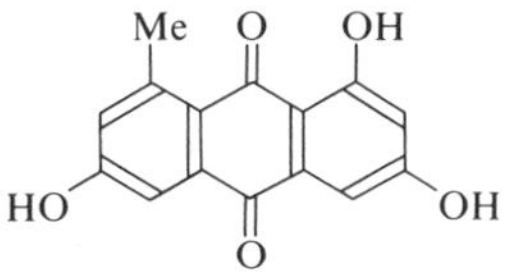 (32) 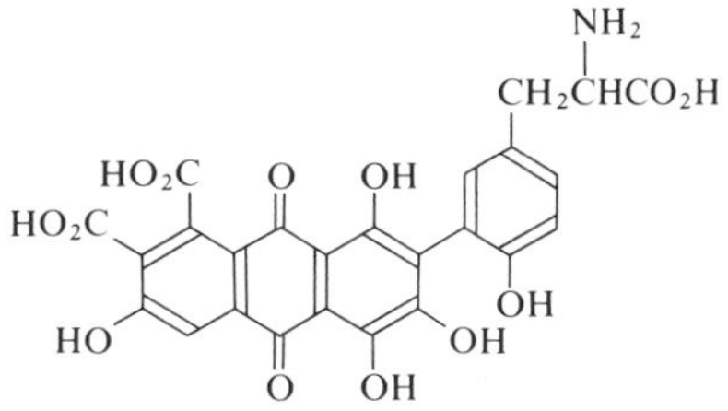 (33)

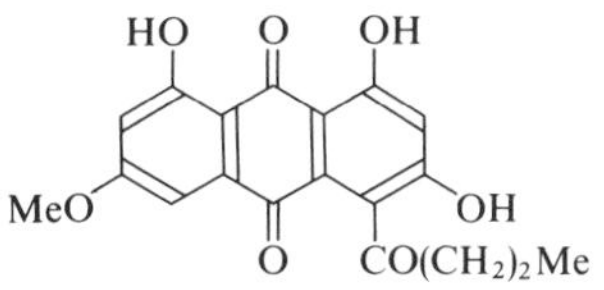

(34) 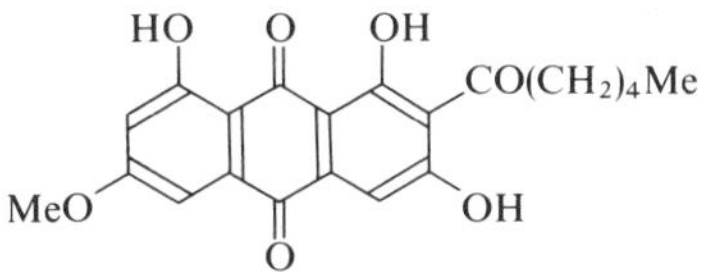 (35) 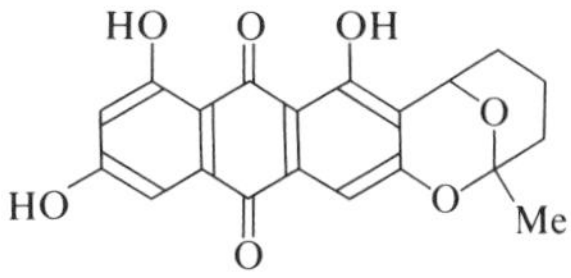 (36)

In the animal kingdom[29] anthraquinones have been isolated from coccid insects and crinoids (sea lilies) and a unique 1,2-anthraquinone, hallachrome (**31**)[9], is present in the skin of the polychaete worm *Halla parthenopeia*. Anthraquinones from animals (in contrast to plants) frequently carry α-side chains and, while emodin (**27**) and related compounds have been isolated from *Eriococcus* insects, *Laccifer lacca* (the lac insect) elaborates the isomer (**32**) and the more complex laccaic acids (*e.g.*, **33**). Similarly α-side chains are present in a number of crinoid anthraquinones such as rhodocomatulin (**34**), which may be compared with the lichen pigment, solorinic acid (**35**). The latter is one of a small group of 1,3,6,8-tetrahydroxyanthraquinones which have β-C_4 or -C_6 side chains. The side chain may be highly modified as in averufin (**36**).

1.5 Anthracyclines

These pigments[1,10], which are all elaborated by streptomycetes, are glycosides, the aglycones (anthracyclinones) being combined with amino-sugars. They have a tetracene polycyclic system and usually one terminal ring is saturated. Consequently they exhibit the infrared and ultraviolet-visible absorption of anthraquinones having two, three of four *peri*-hydroxyl groups. Pyrromycin (**37**) is a typical example which gives rhodosamine and ε-pyrromycinone on hydrolysis. On heating the aglycone it loses the elements of two molecules of water to give the tetracenequinone η-pyrromycinone, which is a copigment. As this has been synthesized, the complete carbon skeleton of ε-pyrromycinone is established and the remaining features of ring A, including the stereochemistry, were deduced from its chemical and spectroscopic properties. While all these pigments have antibiotic properties, daunomycin (rubidomycin) (**38**) whose aglycone differs slightly from the pyrromycinones and rhodomycinones also shows considerable antitumor activity.

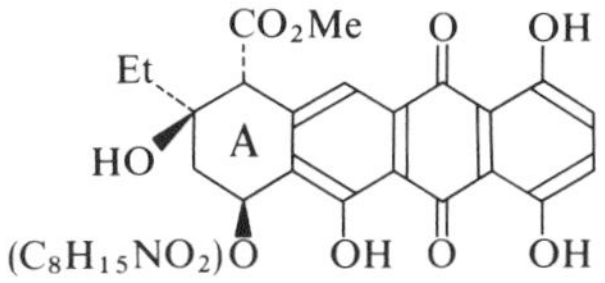

(37)

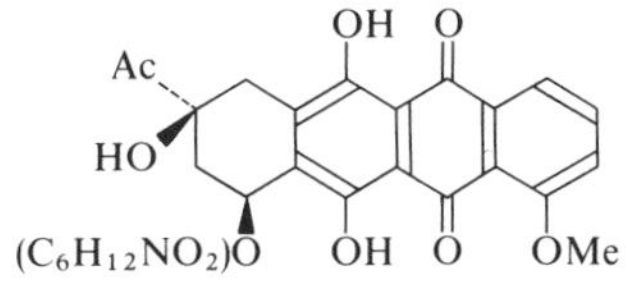

(38)

1.6 Extended Quinones

This small group[1] of highly condensed compounds, in which the quinonoid structure extends through more than one ring, probably arise *in vivo* from simpler precursors by phenolic coupling. The blue quinone (**39**)[11] is related to the *Diospyros* dimers mentioned earlier, while (**40**) (from *Daldinia concentrica*) is essentially a

[8] *N. Tsuji, K. Nagashima*, Tetrahedron *25*, 3007, 3017 (1969); Tetrahedron *26*, 5201, 5719 (1970).

[9] *G. Prota, M. D'Agostino, G. Misuraca*, J.C.S. Perkin Trans. I, 1614 (1972).

[10] *H. Brockmann*, Prog. Chem. Org. Nat. Prods. *21*, 121 (1963).

[11] *O. C. Musgrave, D. Skoyles*, Chem. Commun. 1461 (1970).

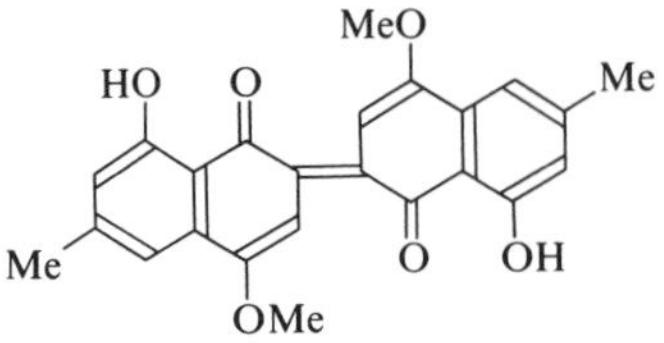

(39)

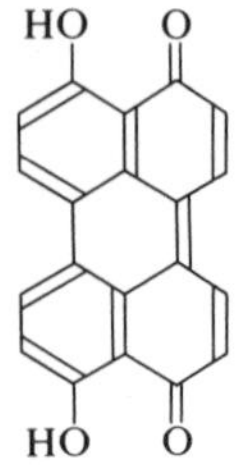

(40)

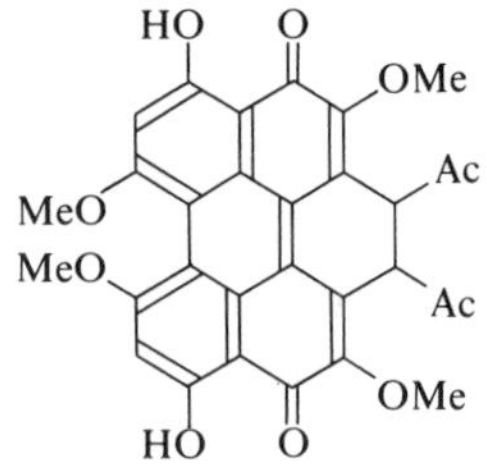

(41)

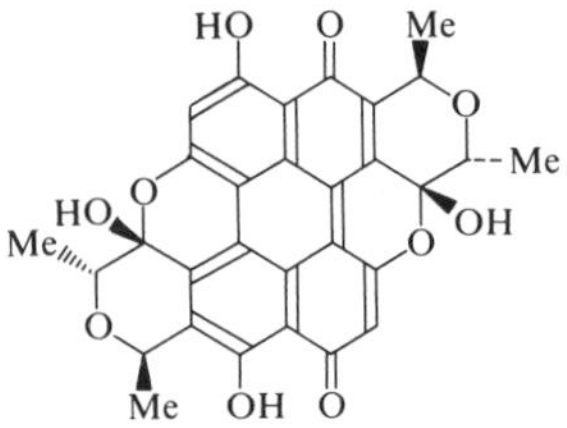

(42)

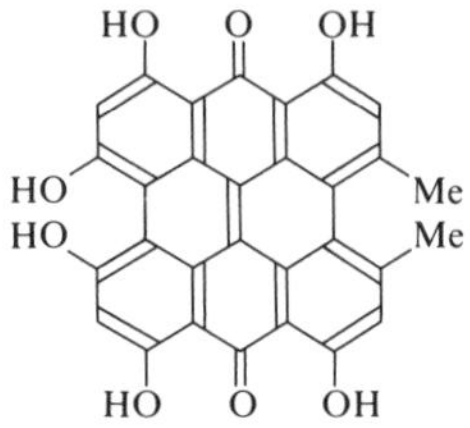

(43)

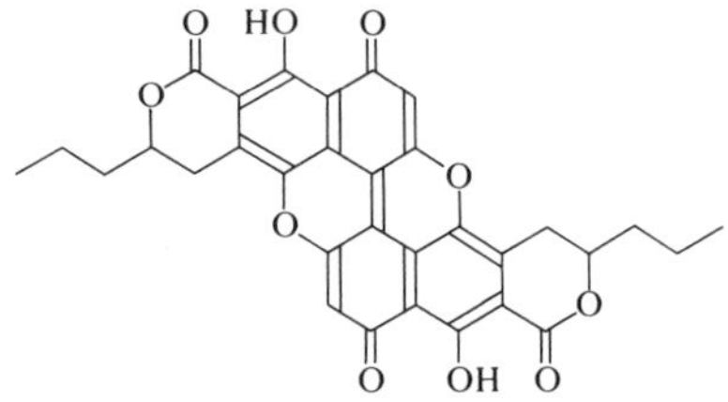

(44)

dimer of 1,8-dihydroxynaphthalene, mono- and dimethyl ethers of which can be isolated from cultures of the fungus. The same chromophore is present in the photodynamic elsinochromes (*e.g.*, **41**) and in the aphid pigment rhodoaphin-*be* (**42**). (Most aphid quinone pigments are complex naphthoquinones which are converted *post mortem* into extended quinones.[12]). The dihydroxyperylenequinone system can be recognized from spectroscopic evidence and by reduction to a leuco-acetate showing the ultraviolet absorption of a perylene derivative.

Hypericin (**43**)[13] is the photodynamic principle in many Hypericaceae which is responsible for the incidence of hypericism in sheep and other animals which feed on these plants. From the flowers of *H. hirsutum*, emodin-9-anthrone has been isolated together with two other dimeric compounds in which the two anthracene nuclei are linked by one and by two C–C bonds, respectively. Both these dimers can be converted into (**43**) by irradiation which clearly indicates the mode of biogenesis. Surprisingly, hypericin and related naphthodianthrones have been identified[14] as the fossil pigments in a Jurassic crinoid (*Apiocrinus* sp.).

Xylindein (**44**) is a remarkable green pigment (λ_{max} 647 nm) produced by the fungus *Chlorociboria aeruginosa* growing on rotting wood. It is a *peri*-xanthenoxanthenequinone and a similar chromophore is present in aphinin, the pigment of greenfly (*Macrosiphium rosae*).

1.7 Miscellaneous Quinones

There are a few natural quinones of varied structure which do not belong to any of the main groups already discussed. These include[1] piloquinone (**45**), tetrangomycin (**46**) and the mitomycins (*e.g.*, **47**); some are entirely terpenoid and

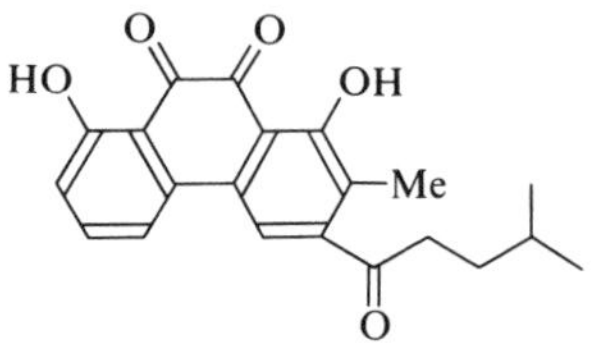

(45)

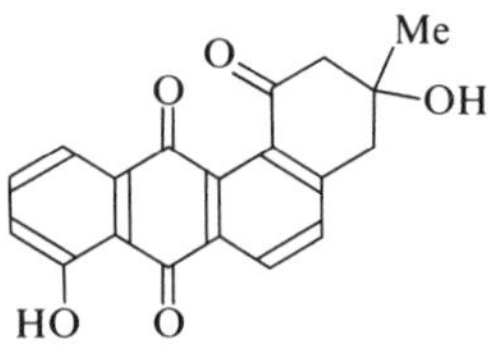

(46)

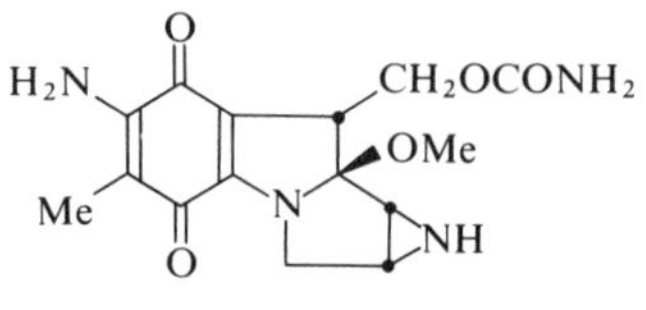

(47)

[12] *D. W. Cameron, Lord Todd, in* Oxidative Coupling, *W. I. Taylor, A. R. Battersby* (Eds.), Arnold, London 1967.

[13] *H. Brockmann*, Prog. Chem. Org. Nat. Prods. *14*, 142 (1957).

[14] *M. Blumer*, Geochim. Cosmochim. Acta *26*, 225 (1962); *M. Blumer*, Science *149*, 722 (1965); *M. Blumer*, personal communication (1968).

in others a quinone ring is part of a macrolide structure as in the streptovaricins and geldanamycin[15]. Mitomycin C (**47**) has antitumor activity and is available commercially. In principle any group of natural aromatic compounds could include quinones and more of these miscellaneous compounds will no doubt be discovered.

1.8 Biogenesis of Quinones[1]

1. *Mevalonate biogenesis.* Terpenoid quinones like thymoquinone, the mansonones (**21**) and the tanshinones (*e.g.*, **48**) are presumably formed from isoprenoid precursors in the usual way although a mevalonate origin has been established experimentally only for the helicobasidins (**49**; R = H and OH)[16]. Others such as lapachol (**17**; R = OH), cyperaquinone (**50**) and the bioquinones are partly terpenoid, the side chain being introduced into a preformed aromatic system by way of the appropriate prenyl pyrophosphate.

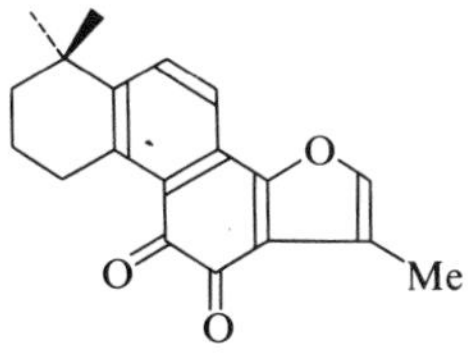

(**48**)

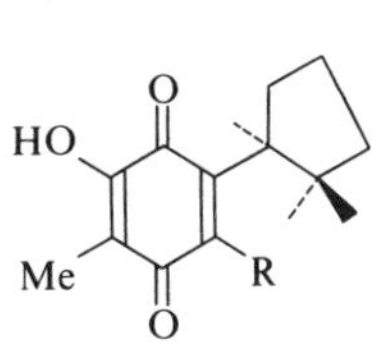

(**49**)

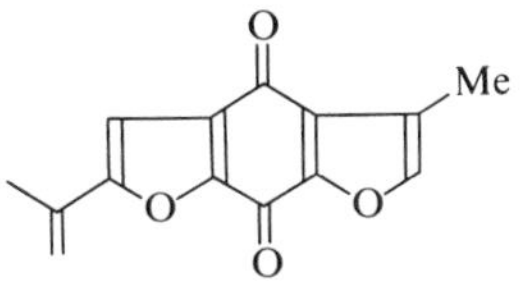

(**50**)

2. *Polyketide biogenesis.* Probably most quinones are formed by the acetate-malonate pathway, especially those produced by microorganisms. This has been established by labelling studies *inter alia* for the benzoquinones fumigatin (**2**; R = OH) and the dimers (**51**; R = H and OH), the naphthoquinones javanicin (**22**; R = CH_2COMe, R′ = OMe) and mollisin (**52**) and the anthraquinones emodin (**27**) and islandicin

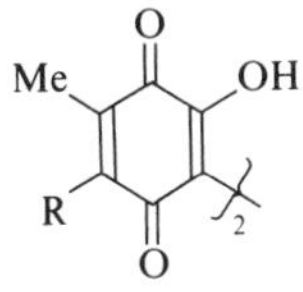

(**51**)

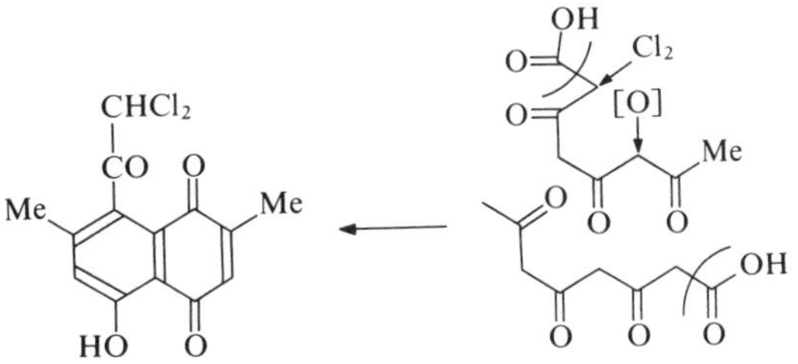

(**52**)

(**53**)

(**55**; R = OH). In higher plants it has also been found experimentally that 7-methyljuglone (**16**; R = Me), its 2-methyl isomer (plumbagin) and chrysophanol (**55**; R = H) are of polyketide origin[17]. These polycyclic systems are normally formed by folding and condensation of a single polyketide chain (*e.g.*, **54**), followed by methylation, side chain oxidation, phenolic coupling and other modifications, but the labelling pattern found in mollisin (**52**) after feeding experiments with *Mollisia caesia* suggests that two polyketide chains are involved, possibly as indicated in (**53**). More highly condensed systems may also be acetate-derived; propionate is the starter unit in the biosynthesis of pyrromycin (**37**) and isobutyrate in the case of piloquinone (**45**). The distribution of ^{14}C in the radioactive elsinochrome A (**41**), obtained from labelling studies, indicated that dimerisation of a "half-molecule" occurred during biogenesis.

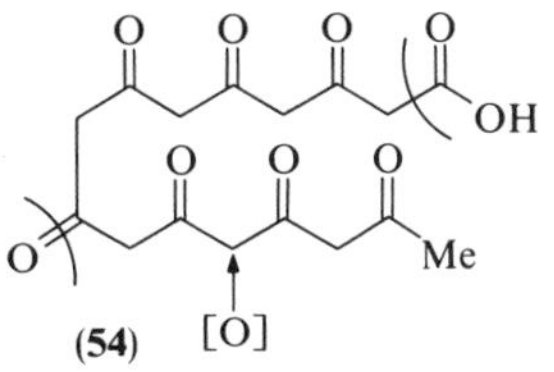

(**54**)

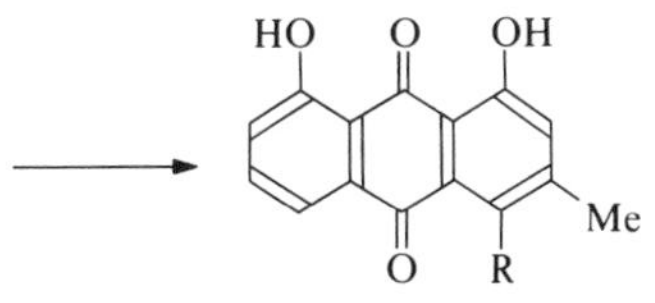

(**55**)

[15] *K. Sasaki, K. L. Rinehart, G. Slomp, M. F. Grostic, E. C. Olson*, J. Am. Chem. Soc. *92*, 7591 (1970).

[16] *S. Natori, Y. Inouye, H. Nishikawa*, Chem. Pharm. Bull. (Tokyo) *15*, 380 (1967); *R. Bentley, D. Chen*, Phytochemistry *8*, 2171 (1969); *S. Nozoe, M. Morisaki, H. Matsumoto*, Chem. Commun. 926 (1970); *P. M. Adams, J. R. Hanson*, J.C.S. Perkin Trans. I, 586 (1972).

[17] *R. Durand, M. H. Zenk*, Tetrahedron Lett. 3009 (1971); *E. Leistner*, Phytochemistry *10*, 3015 (1971).

3. *Shikimate biogenesis.* This route is more important in higher plants (see 4, below) but coprinin (**2**; R = H) is derived from shikimic acid and all three rings of the fungal metabolites volucrisporin (**56**) and phlebiarubrone (**13**) are constructed from two molecules of phenylalanine. There is little doubt that the terphenylquinones found in lichens have the same origin.

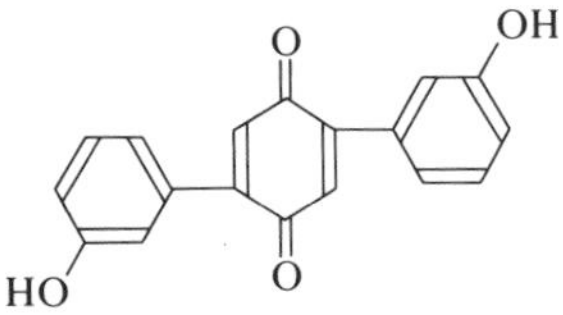

(**56**) 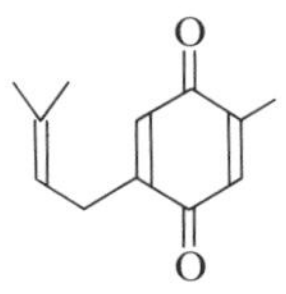 (**57**)

4. *Mixed biogenesis.* In higher plants many quinones probably have a mixed origin. It is known that in *Pyrola media* the quinone ring and the C-2 methyl group of chimaphilin (**14**) are provided by tyrosine (**58**) (via homogentisic acid), while the other ring and methyl group come from mevalonate (**59**). The co-metabolite (**57**) is a likely intermediate (see p. 161). In lawsone (**15**) (*Impatiens balsamina*) and juglone (**16**; R = H) (*Juglans regia*), shikimic acid is incorporated *in toto* to form the benzenoid rings and 50% of the two carbonyl groups; similarly in pseudopurpurin (**26**; R = OH, R′ = CO_2H) (*Rubia tinctorum*) and related anthraquinones, shikimic acid provides ring A and 50% of the quinone carbonyl functions, most of ring C and its side chain being derived from mevalonate. This leaves three carbon atoms in the quinone rings of lawsone and juglone not accounted for, and in (**26**) three carbons in ring B. Recently it has been discovered that *I. balsamina* can utilize glutamate[18] to provide the "missing" carbon atoms of lawsone, and it is suggested that this is transformed via 2-ketoglutarate into the succinyl semialdehyde-thiamine pyrophosphate complex which could attack shikimic[18] or chorismic[19] acid leading to the formation of (**60**) which is well incorporated[19] into lawsone, juglone and pseudopurpurin.

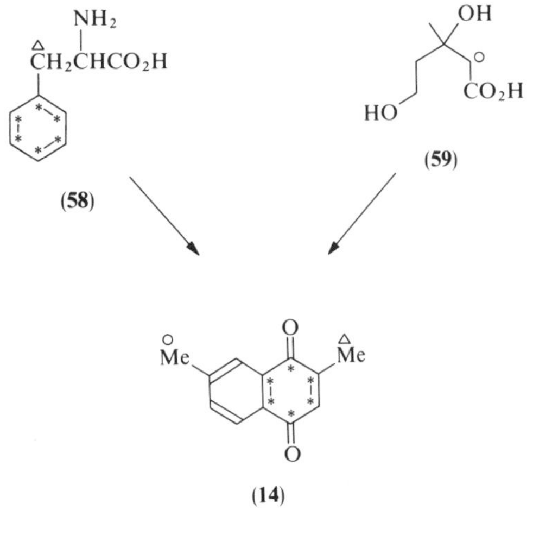

(**58**) (**59**)

(**14**)

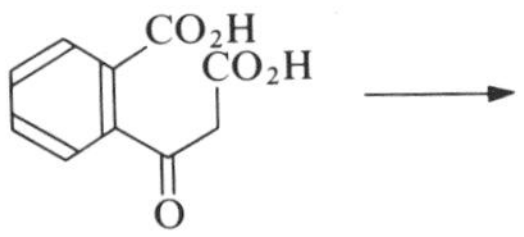

(**60**)

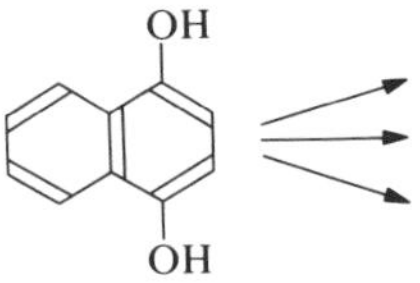

→ (**15**), (**16**: R = H), (**26**)

2 Quinone-Methides[20]

Formally these compounds are derived from quinones by replacing one carbonyl oxygen by a methylene or substituted methylene group. This structural feature appears in about 20 natural, yellow to red, pigments. Examples from higher plants include the flavanoids obtusaquinone (**61**)[21] (*Dalbergia obtusa*) and the "anhydrobase" carajurin (**62**) (*Bignonia chica*), the terpenoids taxodone (**63**)[22] (*Taxodium distichum*) and pristimerin (**64**) (Celastraceae), and the peri-naphthenone lachnanthofluorone (**65**)[23] (*Lachnanthes tinctoria*). Several quinone-methides are mould metabolites (chiefly *Penicillium* spp.) and these are either peri-naphthenones or polyketide structures represented by pulvilloric acid (**66**) (or the *o*-isomer). They can be distinguished from quinones by their resistance to dithionite reduction. However they are easily reduced catalytically and in other ways, and, like quinones, readily undergo addition reactions. These are invariably 1,6-additions.

[18] *I. M. Campbell*, Tetrahedron Lett. 4777 (1969); *E. Grotzinger, I. M. Campbell*, Phytochemistry *11*, 675 (1972).

[19] *P. Dansette, R. Azerad*, Biochem. Biophys. Res. Commun. *40*, 1090 (1970).

[20] *A. B. Turner*, Quart. Rev. (London) *18*, 347 (1964); *A. B. Turner*, Prog. Chem. Org. Nat. Prods. *24*, 288 (1966).

[21] *M. Gregson, W. D. Ollis, B. T. Redman, I. O. Sutherland*, Chem. Commun. 1395 (1968).

[22] *S. M. Kupchan, A. Karim, C. Marcks*, J. Org. Chem. *34*, 3912 (1969).

[23] *U. Weiss, J. M. Edwards*, Tetrahedron Lett. 4325 (1969).

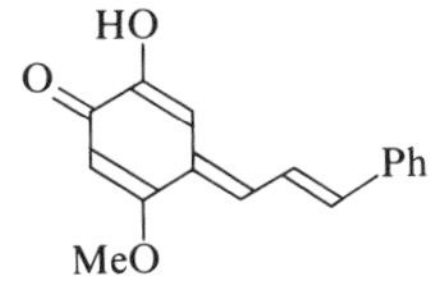

(61) 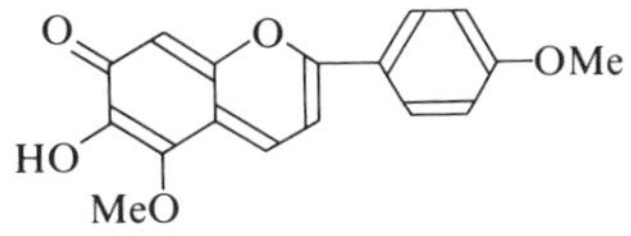 (62) 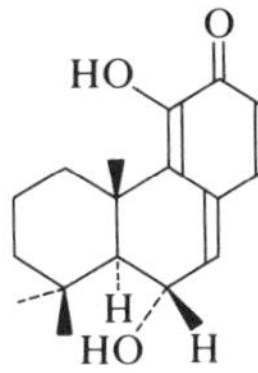 (63)

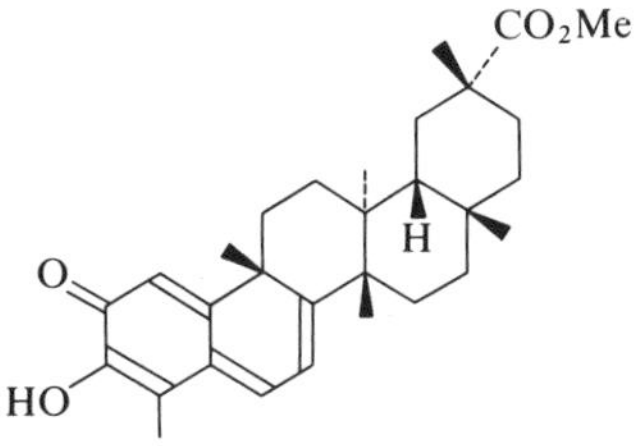

(64) 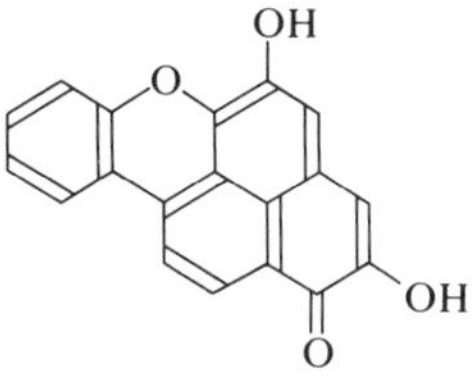 (65) 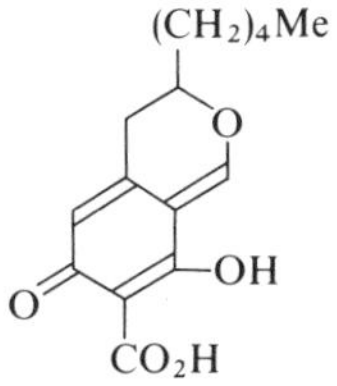 (66)

Thus reduction of pristimerin with borohydride gives (**67**) and pulvilloric acid forms the colorless adduct (**68**) in ethanol which reverts to (**66**) on heating *in vacuo*. Simple quinone-methides are highly reactive and tend to polymerize spontaneously unless the terminal methylene group is substituted. The natural compounds are all stabilized in this way, the terminal methylene group being part of a more extended conjugated system or a vinylogous lactone structure. In many cases intramolecular hydrogen bonding also makes a contribution but the stability of (**63**) is remarkable; in acid, however, it undergoes a dienone-phenol rearrangement[22].

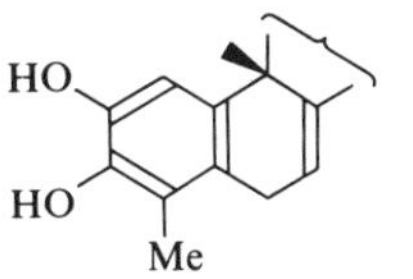

(67) 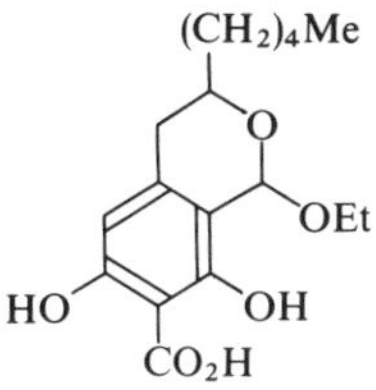 (68)

3 Phenoxazones

The phenoxazones have a quinone-imine structure in which the imino group is part of a heterocyclic ring[24]. This chromophore occurs in a few yellow to red pigments the simplest of which, questiomycin (**69**; R = R′ = H) (from streptomycetes) can be obtained by enzymic oxidation of *o*-aminophenol. Other simple derivatives are produced by woodrotting fungi [*e.g.*, cinnabarin (**69**; R = CH_2OH, R′ = CO_2H)] but the actinomycins (**70**)[25] are of most interest. This group of more than 20 antibiotic, antitumor, chromopeptides, elaborated by streptomycetes, is of considerable biological importance. Actinomycin D inhibits DNA-dependent RNA synthesis although the mechanism is still obscure. They all have the same phenoxazone nucleus but they differ in the sequence of amino-acids in the two pentapeptide lactone rings which can be varied by addition of appropriate amino-acids to the culture medium. It is possible to synthesize symmetrical actinomycins *in vitro* by oxidation of the appropriate peptide (**71**) with ferricyanide and 3-hydroxy-4-methylanthranilic acid and certain peptide derivatives have been biosynthetically incorporated into actinomycins.

In the animal kingdom the phenoxazones are

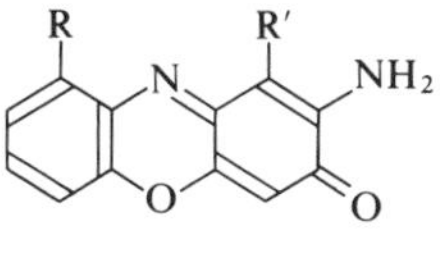

(69) 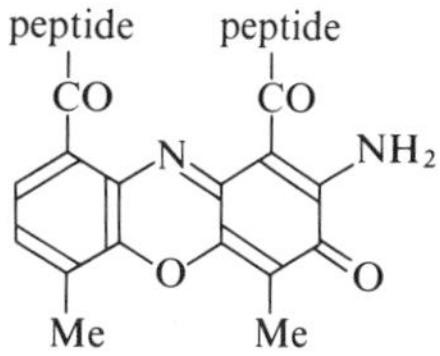 (70)

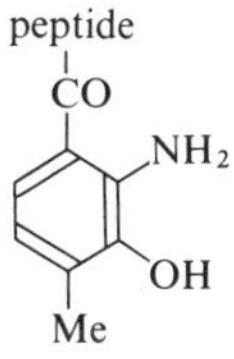

(71)

[24] *W. Schäfer*, Prog. Org. Chem. *6*, 135 (1964); *M. Ionescu, H. Mantsch*, Adv. Heterocyclic Chem. *8*, 83 (1967).

[25] *H. Brockmann*, Prog. Chem. Org. Nat. Prods. *18*, 1 (1960); *H. Lackner*, Tetrahedron Lett. 3189 (1970).

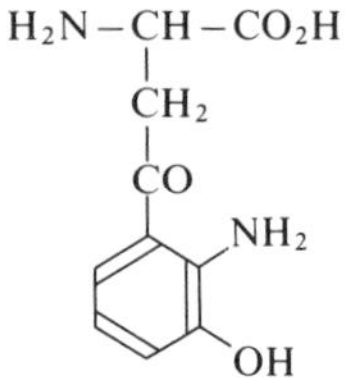

(72) 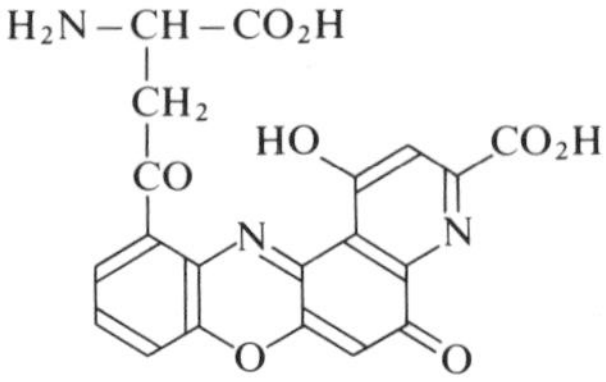 (73) 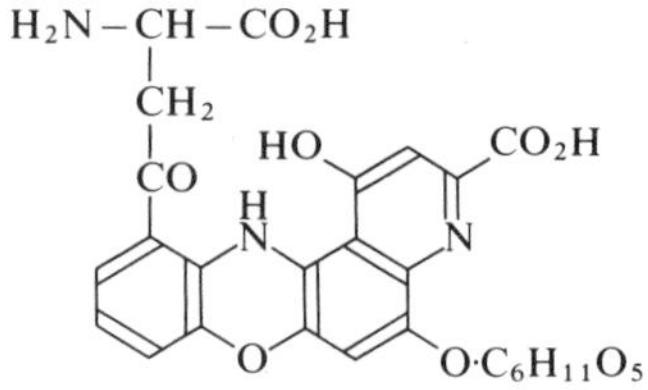 (74)

represented by the ommochromes[26] which are predominantly arthropod pigments, found chiefly in the eyes of insects. Three have been identified. Prior to the chemical investigations it was known from genetic work with *Drosophila* flies that the pigments were products of tryptophan metabolism and were formed by oxidation of 3-hydroxykynurenine (**72**). In fact *in vitro* oxidation of (**72**) with ferricyanide gives the yellow xanthommatin (**73**). Rhodommatin is the glucoside of dihydroxanthommatin (**74**) while ommatin D is the corresponding *O*-sulphate; both these pigments are red.

[26] *A. Butenandt, W. Schäfer, in* Chemistry of Natural and Synthetic Colouring Matters, *T. S. Gore, B. S. Joshi, S. V. Sunthankar, B. D. Tilak* (Eds.), Academic Press, London 1962.

[27] *J. L. Webb*, Enzyme and Metabolic Inhibitors, Vol. III, Academic Press, London 1966.

[28] *J. E. Allebone, R. J. Hamilton, T. A. Bryce, W. Kelly*, Experientia *27*, 13 (1971).

[29] *R. H. Thomson, in* T. R. Seshadri Commemoration Volume, Prentice-Hall of India 1972.

Recommended Literature for Further Reading

a) *R. Bentley, I. M. Campbell, in* The Chemistry of the Quinonoid Compounds, Part 2, *S. Patai* (Ed.), Wiley, London 1974.

b) *R. Bentley, in* Biosynthesis (Specialist Periodical Report—The Chemical Society, London) Vol. 3, 181 (1975).

c) Aromatic and Heteroaromatic Chemistry (Specialist Periodical Report—The Chemical Society, London) Vols. 1–5, (1973–77).

5.4 Polyene Compounds

Kiyoshi Tsukida

Kobe Women's College of Pharmacy, Higashinada-ku, Kobe, 658 Japan

1 New Natural Carotenoids

The carotenoids are one of the most widespread and important classes of natural polyene pigments bearing numerous conjugated carbon-carbon double bonds, and research on carotenoids has been a long story[1,2] since Wackenroder isolated "carotene" in 1831. In addition to the extensive studies in the field of microbiology[3], recent application of spectroscopic analysis[4], especially of nuclear magnetic resonance (NMR) and mass spectral techniques gave impetus to the discoveries of new natural carotenoids and has disclosed some uncertainty in the expediential way of identification which had relied mainly on their ultraviolet spectral information and on analogy with the basic carotenoid such as lycopene, α- or β-carotene. Up to now structural formulae have been established or proposed with considerable certainty for *ca.* 390 natural carotenoids, important groups of which are surveyed in Tables 1 and 2.

Table 1. Chronological development of natural carotenoids*

	–1949	1950 –1959	1960 –1969	1970 –1975
Carotenes**	5	10	13	19
Xanthophylls†	25	21	155	137

* Principally, neither *cis* nor ester type of xanthophyll is not included.

** Carotenoid hydrocarbons.

† Oxygenated carotenoids.

[1] *P. Karrer, E. Jucker*, Carotenoids, Elsevier, Amsterdam 1950; *J. B. Davis*, Rodd's Chemistry of Carbon Compounds, *S. Coffey* (Ed.), 2nd ed., Vol. IIB, p. 231, Elsevier, Amsterdam 1968; *J. B. Davis*, Rodd's Chemistry of Carbon Compounds, *M. F. Ansell* (Ed.), Vol. IIA/B Suppl., p. 191, Elsevier, Amsterdam 1974; *F. H. Foppen*, Chromatogr. Rev. *14*, 133 (1971).

[2] *O. Isler* (Ed.), Carotenoids, Birkhäuser, Basel 1971.

[3] *S. Liaaen-Jensen, A. G. Andrews*, Ann. Rev. Microbiol. *26*, 225 (1972).

Table 2. Structure patterns of natural carotenoids*

Structure pattern**	First example	Structure proposed in
C_{20}	crocetin	1932
C_{30}	β-citraurin	1937
C_{40}	lycopene	1931
C_{45}	2-isopentenyl-3,4-dehydrorhodopin	1969
C_{50}	decaprenoxanthin	1968
Aliphatic in-chain		
normal	lycopene	1931
retro	rhodoxanthin	1933
hydro	phytoene	1956
allenic	fucoxanthin	1964
acetylenic	alloxanthin	1967
methyl-oxidized	lycopenal	1967
tri-nor	peridinin	1971
End group		
acyclic	lycopene	1931
6-ring, alicyclic	β-carotene	1931
″ , aromatic	renieratene	1957
5-ring, C_{40}	capsanthin	1960
″ , nor	actinoerythrin	1968
3,4-dehydro	rhodovibrin	1958
1,2-dihydro	1,2-dihydrolycopene	1970
$=CH_2$, exocyclic	β,γ-carotene	1971
″ , terminal	aleuriaxanthin	1973
methyl-oxidized	torularhodin	1959
5,6-seco	β-carotenone	1968
Xanthophylls		
acid	crocetin	1932
acetate	fucoxanthin	1964
butenolide lactone	peridinin	1971
aldehyde	β-citraurin	1937
O-glycoside, C-1	4-oxo-phleixanthophyll	1967
″ , C-2	myxoxanthophyll	1969
″ , C-3	zeaxanthin rhamnoside	1973
″ , C-4	4-hydroxy-4,4′-diaponeurosporene glycoside	1974
″ , C-4″	sarcinaxanthin monoglucoside	1970
epoxy, 5,8-	flavoxanthin	1942
″ , 5,6-	violaxanthin	1945
″ , 1,2-	phytoene-1,2-oxide	1969
alcoholic OH, C-1	rhodopin	1958
″ , C-2	4-oxophleixanthophyll	1967
″ , C-3	zeaxanthin	1932
″ , C-4	4′-hydroxy-echinenone	1965
″ , C-5	azafrin	1933
″ , C-6	azafrin	1933
″ , C-15	15-hydroxy-octahydro-6′-apo-β-7′-one	1952
″ , C-16	lycoxanthin	1968
″ , C-19	loroxanthin	1969
″ , C-20	rhodopinol	1967
″ , C-8′	8′-hydroxy-7′,8′-dihydrocitranaxanthin	1966
″ , C-3″	bacterioruberine	1970
″ , C-4″	decaprenoxanthin	1968
″ , C-5″	2-(dihydroxy-isopentenyl)-2′-isopentenyl-β-carotene	1970
″ , *prim.*	rhodopinol	1967
″ , *sec.*	zeaxanthin	1932
″ , *tert.*	azafrin	1933
phenolic OH, C-3	astacene	1934
″ , C-9′	pigment 26–5	1971

Table 2. Structure patterns of natural carotenoids* (continued)

Structure pattern**	First example	Structure proposed in
methoxy	spirilloxanthin	1959
keto, C-2	spheroidenone	1961
″ , C-3	rhodoxanthin	1933
″ , C-4	astacene	1934
″ , C-6	capsanthin	1960
″ , C-8	fucoxanthin	1964
″ , C-5′	tangeraxanthin	1962
″ , C-7′	15-hydroxy-octahydro-6′-apo-β-7′-one	1952

* Principally, neither *cis* nor ester type of xanthophyll is not included.
** Numbering.

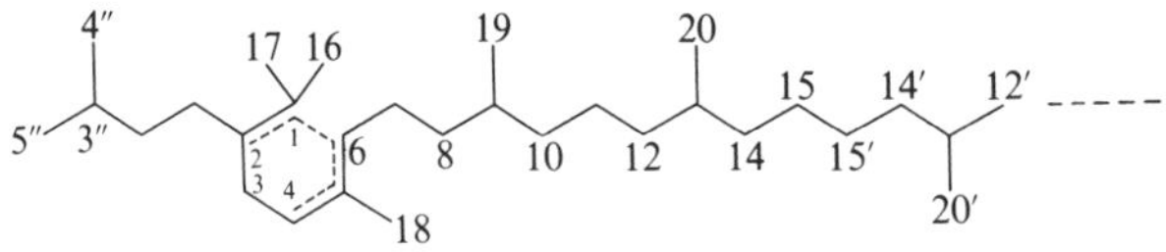

4″ 17 16 19 20 15 14′ 12′ 5″ 3″ 1 2 3 4 6 8 10 12 14 15′ 18 20′

For further details the valuable articles[1,2,5] should be consulted.
It is under the same situation as general natural products that as prompt an isolation of pure crystals as possible is the most important task though it is laborious and troublesome. However, it should also be kept in mind that classical solvent partition procedure and preparative column chromatography are still highly efficient techniques in this field, though preparative thin-layer or high speed liquid chromatographies[6] are getting familiar. While the chemistry of caroteno-proteins and -glycosides as well as biosynthesis and chemotaxonomic features of carotenoids are all largely unexplored and attractive themes, so far accumulated less reliable records should also be re-examined. This is especially desired in the chemistry of the basic chloroplast pigments and formally related carotenoid constituents.

Most stereoisomerization studies on naturally-derived *cis*-polyenes were conducted extensively by Zechmeister in the field of carotenoids and by Wald on the visual pigments[7]. Although natural carotenoids generally preserve the all-*trans* configuration, some preponderant members as exemplified by phytofluene, antheraxanthin, neoxanthin, violeoxanthin and some cross-conjugated pigments (lycopenal, rhodopinal etc.) are known to be genuine *cis* (9-, 13- or 15-*cis*) components. It should be decided carefully whether isolated *cis*-carotenoid is a native component or not.
It is clarified from an X-ray analysis and molecular orbital calculations of some optically inactive carotenoids that the polyene chain of carotenoids is almost coplanar with a slight *S*-shaped bending and the most stable configuration of ring and aliphatic chain is a 6,7-*s-cis* arrangement with a certain degree of deviation. In recent years, stereochemical studies are rapidly extending to the absolute configuration by using the appropriate techniques including optical rotatory dispersion, circular dichroism measurements and NMR (^{1}H and ^{13}C) analysis. It is noteworthy that several important conclusions about the chiralities at C-2, -3, -5 or -6 are now presented[2,5,8].

---

4 *U. Schwieter, H. R. Bolliger, L. H. Chopard-dit-Jean, G. Englert, M. Kofler, A. König, C. v. Planta, R. Rüegg, W. Vetter, O. Isler*, Chimia *19*, 294 (1965); *B. C. L. Weedon*, Chemistry and Biochemistry of Plant Pigments, *T. W. Goodwin* (Ed.), p. 75, Academic Press, London 1965; *B. C. L. Weedon*, Fortschr. Chem. Org. Naturstoffe *27*, 81 (1969); *U. Schwieter, G. Englert, N. Rigassi, W. Vetter*, Pure Appl. Chem. *20*, 365 (1969).

5 *S. Liaaen-Jensen*, Pure Appl. Chem. *14*, 227 (1967), *20*, 421 (1969), *35*, 81 (1973); *S. Liaaen-Jensen*, Experientia *26*, 697 (1970); *B. C. L. Weedon*, Chem. Brit. *3*, 424 (1967); *B. C. L. Weedon*, Pure Appl. Chem. *14*, 265 (1967), *20*, 531 (1969), *35*, 113 (1973).

6 *M. Vecchi, J. Vesely, G. Oesterhelt*, J. Chromatogr. *83*, 447 (1973); *S. K. Reeder, G. L. Park*, J. Assoc. Off. Anal. Chem. *58*, 595 (1975).

7 *L. Zechmeister, Cis-trans* Isomeric Carotenoids, Vitamins A and Arylpolyenes, Springer Verlag, Wien 1962.

8 *G. Britton*, Terpenoids & Steroids *5*, 153 (1975).

2 Allenic Carotenoids and Related Polyenes

Until recently, natural allenes have been rather scarce and only diyne-allenes (mycomycin etc.) and allenic fatty acids (laballenic acid etc.) have been reported. As is foreseen, allenes present poor spectral information except infrared spectra which exhibit a characteristic absorption band around 1950 cm^{-1}. During the past five years, however, allenic isoprenoids have been discovered successively by skillful and patient works. Fucoxanthin is the first example and fucoxanthinol, isofucoxanthin, isofucoxanthinol (=pentaxanthin?), neoxanthin (foliaxanthin; =trollixanthin?), neochrome (foliachrome; = trollichrome?), deepoxyneoxanthin (=trollein?), caloxanthin, dinoxanthin, hexanoyloxy-fucoxanthin, mimulaxanthin, nostoxanthin, paracentrone, hexanoyloxy-paracentrone, peridinin (sulcatoxanthin), peridininol, vaucheriaxanthin and a grasshopper ketol are all involved in this category. Furthermore, there remains another probability that such allenic constituents may have been overlooked in the early literatures.

The last ketol, the simplest allenic terpene yet discovered, was isolated from an ant repellant secretion of over 1,000 adult grasshoppers[9] and was synthesized later[10]. It was supposed that this compound could be derived from neoxanthin by an appropriate oxidative degradation *in vivo*. An X-ray crystallographic analysis of the *p*-bromobenzoate[11] revealed the absolute configuration of this allenic ketol as **I** where the C-5 hydroxy group is *trans* to both other oxygen functions and is *cis* to the allenic hydrogen. Some allenes formally related to **I** were reported as products of the photo-sensitized oxidation of *β*-ionol and related compounds[12]. On the basis of

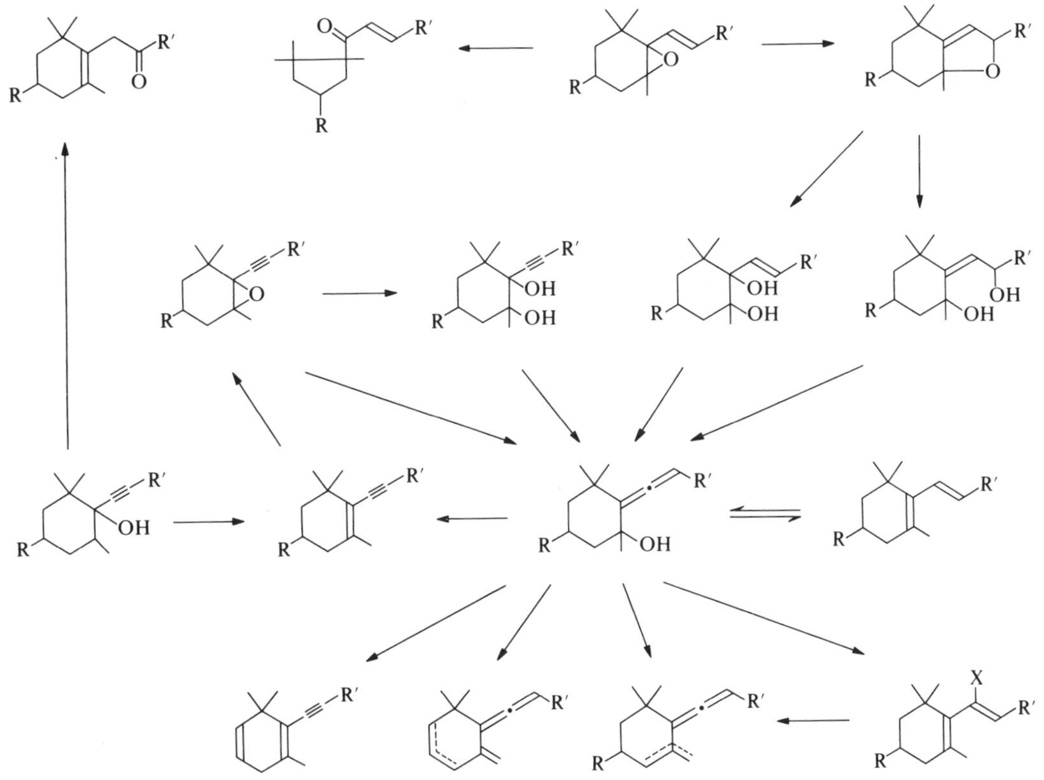

Scheme 1. Some *in vitro* conversions of allenic carotenoids.

[9] *J. Meinwald, K. Erickson, M. Hartshorn, Y. C. Meinwald, T. Eisner*, Tetrahedron Lett. 2959 (1968).

[10] *S. W. Russell, B. C. L. Weedon*, Chem. Commun. 85 (1969); *J. Meinwald, L. Hendry*, Tetrahedron Lett. 1657 (1969); *J. R. Hlubucek, J. Hora, S. W. Russell, T. P. Toube, B. C. L. Weedon*, J. Chem. Soc. Perkin I, 848 (1974); *K. Mori*, Tetrahedron *30*, 1065 (1974).

[11] *T. E. DeVille, M. B. Hursthouse, S. W. Russell, B. C. L. Weedon*, Chem. Commun. 754 (1969).

[12] *C. S. Foote, M. Brenner*, Tetrahedron Lett. 6041 (1968); *S. Isoe, S. B. Hyeon, H. Ichikawa, S. Katsumura, T. Sakan*, Tetrahedron Lett. 5561 (1968), 279 (1969), 1089 (1971); *T. E. DeVille, J. Hora, M. B. Hursthouse, T. P. Toube, B. C. L. Weedon*, Chem. Commun. 1231 (1970).

the reaction mechanism which is explained to proceed by attack of singlet oxygen at C-5 concerted with abstraction of the hydrogen at C-7, the C-5 hydroxy group is believed to be *trans* to the allenic hydrogen. Natural allenic carotenoids[13] conceivably arise by oxidation of zeaxanthin and related pigments, during which intermediates such as biradicals may well be involved. Some *in vitro* conversions are shown in Scheme 1.

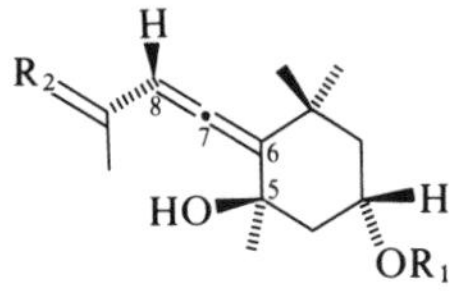

(I) $R_1 = H$, $R_2 = O$
(II) $R_1 = H$
$R_2 =$

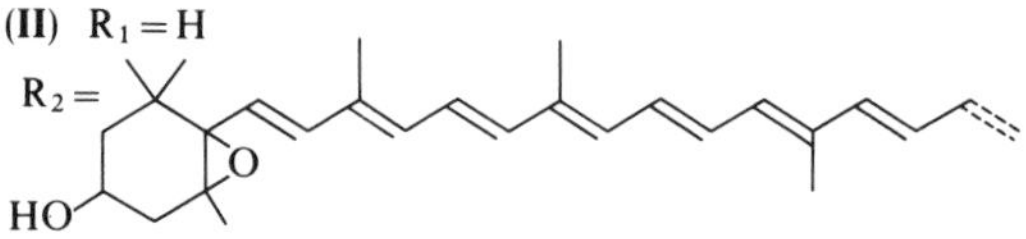

(III) $R_1 = Ac$
$R_2 =$

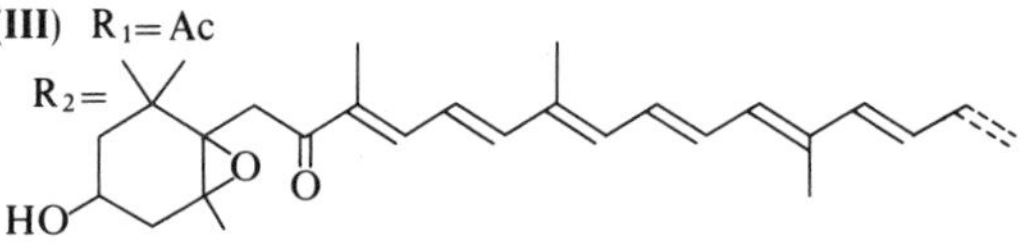

Neoxanthin (**II**) and fucoxanthin (**III**), both are the representative pigment of all green leaves or of common algae, are not yet totally synthesized. It is a question of biological and genetical interest why such structurally peculiar compounds are the most abundant constituents in nature, natural production being estimated up to a million tons per year, or what a distinctive biological significance they might have in the living cells. Clarification of *in vivo* conversions among allenic, acetylenic, and epoxy carotenoids may offer a clue to solve these problems.

Peridinin (**IV**), the principal carotenoid pigment of the dinoflagellates, is another example of the most fascinating carotenoid bearing allenic as well as enol-butenolide ring moieties in the molecule[14]. The suggested biogenesis involves a rhodopinal type of unit for the elimination of three carbon atoms from the central part of a traditional C_{40}-carotenoid skeleton and hydration of an 11′,12′-acetylene to form the butenolide ring (Scheme 2). Further natural polyenic enol-butenolides, (**V**)–(**XIV**), are known[15-21].

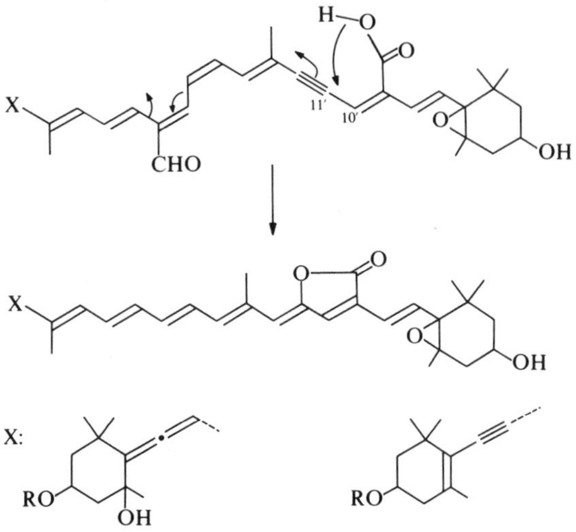

Peridinin (**IV**)[14]
R = Ac

Pyrrhoxanthin (**VI**)[15]
R = Ac

Peridininol (**V**)[15]
R = H

Pyrrhoxanthinol (**VII**)[15]
R = H

Scheme 2. Possible biogenesis of enol-butenolide carotenoids.

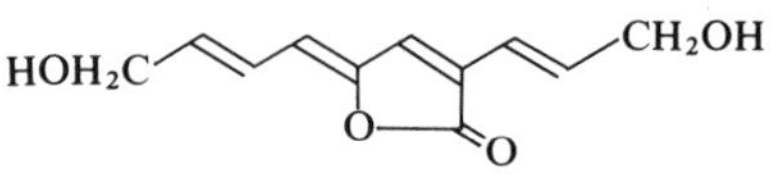

Tetrenolin (**VIII**)[16]

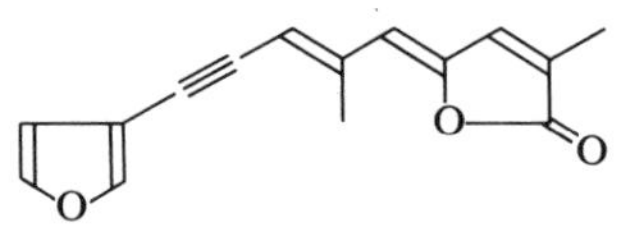

Freelingyne (**IX**)[17]

[13] *T. E. DeVille, M. B. Hursthouse, S. W. Russell, B. C. L. Weedon*, Chem. Commun. 1311 (1969).

[14] *H. H. Strain, W. A. Svec, K. Aitzetmüller, M. C. Grandolfo, J. J. Katz, H. Kjøsen, S. Norgård, S. Liaaen-Jensen, F. T. Haxo, P. Wegfahrt, H. Rapoport*, J. Am. Chem. Soc. *93*, 1823 (1971); *H. H. Strain, W. A. Svec, P. Wegfahrt, H. Rapoport, F. T. Haxo, S. Norgård, H. Kjøsen, S. Liaaen-Jensen*, Acta Chem. Scand. *30B*, 109 (1976); *H. Kjøsen, S. Norgård, S. Liaaen-Jensen, W. A. Svec, H. H. Strain, P. Wegfahrt, H. Rapoport, F. T. Haxo*, Acta Chem. Scand. *30B*, 157 (1976).

[15] *J. E. Johansen, W. A. Svec, S. Liaaen-Jensen, F. T. Haxo*, Phytochemistry *13*, 2261 (1974).

[16] *G. G. Gallo, C. Coronelli, A. Vigevani, G. C. Lancini*, Tetrahedron *25*, 5677 (1969); *H. Pagani, G. Lancini, G. Tamoni, C. Coronelli*, J. Antibiotics *26*, 1 (1973).

[17] *C. F. Ingram, R. A. Massy-Westropp, G. D. Reynolds*, Aust. J. Chem. *27*, 1477, 1491 (1974); *M. J. Begley, D. W. Knight, G. Pattenden*, J. Chem. Soc. Perkin II, 1867 (1975).

[18] *R. Hänsel, A. Pelter*, Phytochemistry *10*, 1627 (1971).

[19] *P. Singh, M. Anchel*, Phytochemistry *10*, 3259 (1971).

[20] *N. Ojima, S. Takenaka, S. Seto*, Phytochemistry *12*, 2527 (1973).

[21] *N. Ojima, I. Takahashi, K. Ogura, S. Seto*, Tetrahedron Lett. 1013 (1976).

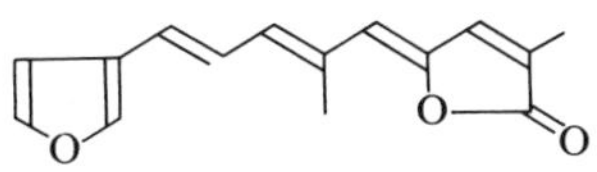

Dihydrofreelingyne (**X**)[17]

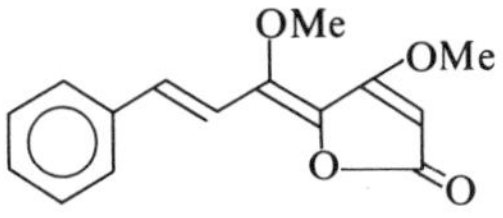

Piperolide (**XI**)[18]

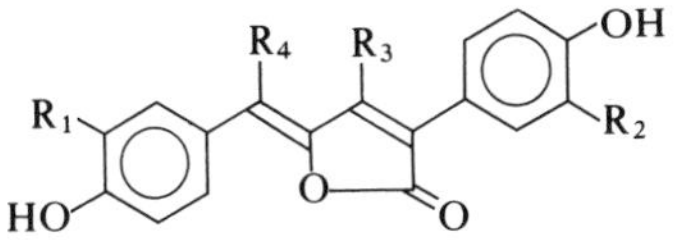

(**XII**)[19] Atromentic acid
$R_1=R_2=H$, $R_3=OH$, $R_4=COOH$

(**XIII**)[20] $R_1=R_2=R_4=H$, $R_3=OH$

(**XIV**)[21] Aspulvinones
$R_1, R_2=H$ or $-CH_2CH=CMe_2$
$R_3=R_4=H$

3 Other Natural Conjugated-Polyenes

Polyenic macrolides are members of another big group. They are tri- to hepta-enes possessing a large-membered (24 to 39) lactone ring and have peculiar antibiotic properties. It is now established that amphotericin B, candidin, and mycoheptin are conjugated heptaenes, dermostatin a conj. hexaenone, chainin and rectilavendomycin conj. pentaenes, and flavofungin (mycotocin) a conj. pentaenone.

Another interesting type of natural conjugated polyene is aryl polyenoic acids such as asperenone (asperyellone) (**XV**), asperrubrol (**XVI**), xanthomonas pigment (**XVII**) or flexirubin (**XVIII**). Polyene acids (**XIX, XX**), δ-lactones (**XXI, XXII**), an acid amide (**XXIII**) as well as polyenones (**XXIV–XXIX**) have also been known.

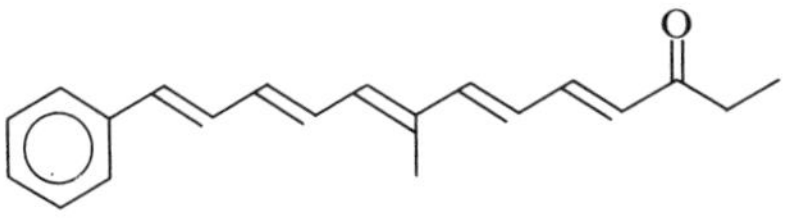

(**XV**)

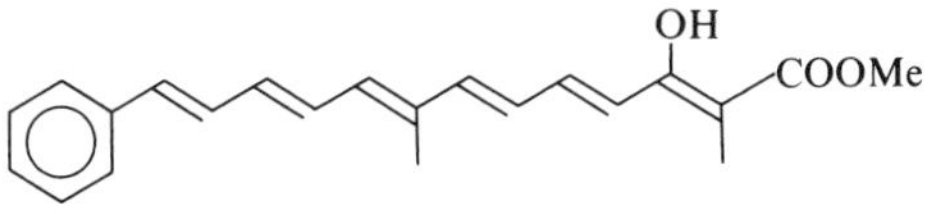

(**XVI**)

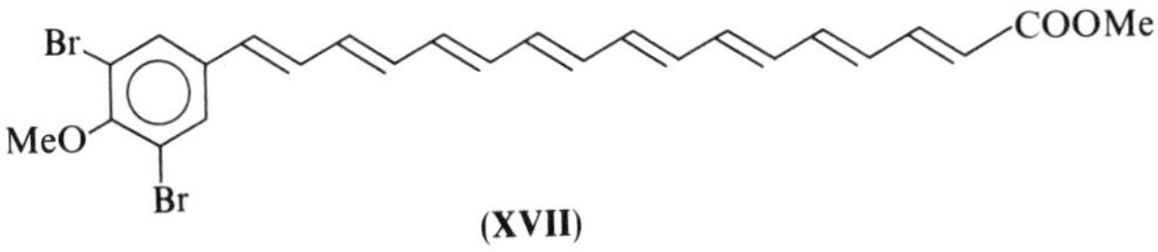

(**XVII**)

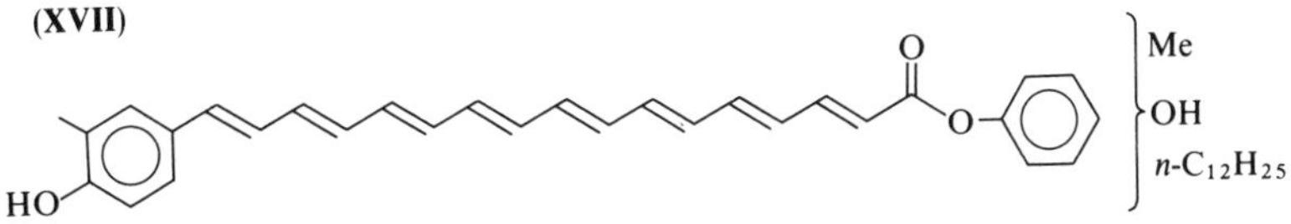

(**XVIII**)

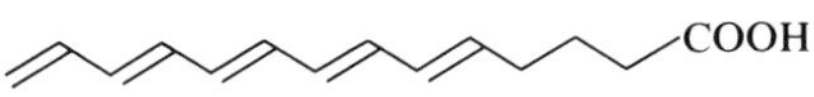

C_{14}-Ajenoic acid (**XIX**)

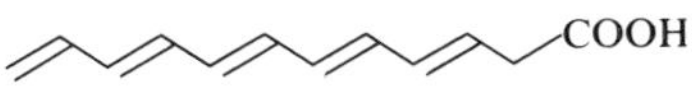

C_{12}-Ajenoic acid (**XX**)

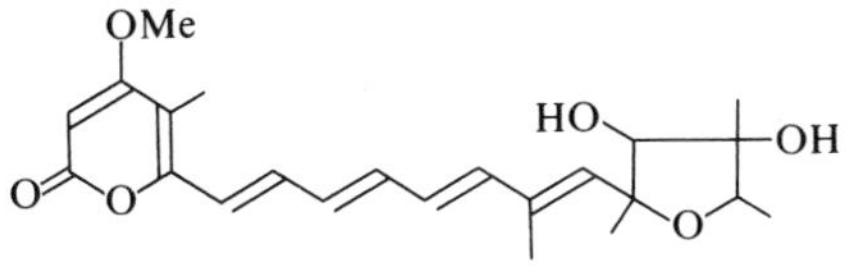

Citreoviridin (**XXI**)

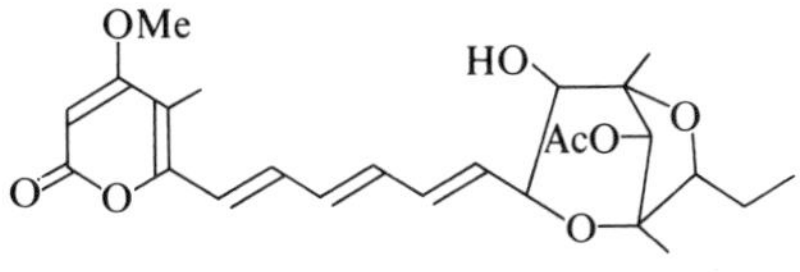

Aurovertin B (**XXII**)

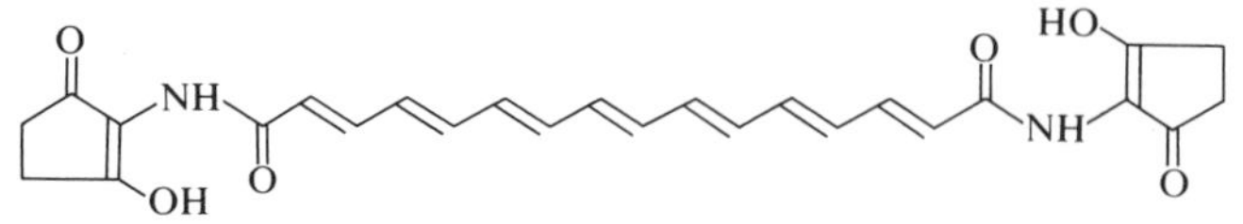

Limocrocin (**XXIII**)

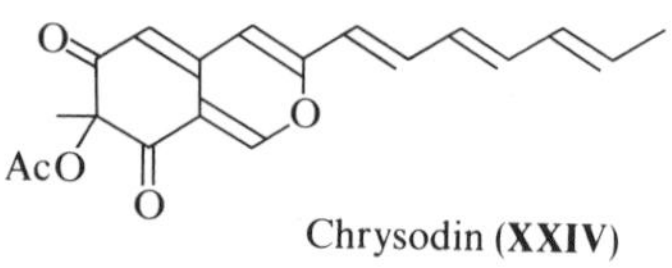

Chrysodin (**XXIV**)

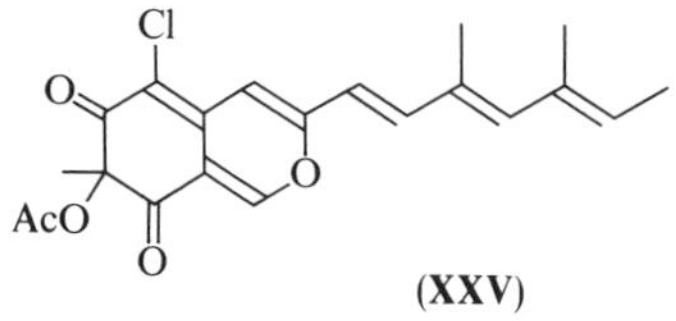

(**XXV**)

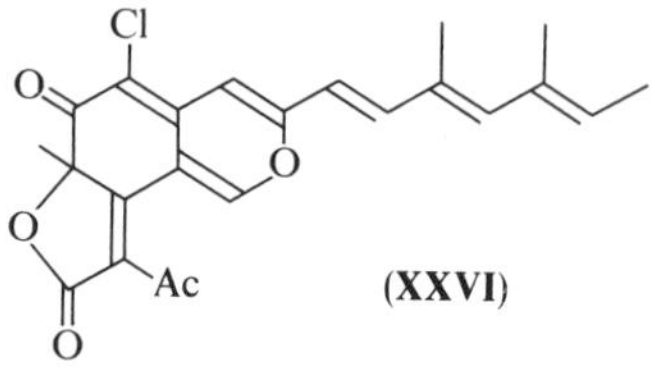

(**XXVI**)

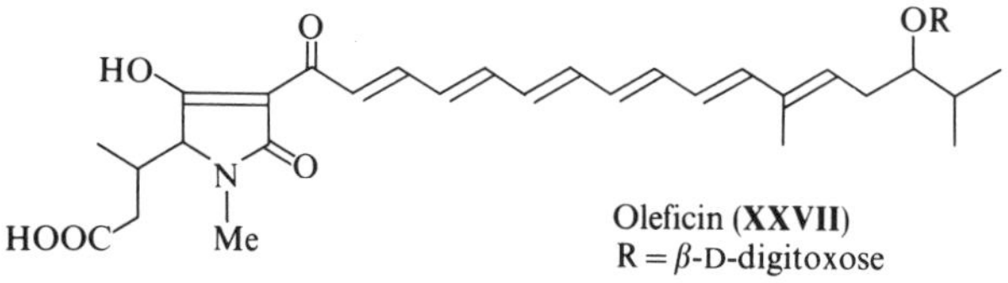

Oleficin (**XXVII**)
R = β-D-digitoxose

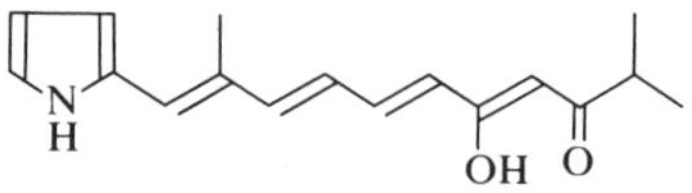

Wallemia A (**XXVIII**)

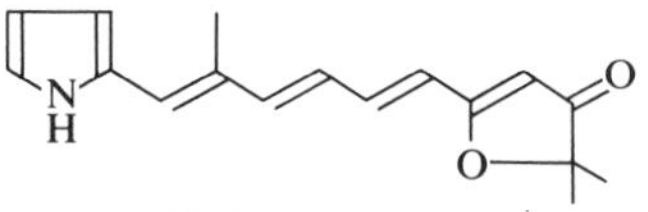

Wallemia C (**XXIX**)

Addendum

During the last few years, a number of important papers on the current interest have appeared. Investigations on the synthesis of carotenoids and vitamin A via sulfones[22], ^{13}C-NMR of *cis-trans* isomeric vitamins A and carotenoids[23], 7-*cis*-retinals[24], application of the NMR shift reagent to isomeric trisubstituted allylic alcohols[25], chemical behavior of conjugated polyenoic acid toward sulfuric acid[26] or chemical inducers of carotenogenesis[27] are only a few instances in this area. In connection with recent developments in carotenoid chemistry, 'Nomenclature of carotenoids (Rules approved 1974)' was issued by IUPAC and IUPAC-IUB[28].

[22] *A. Fischli, H. Mayer*, Helv. *58*, 1492, 1584 (1975); *A. Fischli, H. Mayer, W. Simon, H.-J. Stoller*, Helv. *59*, 397 (1976); *P. S. Manchand, M. Rosenberger, G. Saucy, P. A. Wehrli, H. Wong, L. Chambers, M. P. Ferro, W. Jackson*, Helv. *59*, 387 (1976).

[23] *G. Englert*, Helv. *58*, 2367 (1975).

[24] *V. Ramamurthy, R. S. H. Liu*, Tetrahedron *31*, 201 (1975); *A. E. Asato, R. S. H. Liu*, J. Am. Chem. Soc. *97*, 4128 (1975).

[25] *K. Tsukida, M. Ito, F. Ikeda*, Internat. J. Vit. Nutr. Res. *42*, 91 (1972); *K. Tsukida, M. Ito, F. Ikeda*, Chem. Pharm. Bull. *21*, 248 (1973).

[26] *K. Tsukida, M. Ito, F. Ikeda*, Experientia *29*, 1338 (1973), *30*, 980 (1974).

[27] *W.-J. Hsu, S. M. Poling, C. DeBenedict, C. Rudash, H. Yokoyama*, J. Agr. Food Chem. *23*, 831 (1975).

[28] IUPAC Comm. Nom. Org. Chem., IUPAC-IUB Comm. Biochem. Nom.; Pure Appl. Chem. *41*, 407 (1975).

5.5 Polyacetylene Compounds

Sir Ewart R. H. Jones and Viktor Thaller

The Dyson Perrins Laboratory, Oxford University, South Parks Road, Oxford OX1 3QY, England

1 Introduction

Compounds containing conjugated carbon–carbon triple bonds, polyacetylenes, are formed *in vivo* by dehydrogenation and, in most instances, chain shortening of C_{18} fatty acids. They consist of unbranched chains of carbon atoms which are sometimes part of benzenoid and heterocyclic systems. Monoacetylenes of fatty acid origin and important either as precursors for, or transformation products of, polyacetylenes are also included here.

85% of the natural polyacetylenes known today occur in higher plants and 15% in fungi. Most plant polyacetylenes have been isolated from species of the Compositae and Umbelliferae plant families, in both of which they are widespread; they also occur in species belonging to a dozen other plant families. With the exception of a group of C_{18}-fatty acids which have been found in seed-fats only, polyacetylenes have been isolated from both roots and aerial parts of plants, but individual polyacetylenes usually predominate in one or the other. Fungal polyacetylenes have been isolated from laboratory cultures of Basidiomycete species, but have recently been detected in the sporophores of wild fungi. The large number (~700) of polyacetylenes described[1] comprises a wide range of combinations of differing chain lengths (C_6–C_{18}) and degrees of unsaturation [$-C{\equiv}C-$ to $-(C{\equiv}C)_5-CH{=}CH-$] with a considerable number of functional groups and cyclic systems. A few of these permutations are illustrated by the polyacetylenes **1**–**17**.

[1] For reviews see
F. Bohlmann, T. Burkhardt, C. Zdero, Naturally Occurring Acetylenes, Academic Press, London, New York 1973.
F. Bohlman, H. Bornowski, C. Arndt, Fortschr. Chem. Forsch. *4*, 138 (1962); *F. Bohlmann*, Fortschr. Chem. Forsch. *6*, 65 (1966); *M. Anchel*, Antibiotics *2*, 189, 441 (1967); *Sir Ewart R. H. Jones*, Chem. Brit. *2*, 6 (1966); *J. D. Bu'Lock*, Progr. Org. Chem. *6*, 86 (1964).

$CH_3-C{\equiv}C-$(2,5-thienyl; S)$-CHO$ **1**

$CH_3-CH-CH-[C{\equiv}C]_2-CH_2-CH_2OH$ (epoxide O bridging the two CH) **2**

$HO_2C-[C{\equiv}C]_3-[CH_2]_2-CO_2H$ **3**

$CH_3-[CH_2]_2-[C{\equiv}C]_2-CH\overset{t}{=}CH-CO_2CH_3$ **4**

$CH_3-[C{\equiv}C]_2-CO-C_6H_5$ **5**

$CH_3-CH\overset{t}{=}CH-[C{\equiv}C]_4-CH{=}CH_2$ **6**

$CH_3-[C{\equiv}C]_4-[CH(OH)]_3-CH_2OH$ **7**

$CH_3-CH\overset{t}{=}CH-[C{\equiv}C]_3-CH\overset{t}{=}CH-CH(Cl)-CH_2OCOCH_3$ **8**

$H[C{\equiv}C]_2-CH{=}C{=}CH-CH\overset{c}{=}CH-CH\overset{t}{=}CH-CH_2-CO_2H$ **9**

$CH_3-[C{\equiv}C]_2-CH{=}C$(ring: $HC{=}CH$, O, spiro C; second ring: O, CH_2, $C-CH_2$, H_2) **10**

$CH_3-[C{\equiv}C]_2-CH_2-$(benzene ring with H_3CO_2C and OCH_3) **11** (furan-2-yl)$-C\equiv C-CH_2-C_6H_5$ **12**

$$
\begin{array}{l}
\qquad\qquad\qquad\qquad\qquad\qquad H_2 \\
\qquad\qquad\qquad\qquad\qquad\qquad\; C \\
\qquad\qquad\qquad\qquad\qquad HOHC \quad CH_2 \\
CH_3-[C\equiv C]_3-CH\overset{t}{=}CH-HC \quad CH_2 \\
\qquad\qquad\qquad\qquad\qquad\qquad\; O
\end{array}
$$
13

$HOCH_2-CH\overset{t}{=}CH-[C\equiv C]_2-[CH\overset{t}{=}CH]_2-[CH_2]_2-CH(OH)-[CH_2]_2-CH_3$ **14**

$H_2C=CH-CH(OH)-[C\equiv C]_2-CH_2-CH\overset{c}{=}CH-[CH_2]_6-CH_3$ **15**

$CH_3-[CH_2]_4-C\equiv C-CH_2-CH\overset{c}{=}CH-[CH_2]_7-CO_2H$ **16**

$CH_3-[CH_2]_3-CH=CH-[C\equiv C]_2-[CH_2]_7-CO_2H$ **17**

2 Physical and Chemical Properties

Polyacetylenes are relatively stable in dilute solutions, but are heat and light sensitive in concentrated solutions and condensed phases. They do crystallise (low-temperature crystallisation can often be used for their purification) but the crystalline compounds frequently polymerise and should not be kept. Most polyacetylenes decompose, some very violently, on heating. However, in some instances they can be analysed by gas-liquid chromatography without decomposition. Liquid chromatography (column and thin layer) is invariably used for polyacetylene separations although recoveries are seldom quantitative.

The ultraviolet spectra of the polyacetylenes are their most useful diagnostic feature. With two or more triple bonds present in the chromophore, the spectra are highly characteristic and possess sharp fine structure with a band spacing greater than 2000 cm^{-1}. Two series of bands are generally discernible; they are of very different

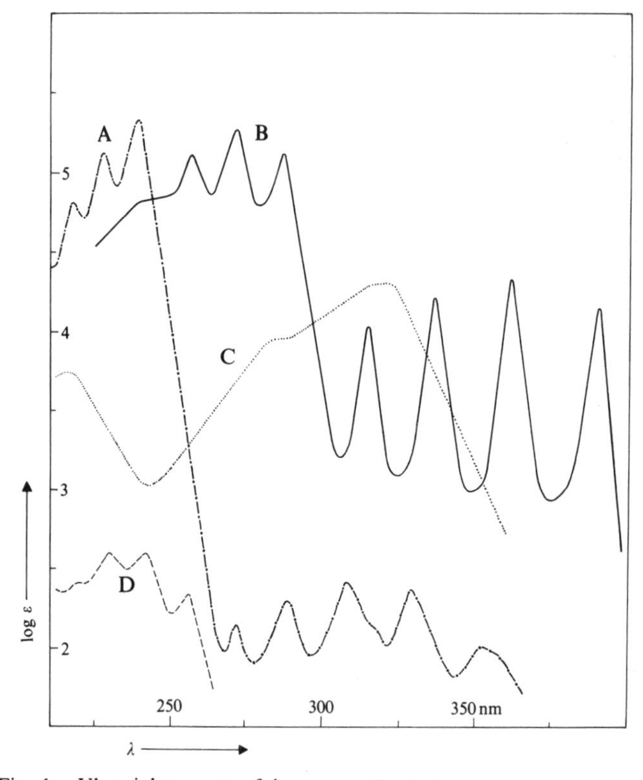

Fig. 1. Ultraviolet spectra of the tetrayne **7** (A), the ene-tetrayn-ene **6** (B), the mono-yne thiophene aldehyde **1** (C), and the diyne **15** (D).

intensities in purely poly-yne chromophores (log $\varepsilon \sim 5$ and 2) in which the short wavelength series become useful for characterisation and detection purposes only with four or more conjugated triple bonds present (Fig. 1, spectrum A). When other unsaturated groups, like the ethylenic or carbonyl group, are in conjugation with the polyyne chain, as is the case with most natural polyacetylenes, the intensities of the two sets of bands are less disparate (log $\varepsilon \sim 5$ and 4) (Fig. 1, spectrum B).

The infared spectra, in addition to giving information about the functional groups present in polyacetylenes, can in favourable cases also be used for their detection. The intensity of the acetylene band (often more than one) at 2250–2100 cm^{-1} varies from very weak [non-existent in crepenynic acid (**16**)] to very strong (*e.g.*, the acetylenic ketone **5**) and can be detected even in crude extracts in the latter case. Characteristic fragmentation patterns can sometimes be recognized in the mass spectra of polyacetylenes; molecular ions are often present thus permitting accurate molecular weight determinations. The nuclear magnetic resonance spectra of polyacetylenes are invaluable in the structure elucidation of the non-acetylenic parts of the molecules. Once the polyacetylenes have been obtained sufficiently pure, their spectral data often suffice to make their identification possible. A few simple chemical transformations, like catalytic hydrogenation to the saturated compounds, manganese dioxide oxidation of allylic and propargylic hydroxyl groups, the sodium borohydride reduction of aldehydes and ketones, the periodate cleavage of diols, the formation of derivatives, etc. readily furnish confirmatory structural information.

3 Detection and Isolation

The concentrations of polyacetylenes in natural sources vary from traces to gram quantities of compound per kilogram of dry plant tissue; the seed lipids of a limited number of species contain C_{18} acetylenic fatty acids as the main fatty acid constituent.

The screening of plant and fungal species for polyacetylenes is effected almost exclusively by ultraviolet spectroscopy. The characteristic and intense absorption common to the majority of natural polyacetylenes permits their detection in crude extracts in spite of the presence of other pigments (*e.g.*, Fig. 2). The detection of very weakly absorbing di-ynes (Fig. 1, spectrum D) and tri-ynes and of polyacetylenes with atypical ultraviolet absorption (Fig. 1, spectrum C) is always difficult; infrared spectroscopy can sometimes be useful in such instances. With the present emphasis on biosynthesis and the recognition of the importance of some of the not easily detectable chromophores, more sophisticated screening methods have been applied to establish their presence. Thus, the ultraviolet screening of chromatographic fractions sometimes reveals chromophores invisible in the crude extract, manganese dioxide oxidation effects a ~20 fold increase in ε-values together with a bathochromic shift of ~25 nm for di-ynes and tri-ynes with propargylic hydroxy groups and alkali treatment converts skipped en-yne systems

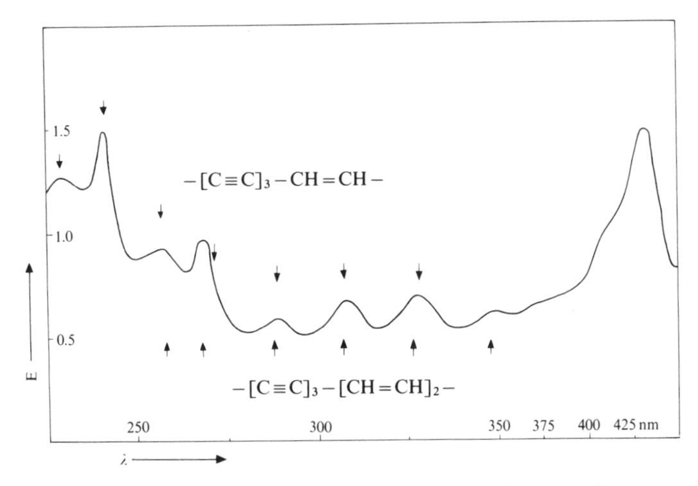

Fig. 2. Assignment of chromophores in the crude extract from leaves of a Dahlia hybrid.

into conjugated polyenes [crepenynic acid (**16**) was detected in this manner[2]].
Polyacetylenes are generally isolated from coarsely cut plant tissues and the aqueous fungal culture media by solvent extraction in the cold. Ether and light petroleum are the most commonly used solvents, but more polar ones have also been employed. The extracts are carefully concentrated (vacuum) and the resulting mixtures, which are often complex and contain very similar compounds, are separated by repeated chromatography (silica gel or alumina) or by counter-current distribution. Thin layer chromatography has recently proved valuable in both the separation of complex mixtures and the final purification of individual polyacetylenes which is completed, whenever possible, by low-temperature crystallisation.

4 Synthesis

A wide variety of reactions, especially those used in acetylene and polyene chemistry have been used in the synthesis of natural polyacetylenes, but two reactions have been particularly valuable in the build-up of a poly-yn-ene chain: the Cadiot-Chodkiewicz reaction permits the asymmetric coupling of acetylenes and the Wittig reaction, the introduction of a double bond into the conjugated system.

$$R{-}[C\equiv C]_n{-}Br + H(C\equiv C)_m{-}R' \longrightarrow R{-}[C\equiv C]_{n+m}{-}R'$$

$$R{-}[C\equiv C]_n{-}CHO + Ph_3P{=}CH{-}R' \longrightarrow R{-}[C\equiv C]_n{-}CH{=}CH{-}R'$$

5 Biosynthesis

Current views on polyacetylene biosynthesis are admirably illustrated[3] by the extensively in-

$$CH_3{-}[CH_2]_7{-}CH\overset{c}{=}CH{-}[CH_2]_7{-}CO_2R$$

$$\downarrow -[H]$$

$$CH_3{-}[CH_2]_4{-}CH\overset{c}{=}CH{-}CH_2{-}CH\overset{c}{=}CH{-}[CH_2]_7{-}CO_2R$$

$$\downarrow -[H]$$

$$CH_3{-}[CH_2]_4{-}C\equiv C{-}CH_2{-}CH\overset{c}{=}CH{-}[CH_2]_7{-}CO_2R \qquad \mathbf{16}$$

$$\downarrow -[H]$$

$$CH_3{-}[C\equiv C]_3{-}CH_2{-}CH\overset{c}{=}CH{-}[CH_2]_7{-}CO_2R$$

$$\downarrow \left.\begin{matrix}1\times\alpha\\2\times\beta\end{matrix}\right\}\text{oxidation}$$

$$CH_3{-}[C\equiv C]_3{-}CH_2{-}CH\overset{c}{=}CH{-}[CH_2]_2{-}CO_2R$$

$$\downarrow \text{reduction}$$

$$CH_3{-}[C\equiv C]_3{-}CH_2{-}CH\overset{c}{=}CH{-}[CH_2]_2{-}CH_2OH \qquad \mathbf{18}$$

$$\downarrow [O]$$

HC=CH OH; CH3—[C≡C]2—C≡C / C(=O) / CH2 / C—CH2 / H2 (bracketed intermediate)

$$\downarrow$$

HC=CH O; CH3—[C≡C]2—CH=C / O / C / CH2 / C—CH2 / H2 **10**

[2] *K. L. Mikolajczak, C. R. Smith, Jr., M. O. Bagby, I. A. Wolff*, J. Org. Chem. *29*, 318 (1964).

[3] *F. Bohlmann, R. Jente, R. Reinecke*, Chem. Ber. *102*, 3283 (1969).

vestigated sequence of intermediates and reactions in the biosynthetic pathway leading from oleate to the spiroketal (**10**) in *Chrysanthemum flosculosum* L. The processes involved in the biogenesis of the triple bond are not yet known. Experiments with leaf homogenates of the above plant indicate that the enzymes required for the dehydrogenation of oleic acid are located within the chloroplasts, whilst the secondary transformations (oxidation, rearrangement and cyclisation) of the C_{13}-alcohol (**18**) can be effected by enzymes outside the plant cell.

6 Physiological Role

The only reported function of polyacetylenes in the organisms in which they are produced appears to be that of self-protection and they act in several cases as systemic fungicides. Some polyacetylenes (*e.g.*, **13** and **14**) are known to be toxic to animals, others (*e.g.*, **9**) show antibiotic activity.

5.6 Polyphenolic Compounds

Takashi Tokoroyama
Faculty of Science, Osaka City University, Sumiyoshi-ku, Osaka, 558 Japan

Takashi Kubota
Medical School, Kinki University, Sayama-cho, Osaka, 589 Japan

1 General Survey of the Compound Classes

Polyphenolic compounds include conveniently groups of complex natural phenols which are mostly polyhydric and are very often oxygen heterocycles. These are fequently colored substances and constitute much of the pigmentation in nature. Several groups of polyphenolic compounds are treated separately in this volume [*e.g.* Chapter 1 (Part 2) Antibiotics; Chapter 6-1 Lichen Substances; Chapter 5-1 Pyran Compounds; Chapter 5-3 Quinoid Compounds]. In this section the natural polyphenolic compounds will be discussed. A number of invaluable surveys dealing with these compounds may be found in comprehensive works[1-7] and reviews (cf. the references of the compound classes below). The important classes of the natural polyphenols are as follows:

a. Simple monocyclic phenols (*e.g.* benzoic acids, acetophenones, cinnamic acids and other simple phenylpropanes)[8,9].
b. Chromones (*e.g.* eugenin and furochromones. Chapter 5-1).
c. Depsides, depsidones and dibenzofurans (Lichen substances, Chapter 6-1).
d. Quinones (Chapter 5-3).
e. Coumarins (isoprenylcoumarins, pyranocoumarins and furocoumarins)[8,10]
f. Lignans[8,11,12]
g. Benzophenones and xanthones[13,14]
h. Flavonoids[15-23]
   1. Stilbenes[24], chalcones and aurones[25]
   2. Flavans (flavones, flavanones, flavonols, anthocyanidins and leucoanthocyanidins).
   3. Isoflavonoids[17,26] (*e.g.* isoflavones, 3-arylcoumarins, pterocarpanes, cumestanes and rotenoids[27]).
   4. Neoflavonoids (4-arylcoumarins)[17,28,29]

[1] *W. Karrer*, Konstitution und Vorkommen der Organischen Pflanzenstoffe, Birkhauser-Verlag, Basel 1958.
[2] *T. K. Devon, A. I. Scott, in* Handbook of Naturally Occurring Compounds, Vol. 1, Acetogenins, Shikimates and Carbohydrates, Academic Press, New York 1975.
[3] *K. Nakanishi, T. Goto, S. Ito, S. Natori, S. Nozoe*, Natural Products Chemistry, Vol. 2, Academic Press, New York 1975.
[4] *F. M. Dean*, Naturally Occurring Oxygen Ring Compounds, Butterworths, London 1963.
[5] *J. B. Harborne* (Ed.), Biochemistry of Phenolic Compounds, Academic Press, London 1964.
[6] *T. A. Geissman, D. H. Crout*, Organic Chemistry of Secondary Metabolism, Freeman Cooper and Co., San Fancisco 1969.
[7] *P. Ribereau-Gayon* (Ed.), Plant Phenolics, Oliver and Boyd, London 1972.
[8] *A. C. Neish, in* Plant Biochemistry, *J. Bonner, J. E. Varner* (Eds.), p. 581, Academic Press, New York 1965.
[9] *J. B. Pridham*, Ann. Rev. Plant Physiol. *16*, 13 (1965).
[10] *F. M. Dean*, Fortschr. Chem. Org. Naturstoffe *9*, 225 (1952).
[11] *K. Weinges, R. Spänig, in* Oxidative Coupling of Phenols, *W. I. Taylor and A. R. Battersby* (Eds.), p. 323, Mercel-Dekker, New York 1967.
[12] *O. R. Gottlieb*, Phytochemistry *11*, 1537 (1972).
[13] *J. C. Roberts*, Chem. Rev. *61*, 591 (1961).
[14] *I. Carpenter, H. D. Locksley, F. Scheinmann*, Phytochemistry *8*, 2013 (1969).

Many polyphenolics occur in living cells in combined forms as glucosides or esters. Lignins[8,30–35] and tannins[5,35–39] are important groups of polymeric polyphenolic materials.

2 Isolation and Identification of Polyphenolic Compounds

Miscellaneous methods are used for the isolation of naturally occurring polyphenols dependent upon the properties of the compounds and the chemical composition of the biological material from which the separation has to be made. For the particular groups of polyphenolic compounds, the isolation procedures have been well described and are rather routine[40–44]. Chromatographic methods are widely and conveniently employed for the separation of polyphenolic compounds[45,46]. Silica gel is probably the most common adsorbent. In some cases deactivation with organic acids or mineral acids is necessary for the separation of sensitive phenols. Silica gel also serves as a support for partition chromatography. Polyamide finds its use in the separation of somewhat polar polyphenolics such as phenolic acids or glucosides[45,47]. Lowering of pH may be pertinent for the chromatography of phenolic acids with this adsorbent in order to prevent ionization[48]. The application of thin layer chromatography and related techniques is extremely efficient and useful for the separation of polyphenolic compounds[49]. It can be widely used not only for the monitoring of isolation processes but also for preparative purposes. Thus the separation of a complex mixture of closely related polyphenols can be achieved in simple manner and in small scale. For mixtures of polyphenolics having extremely close R_f values, the continuous development method[50] may be effective. Nearly twenty anthraquinone pig-

---

15 *T. A. Geissman* (Ed.), The Chemistry of Flavonoid Compounds, Pergamon Press, Oxford 1962.

16 *J. B. Harborne, T. J. Mabry, H. Mabry* (Eds.), The Flavonoids, Chapman and Hall, London 1975.

17 *T. Mabry, A. E. Alston, C. Runeckles* (Eds.), Recent Advances in Phytochemistry, Vol. 1, p. 305, Appleton-Century-Crofts, New York 1968.

18 *H. Wagner*, Fortschr. Chem. Org. Naturstoffe *31*, 153 (1974).

19 *H. D. Locksley*, Fortschr. Chem. Org. Naturstoffe *30*, 207 (1973).

20 *J. B. Harborne in* Recent Advances in Phytochemistry, *T. Mabry, R. E. Alston, V. C. Runeckles* (Eds.), Vol. 4, p. 107, Appleton-Century-Crofts, New York 1972.

21 *T. W. Goodwin* (Ed.), "Chemistry and Biochemistry of Plant Pigments", Academic Press, New York 1965.

22 *J. B. Harborne*, "Comperative Biochemistry of the Flavonoids", Academic Press, New York 1967.

23 *L. Farkas, M. Gabor and F. Kally*, "Topics in Flavonoid Chemistry and Biochemistry", Elsevier, Amsterdam 1975.

24 *G. Billek*, Fortschr. Chem. Org. Naturstoffe *22*, 115 (1964).

25 *L. Farkas and L. Pallos*, Fortschr. Chem. Org. Naturstoffe *25*, 150 (1967).

26 *E. Wong*, Fortschr. Chem. Org. Naturstoffe *28*, 1 (1970).

27 *L. Crombie*, Fortschr. Chem. Org. Naturstoffe *21*, 275 (1963).

28 *W. D. Ollis*, Experientia *22*, 277 (1966).

29 *T. R. Seshadri*, Phytochemistry *11*, 881 (1972).

30 *J. M. Harkin*, in Oxidative Coupling of Phenols, W. I. Taylor, A. R. Battersby (Eds.), p. 243, Mercel-Dekker, New York 1967.

31 *S. A. Brown*, Ann. Rev. Plant Physiol. *17*, 223 (1966).

32 *T. Swain, in* Plant Biochemistry, *J. Bonner, V. E. Varner* (Eds.), p. 552, Academic Press, New York 1965.

33 *K. V. Sarkanen, C. H. Ludwig*, Lignin: Occurrence, Formation, Structure and Reaction, Wiley, New York 1972.

34 *S. M. Siegel, P. Carrol, I. Umeno, C. Corn, in* Recent Advances in Phytochemistry, *T. Mabry, R. E. Alston, V. C. Runeckles* (Eds.), Vol. 4, p. 223, Appleton-Century-Crofts, New York 1968.

35 *K. Freudenberg, A. C. Neish*, Constitution and Biosynthesis of Lignin, Springer, Berlin 1968.

36 *R. D. Haworth, O. T. R. Schmidt, in*, Recent Development in the Chemistry of Natural Phenolic Compounds, *W. D. Ollis* (Ed.), p. 134, Pergamon, Oxford 1961.

37 *K. Weinges, W. Bahr, W. Ebert, K. Goritz, H.-D. Marx*, Fortschr. Chem. Org. Naturstoffe *22*, 115 (1964).

38 *D. G. Roux*, Phytochemistry *11*, 1219 (1972).

39 *E. Haslam*, The Chemistry of Vegetable Tannins, Academic Press, London 1966.

40 *K. Peach, M. V. Tracey* (Eds.), Moderne Methoden der Pflanzenanalyse, Bd. III, Springer, Berlin 1955.

41 *M. K. Seikel, in* Biochemistry of Phenolic Compounds, *J. B. Harborne* (Ed.), p. 33, Academic Press, New York 1964.

42 *J. B. Harborne*, Phytochemical Methods—a Guide to Modern Techniques of Plant Analysis, Chapman and Hall, London 1973.

43 *T. R. Seshadri, M. K. Seikel in* The Chemistry of Flavonoid Compounds, *T. A. Geissman* (Ed.), p. 6, Pergamon, Oxford 1961.

44 *T. J. Mabry, K. R. Markham, M. B. Thomas*, The Systematic Identification of Flavonoids, Springer, Berlin 1970.

45 *J. B. Pridham* (Ed.), Methods in Polyphenol Chemistry, Pergamon, Oxford 1964.

46 *J. B. Harborne, in* Chromatography, *E. Heftmann* (Ed.), p. 677, Reinhold, New York 1967.

47 *H. Enders, H. Hörmann*, Angew. Chem. *75*, 288 (1963).

48 *M. K. Seikel, F. D. Hostettler, D. B. Johnson*, Tetrahedron *24*, 1475 (1968).

49 *K. Egger, in* Thin-Layer Chromatography. A Laboratory Handbook, *E. Stahl* (Ed.), p. 686, Springer, Berlin 1969.

50 *R. D. Bennett, E. Heftmann*, J. Chromatogr. *12*, 245 (1963).

ments (julichromes) have been isolated by this modification[51]. A dry column may be substituted for the thin layer plate in larger scale preparations with comparable efficiency[52,53]. Paper partition chromatography and paper electrophoresis are likewise effective means for the separation and identification of polyphenolics[46] and are used sometimes in preparative scale[54–56]. Thin layer plates coated with cellulose may be more conveniently used for the same purposes. The isolation of water-soluble, highly polar compounds is also achieved by the application of countercurrent techniques[57] and column chromatography using various cellulose preparations or ion-exchange resins[58]. Gel filtration methods are effective in certain cases[59]. Gas-chromatographic techniques may be used conveniently only for the analyses of relatively volatile polyphenols. It is claimed that by this method the principal classes of lignans can be differentiated and the geometrical isomers may usually be distinguished[60]. Acetylation, methylation or, specially, silylation[61] before analysis are pertinent measures for the enhancement of volatility in the case of polyhydric compounds. Combination of gas-chromatographic separation with mass spectrometry (GC-MS) is the highly efficient method which make possible the direct identification of complex mixture on ultramicro scale[62]. Extensive study of cannabis constituents by this procedure is a notable example[63]. High pressure liquid chromatography is also the versatile and potent technique of recent application and specially useful in the separations of nonvolatile and heat sensitive polyphenolics[64]. Identification of polyphenolic compounds is feasible by the application of chromatographic and spectroscopic (Section 3) methods supplemented with classical procedures.

3 Structure Elucidation of Natural Polyphenolic Compounds

Progress in spectroscopic methods has made the identification and structure analysis of polyphenolic compounds considerably easier. In most cases these are possible by several spectroscopic measurements (*i.e.* IR, UV, NMR and MS) on the compound in question and a few of simple derivatives such as the acetate or methyl ether. Even in somewhat complicated cases the problem may be solved by extensive utilization of proton nuclear magnetic resonance techniques, which would enable the complete assignment of all protons. The application of spectroscopic methods to structural problem in polyphenolic compounds is briefly outlined below.

a. *Ultraviolet spectroscopy (UV)*: A. I. Scott has proposed an empirical method to predict the position of the principal electron transfer bands (K-band) in substituted acetophenones, benzaldehydes and benzoic acids[65]. It may have some utility for simple phenols[66,67]. The ultraviolet spectra of the polyphenolics are altered characteristically with change of pH[68] or the presence of added reagents. This increases the value of spectra data in identification and structural analysis. The commonly used reagents for such purposes[69] are aluminum chloride, sodium

[51] *N. Tsuji et al.*, Tetrahedron *25*, 2999 (1967).

[52] *B. Loev, K. M. Snader*, Chemistry and Industry, 15 (1965).

[53] *B. J. Hunt, W. Rigby*, Chemistry and Industry, 1968 (1967).

[54] *S. E. Drewes, D. G. Roux*, Biochem. J. *98*, 493 (1966).

[55] *T. G. Fourie, I. C. du Preez, D. G. Roux*, Phytochemistry, *11*, 1763 (1972).

[56] *K. R. Markham, L. J. Porter*, Phytochemistry *12*, 2007 (1973).

[57] *L. C. Craig, D. Craig, E. G. Scheibel, in* Technique of Organic Chemistry, *Weisberger* (Ed.), Vol. 3, p. 1, Interscience, New York 1951.

[58] *D. Brown, R. O. Asplund, V. A. McMahon*, Phytochemistry *14*, 1083 (1975).

[59] For examples, cf. *D. T. Coxon et al.*, Tetrahedron Lett. 5237, 5241 (1970); *M. Manabe, S. Matsuma*, Agr. Biol. Chem. *35*, 417 (1971); T. K. Kirk, L. F. Lorenz, M. J. Larsen, Phytochemistry *14*, 281 (1975).

[60] *D. C. Ayres, R. B. Chater*, Tetrahedron *25*, 4093 (1969).

[61] For recent examples, cf. *P. D. Collier, T. Bryce, R. Mallows, P. E. Thomas*, Tetrahhedron *29*, 125 (1973); *D. E. Gueffroy, R. E. Kepner, A. D. Webb*, Phytochemistry *10*, 813 (1971); *W. E. Hill, J. H. Hart, Y. Yazaki*, Phytochemistry *13*, 1519 (1974).

[62] *D. E. Games*, Tetrahedron Lett. 3187 (1972); *R. D. Schmid et al.*, Phytochemistry *12*, 2765 (1973); *N. S. P. Rao, L. R. Row, R. T. Brown*, Phytochemistry *12*, 671 (1973); *W. M. Bandaranayake, S. S. Selliah, M. U. S. Sultanbawa, D. E. Games*, Phytochemistry *14*, 265 (1975).

[63] *J. Friedrich-Friechtl, G. Spiteller*, Tetrahedron *31*, 479 (1975).

[64] For example, cf. *R. L. Lyne, L. J. Mulheim, D. P. Leworthy*, J. Chem. Soc., Chem. Commun. 497 (1976).

[65] *A. I. Scott*, Interpretation of the Ultraviolet Spectra of Natural Products, Pergamon, Oxford 1964.

[66] *T. Tokoroyama, T. Kamikawa, T. Kubota*, Tetrahedron *24*, 2345 (1968).

[67] *T. Sassa et al.*, Agr. Biol. Chem. *32*, 1432 (1968).

[68] *A. Erdtman*, Chemistry and Industry, 581 (1955).

[69] *L. Jurd, in* The Chemistry of Flavonoid Compounds, *T. A. Geissman* (Ed.), p. 107, Pergamon, Oxford 1962.

ethoxide, sodium acetate and boric acid-sodium acetate. Systematic studies have been made on flavonoids[44]. In sodium ethoxide solution, all free hydroxyl groups are ionized, whereas the addition of sodium acetate selectively ionizes more acidic phenolic groupings—*i.e.* the 3-, 7- and 4′-hydroxyl groups. With aluminum chloride, flavones and flavonols which contain hydroxyl groups at C-3 or C-5 form acid stable complexes. In addition, aluminum chloride forms acid labile complexes with flavonoids which contain ortho-dihydroxyl systems. In the presence of sodium acetate, boric acid chelates with ortho-dihydroxyl groups at all locations on the flavonoid nucleus, except at C-5, 6. Reflecting these changes, characteristic bathochromic shift of absorption bands will be observed in the spectra. The ultraviolet spectra of the acetylated material also provide useful information being much like those of the corresponding deoxygenated compounds. At present a good deal of the data on the ultraviolet spectra of the polyphenolic compounds are available and the spectral characteristics of many classes can be discerned[44,65,69]. Thus measurement of the ultraviolet spectra is a useful tool for the assignment of the compound class and also the oxygen substitution pattern[44].

b. *Infrared spectroscopy (IR)*: The important absorption bands of the polyphenolics in the infrared region are those due to hydroxyl and carbonyl stretching vibrations. The position of the latter absorption is characteristic of the type of heterocyclic ring present, provided that the influence of the hydrogen bonding can be neglected, and serves for the preliminary assignment among the compound classes[70,71]. On the other hand the specific shift of the carbonyl bands to lower frequencies by intra-molecular hydrogen bonding affords evidence as to the relative positions (*peri* or *ortho*) of the hydroxyl groups. The stretching frequencies of the hydroxyl groups are indicative of various structural features in the polyphenolic compounds, since these affect inter- and intra-molecular hydrogen bondings in many ways. Thus the position and appearance of the hydroxyl and carbonyl stretching bands have been correlated with substitution types in several polyphenolics; flavonoids[72,73], chromones[74] and xanthones[75]. In the structural study of an isoflavan compound, lonchocarpan, the two substitution patterns **I** and **II** were discriminated on the basis of close examination of the hydroxyl absorptions in the model compounds. An intramolecularly hydrogen bonded band with the adjacent methoxyl group was observed in **I** but not in **II**[76]. Lowering of the hydroxyl stretching frequencies due to the OH–π interaction was examined in several natural phenols[77–79].

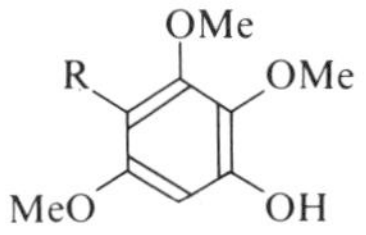

OMe
R OMe
MeO OH

I

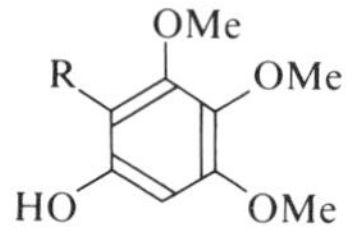

OMe
R OMe
HO OMe

II

c. *Nuclear magnetic resonance spectroscopy (NMR)*: Proton nuclear magnetic resonance spectroscopy (PMR) with additional techniques such as spin decoupling, nuclear Overhauser effect and the solvent shift is a powerful tool for the structural study of the polyphenolic compounds. The hydroxyl proton signals of phenols are usually observed at δ 4–8 ppm, while they are shifted downfield to δ 10.5–16 ppm by the presence of intramolecular hydrogen bonding[80]. The latter fact is good evidence for the presence of a hydroxyl group in chelating position with reference to the carbonyl group. Aromatic proton chemical shifts are useful for defining the

[70] *L. J. Bellamy*, Advances in Infrared Group Frequencies, p. 162, Metheuen, London 1968.

[71] *A. R. Katritzky, A. P. Ambler, in* Physical Methods in Heterocyclic Chemistry, *A. R. Katritzky* (Ed.), Vol. 2, p. 165, Academic Press, New York 1963.

[72] *L. H. Briggs, L. D. Colebrook*, Spectrochim. Acta *18*, 939 (1962).

[73] *J. H. Looker, W. W. Hanneman*, J. Org. Chem. *27*, 381 (1962).

[74] *R. D. H. Murray et al.*, Tetrahedron *25*, 5819, 5839 (1969).

[75] *F. Scheinmann*, Tetrahedron *18*, 853 (1962).

[76] *A. Pelter, P. I. Amenechi*, J. Chem. Soc. (C), 887 (1969).

[77] *R. D. H. Murray, M. M. Ballantyne*, Tetrahedron *26*, 4667 (1970).

[78] *P. H. McCabe, R. McCrindle, R. D. H. Murray*, J. Chem. Soc. (C), 145 (1967).

[79] *N. Tsuji, K. Nagashima*, Tetraheron *25*, 3017 (1969).

[80] *L. M. Jackman, S. Sternhell*, Application of NMR Spectroscopy in Organic Chemistry, 2nd ed., Pergamon, Oxford 1969.

oxygenation patterns in polyphenolic compounds. Ballantine and Pillinger[81] have analyzed the chemical shift data of the aromatic protons in some 450 phenolic compounds in terms of additive substituent shielding values. With some empirical corrections, these values can be used to predict the chemical shift positions and hence the orientation of unknown polysubstituted phenols. This approach has been further refined in xanthone case[82]. In addition to the substitution effect consideration, analysis of the spin-spin coupling and reference to the chemical shift data of related substances are naturally pertinent for allocation of the aromatic protons. A number of relevant data for this purpose available for coumarins[83-85], xanthones[86], flavonoids[42,87-90], isoflavonoids[76], pterocarpins[91] and rotenoids[92,93]. Various types of long-range spin-spin coupling[80] have been also detected in the polyphenolic compounds and are of some utility in the structure arguments[8,93-96]. Long-range coupling between the methoxyl protons and the *ortho* protons is known to cause the broading of both signals[97,98]. The assignment of the aromatic protons is further facilitated by the use of additional techniques. Conversion to the phenolate ion gives rise to an appreciable upfield shift (0.60–0.84 ppm) at the *ortho* and *para* positions with only a smaller upfield shift at the *meta* position[99]. Acetylation generally leads to a downfield shift of the *para* protons (0.35–0.50 ppm), whereas on *ortho* proton signal is shifted downfield by 0.20–0.25 ppm[88,99-104]. The deshiedling effect by the acetoxyl group is recognizable at the *peri*-proton and may be applied for the differentiation of the substitution types in hydroxychromones[94]. The nuclear Overhauser effect, observable among two or more protons in close proximity, has been examined in a number of polyphenolic compounds. The interactions of hydroxyl[105], methoxyl[98,106-108,110] or methyl[96,108-110] protons with *ortho*[98,108-110] or *peri*[105-107,110] standing aromatic protons are the most commonly observable examples. Solvent effects are also used frequently as an aid in the analysis of the NMR spectra of polyphenolics[80,111]. The systematic application of the solvent effect is the use of the solvent shift values ($\Delta = \delta_{C_6D_6} - \delta_{CDCl_3}$). The studies in this direction have been made on a number of polyphenolic compounds; flavones[112-114], quinones[98,115], xanthones[116,117] and coumarins[118]. The solvent shift of the methoxyl signal is employed most prominently. A

81 *J. A. Ballantine, C. T. Pillinger*, Tetrahedron *23*, 1691 (1967).
82 *D. Barraclough et al.*, J. Chem. Soc. (C), 603 (1970).
83 *J. F. Fischer, H. E. Nordby*, Tetrahedron *22*, 1489 (1966).
84 *E. V. Lassak, J. T. Pinhey*, J. Chem. Soc. (C), 2000 (1967).
85 *A. I. Gray, R. D. Waigh, P. G. Waterman*, J. Chem. Soc., Chem. Commun. 632 (1974) and references cited therein.
86 *G. H. Stout et al.*, Tetrahedron *25*, 1961 (1969).
87 *L. M. Jackman*, Fortschr. Chem. Org. Naturstoffe *23*, 349 (1965).
88 *T. T. Batterham, R. J. Highet*, Aust. J. Chem. *17*, 428 (1964).
89 *J. Massicot et al.* Bull Soc. Chim. Fr., 1962 (1962), 2719 (1963).
90 *T. J. Mabry, in* Perspective in Phytochemistry, *J. B. Harborne, T. Swain* (Eds.), p. 1, Academic Press, New York 1969.
91 *K. G. R. Pacher, W. G. E. Underwood*, Tetrahedron *23*, 1817 (1967).
92 *L. Crombie, J. W. Lown*, J. Chem. Soc. 775 (1962).
93 *D. G. Carlson, D. Weisleder, W. H. Tallent*, Tetrahedron *29*, 2731 (1973).
94 *A. Aron et al.*, Tetrahedron Lett. 4201 (1967).
95 *C.-L. Chen, F. D. Hostettler*, Tetrahedron *25*, 3223 (1969).
96 *W. T. L. Sidwell, H. Fritz, Ch. Tamm.*, Helv. Chim. Acta *54*, 207 (1971).
97 *C. Farid*, Tetrahedron *24*, 2121 (1968).
98 *R. J. J. Ch. Lousberg et al.*, J. Chem. Soc. (C), 2154 (1970).
99 *R. J. Highet, P. F. Highet*, J. Org. Chem. *30*, 902 (1965).
100 *R. G. Cooke, I. D. Rae*, Austral. J. Chem. *17*, 379 (1964).
101 *A. Arnone, G. Cardillo, L. Merlini, R. Mondelli*, Tetrahedron Lett. 4201 (1967).
102 *Dun-Mei Yang, N. Takeda, Y. Iitaka, U. Sankawa, S. Shibata*, Tetrahedron *29*, 519 (1973).
103 *R. Vleggaar, T. M. Smalberger, A. J. von den Berg*, Tetrahedron *31*, 2571 (1975).
104 *M. Aritomi, T. Kawasaki*, Chem. Pharm. Bull. *18*, 2224 (1970).
105 *R. A. Archer et al.*, J. Am. Chem. Soc. *92*, 5200 (1970).
106 *I. Morimoto et al.*, Chem. Commun. 550 (1970).
107 *R. D. H. Murray, M. M. Ballantyne*, Tetrahedron *26*, 4473 (1970).
108 *K. Yoshihara, M. Tezuka, S. Natori*, Tetrahedron Lett. 7 (1970).
109 *L. A. Mitscher et al.*, J. Am. Chem. Soc. *92*, 6070 (1970).
110 *M. Sezaki et al.*, Tetrahedron *26*, 5171 (1970).
111 For example, cf. *D. L. Dreyer, D. J. Berteilli*, Tetrahedron *23*, 4607 (1967).
112 *R. G. Wilson, J. H. Bowie, D. H. Williams*, Tetrahedron *24*, 1407 (1968).
113 *P. J. Garrat et al.*, Tetrahedron *23*, 2413 (1967).
114 *S. M. Kupchan et al.*, Tetrahedron *25*, 1603 (1969).
115 *J. H. Bowie et al.*, Tetrahedron *22*, 1771 (1966).
116 *H. D. Locksley, I. G. Murray*, J. Chem. Soc. (C), 392 (1970).
117 *B. Jackson, H. D. Locksley, F. Schelmann*, J. Chem. Soc. (C), 2500 (1967).
118 *R. Grigg, J. A. Knight, P. Roffey*, Tetrahedron *22*, 3301 (1966).

large positive value (30 Hz) is obtained in the presence of at least one free *ortho* position, while a small positive (0–20 Hz) or negative value results for the methoxyl group flanked by substituents. The addition of small quantity (3% v/v) of trifluoroacetic acid may enhance the shift[119]. However precautions are necessary for the application of the above mentioned rule since in certain cases, especially when a free hydroxyl group is present in the molecule, a large solvent shift is observed in spite of the absence of the free *ortho* position and may lead to an incorrect result[76,120,121]. In this respect the measurement of the solvent induced shift as trimethylsilyl ether may be more pertinent and the locatings of methoxyl groups by this procedure have been reported in flavonoids[122] and coumarins[123]. The use of lanthanide shift reagents[124] is also very helpful in the analysis of complex spectra. The utility of the method is proved in a number of examples, flavonoids[125], coumarins[126] and others[127]. Even when the NMR spectra at 60 or 100 MHz are not sufficiently resolved for first-order analysis, the measurement of spectra at 220 MHz may enable such analysis and are sometimes sufficient for the solution of the problem[82,105,116,117,128]. The application of carbon-13 magnetic resonance spectroscopy (CMR) to the structure problems of polyphenols are the field of current interest[129] and, the relevant chemical shifts and other data are accumulating for *e.g.* 2-pyrones and coumarins[130–133], tocopherols and 2,2-dimethylchromanols[134], lignans[135,136], flavonoids[137–139] and rotenoids[140].

d. *Mass Spectrometry (MS)*: Mass spectrometry enables not only the determination of precise molecular weight but the characterization of structure types by the analysis of the fragmentation patterns, and is also a powerful means in structure studies of polyphenolic compounds. The basic fragmentation patterns of the main oxygen heterocycles have been summarized in a monograph[141,142]. Other relevant works on the mass spectra of polyphenolic compounds deal with lichen metabolites (depside, depsidone and dibenzofuran)[143], benzofuran[144,145], benzophenones[146], chromenes and chromones[144,147–149], phenylpropanes[150], lignans[151,152], couma-

[119] *R. G. Wilson, D. H. Williams*, J. Chem. Soc. (C), 2477 (1968).

[120] *G. H. Stout et al.*, Chem. Commun. 211 (1968).

[121] *C. K. Atal, P. N. Moza, A. Pelter*, Tetrahedron Lett. 1397 (1968).

[122] *E. Rodriguez, N. J. Carman, T. J. Mabry*, Phytochemistry *11*, 409 (1972).

[123] *P. Brown, R. Owen Asplund, V. A. McMahon*, Phytochemistry *13*, 1923 (1974).

[124] *B. C. Mayo*, Chem. Soc. Rev. *2*, 49 (1973).

[125] *M. Okigawa, N. U. Khan, N. Kawano*, J. Chem. Soc., Perkin I, 1563 (1976); *M. Okigawa, N. U. Khan, N. Kawano, W. Rahman*, J. Chem. Soc., Perkin I, 1563 (1975), cf. also *F-C. Chen, Y-M Liu, L-C. Wu*, Phytochemistry *13*, 1571 (1974).

[126] *A. I. Gray, R. D. Waigh, P. G. Waterman*, J. Chem. Soc., Chem. Commun., 632 (1974); *A. I. Gray, R. D. Waigh, P. G. Waterman*, J. Chem. Soc., Perkin I, 488 (1975).

[127] *S. G. Levine, R. S. Hicks*, Tetrahedron Lett. 311 (1971); *K. K. Chexal, J. S. E. Holker, T. J. Simpson, K. Young*, J. Chem. Soc., Perkin I, 543 (1975); *F. Fish, A. I. Gray, R. D. Waigh, P. G. Waterman*, Phytochemistry *15*, 313 (1976); *A. Arnone, L. Camarda, L. Merlini, G. Nasini*, J. Chem. Soc., Perkin I, 186 (1976); *R. B. Fo, J. C. Mourão, O. R. Gottlieb, T. G. S. Maia*, Tetrahedron Lett. 1157 (1976).

[128] *L. R. Row, P. Satyanarayana, C. Scrinivasula*, Tetrahedron *26*, 3051 (1970).

[129] *W. Wenkert, B. L. Buckwalter, I. R. Burfitt, M. J. Gašic, H. E. Gottlieb, F. W. Hagaman, F. M. Schell, P. M. Wovkulich, in* Topics in Carbon-13 NMR Spectroscopy, *G. C. Levy* (Ed.), Vol. 2, Wiley-Interscience, New York 1976.

[130] *K. Tori, T. Hirata, O. Koshitani, T. Suga*, Tetrahedron Lett. 1311 (1976) and references cited therein.

[131] *W. V. Turner, W. H. Pirkle*, J. Org. Chem. *39*, 1935 (1974).

[132] *S. A. Sojka*, J. Org. Chem. *40*, 1175 (1976).

[133] *N. J. Cussans, T. N. Huckerby*, Tetrahedron *31*, 2587, 2591, 2719 (1975).

[134] *M. Matsuo, S. Urano*, Tetrahedron *32*, 229 (1976).

[135] *C. J. Kelly, R. C. Harruff, M. Carmack*, J. Org. Chem. *41*, 449 (1976).

[136] *E. Wenkert et al.*, Phytochemistry *15*, 1547 (1976).

[137] *C. A. Kingsbury, J. H. Looker*, J. Org. Chem. *40*, 1120 (1975).

[138] *H. Wagner, V. M. Chari, J. Sonnenbichler*, Tetrahedron Lett. 1799 (1976).

[139] *B. Ternai, K. R. Markham*, Tetrahedron *32*, 565, 2607 (1976).

[140] *L. Crombie, G. W. Kilbee, D. A. Whiting*, J. Chem. Soc., Perkin I, 1497 (1975).

[141] *H. B. Budzikiewicz, C. Djerassi, D. H. Williams*, Structure Elucidation of Natural Products by Mass Spectrometry, Vol. 2, p. 254, Holden-Day, San Franscisco 1964.

[142] *Q. N. Porter, J. Baldas*, Mass Spectrometry of Heterocyclic Compounds, Wiley-Interscience, New York 1971.

[143] *S. Huneck et al.*, Tetrahedron *24*, 2707 (1968).

[144] *C. S. Barnes, J. L. Occolowitz*, Australian J. Chem. *17*, 975 (1969).

[145] *E. N. Givens, L. G. Alexakos, P. B. Venuto*, Tetrahedron *25*, 2407 (1969).

[146] *J. A. Ballantine, C. T. Pillinger*, Org. Mass Spectrometry *1*, 425 (1968).

[147] *R. Willhalm, A. Thomas, F. Gautschi*, Tetrahedron *20*, 1185 (1964).

rins[144,153–157], flavonoids[158–164], flavonoid glycosides[165–167] and isoflavonoids[76,168,169].

4 Reactions of Natural Phenolic Compounds

Recently it seems to be a common trend that the application of chemical reactions in structure determination of polyphenolic compounds has become very limited due to the development of the spectroscopic methods outlined in the previous Section 3. For that purpose a few reactions like esterification (Houben-Weyl Vol. VIII) and alkylation (Houben-Weyl Vol. VI/3) with some correlative or synthetic reactions would generally be sufficient. On the other hand the multifarious reactions of the natural polyphenolics have aroused new stimulus for the study in connection with the metabolism, biosynthesis and biological activity. Mention will be made of some of these reactions in the following.

a. *Photo-induced reaction*: The well-known dimerization of cinnamic acid to α-truxillic and β-truxinic acids on ultraviolet irradiation[170] has been studied in the light of recent concept[171–173]. The photo-dimerization of coumarin is effected with or without sensitizer and has been investigated mechanistically[174]. The dimerizations of some natural coumarins including several furocoumarins were achieved in solid state[170]. The photo-addition reactions of coumarin[175] and chromone[176] with olefins, and of flavone[177] with diphenylacetylene are reported. A photosensitizer is required for the reaction of coumarin. Photo-Fries rearrangement of phenyl cinnamates leads to the formation of 2′-hydroxychalcones[178–182]. Photo-Claisen rearrangement of allyl phenyl ether, which afford allylphenols, also reported[183]. Photolysis of flavanone produces 2′-hydroxychalcone (20%),

III → (hν) [IV] → V

[148] *M. M. Baldawi, M. B. E. Fayez*, Chemistry and Industry, 498 (1966).
[149] *S. E. Drewes, in* Progress in Mass Spectrometry, Vol. 2, p. 25, Verlag Chemie, Weinheim 1974.
[150] *V. Kovacik et al.*, Chem. Ber. *102*, 1513 (1969).
[151] *A. Pelter, P. Stainton, M. Baker*, J. Heterocyclic Chem. *3*, 191 (1966).
[152] *A. Pelter*, J. Chem. Soc. (C), 1376 (1967), 74 (1968).
[153] *R. A. W. Johnston et al.*, J. Chem. Soc. (C), 1712 (1966).
[154] *F. M. Dean et al.*, J. Chem. Soc. (C), 2232 (1967).
[155] *J. P. Kutney et al.*, Org. Mass Spectrometry *5*, 249 (1971).
[156] *L. Crombie, D. E. Games, A. McCormick*, J. Chem. Soc. (C), 2545, 2553 (1967).
[157] *R. H. Shapiro, C. Djerassi*, J. Org. Chem. *30*, 955 (1965).
[158] *D. G. I. Kingston*, Tetrahedron *27*, 2691 (1971); *29*, 4083 (1973).
[159] *A. Pelter, P. Stainton, M. Barber*, J. Heterocyclic Chem. *2*, 262 (1965).
[160] *A. Pelter, P. Stainton*, J. Chem. Soc. (C), 1933 (1967).
[161] *Y. Itagaki et al.*, Bull. Chem. Soc. Jap. *39*, 538 (1966).
[162] *H. Audier*, Bull. Soc. Chim. Fr. 2892 (1966).
[163] *R. I. Reed, J. M. Wilson*, J. Chem. Soc. (C), 5949 (1963).
[164] *S. E. Drewes*, J. Chem. Soc. (C), 1140 (1968).
[165] *R. D. Schmid*, Tetrahedron *28*, 3259, 5037 (1972).
[166] *H. Wagner, O. Seligmann*, Tetrahedron *29*, 3029 (1973).
[167] *M-L. Bouillant, J. Favre-Bouvin, J. Chopin*, Phytochemistry *14*, 2267 (1975).
[168] *A. Pelter et al.*, J. Heterocyclic Chem. *2*, 256 (1965).
[169] *A. P. Johnson, A. Pelter, P. Stainton*, J. Chem. Soc. (C), 192 (1966).
[170] *K. Schaffner*, Fortschr. Chem. Org. Naturstoffe *22*, 82 (1964).
[171] *M. D. Cohen, G. M. T. Schmidt, F. I. Sonntag*, J. Chem. Soc. 2000 (1964); *G. M. T. Schmidt*, J. Chem. Soc. 2014 (1964).
[172] *J. Bergman et al.*, J. Chem. Soc. 2021 (1964).
[173] *D. Rabinovich, G. M. T. Schmidt*, J. Chem. Soc. 2030 (1964).
[174] *G. S. Hammond, C. A. Stout, A. A. Lamola*, J. Am. Chem. Soc. *86*, 3103 (1964).
[175] *J. W. Hanifin, E. Cohen*, Tetrahedron Lett. 1419 (1966).
[176] *J. W. Hanifin, E. Cohen*, Tetrahedron Lett. 5421 (1966).
[177] *A. Schönberg, G. D. Khandelwal*, Chem. Ber. *103*, 2780 (1970).
[178] *H. Obara, H. Takahashi, H. Hirano*, Bull. Chem. Soc. Jap. *42*, 560 (1969).
[179] *V. K. Bhatia, J. Kagan*, Chemistry and Industry 1203 (1970).
[180] *V. T. Ramakrishnan, J. Kagan*, J. Org. Chem. *35*, 2898, 2901 (1970).
[181] *J. Onodera, H. Obara*, Bull. Chem. Soc. Jap. *47*, 240 (1974).
[182] *C. E. Kalmus, D. M. Hercules*, J. Am. Chem. Soc. *96*, 449 (1974).
[183] *N. Shimamura, A. Sugimori*, Bull. Chem. Soc. Jap. *44*, 281 (1971).

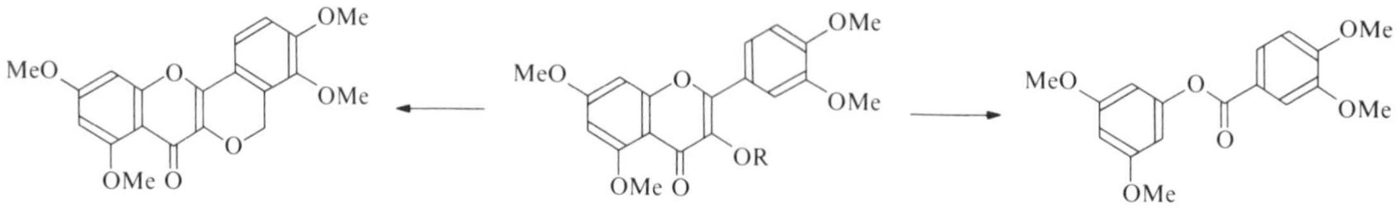

VII VI(R = Me) VIII(R = H) IX

4-phenyldihydrocoumarin (13%) and salicylic acid (4%)[184]. Irradiation of 3-hydroxy-2-methoxyflavone (**III**) affords a photo-rearranged product (**V**) which may be derived from the intermediate (**IV**) formed by a formal $_{\sigma}2+_{\pi}2$ cycloaddition[185].

Photo-induced oxygenation has given rise to a number of interesting results in relation to the biological process. Sensitized photo-oxidation of cis and trans isoeugenol yields dehydrodiisoeugenol with other coupling products. This fact suggests that singlet oxygen may take part in lignin formation[186]. Biologically patterned sensitized photo-oxygenation of chalcones to produce flavones has recently been reported[187]. The investigation of the dye-sensitized photo-oxygenation of α-tocopherol is motivated by the implication of singlet oxygen as a possible *in vivo* oxidant[188]. Irradiation of 3,4-dihydroxycinnamic acid in the presence of oxygen produces 6,7-dihydroxycoumarin (esculetin) *albeit* in low yield[189]. 2-Hydroxyxanthone is formed by photo-induced oxidative coupling of 2,3-dihydroxybenzophenone[190]. Sensitized allylic photo-oxidation is applied to the cyclization of isopentenylcoumarin (mameisine) to a chromenocoumarin (mameigine)[191]. A somewhat different kind of photo-oxygenation is used for the preparation of cannabielsoic acid[192]. Dehydrorotenones are converted to rotenones by photo-induced oxidations[193]. Irradiation of quercetin pentamethylether (**VII**) gives four tetracyclic derivatives of which (**VI**) is the major product[194]. On the other hand the photo-sensitized oxygenation of quercetin-5,7,3′,4′-tetramethylether (**VIII**) effects a cleavage of the bond between the carbon atoms 2 and 3 to produce the depside (**IX**) together with carbon dioxide and carbon monoxide[195].

b. *Oxidative coupling*: This important reaction of phenols has been extensively studied in connection with the biogenesis of certain classes of the polyphenolic compounds and alkaloids. The subject is fully discussed in a excellent book[196] and only brief mention will be made here. Oxidative coupling of the phenolic compound can be effected by the use of one-electron oxidizing agents[197–199] or enzyme systems such as laccase or peroxydase[197]. Potassium ferricyanide or peroxidase oxidation of 2,3-dihydroxybenzophenone gives mainly 2-hydroxyxanthone together with some 4-hydroxyxanthone by intramolecular C–O coupling[116,200]. This reaction gives support for a biosynthetic proposal of xanthones (Section 5). The naturally occurring bisflavonyls are constructed from C–C and C–O bonding between the flavonoid units and their formations through oxidative coupling of the latter are reasonably expected. However the oxidative coupling of apigenin (4,5,7-trihydroxyflavone) using potassium ferricyanide lead to the formation of 3,3- and 3′,3-biapigenyls, linkages which are not observed in known natural bisflavonyls. From this result it is suggested that the biogenesis of the natural

[184] *P. O. L. Mack, J. T. Pinhey*, Chem. Commun. 451 (1972).
[185] *T. Matsuura, T. Takemoto, R. Nakashima*, Tetrahedron Lett. 1539 (1971).
[186] *K. Eskins et al.*, Tetrahedron Lett. 861 (1972).
[187] *H. M. Chawla, S. S. Chibleer*, Tetrahedron Lett. 2171 (1976).
[188] *G. W. Grams, K. Eskins, G. E. Inglett*, J. Am. Chem. Soc. *94*, 866 (1972).
[189] *J. Kagan*, J. Am. Chem. Soc. *88*, 2617 (1966).
[190] *A. Jefferson, F. Scheinmann*, Nature *207*, 1193 (1965).
[191] *J. L. Fourrey, J. Rondest, J. Polonsky*, Tetrahedron *26*, 3839 (1970).
[192] *A. Shani, R. Mechoulam*, Chem. Commun. 273 (1970).
[193] *H. Suginome, T. Yonezawa, T. Masamune*, Tetrahedron Lett. 5079 (1968).
[194] *A. C. Waiss, Jr. et al.*, J. Am. Chem. Soc. *89*, 6213 (1967).
[195] *T. Matsuura, H. Matsushima, H. Sakamoto*, J. Am. Chem. Soc. *89*, 6370 (1967); *T. Matsuura, H. Matsushima, R. Nakashima*, Tetrahedron *26*, 435 (1970).
[196] *W. I. Taylor, A. R. Battersby* (Eds.), Oxidative Coupling of Phenols, Mercel-Dekker, New York 1967.
[197] *A. I. Scott*, Quart. Rev. *19*, 1 (1965).
[198] *T. Kametani, K. Fukumoto*, Synthesis 657 (1972).
[199] *S. Tobinaga*, Bioorganic Chemistry *4*, 110 (1975).
[200] *J. E. Atkinson, J. R. Lewis*, J. Chem. Soc. (C), 281 (1969) and references cited therein.

compound may not involve direct radical coupling[201]. Teaflavin, polyphenol pigment of black tea, is known to arise by the oxidation of a mixture of epicatechin and epigallocatechin, catalyzed by the polyphenoloxidase of tea[196,202]. The enzymic oxidation of coniferyl alcohol has been investigated in great detail in connection with lignin biosynthesis and provides an example of wide exploration of the subject[196].

c. *Other biosynthetic type reactions*: The biogenetic type syntheses polyketides have appeared using skillfully prepared poly-β-keto compounds or their masked equivalents of various types[203,204]. Various isoprenylation methods of phenols including biogenetic condition have been investigated for the purpose of their applications to the syntheses of natural products[205-208]. One step synthesis of some tetrahydrocannabinols has been achieved by electrophilic substitution of 5-pentylresorcinol with various terpenic components[209-211]. The base-catalized autooxidation or metal catalyzed oxygenation of flavonols have been found to afford the depsides through the cleavage of $\Delta^{2,3}$-double bond and these transformations provide the nonenzymic model for the reaction of dioxygenase like quercetinase[212]. The heating of cinnamyl pyrophosphate with resorcinol in ammonium acetate buffer pH 7.2 gave *trans*-3-(2,4-dihydroxyphenyl)-1-phenylpropene and 3-(2,4-dihydroxyphenyl)-3-phenylpropene in 3:1 ratio[213]. Interconversion reactions of neoflavonoids also have been reported[214]. The conversion of 2′-hydroxychalcones to isoflavones involving the oxidative rearrangement reaction with thallic salts represents the first chemical analogy for isoflavone biosynthesis[215,216]. The synthesis of isorotenone from isoderritol is effected by th reaction using dimethyl sulphoxonium methylide as a methylating agent[217].

5 Biosynthesis of Polyphenolic Compounds[218-223]

Polyphenolic compounds arise essentially by one of three pathways: a. the acetate-malonate pathway; b. the shikimic acid pathways; c. the combination of the both (the mixed pathway). Feeding experiments with labelled precursors indicate which of these pathways is operative in a given case. Recently the individual steps of the biosynthesis have been elaborated in a number of polyphenols by the realization of respective transformations by the isolated enzymes. It should be noted that the various pathways could operate for the production of a class of polyphenols or even a single compound up to the organisms concerned, as seen in the case of certain quinones, benzoquinones and xanthones. From comperative viewpoint, the acetate-malonate pathway is more common on the

201 *R. J. Molyneux, A. Waiss, Jr., W. F. Haddon*, Tetrahedron *26*, 1409 (1970).
202 *D. T. Coxon, A. Holmes, W. D. Ollis*, Tetrahedron Lett. 5247 (1970).
203 *T. Money*, Chem. Rev. *70*, 553 (1970).
204 *T. M. Harris, C. M. Harris, K. B. Hindley*, Fortschr. Chem. Org. Naturstoffe *31*, 217 (1974).
205 *A. C. Jain, P. Pal, T. R. Seshadri*, Tetrahedron *26*, 1977, 2631 (1970); *A. C. Jain, V. K. Khanna, T. R. Seshadri*, Tetrahedron *25*, 2787 (1969).
206 *R. D. H. Murray, M. M. Ballantyne*, Tetrahedon *26*, 4667 (1970); *R. D. H. Murray, M. M. Ballantyne, K. P. Mathai*, Tetrahedron Lett. 243 (1970).
207 *R. J. Molyneux, L. Jurd*, Tetrahedron *26*, 4743 (1970).
208 *G. Casnati, A. Guareschi, A. Pochiui*, Tetrahedron Lett. 3737 (1971).
209 *R. K. Razdam, G. R. Handrick*, J. Am. Chem. Soc. *92*, 6061 (1970).
210 *E. C. Taylor, K. Lenard, Y. Sho*, J. Am. Chem. Soc. *88*, 367 (1966).
211 *U. Claussen, P. Mummenhoff, F. Korte*, Tetrahedron *24*, 2897 (1968).
212 *A. Nishinaga, T. Matsuura*, Chem. Commun. 9 (1973); *A. Nishinaga, T. Tojo, T. Matsuura*, Chem. Commun. 896 (1974).
213 *J. Larkin, D. C. Nonhebel, H. C. S. Wood*, Chem. Commun. 455 (1970).
214 *D. M. X. Donnelly et al.*, J. Chem. Soc., Perkin I, 965 (1973); *L. Jurd, K. Stevens, G. Manners*, Tetrahedron *29*, 2347 (1973).
215 *W. D. Ollis et al.*, Chem. Commun. 1237 (1968); *W. D. Ollis et al.*, J. Chem. Soc. (C), 119, 125 (1970).
216 *L. Farkas et al.*, J. Chem. Soc., Perkin I, 305 (1974).
217 *L. Crombie, P. W. Freeman, D. A. Whiting*, Chem. Commun. 563 (1970). For the different reactivities of flavones and isoflavones to this reagent, cf. *G. A. Calpin, W. D. Ollis, I. O. Sutherland*, Chem. Commun. 575 (1967).
218 *D. H. G. Crout, in* Topics in Carbocyclic Chemistry, *D. Lloyd* (Ed.), Vol. 1, p. 63, Logos Press, London 1969.
219 *M. Luckner*, Der Sekundärstoffwechsel in Pflanzen und Tieren, VEP Gustav Fischer Verlag, Jena 1969.
220 *G. Billek* (Ed.), Biosynthesis of Aromatic Compounds, Pergamon, Oxford 1966.
221 *P. Bernfeld* (Ed.), Biogenesis of Natural Compounds, 2nd ed., Pergamon, Oxford 1967.
222 *J. H. Richards, J. B. Hendrickson*, The Biosynthesis of Steroids, Terpenes and Acetogenins, W. A. Benjamin, New York 1964.
223 *J. B. Pridham, T. Swain* (Eds.), Biosynthetic Pathways in Higher Plants, Academic Press, London 1965.

(1)

$RCH_2COCH_2COCH_2COCH_2COSCoA$

(2)

CH_2R, CO_2H, HO, OH

(X)

$COCH_2R$, HO, OH, OH

(XI)

biosynthesis in micro-organisms and the shikimate pathways is more popular for higher plants.
a. *Acetate-malonate pathway*[224,225]: The primary building block in this pathway is a linear poly-β-keto compound, produced from an acetate and several malonates by a process analogous to fatty acid biosynthesis[226]. For the formation of poly-β-keto compounds, the intermediate reduction steps before the chain extension, which are requisite in fatty acid biosynthesis, must be partly omitted. The presence of poly-β-keto compounds as precursors in polyphenolic compound biosynthesis is greatly strengthened by the isolation of the acyclic polyketides, triacetic and tetraacetic lactones from a strain of *Penicillium stiptatum*[227]. The cyclization of the β-polyacetyl chain occurs in two ways. The aldol type condensation (1) affords orsellic acid type derivatives (**X**) whereas the Claisen type condensation (2) gives rise to acylphloroglucinol derivatives (**XI**). These initial cyclization products are subject to various secondary transformations (*i.e.*, *C*- and *O*-alkylation, hydroxylation and further cyclization etc.) to form the final products. Both types of cyclization have been reproduced *in vitro* reactions using various model poly-β-keto compounds[203;204]. Another mode of β-polyacetyl chain cyclization is demonstrated on the biosyntheses of several polyketides: citromycetin[228], rotiorin[229], mollisin[230,231] and sclerin[232]. In these instances the molecules are constructed from the separate β-polyacetyl chains and this fact poses an interesting problem for the mechanism of polyketide cyclization. Timing of the secondary transformation relative to the aromatic ring formation is an outstanding problem in the polyketide biosynthesis. More evidence seems to support the view that the *C*-methylation precedes the aromatization step[233]. On the basis of the incorporation experiment using labelled intermediates it is proposed for the biosynthesis of mycophenolic acid that *C*-methylation occurs before aromatization and the attachment of the isoprenoid residue follows the latter step[225,234,235].

b. *Shikimic acid pathway*[236,237]: A wide range of the polyphenolic compounds originate from the

[224] *W. B. Turner*, Fungal Metabolites, Academic Press, London 1971.

[225] *T. Money, in* Biosynthesis, A Special Periodical Report, The Chemical Society, Vol. 2, p. 183 (1973).

[226] *F. Lynen*, Pure Appl. Chem. *14*, 137 (1967); *F. Lynen*, Biochem. J. *102*, 381 (1967).

[227] *T. E. Acker, P. E. Breneisen, S. W. Tamenbaum*, J. Am. Chem. Soc. *88*, 834 (1966); *R. Bentley, P. M. Zwitkowitz*, J. Am. Chem. Soc. *89*, 676 (1967).

[228] *S. Gatenbeck, K. Mosbach*, Biochem. Biophys. Res. Commun. *11*, 166 (1963).

[229] *J. S. E. Holker, J. Staunton, W. B. Whalley*, J. Chem. Soc. 16 (1964).

[230] *R. Bentley, S. Gatenbeck*, Biochemistry *4*, 1150 (1965).

[231] *M. Tanabe, H. Seto*, Biochemistry *9*, 4851 (1970); *H. Seto, L. W. Cary, M. Tanabe*, J. Chem. Soc., Chem. Commun. 867 (1973); *M. L. Casey, R. C. Paulick, H. W. Whitlock, Jr.*, J. Am. Chem. Soc. *98*, 2636 (1976).

[232] *T. Tokoroyama, T. Kubota*, J. Chem. Soc. (C), 2703 (1971); cf. also *M. J. Garson, J. Staunton*, J. Chem. Soc. Chem. Commun. 928 (1976); *M. Yamazaki, Y. Maebayashi, I. Tokoroyama*, Tetrahedron Lett. 489 (1977).

[233] *E. Lederer*, Quart. Rev. *23*, 453 (1969), cf. also *H. Taguchi, U. Sankawa, S. Shibata*, Chem. Pharm. Bull. *17*, 2054 (1969).

[234] *L. Canonica et al.*, J. Chem. Soc., Perkin I, 2639 (1972).

[235] *C. T. Bedford et al.*, Canad. J. Chem. *51*, 694 (1973).

[236] *E. Haslam*, The Shikimate Pathway, Butterworth, London 1974.

essential amino acids, phenylalanine, tyrosine and tryptophane or the intermediate involved in their biosyntheses from glucose, in which shikimic acid plays a key role.

1. C_6–C_1, C_6–C_2 and C_6–C_3 compounds[238–240]: C_6–C_3 and C_6–C_1 compounds oxygenated at 4-, 3,4- or 3,4,5-positions (*e.g.*, *p*-hydroxycinnamic acid, sinapic acid, protocatechuic acid, syringic acid) are very common constituents especially in higher plants. C_6–C_2 compounds (*e.g.*, acetophenones and phenylacetic acids) occur less commonly. Most of these compounds derive from phenylalanine and tyrosine through deamination followed by hydroxylation on the ring or degradation of the side chain or other modifications. These processes have been studied closely from the ground of enzymology[237b,c,241]. In several cases the direct production of some of these compounds from shikimic acid has been demonstrated[237b,242]. The formation of protocatechuic acid and gallic acid is closely related to the biogenesis of some tannins[32].

2. Coumarins[243,244]: It has been generally accepted that coumarin is biosynthesized from *trans*-cinnamic acid *via* the sequence of the reactions: *ortho*-hydroxylation, glycosylation, stereomutation and cyclization. In the case of 7-oxygenated coumarins, *para*-hydroxylation precedes the *ortho*-hydroxylation. More direct pathways involving free *cis*-cinnamic acid as an intermediate seem to be less likely. However Ourisson has established in a feeding experiment using tabacco tissue cultures that the following sequence is the main pathway in the biosynthesis of scopoletin[245]: phenylalanine → free cinnamic acid → free *p*-coumaric acid → free caffeic acid → free ferulic acid → scopoletin → scopolin. Isopentenylation of the coumarin nuclei gives rise to further classes of coumarins—isopentenylcoumarins, pyranocoumarins and furocoumarins[237a,c]. The oxidative fission of the isopropyl residue is involved in the production of the last type of the compound[246].

3. Lignins[6,24,25,27,247,248] and lignans[9]: The biosynthesis of lignins is generally pictured as the radical polymerisation of cinnamic alcohols like *trans-p*-hydroxycinnamic alcohol, *trans*-coniferyl alcohol and *trans*-sinapyl alcohol, which are derived from the corresponding cinnamic acids. The relative ratio of the incorporation of these monomer phenylpropanoid units is variable in the plants. This view of lignin biosynthesis is supported by the polymerization of phenylpropane monomers by peroxidase or laccase. Lignans comprise the dimers of all types formed by the oxidative coupling of *p*-hydroxyphenylpropane units[249]. Most common (and in the original sense) lignans are those derived by 2′–2′ coupling. The other types of compounds which have been found recently include the product of 1–2′, 2′–3 and 3–3 couplings[197,249–251] and they have been named as neolignan[252].

c. *Mixed pathway*: 1. Xanthones; The results of labelling experiments[253] seem to favor the proposal that xanthones in higher plants are biosynthesized by direct oxidative coupling of the hydroxybenzophenones derived from acetate and shikimate[25]. The attainment of this coupling reaction by *in vitro* processes is a further support[116,200].

---

[237] *T. Harborne, in* Biosynthesis, A Specialist Periodical Report: (a), vol. 1, p. 119 (1972); (b), vol. 2, p. 215 (1973); (c), vol. 3, p. 88 (1974).

[238] *J. B. Pridham*, Ann. Rev. Plant Physiol. *16*, 13 (1965).

[239] *M. H. Zenk, in* Biosynthesis of Aromatic Compounds, *G. Billek* (Ed.), p. 45, Pergamon, Oxford 1966.

[240] *A. C. Neish, in* Biochemistry of Phenolic Compounds, *J. B. Harborne* (Ed.), p. 313, Academic Press, London 1964.

[241] *K. R. Hanson, E. A. Havir, in* Recent Advances in Phytochemistry, *V. C. Runeckles, J. F. Watkins* (Eds.), Vol. 2, p. 48, Appleton-Century-Crofts, New York 1972.

[242] *P. M. Dewick, E. Haslam*, Chem. Commun. 673 (1968) and references cited therein.

[243] *S. A. Brown, in* Biosynthesis of Aromatic Compounds, *G. Billek* (Ed.), p. 15, Pergamon, Oxford 1966.

[244] *A. Neish, in* Plant Biochemistry, *J. Bonner and V. E. Varner* (Ed.), p. 581, Academic Press, New York 1965.

[245] *B. Fritig, L. Hirth, G. Ourrison*, Phytochemistry *9*, 1963 (1970).

[246] *S. A. Brown*, Phytochemistry *9*, 2471 (1970).

[247] *F. A. Isherwood, in* Biosynthetic Pathways in Higher Plants, *J. B. Pridham, T. Swain* (Eds.), p. 133, Academic Press, New York 1965.

[248] *F. F. Nord, W. J. Schubert, in* Biogenesis of Natural Compounds, *P. Bernfeld* (Ed.), 2nd ed., Pergamon, Oxford 1967.

[249] *R. S. McCredie, E. R. Ritchie, W. C. Taylor*, Austral. J. Chem. *22*, 1011 (1969).

[250] *A. Ogiso et al.*, Tetrahedron Lett. 2003 (1968).

[251] *A. Stoessl*, Tetrahedron Lett. 2287, 2849 (1966).

[252] *O. R. Gottlieb*, Phytochemistry *11*, 1537 (1972).

[253] *P. Gupta, J. R. Lewis*, J. Chem. Soc. (C), 629 (1971).

[254] *I. Carpenter, H. D. Locksley, F. Sceinmann*, Phytochemistry *8*, 2013 (1969). For another proposal as to the xanthone biosynthesis, cf. *O. R. Gottlieb*, Phytochemistry *7*, 411 (1966).

[255] *H. Grisebach et al., in* Biosynthesis of Aromatic Compounds, *G. Billek* (Ed.), p. 25, Pergamon, Oxford 1966.

2. Flavonoids[22,255–258]: The key compounds in the biosyntheses of flavonoids are chalcones which are constructed by the condensation of β-polyacetyl intermediates composed of a chain initiating cinnamic acids and three malonyl-CoA units. The interconversion between chalcones and flavanones occurs easily and these are converted to various flavonoid compounds by oxidation, reduction or numerous other secondary transformations. Extensive enzymological studies have been directed to these processes. Polymerization of catechins and flavan-3,4-diols lead to the formation of some sort of tannins[37].

3. Isoflavonoids[17,22,255–257]: It is established that the isoflavones are formed by the 1,2-aryl migration of the corresponding flavones although the exact mechanism of the rearrangement are not yet disclosed[258]. 3-Arylcoumarins, pterocarpanes[259], cumestanes[260] and rotenoid are all related biogenetically to the isoflavones. The extra carbon in rotenoid has been shown to derive from methionine[261,262].

4. Neoflavonoids[17,28,29,218]: A novel group of interesting compounds which have the carbon skeleton of 4-phenylcoumarin have been demonstrated to arise by nucleophilic attack of a phloroglucinol nucleus on a cinnamoyl residue at the α-carbon of the side chain with elimination of a suitable leaving group (a phosphate or pyrophosphate anion)[263].

5. Stilbenes[264]: The same β-polyketo intermediates as included in chalcone formation may be subject to another mode of cyclization (aldol type) with subsequent decarboxylation to give stilbenes. The retainment of the carboxyl group is found in a few examples, such as hydrangenol. This scheme of stilbene biosynthesis has been substantiated by feeding experiments and also in the model reaction[203,204].

[256] *H. Grisebach, in* Chemistry and Biochemistry of Plant Pigments, *T. W. Goodwin* (Ed.), p. 279, Academic Press, London 1965.

[257] *H. Grisebach, in* Recent Advances in Phytochemistry, *V. C. Runeckles* (Ed.), Vol. 9, Plenum Press, New Nork 1975.

[258] *H. Grisebach, H. Zily, Z. Naturforsch. 23B*, 494 (1968).

[259] *N. T. Keen, A. I. Zaki, J. J. Sims*, Phytochemistry *11*, 1031 (1972).

[260] *P. M. Dewick, W. Barz, Grisebach*, Phytochemistry *9*, 775 (1970).

[261] *L. Crombie, C. L. Green, D. A. Whiting*, J. Chem. Soc. (C), 3029 (1968).

[262] *L. Crombie, P. M. Dewick, D. A. Whiting*, Chem. Commun. 1469 (1970).

[263] *G. Kunesch, J. Polonsky*, Chem. Commun. 317 (1967); *G. Kunesch, J. Polonsky*, Phytochemistry *8*, 1221 (1969).

[264] *G. Billek, A. Shimpl, in* Biosynthesis of Aromatic Compounds, *G. Billek* (Ed.), p. 37, Pergamon, Oxford 1966.

5.7 Melanins

Rodolfo A. Nicolaus
Faculty of Pharmacy, University of Naples, Via L. Rodinò, 22
80138 Naples, Italy

1 Introduction

Melanins are a widely distributed class of pigments[1–9] found in animals and plants. On the basis of their chemical precursors, melanins can be classified into three groups, namely:

Eumelanins which are DOPA (3,4-dihydroxyphenylalanine) derivatives;

Pheomelanins which are 5-*S*-cysteinyldopa and 2-*S*-cysteinyldopa derivatives;

Allomelanins which are derivatives of nitrogen-free precursors such as pyrocatechol and 1,8-dihydroxynaphthalene.

In most cases a definite chemical structure cannot be assigned to these dyes as they form part of solid, insoluble particles or granules. In polymerization, which is an oxidative process of a radical type, the generated quinones play an important role; the final product is always a heterogeneous macromolecule with respect to the monomeric units present and the type of linkages. Both the intermediate products and other substances present can, in fact, either be enclosed in or chemically bound to the main

[1] *A. Quilico*, I. pigmenti neri animali, Tipografia Fusi, Pavia 1937.

[2] *D. L. Fox*, Animal Biochromes and Structural Colours, Cambridge University Press, London 1953.

[3] *M. Thomas, in* Modern Methods of Plant Analysis, Vol. IV, Springer, Berlin 1955.

[4] *T. B. Fitzpatrick, P. Brunet, A. Kukita, in* The Biology of Hair Growth, *W. Montagna* (Ed.), Academic Press, New York 1958.

polymer. This phenomenon becomes apparent with allomelanins which are formed by the non-specific action of various polyphenoloxidases and in which can be detected the presence of different residues such as amino acids, polypeptides, sugars, phenols, or transformation products with furan, pyrrole, and indole structures. Eumelanins are responsible for the brown and black coloration of the skin, hair, and eyes of mammals, the brown and black coloration of feathers, skin, and eggs of birds, of the scales and eggs of reptiles, amphibious animals, and fish, of the cuticula of insects, and of the skin and the so-called ink of cephalopoda. In addition, eumelanins form the pigment and, in some cases, the major component of many tumors (melanoma) frequently encountered with mammals, birds, fishes, etc. Finally, eumelanins are partly responsible for some characteristic blue or green colorations (Tyndall effect) in animals.

Eumelanins are generated in the so-called melanocytes which originate from nerve cells of the central nervous system during the embryonic development. Melanocytes contain organellae in different stages of development. The last stage, the melanosome, is a pigment granule with a characteric shape and size devoid of any chemical activity. The melanosomes produced by the melanocytes are transferred to other cells, *e.g.* to the Malpighian cells, in order to effect pigmentation of the skin. Melanocytes and keratinocytes are presumed to form a single, harmonic system exhibiting a particular biological structure which completes the individual functions of the two cell types.

Size, type, and distribution of the melanosomes are the main factors which determine the coloration of animals. For example, the red-brown coloration of the coat of certain lambs (argentinian domestic goat) is not due to pheomelanin-containing granules but to granules which have different sizes as compared to those of black lambs. In the case of chestnut brown hair, melanosomes have been, in contrast to those of blond hair, transferred groupwise, under genetic control from melanocytes into the hair. Melanosomes of races with blond hair are less numerous and not completely filled with pigments, as is true of races with black hair, but exhibit a lamellar structure similar to that of immature melanosomes (premelanosomes) of melanocytes.

With some animals, a further cell type is encountered, the melanophore in which the melanosomes can migrate from the center of cell to the periphery or vice versa under the control of the central nervous system or of a hormonal system or of both systems. In this way, reptiles may, for example, change the coloration of their skin and their pigmented patterns such that the appearance of the animal is completely change in certain extreme cases.

Pheomelanins are responsible for the characteristic red and red-brown coloration of the hide of mammals and the plumage of birds. With the exception of man, eumelanins and pheomelanins may occur together and thus give rise to dappling of animals. The granules generated from pheomelanin seem to be smaller than those generated from eumelanin; however, in certain cases, a diffuse distribution of the dye within the cell cannot be excluded.

Allomelanins are black or brown dyes of fungi, bacteria, and mold. The coloration of seed and fruit of higher plants and of some timbers, such as ebony, can be attributed to allomelanins. The allomelanins are, under the term "humic acid", responsible to the characteristic coloration of the soil.

2 Eumelanins

2.1 Biogenesis of Eumelanins

Tyrosine may be transformed into a black substance by oxydases of animal and vegetable origin[10]. In the course of this process, dopa (**1**), dopaquinone (**2**), dopachrome (**3**), 5,6-

[5] *H. Munro Fox, G. Vevers*, The Nature of Animal Colours, Sidgwick and Jackson, London 1960.

[6] *R. H. Thomson, in* Comparative Biochemistry, *Florkin-Mason* (Ed.), Vol. III, Academic Press, New York 1962.

[7] *W. Montagna, F. Hu*, Advances in Biology of Skin, Pergamon Press, New York 1967.

[8] *R. A. Nicolaus, in* Melanins, *E. Lederer* (Ed.), Hermann, Paris 1968.

[9] *R. Novales*, Cellular Aspects of the Control of Color Changes, American Zoologist Vol. 9, No. 2, p. 427–549, T. J. Griffiths, Utica 1969.

[10] *H. S. Raper*, Biochem. J. *17*, 454 (1923); *H. S. Raper, H. B. Speakman*, Biochem. J. *20*, 69 (1926); *C. E. Pugh, H. S. Raper*, Biochem. J. *21*, 1370 (1927); *H. S. Raper*, Biochem. J. *21*, 89 (1927); *H. S. Raper*, Physiol. Rev. *8*, 245 (1928); *H. S. Raper*, Biochem. J. *26*, 2000 (1932); *H. S. Raper*, Biochem. J. *31*, 2155 (1937); *H. S. Raper*, J. Chem. Soc. 125 (1938); *H. S. Raper, in Bamann-Myrback*, Methoden der Fermentforschung, 2476, Leipzig 1941.

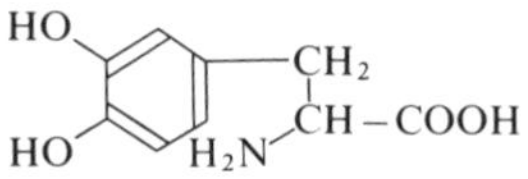

(1) 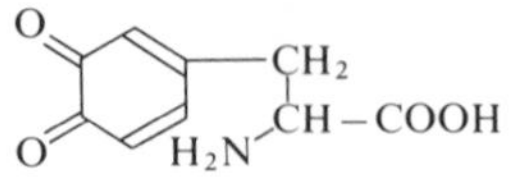 (2) 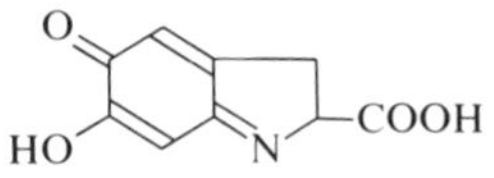 (3)

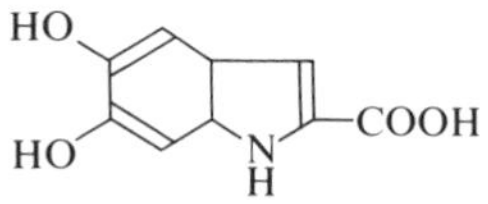

(4) 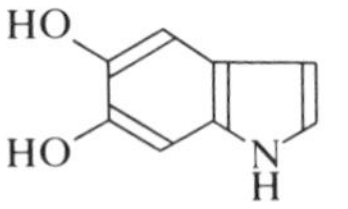 (5) 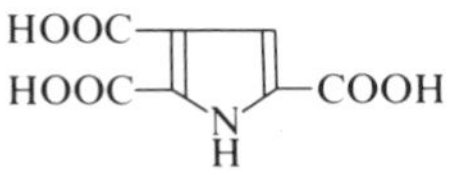 (6)

dihydroxyindole-2-carboxylic acid (**4**) and 5,6-dihydroxyindole (**5**) are formed. The same compounds can also be obtained through chemical oxidation of tyrosine[11]. In the processes which lead to melanin, consumption of oxygen and liberation of carbon dioxide take place, both of which vary depending on the experimental conditions. The process is accompanied by degradation reactions[8].

2.1.1 Isolation and Identification of 5,6-Dihydroxyindole[12]

To a solution of dopa (25 mg) in M20 phosphate buffer (35 ml) at pH 6.8 is added a solution of tyrosinase of fungi (350E) in water (1 ml). A rapid stream of air is passed through the solution kept at 35° for 15 min; the red solution is extracted twice with peroxide-free ether (5 ml each). The combined ether extracts are dried with Na_2SO_4 in the presence of little $Na_2S_2O_4$ and evaporated to dryness *in vacuo*. The residue is taken up in the smallest possible amount of water and subjected to paper chromatography (Whatman No. 1). Using butanol / acetic acid / water (60 : 15 : 25) as eluent, 5,6-dihydroxyindole appears on treatment with sulfanilic acid, as a violet spot, with iron(III) chloride as a blue-green spot, and with sodium hydroxide as a black spot ($R_f = 0.7$).

2.1.2 Polymerization of 5,6-Dihydroxyindole

In an alkaline medium, 5,6-dihydroxyindole is rapidly transformed into a black, insoluble substance both enzymatically and on the action of oxygen. In this transformation process which is described in Scheme 1, hydrogen peroxide, carbon dioxide, and minor amounts of pyrrolecarboxylic acids are formed. The type and position of the bonds in the polymer have not been determined. The number X of 5,6-dihydroxyindole units is high at the end of the process but it cannot, *a priori*, be exactly calculated (Scheme 1). Starting from tyrosine or dopa, the black granule formed is not composed of the substance which can be obtained from 5,6-dihydroxyindole but also contains, depending on the experimental condition, different amounts of open units (dopa) and units containing carboxy groups (dopachrome, 5,6-dihydroxyindole-2-carboxylic acid). This also accounts for the structural differences which are observed with eumelanins of different origin.

2.2 Chemistry of Eumelanins

After purification and drying, eumelanins exist as a black insoluble, amorphous, hygroscopic powder. Some eumelanins such as those of sepia occur as salts. These substances display some properties of free radicals which are presumed to be due to the presence of a limited number of phenoxy units trapped in the macromolecules which form the granule. The modern spectroscopic techniques are only of limited value in this field both for qualitative determinations and for structure elucidation. As the pigments forming the granule are completely insoluble, a purification cannot be carried out in the conventional manner. The soluble portion is usually removed with ether and acetone and the hydrolyzable substances such as proteins are separated by means of hydrochloric acid. The analysis of a typical elumelanin as obtained from the sac of the sepia affords, after purification, the following values:

C: 63.9 H: 2.8 N: 8.9 O: 24.3 S: 0.2%.

---

[11] *S. Dukler, M. Wilchek, D. Lavie*, Tetrahedron *27*, 607 (1971).

[12] *M. Piattelli, E. Fattorusso, S. Magno*, Rend. Acc. Sci. Fis. Mat. Napoli *28*, 168 (1961); Chem. Abst. *62*, 5510 (1965).

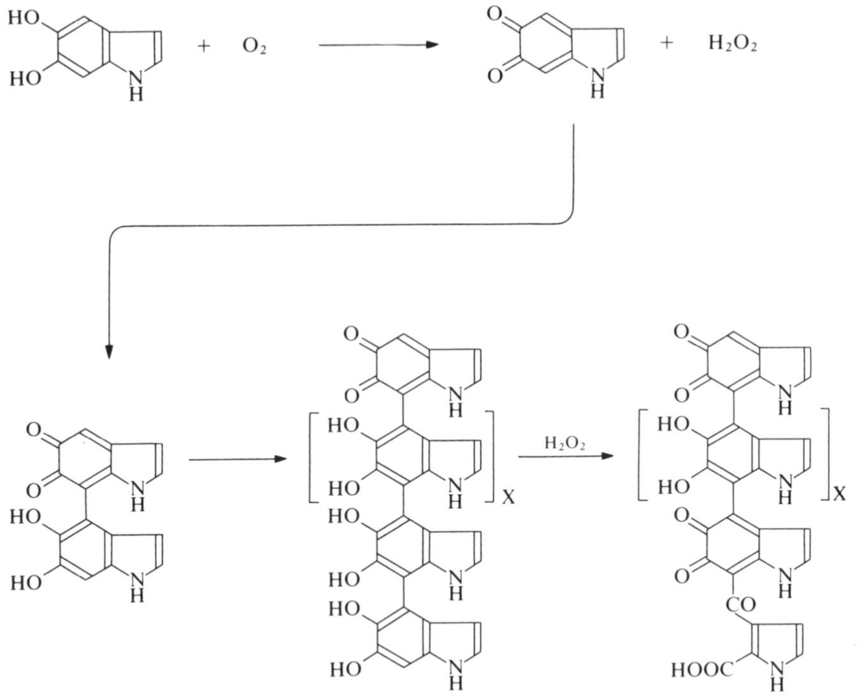

Scheme 1. Polymerization of 5,6-Dihydroxyindole.

Calculated for an indole-quinone polymer $(C_8H_3O_2N)_x$:

C: 66.2 H: 2.0 N: 9.6 O: 22.0%.

This eumelanin has a neutralization equivalent of 306 and liberates 9.1% CO_2 upon heating to 220°. It reacts with $CH_3OH + HCl$ ($OCH_3 = 6.0\%$) and with diazomethane ($OCH_3 = 18.8\%$). Both oxidation and reduction of eumelanins may afford different degradation products, generally in poor yields. A characteristic oxidation product is 2,3,5-pyrroletricarboxylic acid (**6**). This acid can also be obtained by boiling the pigment with an aqueous 4% NaOH solution. When the pigment is oxidized mildly with hydrogen peroxide and subsequently again treated with alkali, the acid (**6**) is obtained. In the case of sepiomelanin, with approximately 25 of such operations afford a total yield of 6.5% of (**6**). Compounds **4** and **5** are formed by alkaline fusion. Degradation experiments do not give information on the structural formula, which is understandable when one considers that the substance concerned is a solid, insoluble particle, *i.e.* the results are not transferrable to individual molecules or macromolecules. The degradation experiments are, however, important both in the determination of partial structures and in the detection of some structural differences of the various eumelanins.

Thus, for example, the direct chemical investigation of eumelanins has shown that the melanin of melanoma contains non-cyclized units (dopa) which do not occur in sepiomelanin. This has also been confirmed *in vivo* by use of labeled precursors[7,13].

3 Pheomelanins

3.1 Biogenesis of Pheomelanins

Pheomelanins are formed from dopa and cysteine in the presence of a mild oxidizing agent or of a polyphenoloxidase. Scheme 2 describes the different types of conversions of the precursors; it is based on the *in vitro* isolation or characterization of the intermediate products and on *in vivo* investigations with labeled precursors which have been carried out with skin preparations of New Hampshire chicken embryos[14]. The first step of the reaction is the coupling of cysteine and dopaquinone (**2**) which, as with the eumelanins, is obtained from dopa, which, in turn, is derived

[13] *K. Hempel, H. F. K. Männl*, Biochim. Biophys. Acta *124*, 192 (1966).

[14] *E. Fattorusso, L. Minale, S. De Stefano, G. Gimino, R. A. Nicolaus*, Gazz. Chim. Ital. *99*, 969 (1969); *G. Misuraca, R. A. Nicolaus, G. Prota, G. Ghiara*, Experientia *25*, 920 (1969); *G. Prota, S. Crescenzi, G. Misuraca, R. A. Nicolaus*, Experientia *26*, 1058 (1970).

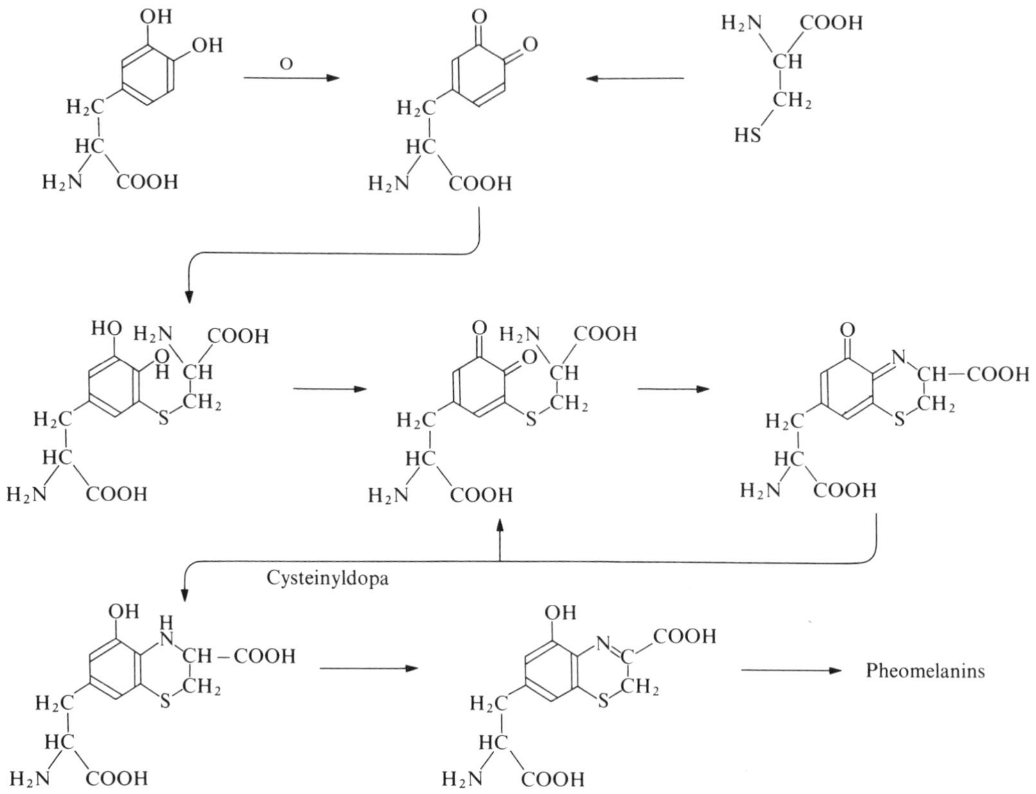

Cysteinyldopa

Pheomelanins

Scheme 2. Biogenesis of Pheomelanins.

from tyrosine; thus it may be assumed that the enzyme, the so-called tyrosinase, is common to the first stages of the processes leading to eumelanins and pheomelanins.

Considering the fact that both the formation and cyclization of 5-*S*-cysteinyldopa (which is formed in addition to the isomeric 2-*S*-cysteinyldopa in a ratio of 95 : 5) proceed rapidly and quantitatively, it must be assumed that autocatalytic processes involving redox systems are required for the formation of dihydrobenzothiazines.

The cyclization reaction presumably starts with the formation of catalytic amounts of 5-*S*-cysteinyldopaquinone which loses one molecule of water to form the *o*-quinonimine; the latter can be reduced to the dihydrobenzothiazine by 5-*S*-cysteinyldopa which is simultaneously oxidized to the quinone and the reaction can proceed spontaneously.

The last stage of the process shows how trichosiderins may be formed whereas the formation of gallopheomelanins has not yet been elucidated.

3.2 Chemistry of Pheomelanins

Although the study of these pigments is simpler than that of eumelanins especially due to their solubility in alkali, their structure has only recently been investigated[15]. The method of isolation depicted in Scheme 3 was first used in the investigation of dyes of the plumage of New Hampshire chickens. This method can, however, also be applied to the isolation of other compounds of biological origin. Yields and spectroscopic properties of the pigments are listed in Table 1.

3.2.1 Gallopheomelanins

These red-brown dyes, extractable from chicken feathers, are amorphous, non-melting powders which are insoluble in water and organic solvents. The average molecular weight of gallopheomelanin 1 has been found to be about 2000. The structure of this dye has not yet been completely elucidated as each fraction is presumed to contain very similar compounds whose

[15] *R. A. Nicolaus et al.*, Struttura e biogenesi delle feomelanine, Note I–XIII, Gazz. Chim. Ital. 1968–1970.

Table 1 Yields and Spectroscopic Properties of Pheomelanins

Pigment	Yield mg/100 g Feathers	λ_{max} (nm)	(log ε)
Gallopheomelanin 1	24	no maximum determined	
Gallopheomelanin 2*	140	no maximum determined	
Gallopheomelanin 3*	25	no maximum determined	
Gallopheomelanin 4	7	no maximum determined	
Trichosiderin A	traces	440 (non-purified)	
Trichosiderin B	2.6	454, 329, 243	(3.98; 4.01; 4,60)
Trichosiderin C	17	452, 327, 240	(4.13; 4.04; 4.56)
Trichosiderin E	traces	482, 312, 247	(4.33; 4.11; 4.34)
Trichosiderin F	0.8	482, 313, 346	—

* Isolated as proteín complex.

molecular weights change continuously.
Gallopheomelanin 1 may undergo degradation upon the action of acids, oxidizing and reducing agents to afford a series of substances (**7**–**21**) which possess benzothiazole, benzothiazine, and isoquinoline structures, whose degree of oxidation has not yet been completely established. The partial structures **22**–**25** have been proposed on the basis of these results and biogenetic considerations.
The other gallopheomelanins afford the same degradation products.

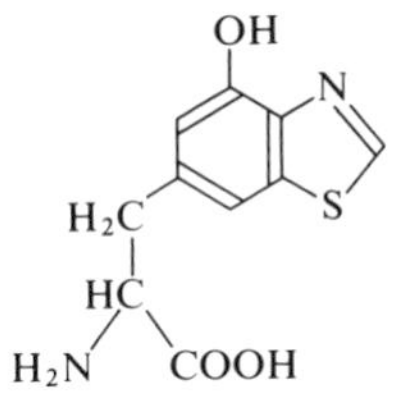

(**7**) 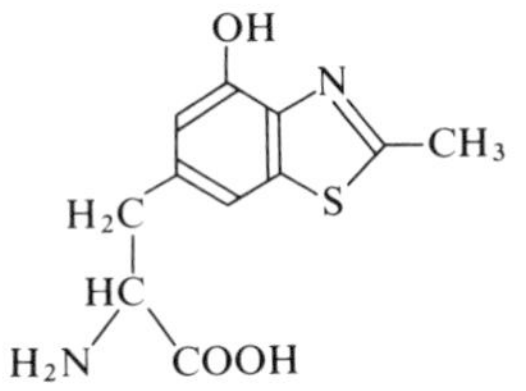 (**8**) 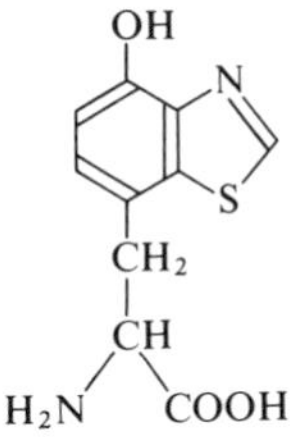 (**9**)

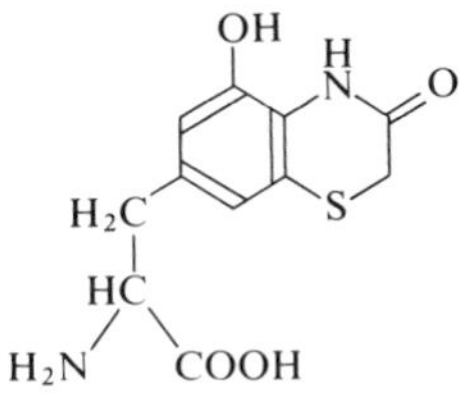

(**10**) 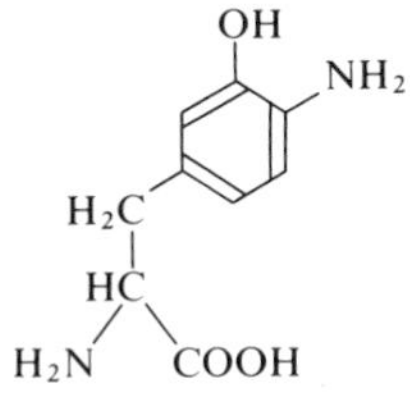 (**11**) 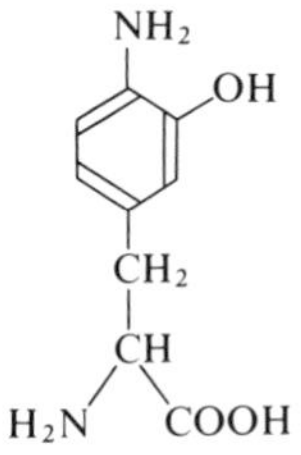 (**12**)

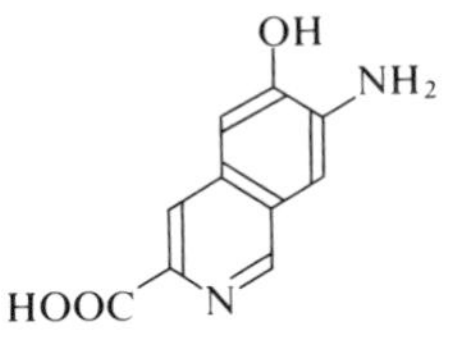

(**13**) 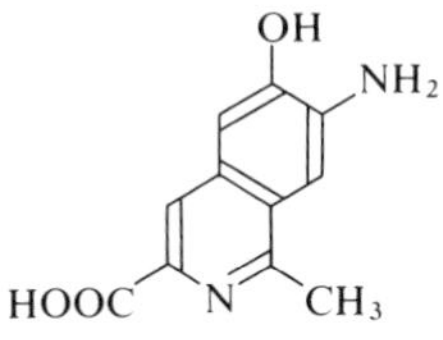 (**14**) 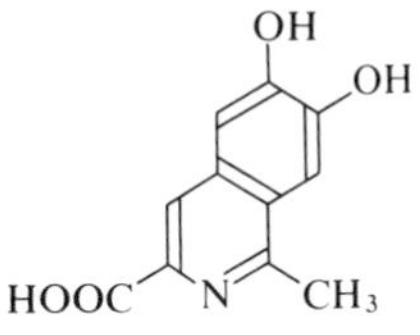 (**15**)

Compounds obtained by Degradation of Gallopheomelanin 1 with Hydrochloric Acid or Hydriodic Acid.

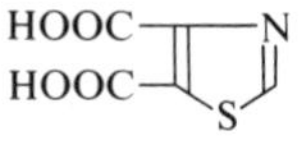

(16) 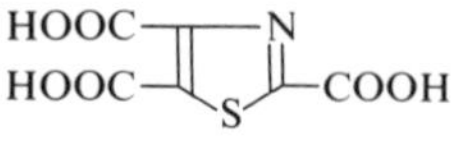 (17) 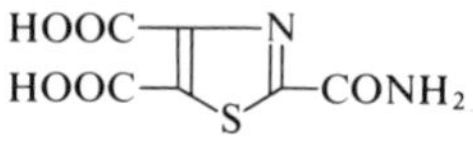 (18)

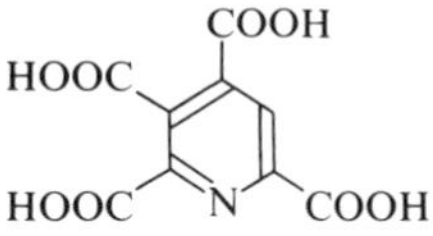

(19) 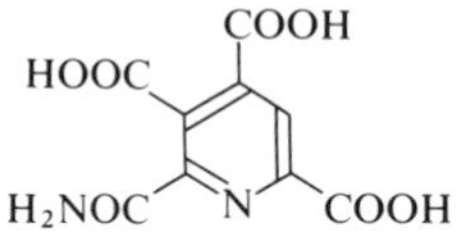 (20) 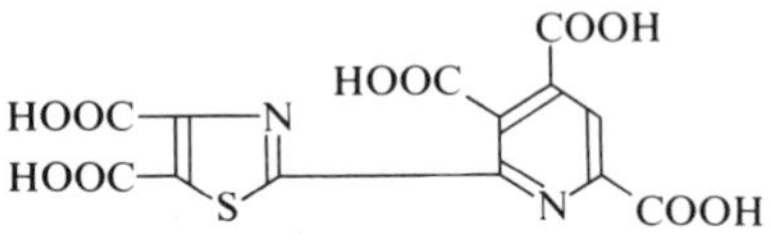 (21)

Compounds obtained by Permanganate Oxidation of Gallopheomelanin 1.

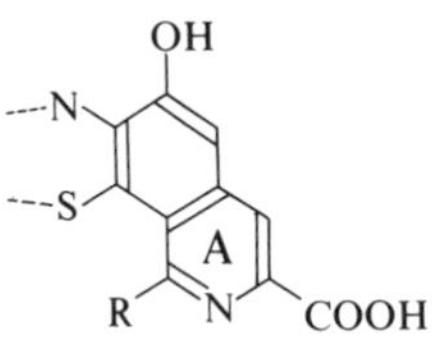

(22) 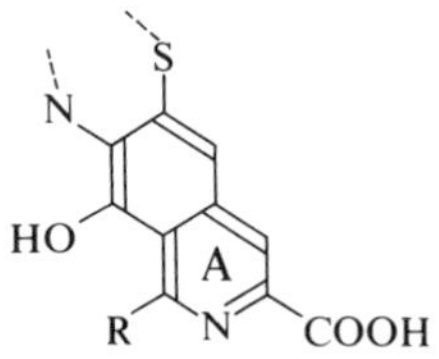 (23) 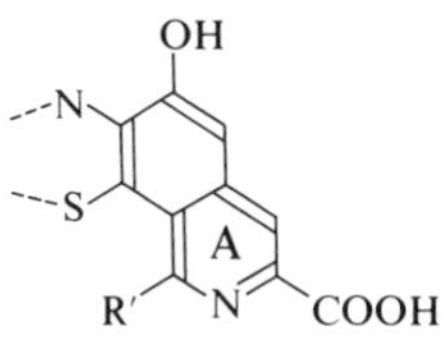 (24)

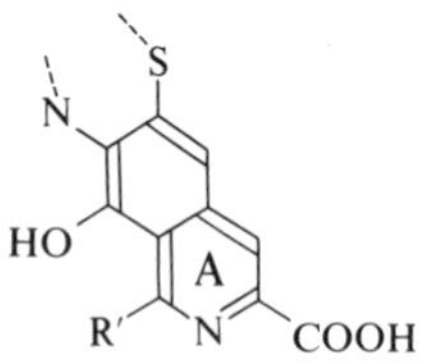

(25)

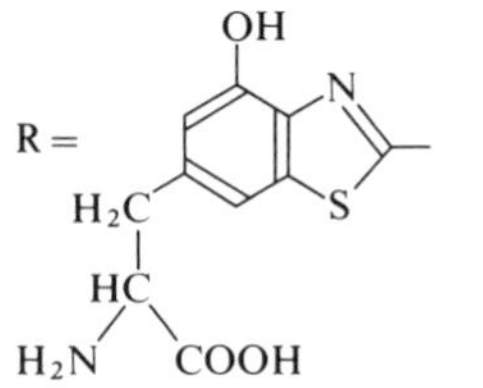

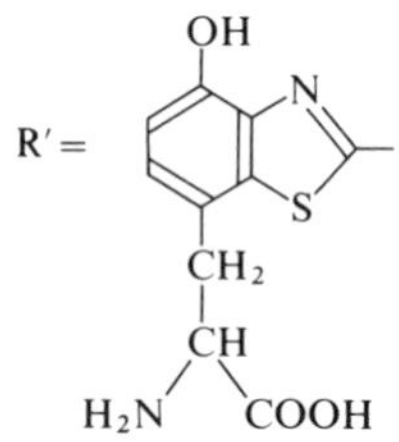

Partial structures contained in Gallopheomelanin 1 (The A ring may occur in different degree of oxidation).

3.2.2 Trichosiderins

The term "trichosiderin", which first appeared in the literature 30 years ago[16], has been temporarily retained to designate pigments which exhibit a characteristic absorption spectrum (Table 1) and which belong to the substances soluble in alkali and acids (Scheme 3).

It is still doubtful whether trichosiderins occur in the free state in tissues or whether they are linked to gallopheomelanins by means of readily hydrolyzable bonds.

On the basis of chemical and physical properties, the following structures have been proposed:

- structure **26** for trichosiderin B,
- structure **27** for trichosiderin C,
- structure **28** for trichosiderin E,
- structure **29** for trichosiderin F.

Trichosiderins may be found in *cis* or *trans* configuration as pure geometrical isomers or as an equilibrium mixture.

The ready decarboxylation with elimination of the carboxy group near the nitrogen of the 1,4-thiazine ring and conversion into pigments whose absorption spectra vary considerably according to the pH value (Table 2) are characteristic of structures **26** and **27**. This variation of the absorption spectra is due to the conjugated system which is positively charged with formation of a mesomeric ammonium-thionium cation which accounts for the strong bathochromic shift observed in acid medium.

Naturally, decarboxytrichosiderins are formed spontaneously upon warm-acid extraction of feathers or hair. For biological studies on pigmentation, it is frequently of importance to know whether the coloration is caused by eumelanins or pheomelanins. For this purpose, warm-acid

[16] *P. Flesch*, J. Invest. Derm. *6*, 257 (1946).

Table 2 Spectroscopic Properties of Decarboxytrichosiderins

Compounds	pH 1	pH 13
	λ_{max} (log ε)	λ_{max} (log ε)
Decarboxytrichosiderin B	525(4.18), 360(3.79), 296(4.06)	458(4.21), 312(4.16), 246(4.49)
Decarboxytrichosiderin C	525(4.20), 356(3.80), 295(4.08)	462(4.19), 313(4.15), 245(4.49)

extraction (0.2 *N* HCl) and analysis of trichosiderins and decarboxytrichosiderins are recommended[7,8].

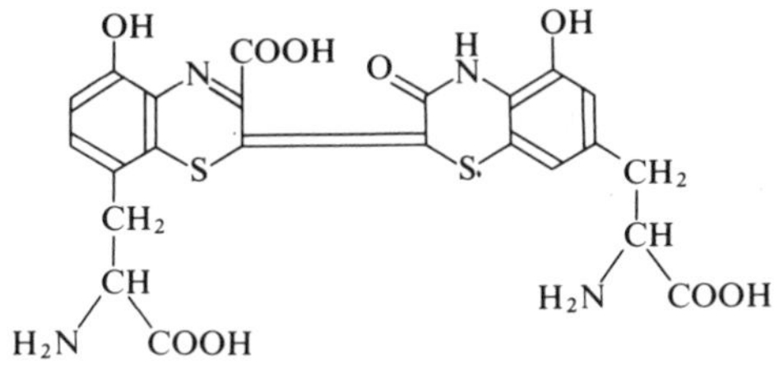

(**26**) Trichosiderin B

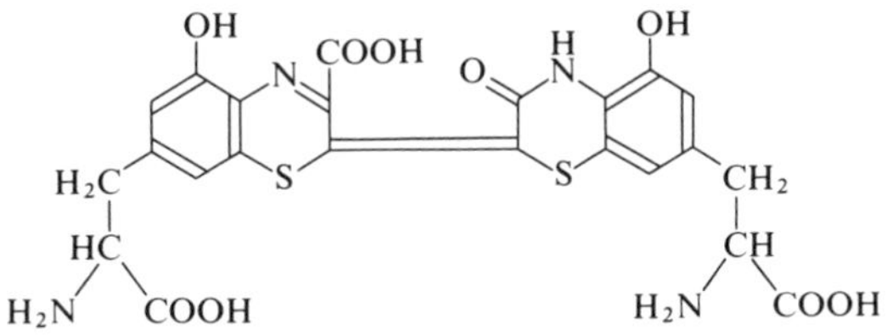

(**27**) Trichosiderin C

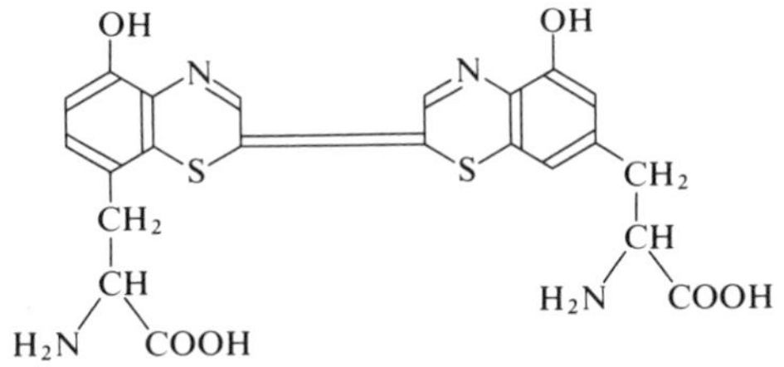

(**28**) Trichosiderin E

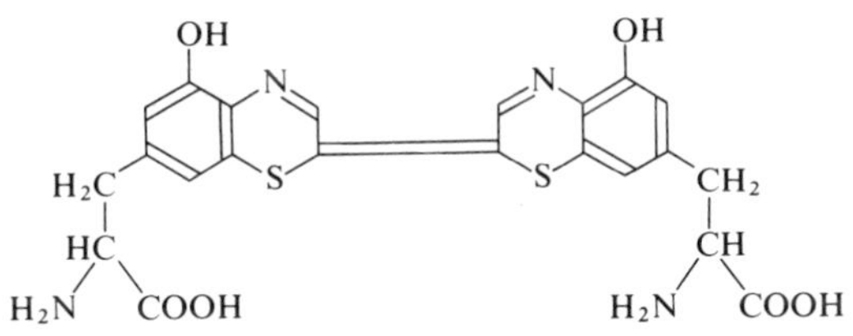

(**29**) Trichosiderin F

3.2.2.1 Effect of Acids on Trichosiderins B or C

The pigment (20 mg) is dissolved in 0.5 *N* HCl (20 ml) and the yellow solution is heated to boiling. After several minutes, the color turns from yellow to red with the formation of carbon dioxide (1 mol) which is determined through absorption in 0.05 *N* $Ba(OH)_2$ and subsequent titration of the excess alkali with 0.1 *N* HCl (indicator: phenolphthalein). The reaction mixture is passed through a column (Dowex 50 W, 2 × 5 cm; $H^{\oplus}$-form). After flushing with 2 *N* HCl, water, and 0.2 *M* sodium acetate, the decarboxylation product is eluated from the ion exchanger with 0.1 *N* NaOH. The eluate is then brought to pH 3 by acidification and the red precipitate (15 mg) is collected and washed by centrifugation.

4 Allomelanins

Allomelanins are produced predominantly by plants, fungi, molds, and bacteria. Black pigments are also formed after damage of plant tissue (blackening of potatoes, bananas, fungi, etc.) in an oxidation process in which the polyphenols contained in the plant are involved. In addition to humic acids, this compound class also includes, with respect to their origin, melanoidins which are prepared by polymerization and copolymerization of various compounds (amino acids, phenols, sugars, etc.) in the presence of oxygen. They are notorious because of the difficulties and damages which they cause in the cosmetic, tea, beer, and canning industries as well as in protein hydrolysis. Allomelanins are formed by oxidation of nitrogen-free precursors although they may themselves be azotized and the structures appear more complicated than expected as amino acids, polypeptides, and sugar are frequently involved in the polymerization process.

Owing to the complex structure of allomelanins, little is known of their chemistry. The most interesting results have been obtained in the investigation of allomelanins produced from *Aspergillus niger*[1,17], *Daldinia concentrica*[18], and

[17] *M. Barbetta, G. Casnati, G. Ricca*, Rend. Acc. Lincei *40*, 450 (1966). Chem. Abst. *65*, 10543 (1966).

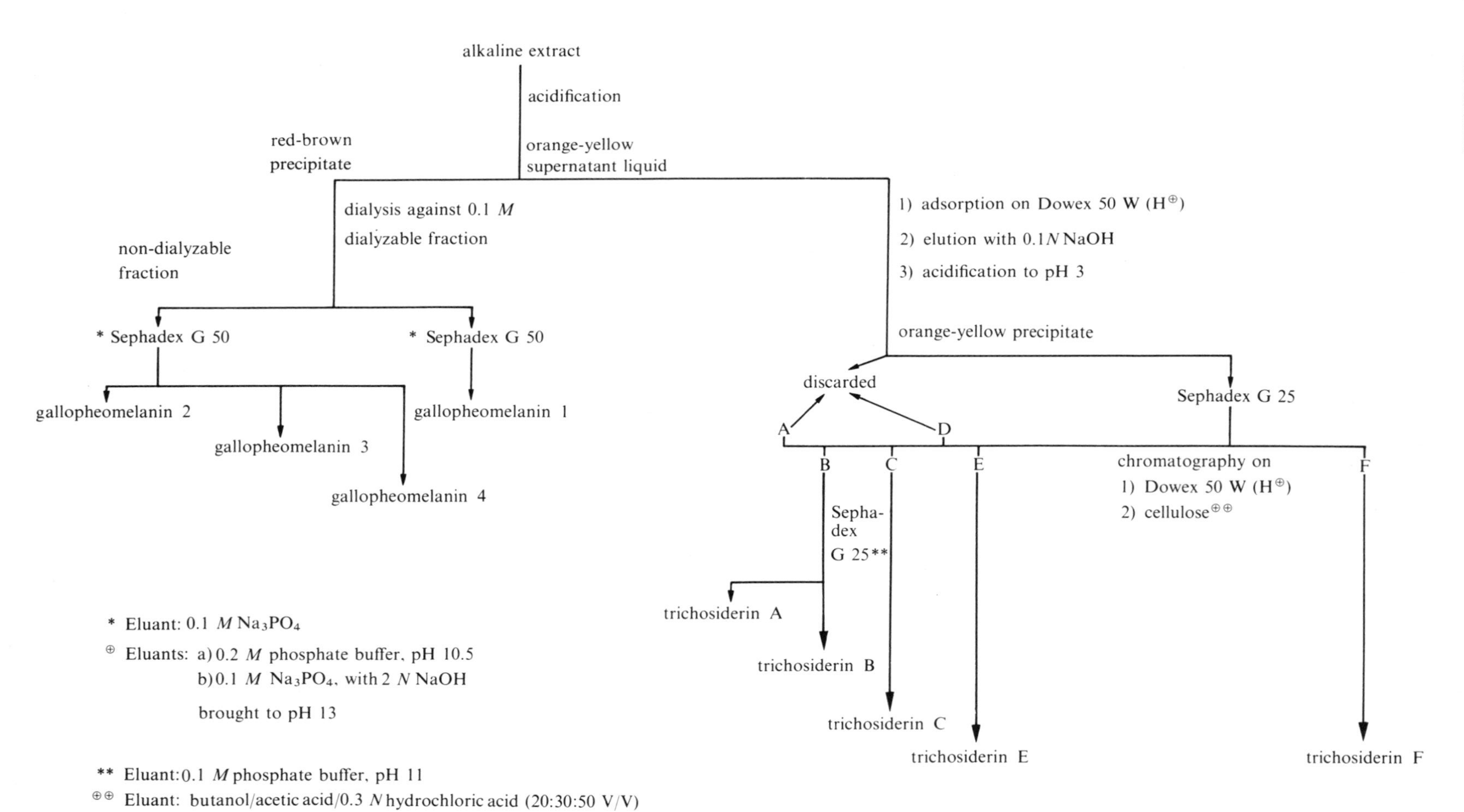

* Eluant: 0.1 M Na_3PO_4

⊕ Eluants: a) 0.2 M phosphate buffer, pH 10.5
b) 0.1 M Na_3PO_4, with 2 N NaOH brought to pH 13

** Eluant: 0.1 M phosphate buffer, pH 11

⊕⊕ Eluant: butanol/acetic acid/0.3 N hydrochloric acid (20:30:50 V/V)

Scheme 3. Extraction of Pheomelanins from Feathers of New-Hampshire Chickens

Ustilago maydis[19] and in the study of humic acids.

On the basis of chemical investigations[1,8] it can be concluded that various allomelanins (*Aspergillus, Daldinia*, humic acids) are characterized by the presence of perylenic units substituted by various oxidized functions. The precursor of these units is presumed to be 1,8-dihydroxynaphthalene (**30**). It is also possible that the perylenic units are formed in nature from compound **30** via **31** and **32** by an oxidation process which is already known in the laboratory.

On the other hand, the pigment granulae produced from *Ustilago maydis* and various black plant seeds are composed of polymers formed from pyrocatechol in different degrees of oxidation. The brown or black substance which can be extracted from these compounds is, in fact, with respect to its chemical properties, very similar to that which can be obtained in laboratory by oxidation of pyrocatechol[8,19].

2 (**30**) → (**31**) → (**32**)

[18] *D. C. Allport, J. D. Bu'Lock*, J. Chem. Soc. 4090 (1958); 654 (1960).

[19] *M. Piattelli, E. Fattorusso, R. A. Nicolaus, S. Magno*, Tetrahedron *21*, 3229 (1965).

Recommended Literature for Further Reading

G. Prota, R. H. Thomson, Melanin pigmentation in mammals, Endeavour *35*, 32 (1976).

R. A. Nicolaus, The Nature of Mammalian Colors, Chim. Ind. *54*, 427 (1972).

S. P. Lioh, E. L. Ruban, Microbnye Melaniny, Akad. Nauk SSSR, Institut Mikrobiologii, Izd, Nauka Moskva 1972.

G. Prota, M. D'Agostino, G. Misuraca, The Structure of Hallachrome: 7-Hydroxy-8-methoxy-6-methyl-1,2-anthraquinone, J. Chem. Soc. Perkin I, 1614 (1972).

H. Rorsman, A. M. Rosengren, E. Rosengren, A Sensitive Method for Determination of 5-*S*-cysteinyldopa, Acta Dermatovener (Stockholm) *53*, 248 (1973).

R. A. Nicolaus, E. Novellino, G. Prota, Origine e significato del colore negli animali, Acc. Sci. Fis. Mat. Napoli, Serie IV, Vol. XLII (1975).

G. Agrup, B. Falck, B. M. Kennedy, H. Rorsman, A. M. Rosengren, E. Rosengren, Formation of Cysteinyldopa from Glutathionedopa in Melanoma, Acta Dermatovener (Stockholm) *55*, 1 (1975).

G. Prota, H. Rorsman, A. M. Rosengren, E. Rosengren, Phaeomelanic Pigments from a Human Melanoma, Experientia *32*, 970 (1976).

G. Prota, H. Rorsman, A. M. Rosengren, E. Rosengren, Occurrence of Trichochromes in the Urine of a Melanoma Patients, Experientia *32*, 1122 (1976).

R. H. Thomson, G. Prota, Melanin Pigmentation in Mammals, Endeavour *35*, 32 (1976).

K. Jimbow, W. C. Quevedo, T. B. Fitzpatrick, G. Szabo, Some Aspects of Melanin Biology: 1950–1975, J. Invest. Dermat. *67*, 72 (1976).

A. A. Bell, R. D. Stipanovic, J. E. Puhalla, Pentaketide Metabolites of *Verticillium dahliae*. Identification of (+)-scytalone as a Natural Precursor to Melanin, Tetrahedron *32*, 1353 (1976); J. Org. Chem. *41*, 2468 (1976).

M. S. Blois, P. W. Banda, Detection of occult metastatic melanoma by urine chromatography, Cancer Research *36*, 3317 (1976).

D. L. Williams-Smith, L. J. Dunne, S. Evans, R. G. Prtchard, E. L. Evans, X-Ray Photoelectron Spectroscopy and the Structure on Melanins, Febs Letters *69*, 291 (1976).

T. Sarna, C. Mailer, J. S. Hyde, H. M. Swartz, B. M. Hoffman, Electron-Nuclear Double Resonance in Melanins, Biophys. J. *16*, 1165 (1976).

T. Sarna, J. S. Hyde, H. M. Swartz, Ion-Exchange in Melanin: An Electron Spin Resonance Study with Lanthanide Probes, Science *192*, 1132 (1976).

J. Filatovs, J. McGinness, P. Corry, Thermal and Electronic Contributions Switching in Melanins, Biopolymers *15*, 2309 (1976).

5.8 Visual Pigments

Christian Baumann
Dept. of Physiology, Justus-Liebig-University, Aulweg 129, D-63 Giessen, F. R. Germany

1 Introduction

Visual pigments are found in photoreceptors which convert electromagnetic radiation ($\lambda \approx 300$–700 nm) into bioelectric signals. This requires the absorption of light by the chromophores of the pigments. In vertebrates the pigment molecules are part of the outer receptor limbs; in cephalopods and arthropods, part of the rhabdomeres.

The visual pigments can be studied in their natural condition in photoreceptors. The cells or tissues containing the pigment are isolated and maintained as single cells[1,2], cell suspensions[3] or isolated retinae[4] in appropriate incubation media. The pigments themselves are identified on the basis of their absorption or photolabile properties. In the case of the intact human eye, the pigments can be studied by analyzing light reflected from the back of the eye (fundus reflectometry)[5,6]; this technique can be considered as a modified form of reflectance spectroscopy.

The pigments can be extracted from isolated retinae or cell suspensions. Since they are not soluble in water, they generally have to be prepared as micelles in a colloidal solution. The preparation involves the use of surface active agents such as bile salts[7], digitonin[8], quarternary ammonium salts[9] or triton X-100[10]. The micelles formed by digitonin and the pigment of *Bos taurus* have been characterized in detail[11]: associated with each pigment molecule are 180–200 digitonin molecules. The molecular weight of this complex amounts to 260–290,000 of which the pigment alone contributes about 40,000.

2 Structure

Visual pigments are proteins. In their natural state in the membrane they are in close contact with lipids, from which they cannot be completely separated by extraction; thus visual pigments in an aqueous micellar suspension contain some phospholipids (*e.g.* phosphatidyl ethanolamine) and are accordingly classified as lipoproteins[12]. Nonetheless, it is possible that the pigment could be purified from lipids without the loss of those properties essential for its function (absorption spectrum, photosensitivity)[13–15]. The number of amino acids per molecule of visual pigment of *Bos taurus* is said to be 305[16], of which 10 are lysin residues to one of which the prosthetic group is bound. (The molecular weight lies at 36,000 according to these investigations). Similar counts have been made on other kinds of animals (frogs, rats). The protein part of the visual pigment is labelled opsin. Its polypeptide chain is covalently bound to an oligosaccharide[17]. Opsin should therefore be termed a glycoprotein.

Opsin absorbs radiation of wavelengths $\lambda \leqq 300$ nm. The characteristic absorption of the visual pigments in the visible and near ultraviolet regions comes into existence when a prosthetic group is incorporated into a pigment molecule as a chromophore. The prosthetic groups of the pigments are derived from retinal or 3-dehydroretinal, the aldehydes of vitamin A_1 (retinol) and A_2 (3-dehydroretinol)[18]. The conjugated double bonds (cf. Fig. 1) imply that in theory 16 isomers are possible. In fact, 6 isomers of the two vitamins and their aldehydes exist: all-*trans*, 9-*cis*, 11-*cis*, 13-*cis* as well as 9,13-di-*cis* and 11,13-

1 *P. K. Brown*, J. Opt. Soc. Am. *51*, 1000 (1961).
2 *H. Langer, B. Thorell*, The Functional Organization of the Compound Eye, p. 145, Pergamon Press, Oxford, New York 1966.
3 *G. B. Arden*, J. Physiol. *123*, 377 (1954).
4 *E. J. Denton, M. A. Walker*, Proc. Roy. Soc. B *148*, 257 (1958).
5 *W. A. H. Rushton*, J. Physiol. *117*, 47 P (1952).
6 *R. A. Weale*, J. Physiol. *122*, 322 (1953).
7 *W. Kühne*, Unters. Physiol. Inst. Heidelberg *1*, 15 (1878).
8 *K. Tansley*, J. Physiol. **71**, 442 (1931).
9 *C. D. B. Bridges*, Biochem. J. *66*, 375 (1957).
10 *F. Crescitelli*, Vision Res. *7*, 685 (1967).
11 *R. Hubbard*, J. Gen. Physiol. *37*, 381 (1954).
12 *N. I. Krinsky*, Arch. Ophthal. (Chic.) *60*, 688 (1958).
13 *J. Heller*, Biochemistry *7*, 2906 (1968).
14 *M. O. Hall, A. D. E. Bacharach*, Nature *225*, 637 (1970).
15 *J. M. P. M. Borggreven, F. J. M. Daemen, S. L. Bonting*, Arch. Biochem. Biophys. *151*, 1 (1972).
16 *T. G. Ebrey, B. Honig*, Q. Rev. Biophys. *8*, 129 (1975).
17 *J. Heller, M. A. Lawrence*, Biochemistry *9*, 864 (1970).
18 *R. A. Morton, in* Handbook of Sensory Physiology, Vol. VII/1, Chapter 3, Springer-Verlag, Heidelberg 1972.

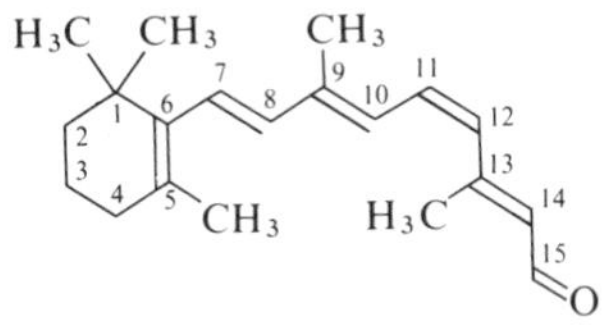

Fig. 1. 11-*cis*-retinal (conventional numbering of carbon atoms).

di-*cis*[19]. Of these only the sterically hindered 11-*cis* form is found in naturally occurring pigments. This is certainly true of retinal, the 11-*cis* isomer of which is released as a result of the thermal decomposition of the pigment[20,21] and is incorporated into the pigment molecule during its synthesis from opsin and retinal[22]. Similar is true for 3-dehydroretinal; in this case also the opsin must be incubated with a *cis*-form (probably likewise 11-*cis*) if a natural pigment is to be obtained[23]. A photolabile pigment can also be synthetized out of 9-*cis*-retinal and opsin[22]. The iso-pigments produced in this manner do not occur *in vivo*. However, the natural pigments can be converted into iso-pigments by flash photolysis. This reaction can also take place in intact photoreceptors. The yield of iso-pigment is especially great when the photolysis is carried out with an extremely short flash of a few nanoseconds duration[24]. The formation of the iso-pigments results from even numbered multiple absorptions. Pigment molecules which have absorbed single photons (or an odd number of quanta) become unstable and decay to opsin and retinal, which is released in the all-*trans* form[22]. Retinal ($C_{19}H_{27}CHO$) condenses with many compounds which contain amino groups to form Schiff's bases

$$\underset{\text{Retinal}}{C_{19}H_{27}CHO} + H_2NR \longrightarrow \underset{\text{Schiff's base}}{C_{19}H_{27}CH{=}NR} + H_2O$$

It is generally accepted that opsin and retinal are also covalently bound together as a Schiff's base. The aldimine linkage is protonated as was recently shown with the aid of resonance Raman spectroscopy[25,26]. In the equation above R stands for opsin which combines with retinal to form *N*-retinylidene opsin (NRO). The latter is not identical with the intact visual pigment but is encountered during the photolysis of the pigment. The nitrogen atom involved in the bond comes from the ε-amino group of a lysin residue[27,28]. Other non-covalent interactions between retinal and opsin are postulated but their nature is still largely unclear.

3 Nomenclature

The large number of known visual pigments requires the introduction of a convenient nomenclature which has its basis in the chemical nature of the prosthetic group and the peak absorption wavelength. Since there are only two different prosthetic groups, there are only two classes of visual pigments: A_1-pigments and A_2-pigments, *i.e.*, those derived from vitamin A_1 and those from vitamin A_2. Within each class classifications are made according to the λ_{max} of the pigments. Thus the pigment P 499_1 has 11-*cis*-retinal as its prosthetic group and absorbs maximally at 499 nm. The prosthetic group of the pigment P 523_2 is 11-*cis*-dehydroretinal and the peak absorption is at 523 nm. The A_1-pigments are frequently termed rhodopsins and the A_2-pigments porphyropsins. The name rhodopsin on its own usually refers to P 502_1 (frog) or P 499_1 (bovine)—the two pigments which have been most exactly investigated.

4 Absorption Properties

Figure 2 shows the absorption spectra of the pigments P 499_1 and P 523_2. Rhodopsin has in addition to its main absorption band (α) with its characteristic peak (λ_{max}) a subsidiary band (β) with a maximum at 350 nm. The minimum between the α and β bands at a wavelength of 400 nm can be used as a measure for the purity of the pigments; the absorption there should not be greater than 21–24% relative to the peak of the α band[18]. A third absorption band (γ—not shown)

19 *L. Zechmeister*, *Cis-Trans* Isomeric Carotenoids, Vitamins A and Arylpolyenes, Springer-Verlag, Wien 1962.

20 *R. Hubbard*, Nature *181*, 1126 (1958).

21 *R. Hubbard*, J. Gen. Physiol. *42*, 259 (1959).

22 *R. Hubbard*, *G. Wald*, J. Gen. Physiol. *36*, 269 (1952–1953).

23 *G. Wald*, Fedn. Proc. Fedn. Am. Socs. Exp. Biol. *12*, 606 (1953).

24 *Ch. Baumann*, *W. Ernst*, J. Physiol. *210*, 156 (1970).

25 *A. Lewis*, *R. Fager*, *E. Abrahamson*, J. Raman Spect. *1*, 465 (1973).

26 *A. Oseroff*, *R. Callender*, Biochemistry *13*, 4243 (1974).

27 *M. Akhtar*, *P. T. Blosse*, *P. B. Dewhurst*, Chem. Commun. *13*, 631 (1967).

28 *D. Bownds*, Nature *216*, 1178 (1967).

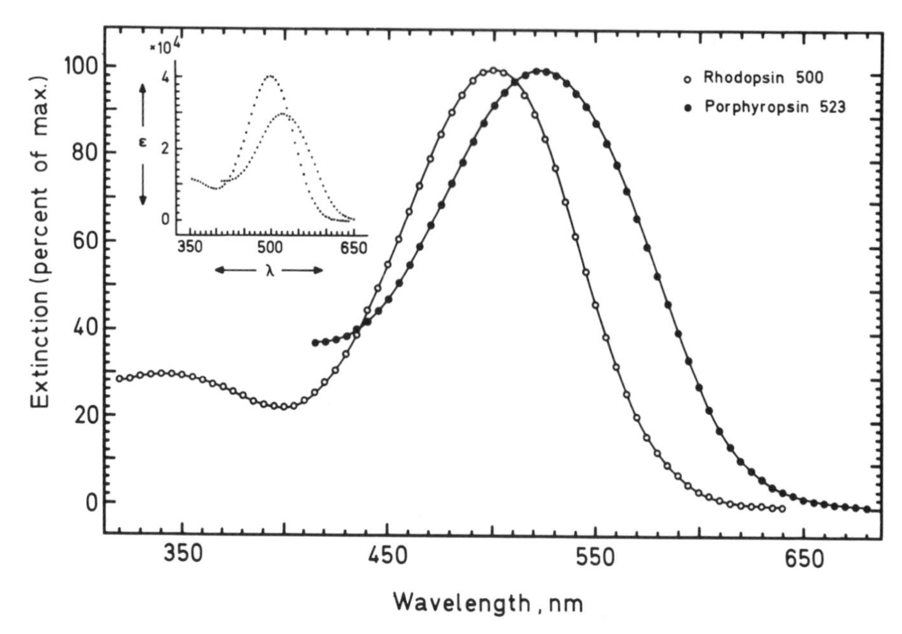

Fig. 2. The absorbance spectra of the pigments P 499_1 (or P 500_1) (rhodopsin) and P 523_2 (porphyropsin), from Bridges[67]. Inset: the two curves plotted in terms of their molar absorbances (ε)[31,32].

has its maximum at 278 nm; its peak is approximately twice that of the α band. (However, the description λ_{max} always refers to the absorption of visible light by the α band). The γ band is mainly caused by tyrosine and tryptophan residues of opsin.

The second pigment illustrated in Fig. 2 is porphyropsin (P 523_2) which commonly occurs in freshwater fishes. In a number of species it is found together with rhodopsin[29]. The distinctive feature of such a pigment pair is that apparently both pigments contain the same opsin and the only difference between them is in the prosthetic group. The latter can be exchanged for one another. The introduction of a further double bond in the transition from an A_1 to an A_2 pigment has three sorts of consequences[30]: a) a bathochromic effect, *i.e.*, the displacement of λ_{max} by about 20 nm towards longer wavelengths, b) the reduction of the molar extinction coefficient from 40,600 (rhodopsin)[31] to about 30,000 (porphyropsin)[32] and c) an increase in the absorption on the shortwave side of the α band.

The absorption spectra of the vertebrate visual pigments correspond in general to the type of curve shown in Fig. 2. The position of the maximum, however, can vary considerably. The λ_{max} of the pigments identified up to now lie between 433 and 575 and those of the A_2-pigments between 438 and 620 nm[33]. They are not equally distributed over the respective wavelength ranges but form clusters at certain discrete points the distance between which is nearly constant. For the A_1-pigments the points recur at approximately 7 nm intervals between 467 and 528 nm[29]. Likewise for the A_2-pigments it has been shown that the points recur regularly at the wavelengths 512, 523, 534 and 543[34]. Moreover there is a fixed relation between the λ_{max} of the A_1 and A_2-pigments in those species in which they occur as a pair. A mutual substitution of the chromophoric groups of such pairs may occur. The longer the wavelength of the light absorbed by the pigment pair, the greater is the resulting spectral displacement (bathochromic in the case of $A_1 \rightarrow A_2$; hypsochromic in the case of $A_2 \rightarrow A_1$). During the metamorphosis of *Rana pipiens*, for example, the pigments undergo the following changes: P $438_2 \rightarrow$ P 433_1; P $527_2 \rightarrow$ P 502_1; P $620_2 \rightarrow$ P 575_1[33]. The mechanism of these shifts is

[29] *H. J. A. Dartnall, J. N. Lythgoe*, Vision Res. *5*, 81 (1965).
[30] *C. D. B. Bridges*, Comprehensive Biochemistry, Vol. 27, p. 31, Elsevier, Amsterdam 1967.
[31] *G. Wald, P. K. Brown*, J. Gen. Physiol. *37*, 189 (1953).
[32] *P. K. Brown, I. R. Gibbons, G. Wald*, J. Cell. Biol. *19*, 79 (1963).
[33] *P. A. Liebman, G. Entine*, Vision Res. *8*, 761 (1968).
[34] *C. D. B. Bridges*, Cold Spring Harbor Symp. Quant. Biol. *30*, 317 (1965).

not fully understood and there is no generally accepted theory which can explain the spectra of the visual pigments and their variability in terms of their molecular structure.

5 Photolysis

In the following description of the photolysis of visual pigments, rhodopsin will be used as an example, since most of the data were obtained for this pigment. Fig. 3 shows schematically the sequence of reactions. Most of the substances involved are unstable at room or body temperature; in general the earlier they occur in the sequence the shorter is their life time. After the absorption of a photon, rhodopsin is first of all converted to prelumirhodopsin[35]. At room temperature, the transition of rhodopsin to prelumirhodopsin takes less than 6×10^{-12} sec[36]. The first step in the photolysis is most likely an isomerization of the rhodopsin chromophore from 11-*cis* to all-*trans*[37,38]. The excited state of the rhodopsin molecule which must necessarily occur during the photochemical reaction is unknown. The reactions which follow occur spontaneously and without the further participation of light; they are therefore frequently termed dark reactions. The steps from pre-lumi- to lumi- to metarhodopsin I have mainly been investigated *in vitro* at low temperatures[39]. On the other hand the conversion of metarhodopsin I into metarhodopsin II can be measured in eyes at 37°[40]. The rate constants of this reaction have an order of magnitude of $10^3\,\mathrm{sec}^{-1}$[41]. The conversion of meta I to meta II which involves the uptake of a proton must correctly be described as an equilibrium reaction[42]. *Acid* and higher temperatures favour the formation of meta II.

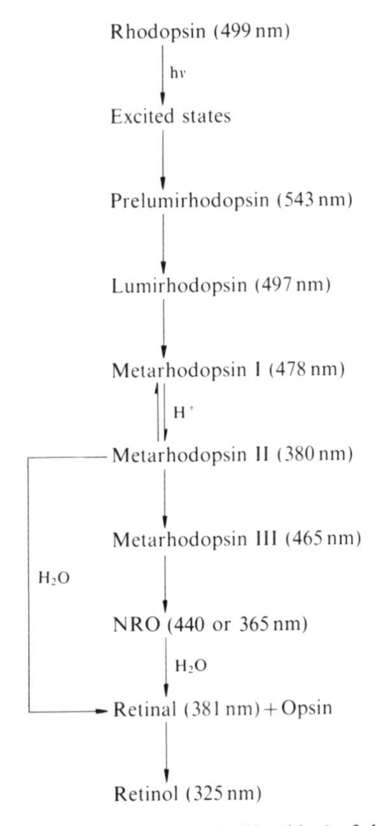

Fig. 3. Scheme of the photolysis (bleaching) of the visual pigment P 499_1 (rhodopsin).
In brackets: λ_{max} of the substances involved.

In cephalopods and arthropods the photolysis ends with a stable metarhodopsin[43,44]. This occurs in one of two forms which can be readily converted to the other by pH changes. The λ_{max} of the acid metarhodopsin is at 500 nm, that of the alkaline form at 380 nm. On illumination, the stable metarhodopsins of invertebrates can easily revert to rhodopsin.

In vertebrates a series of very slow dark reactions begins with metarhodopsin II and ends with the stable end products—opsin and retinol. Since neither of these substances absorb light, rhodopsin containing preparations, *e.g.* the isolated retina, lose their colour in the course of the slow dark reactions. The photolysis described here is also, therefore, called bleaching of the visual pigment. Apart from the back reaction to meta I, metarhodopsin II may be involved in two possible reaction sequences which can occur simultaneously *in vivo*[45,46]. In one it may be directly hydrolyzed to all-*trans*-retinal and opsin. Under

35 *T. Yoshizawa, Y. Kito*, Nature *182*, 1604 (1958).

36 *G. E. Busch, M. L. Applebury, A. A. Lamola, P. M. Rentzepis*. Proc. Nat. Acad. Sci. U.S. *69*, 2802 (1972).

37 *T. Yoshizawa, G. Wald*, Nature *197*, 1279 (1963).

38 *T. Yoshizawa, G. Wald*, Nature *201*, 340 (1964).

39 *R. Hubbard, D. Bownds, T. Yoshizawa*, Cold Spring Harbor Symp. Quant. Biol. *30*, 301 (1965).

40 *R. A. Cone, W. L. Pak, in* Handbook of Sensory Physiology, Vol. 1, Chapter 2, Springer-Verlag, Heidelberg 1971.

41 *Ch. Baumann*, J. Physiol. *259*, 357 (1976).

42 *R. G. Matthews, R. Hubbard, P. K. Brown, G. Wald*, J. Gen. Physiol. *47*, 215 (1963).

43 *R. Hubbard, R. C. C. St. George*, J. Gen. Physiol. *41*, 501 (1958).

44 *H. Langer*, Verh. Dtsch. Zool. Ges. Göttingen 1966, 195 (1967).

45 *T. G. Ebrey*, Vision Res. *8*, 965 (1968).

46 *Ch. Baumann*, J. Physiol. *222*, 643 (1972).

other conditions meta II gives rise to a relatively stable metarhodopsin III and subsequently to *N*-retinylidene opsin (NRO), which is then hydrolyzed to retinal and opsin[47]. NRO is a pH indicator which can occur as a Schiff's base (pH 7.7; $\lambda_{max}=365$ nm) or as the conjugate acid (pH 5.5; $\lambda_{max}=440$ nm)[18]. At a neutral pH it decays very quickly; whether it is to be found as a photoproduct in intact receptor cells is unknown. The reduction of retinal to retinol is catalyzed by an enzyme[48] which is either inactivated or destroyed by extraction procedures; so the photolysis ends with retinal in pigment solutions.

Lumi and metarhodopsin (I and II) like prelumirhodopsin contain all-*trans*-retinal as a prosthetic group. If these substances absorb light, the original pigment can be reformed. The basis of this photochemical reaction is the reisomerization of the chromophore (retinal) from all-*trans* to the 11-*cis* configuration[39,42]. Such a photoregeneration of pigments leads not only to the formation of the normal pigment (11-*cis*) but also to the isopigment (9-*cis*). Irradiation of metarhodopsin III *in vitro* results in a mixture of rhodopsin (P 499_1), isorhodopsin (P 487_1) and *cis*-isomers of retinal[47]. It is a point of dispute whether meta III is an all-*trans*-chromoprotein, for when the all-*trans*-metarhodopsin II is irradiated with ultraviolet light a mixture of rhodopsin and metarhodopsin III is obtained[49], so that it is not altogether improbable that a 13-*cis*-chromophore exists in meta III. According to a more recent study with frog rhodopsin (P 502_1) the chromophore of metarhodopsin III is an all-*trans*- conformer of retinal[50].

The extent to which a pigment is bleached is determined by its photosensitivity[51] and the amount of incident light ($I\times t$). The photosensitivity is the product of the molar extinction coefficient, ε_{max}, and the quantum efficiency, γ. For dilute pigment solutions, pigment concentration can be obtained from the equation:

$$c=c_0\exp(-\varepsilon_{max}\gamma It),$$

provided multiple absorptions which may occur during flash irradiation are ignored[52]. c_0 is the initial concentration, while c is the value to which it decays after time t. The method of photometric curves allows the determination of the product $\varepsilon_{max}\gamma$[53,54]. For a known ε_{max} the quantum efficiency of bleaching can be determined, 6 A_1-pigments and 5 A_2-pigments give values of γ between 0.64 and 0.67[55], which implies that about 1/3 of the pigment molecules absorb quanta without undergoing photolysis.

6 Resynthesis (Regeneration)

In the living eye light constantly breaks down a part of the pigment. However, this bleaching does not lead to a permanent reduction of the amount of pigment. Regeneration (resynthesis) compensates for the bleaching. In this process the pigment is not resynthesized *de novo*; the 11-*cis*-form of retinal is produced and combines with the available opsin. 11-*cis*-retinal originates either from the isomerization of the all-*trans*-form[56] or from the oxidation of the corresponding alcohol (11-*cis*-retinol)[48]. Both reactions are catalyzed by enzymes. Retinol is stored as a fatty acid ester in the eyes of many species[57]. The following scheme shows the reactions which are relevant for pigment regeneration.

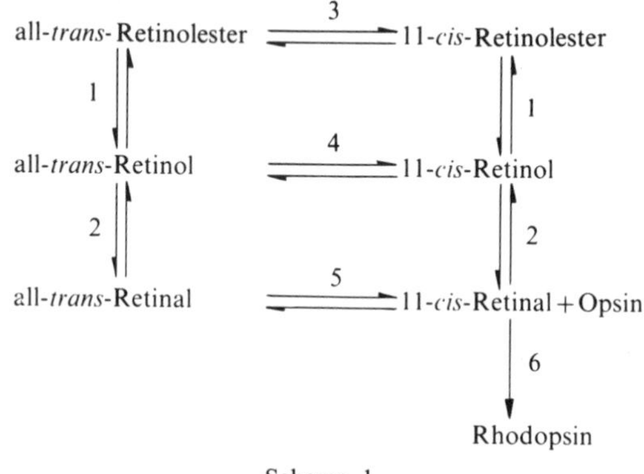

Scheme 1.

Reaction 1 takes place in the living eye in two different tissues: the esterification (like the storage of retinol) occurs in the pigment epithelium

---

[47] *S. E. Ostroy, F. Erhardt, E. W. Abrahamson*, Biochim. Biophys. Acta *112*, 265 (1966).

[48] *A. L. Koen, C. R. Shaw*, Biochim. Biophys. Acta *128*, 48 (1966).

[49] *T. P. Williams*, Vision Res. *8*, 1457 (1968).

[50] *T. Reuter*, Vision Res. *16*, 909 (1976).

[51] *C. F. Goodeve, L. J. Wood*, Proc. Roy. Soc. A *166*, 342 (1938).

[52] *H. J. A. Dartnall*, The Visual Pigments, Methuen & Co., Ltd., London 1957.

[53] *H. J. A. Dartnall, C. F. Goodeve, R. J. Lythgoe*, Proc. Roy. Soc. A *156*, 158 (1936).

[54] *H. J. A. Dartnall, C. F. Goodeve, R. J. Lythgoe*, Proc. Roy. Soc. A *164*, 216 (1938).

[55] *H. J. A. Dartnall*, Vision Res. *8*, 339 (1968).

[56] *R. Hubbard*, J. Gen. Physiol. *39*, 935 (1956).

[57] *S. Futterman, J. S. Andrews*, J. Biol. Chem. *239*, 81 (1964).

adjacent to the retina; on the other hand, the hydrolysis of the ester requires the retinal tissue[58]. It is not known for certain to what extent the two isomers of retinol can be distinguished in these two reactions. Reaction 2 is a reversible reduction of retinal to retinol which is catalysed by an alcohol dehydrogenase localized in the outer segments[48,59–61]. The reaction shows only a limited enzyme specificity. Reaction 3, a slow isomerization, presumably occurs in the pigment epithelium where from 55% (frog)[62] up to 65% (bovine eye)[58] of the retinol ester is found in the 11-*cis*-configuration. Reaction 4 is hypothetical; it has been postulated because the other isomerizations are either very slow (reaction 3) or not very effective (reaction 5) in producing the 11-*cis*-form. The enzyme taking part in reaction 5, a retinal isomerase, has been isolated and examined in detail *in vitro*[56]. The enzyme can catalyze the conversion of the all-*trans*-form into the 11-*cis*-form. The equilibrium mixture consists of 5% 11-*cis* and 95% all-*trans*-retinal (in the dark). If the enzyme substrate complex is irradiated with orange light ($\lambda > 520$ nm) the equilibrium is shifted in favour of 11-*cis*. This effect is difficult to interpret because retinal can absorb only a very limited amount of light of these wavelengths. In the frog retina, the isomerase is likely to be specific for retinyl ester, and reaction 3 occuring in photoreceptors is possibly the main source of the 11-*cis*-conformer[62]. Reaction 6 is highly specific for 11-*cis*-retinal which combines spontaneously with opsin to bring the resynthesis to an end[22].

58 *N. I. Krinsky*, J. Biol. Chem. *232*, 881 (1958).
59 *G. Wald, R. Hubbard*, J. Gen. Physiol. *32*, 367 (1949).
60 *S. Futterman*, J. Biol. Chem. *238*, 1145 (1963).
61 *S. Futterman, L. D. Saslaw*, J. Biol. Chem. *236*, 1652 (1961).
62 *C. D. B. Bridges*, Exp. Eye Res. *22*, 435 (1976).

7 Other Protein-Polyene Complexes

Visual pigments are not the only proteins which employ a retinyl polyene as prosthetic group. A chromoprotein recently found in the outer membrane of the bacterium *Halobacterium halobium* contains retinal in two isomeric configurations[63,64]. The chromophore of the dark form is 13-*cis*-retinal which, on light exposure, is converted to the all-*trans*-configuration. The pigment is very photosensitive and has been named bacteriorhodopsin. It is involved in the transport of protons across the membrane of the bacterium.

Another complex closely related to, but not identical with, the visual pigments was found in retinae of various cephalopods. Cephalopod retinochrome[65] includes all-*trans*-retinal as prosthetic group. On illumination, the chromophore is converted into the 11-*cis*-configuration and is then released from the binding site. The physiological significance of this photochemical reaction is not yet fully understood.

In human blood, retinol is carried by a transporting complex called retinol binding protein (RBP)[66]. The linkage between retinol and the protein moiety is non-covalent (thus differing from the other chromoproteins mentioned here), and no photochemical reaction of this complex has been reported.

63 *D. Oesterhelt, W. Stoeckenius*, Nature New Biol. *233*, 149 (1971).
64 *L. Y. Jan*, Vision Res. *15*, 1081 (1975).
65 *T. Hara, R. Hara, in* Handbook of Sensory Physiology, Vol. VII/1, Chapter 18, Springer, Heidelberg 1972.
66 *M. Kanai, A. Raz, D. S. Goodman*, J. Clin. Invest. *47*, 2025 (1968).
67 *C. D. B. Bridges*, Vision Res. *7*, 349 (1967).

6 Miscellaneous

Contributions by

A. Kawaguchi, Sendai
S. Nozoe, Sendai
S. Shibata, Tokyo

6.1 Lichen Substances

Shoji Shibata

Faculty of Pharmaceutical Sciences, University of Tokyo, Bunkyo-ku, Tokyo, 113 Japan

1 Introduction

Lichens are characteristic symbiotic organisms consisting of algae and fungi. The secondary metabolites of lichens, lichen substances, which are regarded mostly to be produced by the fungal symbionts are structurally unique among the natural products. As the morphological differentiation of lichens is sometimes very difficult, they can often only be distinguished by their secondary metabolites and the thalline color reactions and microcrystallization tests have been employed as useful tools.

Further details of lichen substances are described in books and articles listed in the Bibliography (Section 10).

2 Biogenetical Classification of Lichen Substances

The lichen substances provide a good example of natural regularity in the structures from which Birch's biogenetical acetate theory[1] would have been deduced. The lichen substances hitherto known are classified on the basis of the biogenetical pathways[2] which are now generally recognized:

Carbohydrates (2.1): Polyalchohols (meso-erythritol, D-arabitol, D-mannitol, D-volemitol etc.), disaccharides and polysaccharides (lichenan, isolichenan, pustulan etc.)

Polyketides (2.2) of acetate-malonate origin:

- Higher fatty acids and their lactones (Scheme 1)
- Phenol carboxylic acid derivatives (Scheme 2)
  - a) Orcinol-type compounds (**3**): Depsides, depsidones, depsone, dihydrochromone and dibenzofuranes;
  - b) β-Orcinol-type compounds (**4**): Depsides and depsidones;
  - c) Usnic acid (**8**) and isousnic acid (**9**)
- Quinonoids....Anthraquinones, bianthronyls, bianthraquinone, and modified bianthraquinone.
- Xanthones....Xanthones and modified bixanthones.

Shikimic acid origin:

- Terphenylquinones.
- Diphenylbutadienes (Pulvinic acid derivatives)

Mevalonic acid origin:

- Diterpene, triterpenes (ursane, taraxerane, friedelane, hopane and migrated hopane types), carotenoids (β-carotene, γ-carotene, xanthophyll), steroids (β-sitosterol, ergosterol).

Amino acid origin: Diketopiperazine and peptide.

2.1 Carbohydrates

Simple sugar alcohols, such as erythritol and D-mannitol are contained generally, while D-siphuritol and D-volemitol are found in some special lichens.

Water-soluble polysaccharides show a wide distribution among lichens with some chemotaxonomical correlations. The structures of lichenan and isolichenan have been studied extensively and they have been shown to be (1→3)(1→4)β-glucan (the ratio: about 27:73) and (1→3)(1→4)α-glucan (the ratio: 56.5:43.5), respectively. Pustulan which is contained in *Gyrophora, Umbilicaria* and *Lasallia* spp. was recognized to be a β(1→6) glucan, but recently it has been found to occur in lichens as a partly acetylated form[3].

2.2 Polyketides

Lichen fatty acids like caperatic acid (**2**) like protolichesteric acid (**1**) and γ-lactonic acids are regarded to be biosynthesized following the pathway in Scheme 1.

Phenol carboxylic acids, their condensation products, depsides and depsidones are characteristic metabolites of lichens. A structural regularity is observed among these compounds, *e.g.*, the orcinol type compounds **3** possess alkyl side chains of odd number carbon atoms, while β-orcinol type compounds **4** have a C_1-unit group of a different stage of oxidation, such as CH_3,

---

[1] *A. J. Birch, F. W. Donovan*, Austl. J. Chem. *6*, 373 (1953); cf. *A. J. Birch*, Science *156*, 202 (1967).

[2] *S. Shibata*, Beiträge zur Biochemie u. Physiologie v. Naturstoffen p. 451–465, Gustav Fischer, Jena 1965.

[3] *Y. Nishikawa, T. Takeda, S. Shibata, F. Fukuoka*, Chem. Pharm. Bull. (Tokyo) *17*, 1910 (1969); *18*, 1431 (1970).

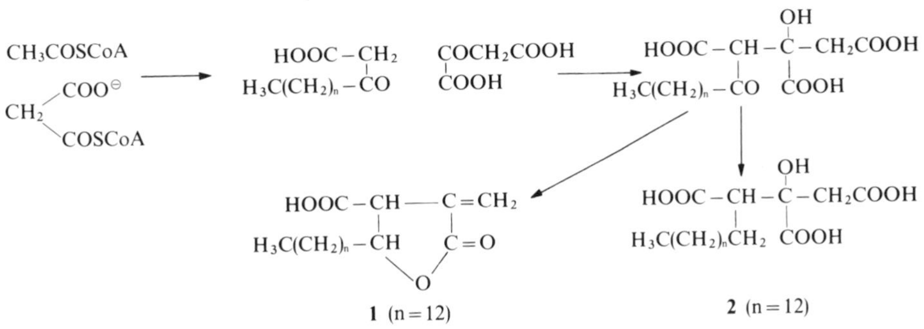

$CH_3COSCoA$ + $CH_2(COO^{\ominus})COSCoA$ → $HOOC-CH_2$, $H_3C(CH_2)_n-CO$ + $COCH_2COOH$, $COOH$ → $HOOC-CH-C(OH)(COOH)-CH_2COOH$, $H_3C(CH_2)_n-CO$

1 (n = 12) $HOOC-CH—C=CH_2$, $H_3C(CH_2)_n-CH$, $C=O$, O

2 (n = 12) $HOOC-CH-C(OH)(COOH)-CH_2COOH$, $H_3C(CH_2)_nCH_2$

Scheme 1.

R, CO, CH_2 CH_2-COOH, CO CO, CH_2 → **3** (HO, COOH, OH, R)

$R = CH_3$, C_3H_7, C_5H_{11}, C_7H_{15}, $CH_2COC_3H_7$, $CH_2COC_5H_{11}$, COC_4H_9

C_1

4 (CH_3, COOH, OH, HO, CH_3)

$CH_3 \rightarrow CH_2OH \rightarrow CHO \rightarrow COOH \rightarrow COOCH_3$

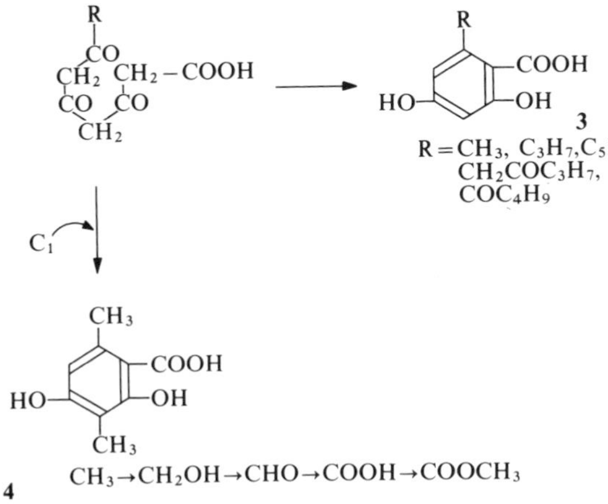

Scheme 2.

CH_2OH, CHO, COOH or $COOCH_3$, at the 1,4-positions.

This regularity can be explained biogenetically by the cyclization of a polyketomethylene chain derived from acetyl-CoA and malonyl CoA and by the substitution of a C_1-fragment.

Isotopic tracer experiments using living lichen thalli provided the evidence for this type of cyclization from acetate and malonate and for C_1-fragment incorporation, if any, prior to aromatization[4]. Nuclear hydroxylation or chlorination occasionally occurs to form polyhydroxy or chloro derivatives of depsides or depsidones.

The ester condensation of two or sometimes three molecules of orcinol or β-orcinol type carboxylic acids forms the corresponding depsides. The intramolecular phenol oxidative coupling of depside (**5**) provides depsidone (**6**). Only one depsone, picrolichenic acid (**7**), is known, which is an oxidative coupling product possessing a spiran-type structure[5].

Another type of cyclization of the polyketomethylene chain provides a phloroglucinol type unit. The experiments using ^{14}C-labeled acetate, malonate and formate revealed that usnic acid (**8**) is biosynthesized in lichens by Scheme 4 where the methyl group is also introduced prior to ring closure[6]. Isousnic acid (**9**)[7] which occurs along with usnic acid in some lichens would be formed by secondary ring closure in a different way after the phenol oxidative coupling of methylphloroacetophenone units.

Dibenzofurane derivatives **10** such as didymic acid (**10**, $R=CH_3$) occurring in lichens would be produced also by the phenol oxidative coupling[8,9].

The single polyketomethylene chain can be cyc-

[4] *M. Yamazaki, S. Shibata*, Chem. Pharm. Bull. (Tokyo) *14*, 96 (1966).

[5] *H. Erdtman, C. A. Wachtmeister*, Chem. & Ind. 1042 (1957).

[6] *H. Taguchi, U. Sankawa, S. Shibata*, Tetrahedron Lett. 5211 (1966).

[7] *S. Shibata, H. Taguchi*, Tetrahedron Lett. 4867 (1967).

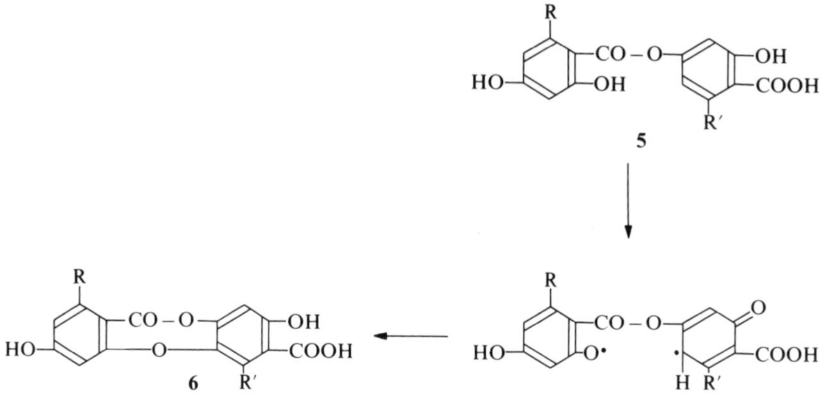

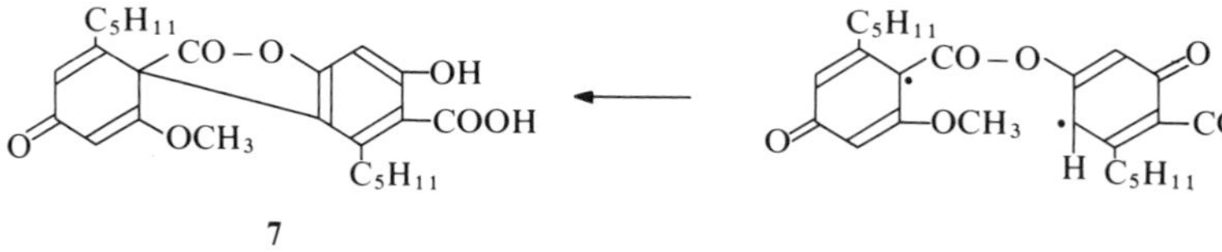

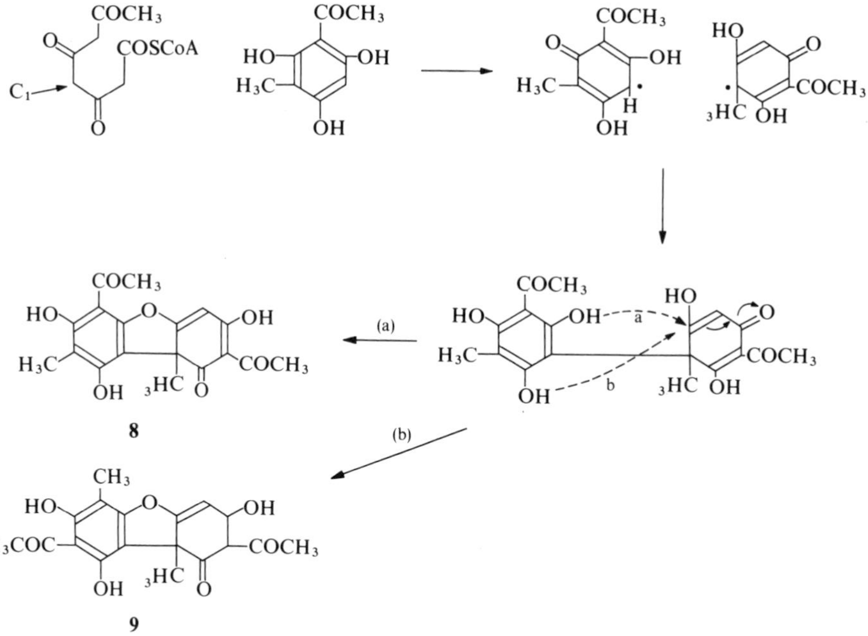

5

6

7

8

9

lized to form polynuclear aromatic compounds, such as anthraquinones (**12**) and xanthones (**11**). It has been proved by isotopic tracer experiments that the polyhydroxyanthraquinones also occur in fungi by the same biosynthetic scheme[10].

Endocrocin (**12**, R = COOH) which was isolated first from the lichen, *Cetraria endocrocea*[11], and thereafter from the fungi, *Aspergillus amstelodami*[12], *Penicillium islandicum*[13] and *Claviceps purpurea*[14], is a key compound in the biosyn-

[8] *D. H. R. Barton, T. Cohen*, Festschrift Prof. A. Stoll p. 117–143, Birkhäuser, Basel 1957.

[9] *H. Erdtman, C. A. Wachtmeister*, Festschrift Prof. A. Stoll p. 144–165, Birkhäuser, Basel 1957.

[10] *S. Shibata, T. Ikekawa*, Chem. Pharm. Bull. (Tokyo) *11*, 368 (1963).

[11] *Y. Asahina, F. Fujikawa*, Ber. *68*, 1558 (1935).

[12] *S. Shibata, S. Natori*, Pharm. Bull. (Tokyo) *1*, 160 (1953).

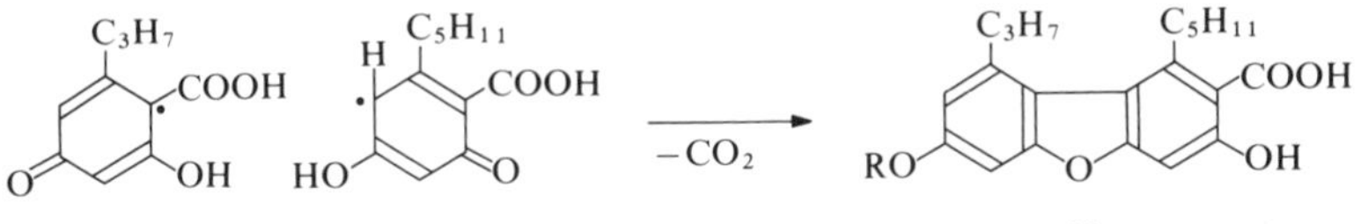

10

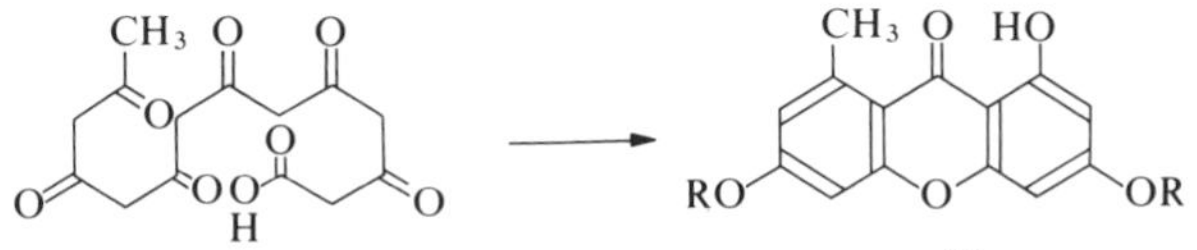

11

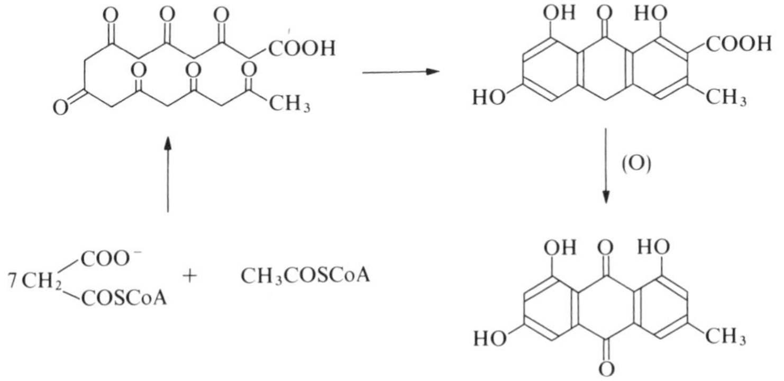

12

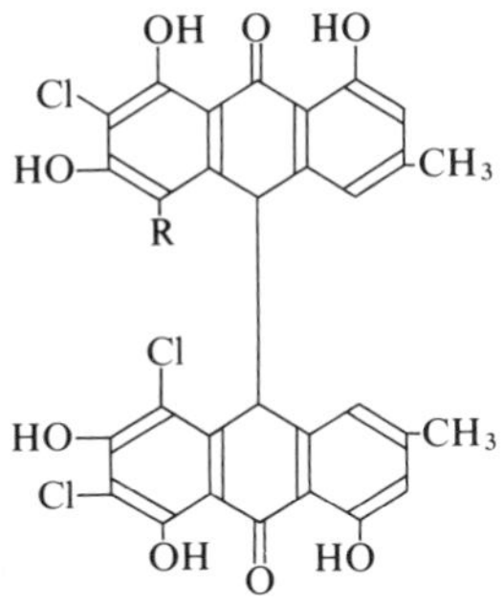

13

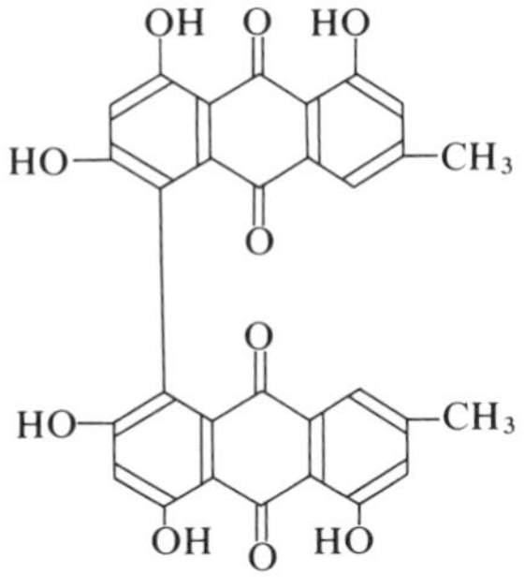

14

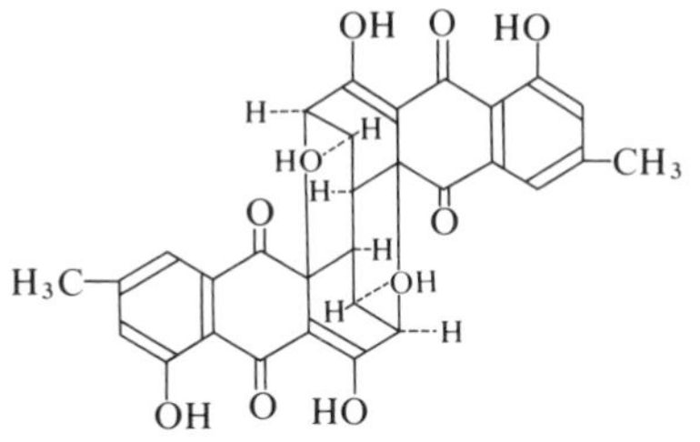

15

thesis of various anthraquinones of lichens, fungi and higher plants. Bianthronyl-10,10′ (**13**), named flavoobscurins A (R=H), B_1 and B_2 (R=Cl), were isolated from *Anaptychia obscùlata* Vain. by Yosioka *et al.*[15] It is noted that emodin (a bimolecular **12**, connected in (8,8′), skyrin (**14**) and its related modified bianthraquinone, rugulosin (**15**), which were originally isolated from *Penicillium* spp. have also been obtained from the lichen, *Acroscyphus sphaerophoIoides*[16].

The structure of rhodocladonic acid which was formulated previously by Shibata[17] as an polyhydroxyanthraquinone has now been revised by the renewed investigation by Baker *et al.*[18] to be a furonaphthoquinone (**16**).

[13] *S. Gatenbeck*, Acta Chem. Scand. *13*, 386 (1959).

[14] *B. Franck, T. Reschke*, Chem. Ber. *93*, 347 (1960).

[15] *I. Yosioka, H. Yamauchi, K. Morimoto, I. Kitagawa*, Journ. Jap. Bot. *43*, 343 (1968); Tetrahedron Lett. 3749 (1968).

[16] *S. Shibata, O. Tanaka, U. Sankawa, Y. Ogihara, R. Takahashi, S. Seo, D. M. Yang, Y. Iida*, Journ. Jap. Bot. *43*, 335 (1968).

[17] *S. Shibata*, Yakugaku-zasshi (J. Japan. Pharm. Soc.) *61*, 320 (1941).

[18] *P. M. Baker, E. Bullock*, Can. J. Chem. *47*, 2733 (1969).

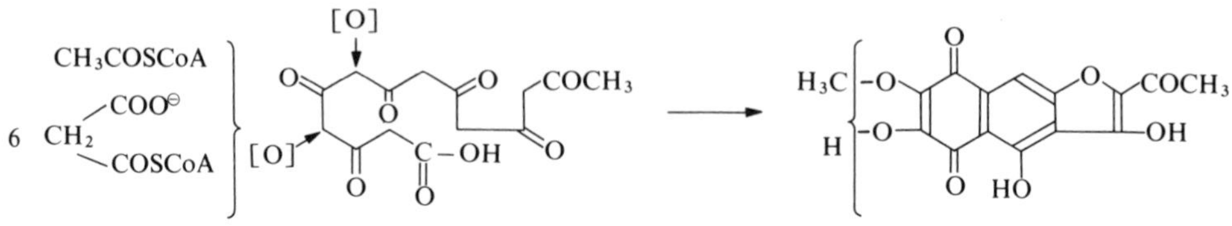

16

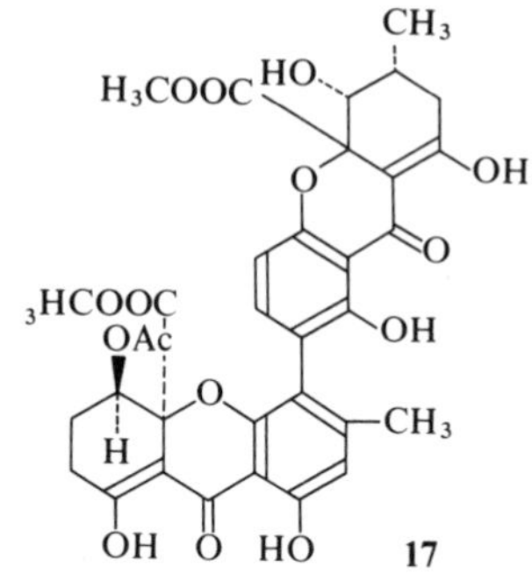

17

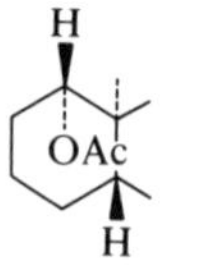

18

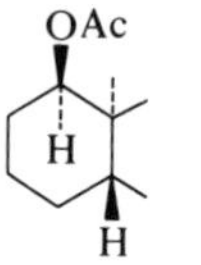

19

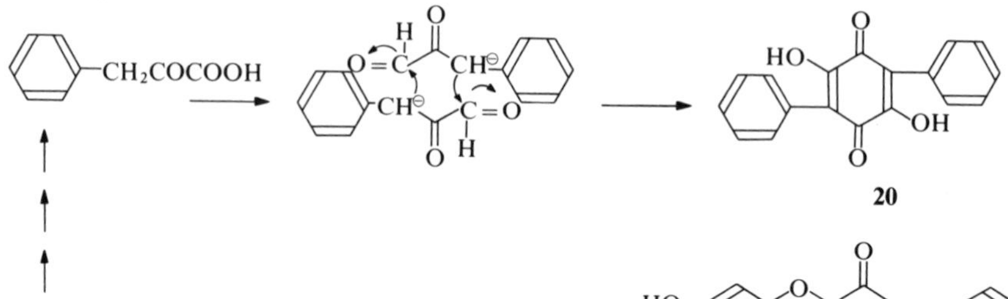

20

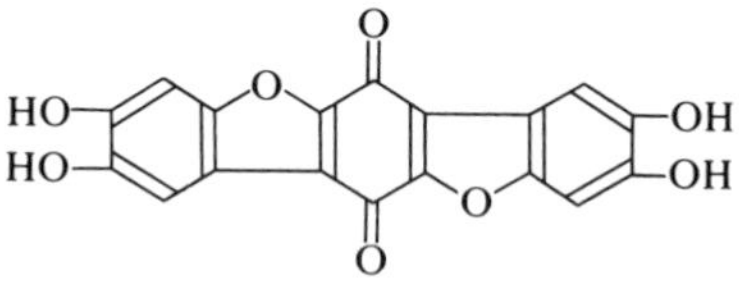

21

Modified bixanthones, eumitrins A (**17**), B (**18**), and C (**19**), were isolated from *Usnea beyleyi*[19]. Correlating with the pigments of Ergot, ergochromes[20], eumitrins are regarded to be derived from an acetate-polymalonate precursor via an anthraquinone intermediate.

2.3 Shikimate Metabolites

Fungal terphenylquinones have been shown to be derived from 2 molecules of C_6–C_3 units which are biosynthesized by the shikimic acid pathway. A terphenyl quinone type compound, polyporic acid **(20)**, was isolated from the lichen, *Sticta coronata*[21]; thelephoric acid (**21**), which was originally isolated from the basidiomycete,

[19] *D. M. Yang, N. Takeda (née Kobayashi), Y. Iitaka, U. Sankawa, S. Shibata*, Tetrahedron *29*, 519 (1973).

[20] cf. *B. Franck*, Angew. Chem. *8*, 251 (1969).

[21] *J. Murray*, J. Chem. Soc. 1345 (1952).

20 → 22

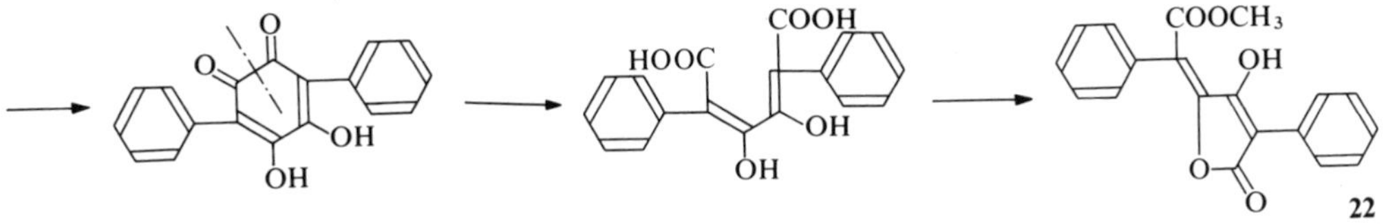

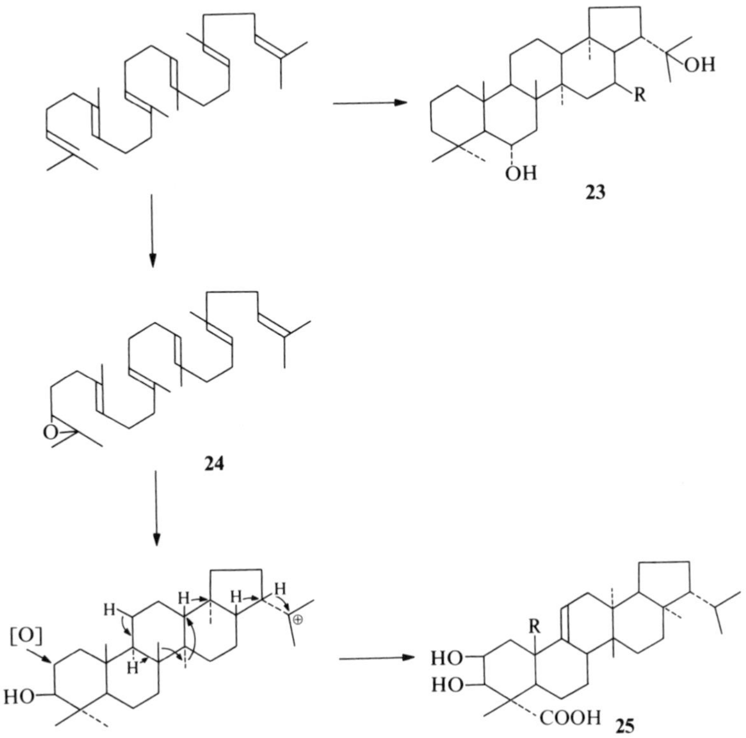

23

24

25

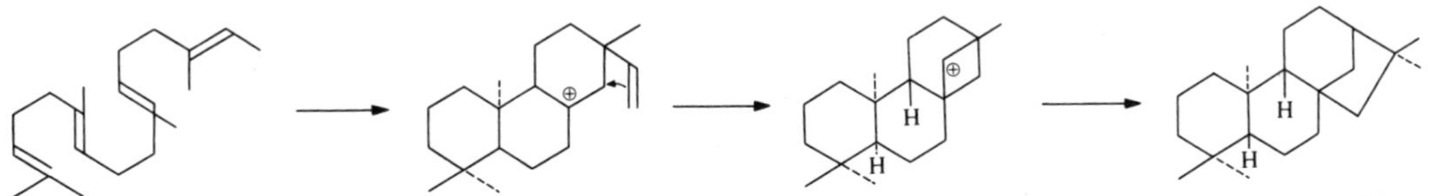

26

Thelephora palmata, was found in the rhizines of the lichens of the *Lobaria retigera* group[22]. The chemical oxidation of polyporic acid with hydrogen peroxide afforded a pulvinic acid type compound which is an unique metabolite of lichens. Feeding phenylalanine-1-^{14}C to *Letharia vulgaris*, Mosbach[23] demonstrated the incorporation of ^{14}C into vulpinic acid (**22**).

2.4 Isoprenoids

Zeorin (**23**), a hopane type triterpene, is distributed widely in various lichens. It has generally been established that lanosterol and other triterpenes having a hydroxyl at the 3-position are biosynthesized from 2,3-oxidosqualene (**24**), but it should be noted that most hopane type triterpenes occurring in lichens have no hydroxyl at the 3-position of the A-ring. In this case protonation initiates the ring closure of squalene. However the migrated hopanes, retigeric acids A (**25**, R=CH_3) and B (**25**, R=COOH)[24], which have recently been isolated from the lichens of

22 *Y. Asahina S. Shibata*, Ber. *72*, 1531 (1939).

23 *K. Mosbach*, Biochem. Biophys. Res. Commun. *17*, 363 (1964).

24 *R. Takahashi, H. C. Chiang, N. Aimi, O. Tanaka, S. Shibata*, Phytochem. *11*, 2039 (1972).

the *Lobaria retigera* group possess hydroxyls at the 2 and 3-positions, which would be biosynthesized in the usual way accompanying the migration of methyl groups and a hydride anion.

A diterpene, (−)16-hydroxykauren (=ceruchinol) has been obtained from *Ramalina ceruchis* and other lichens of *Ramalina* spp[25].

3 Microchemical Identification of Lichen Substances

The classical thallus color reaction has been improved by Asahina[26] for precise identification of the lichens based on their chemical constituents. Asahina[26] has also developed a microcrystallization technique for lichenological investigations. Minute lichen fragments are extracted with acetone in a micro-extractor and the extract is observed under a microscope, using suitable reagents to identify the microcrystals of lichen substances. The following abbreviations for the reagents are used in descriptions:

K: aq. KOH, C: aq. $Ca(OCl_2)_2$, KC: aq. KOH followed by aq. $Ca(OCl_2)_2$, PD: alcoholic *p*-phenylenediamine, FeCl: 1% aq. or alcoholic $FeCl_3$, GE: acetic acid–glycol mixture for microcrystallization, G Ao-T: glycerol–ethanol-*o*-toluidine (2:2:1), GAQ: glycerol–ethanol–quinoline (2:2:1) for the microcrystal test.

Paper chromatography[27-29] and thin layer chromatography have been a useful tools[30], the latter being standardized[31] for the identification of lichen substances in three solvent systems.

The unknowns can be identified by sorting punched cards summarizing R*f* values and microchemical data. Chromatograms are developed on Merck Silica gel F to a height of 10 cm using solvent A (benzene–dioxane–acetic acid (90:25:4), solvent B (hexane–ethyl ether– formic acid (5:4:1), solvent C (toluene – acetic acid (85:15), and marker controls of two lichen substances, atranorin and norstictic acid.

Gas-liquid chromatography is also applicable for the microdetection of lichen substances after tetramethylsilylation[32].

4 Extraction and Isolation

The dried lichen thalli are extracted with organic solvents, hexane, ether, acetone and sometimes benzene for the isolation of most lichen substances. Alcohols are not suitable for extraction of phenolic lichen substances, since the depside linkage is readily cleaved by alcoholysis. Column chromatography and preparative thin layer chromatography with silica gel or silicic acid using various solvent systems, such as benzene–acetone, chloroform–acetone mixtures etc. are employed for the separation of lichen metabolites.

Polysaccharide constituents of lichens are extracted from lichen thalli with water and precipitated by the addition of ethanol. The precipitates are purified by a freezing and thawing method. Dialysis is occasionally used and Sephadex gel filtration is employed for separation of fractions of different molecular weight.

5 Chemical Methods for Structural Determination

Degradation reactions have been used for the classical investigations of the structures of lichen substances. Depsides are readily cleaved to the phenolic components by the action of cold conc. sulfuric acid or by methanolysis with methanolic potassium hydroxide. By these reactions depsidones afford biphenyl ether derivatives, which are cleaved under more drastic conditions by the action of oxidation reagents or potash fusion.

Extensive studies of degradations were reported for the determination of the structure of usnic acid (**8**), in which ozonolysis[33] of the *O*-diacetyl product gave the most reliable evidence for the structure.

6 Physical Methods for Structural Determination

6.1 Ultraviolet Spectroscopy

According to Hale[34], the depsides and depsidones of orcinol and β-orcinol types can be

[25] *S. Huneck, G. Follmann,* Z. Naturforsch. *20b*, 611 (1965); *J. M. Lehn, S. Huneck,* Z. Naturforsch. *20b*, 1013 (1965).

[26] *Y. Asahina,* Journ. Jap. Bot. *12*, 516, 859 (1936); *13*, 529, 855 (1937); *14*, 39, 244, 318, 650, 767 (1938); *15*, 465 (1939); *18*, 622 (1942).

[27] *C. A. Wachtmeister,* Acta Chem. Scand. *6*, 818 (1952).

[28] *M. Mitsuno,* Pharm. Bull. (Tokyo) *1*, 170 (1953); *3*, 60 (1955).

[29] *D. Hess,* Planta *52*, 65 (1958).

[30] *J. Santesson,* Acta Chem. Scand. *19*, 2254 (1965).

[31] *C. F. Culberson, H. D. Kristinsson,* J. Chromatogr. *46*, 85 (1970).

[32] *S. Shibata, T. Furuya, H. Iizuka,* Chem. Pharm. Bull. (Tokyo) *13*, 1254 (1965).

[33] *Cl. Schöpf, X. Ross,* Ann. Chem. *546*, 1 (1941).

[34] *M. E. Hale,* Science *123*, 671 (1956).

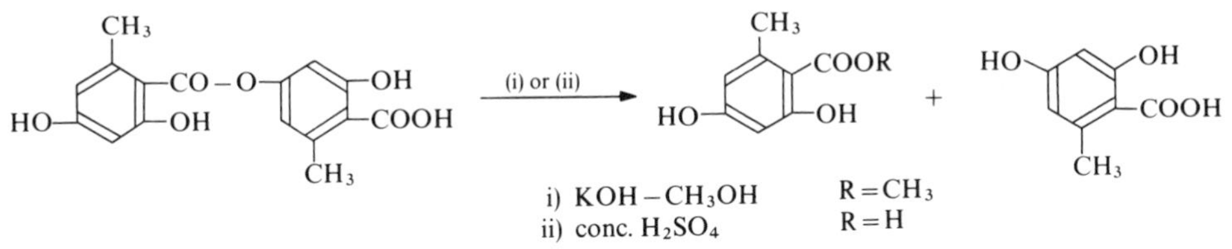

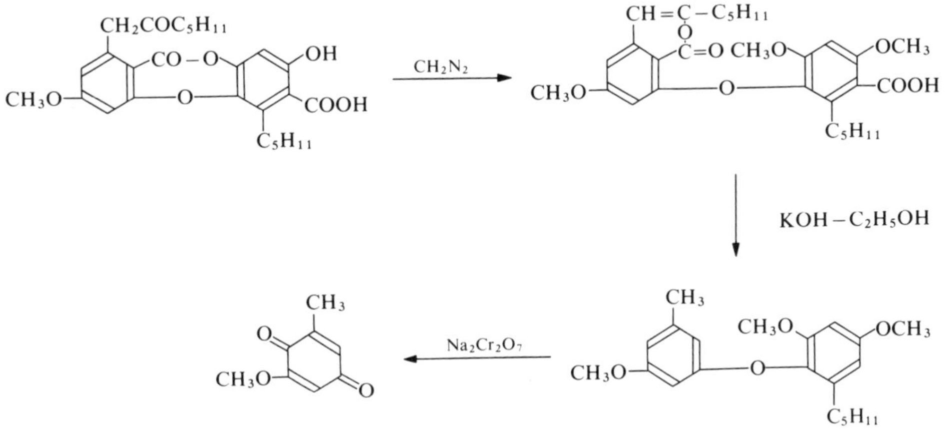

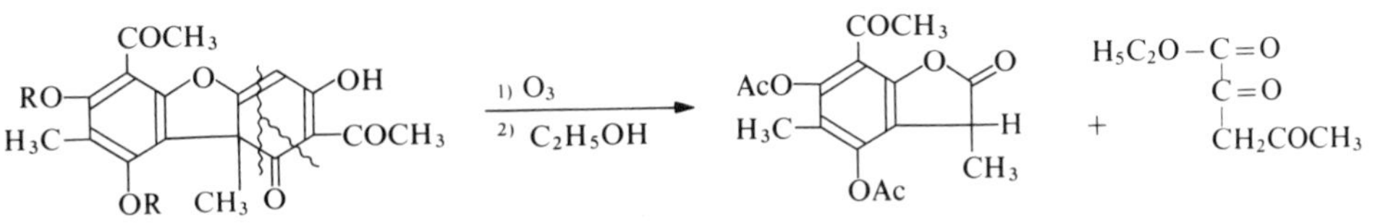

i) $KOH-CH_3OH$ $R=CH_3$
ii) conc. H_2SO_4 $R=H$

8: R=H, Ac

distinguished from each other by their ultraviolet (UV) absorption curves:

Orcinol-type depsides (λ_{max} 270, 307 nm),
orcinol-type depsidones (λ_{max} 245–255, 310–320 nm (sh)),
β-orcinol-type depsides (238, 312 nm),
β-orcinol type depsidones (238, 312 nm).
The depsides having a methoxyl at the *ortho* position of the depside linkage on the S-ring show a different type of UV-curve.

6.2 Infrared (IR) Spectrometry

The infrared (IR) spectral absorptions in the carbonyl (C=O) region give the most useful information for the structural elucidation of lichen substances. Some examples of the IR spectral absorptions [cm^{-1}] of –CO–O–, –COOH, $-COOCH_3$, –CHO, ring C=O in the molecules of lichen substances (in Nujol, unless stated otherwise) are shown in Scheme 3:

IR spectrometry is also used for the determination of configurations of the linkage (α or β) of lichen polysaccharides:

α-Glycan ($844 \pm 8\ cm^{-1}$)
β-Glycan ($891 \pm 7\ cm^{-1}$)
This absorption maximum is shifted to $910\ cm^{-1}$ in 1,6-β-glycan[3].

The presence of acetyl group in the native pustulan of *Gyrophora* spp. was revealed by the IR absorption at $1735\ cm^{-1}$.[3]

6.3 Nuclear Magnetic Resonance (NMR) Spectrometry

Nuclear magnetic resonance (NMR) spectro-

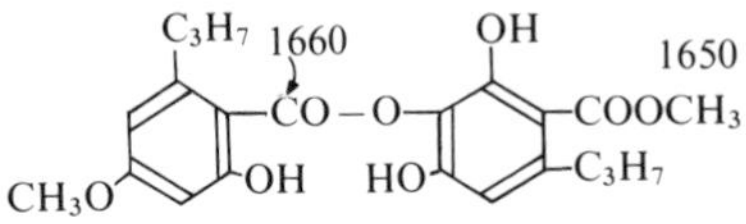

Scrobiculin (Nujol)

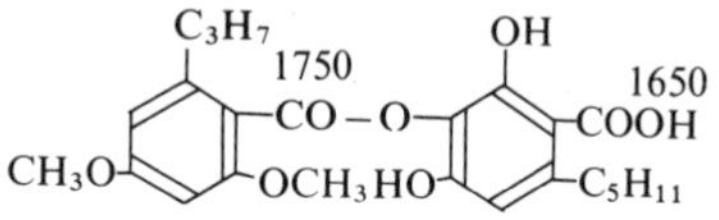

Merochlorophaeic acid **27** (KBr)

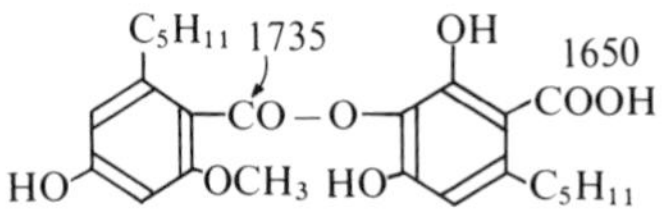

Cryptochlorophaeic acid (Nujol)

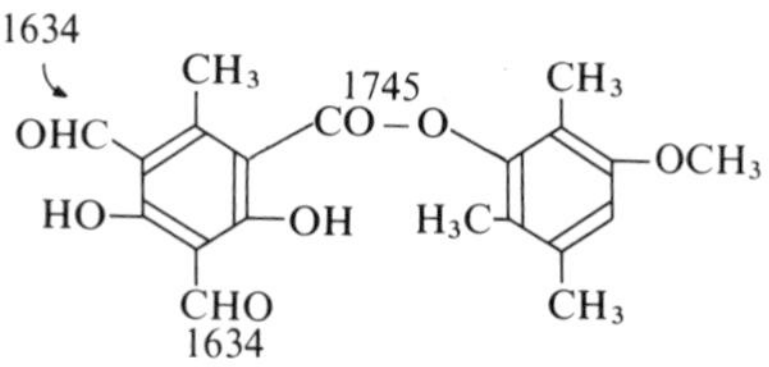

Nephroarctin **28**

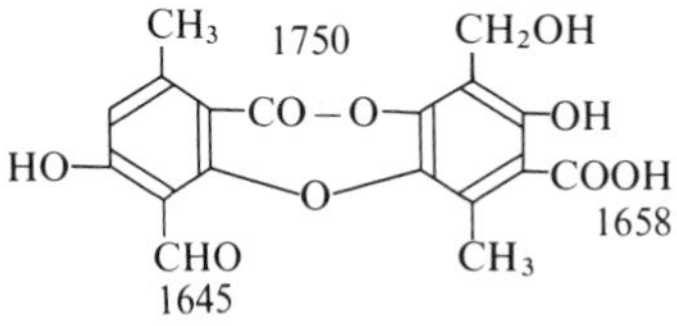

Protocetraric acid **29**

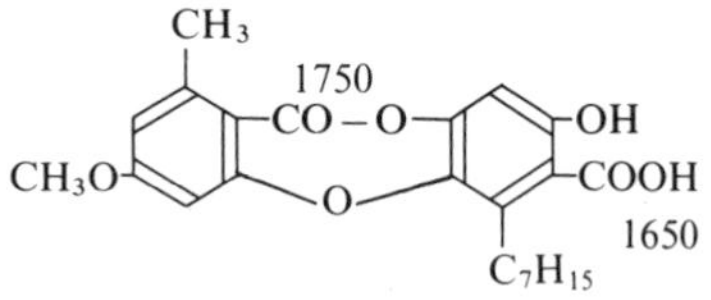

Grayanic acid **30**

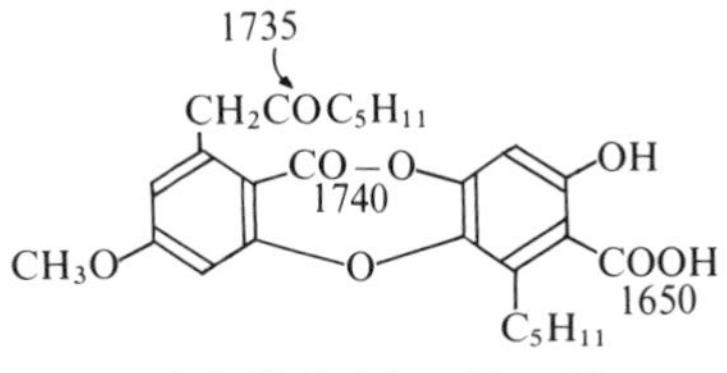

4-O-Methylphysodic acid

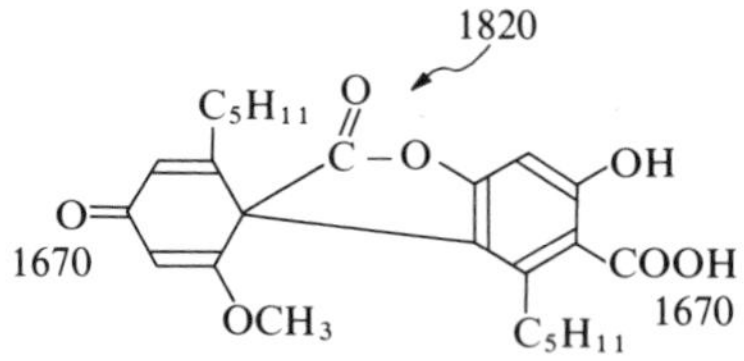

Picrolichenic acid **7**

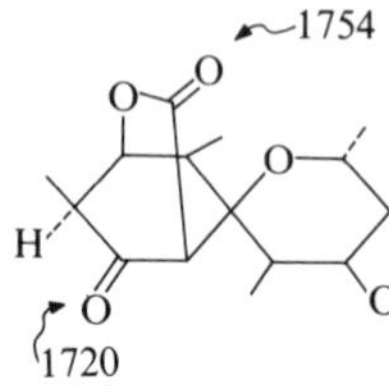

Portentol **31** (KBr)

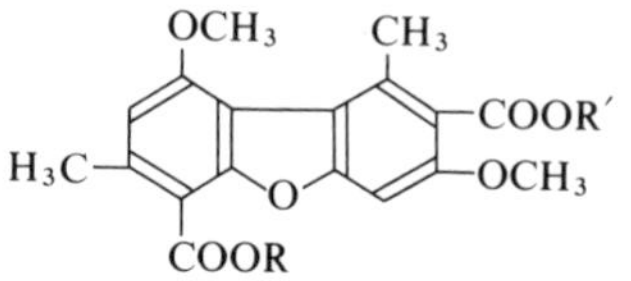

R,R′ = CH_3: 1725 Ester C=O

R,R′ = H: 1680 Carboxyl C=O

Schizopeltic acid **32**

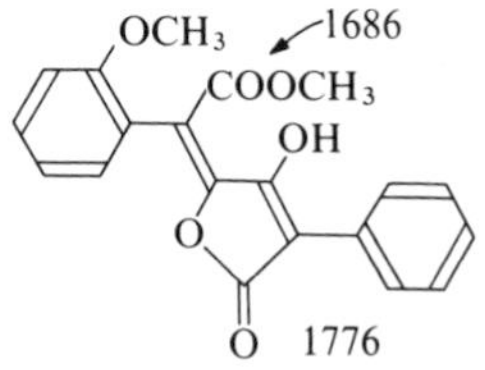

Leprapinic acid **33** (KBr)

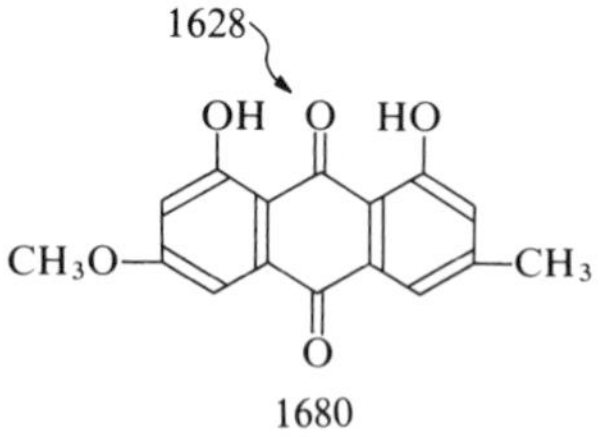

Parietin **34**

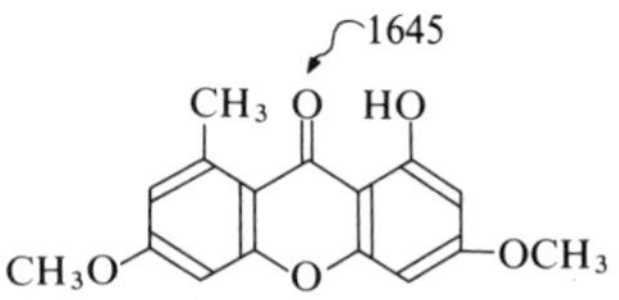

Lichexanthone **11**

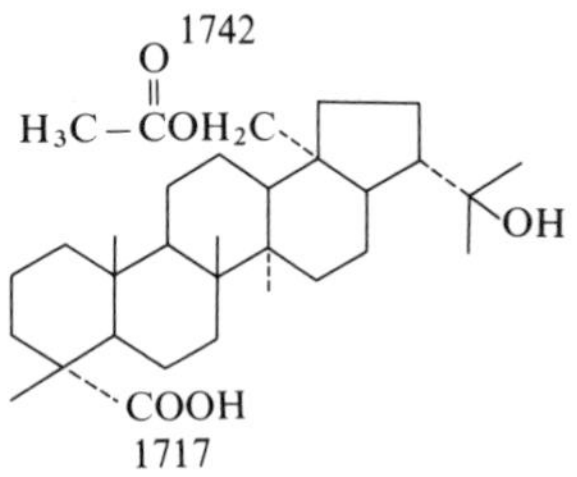

Phlebic acid A **35** (KBr)

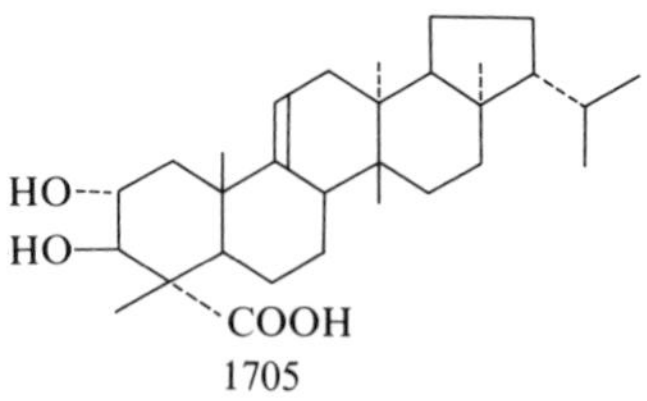

Retigeric acid A **25** (KBr)

Scheme 3.

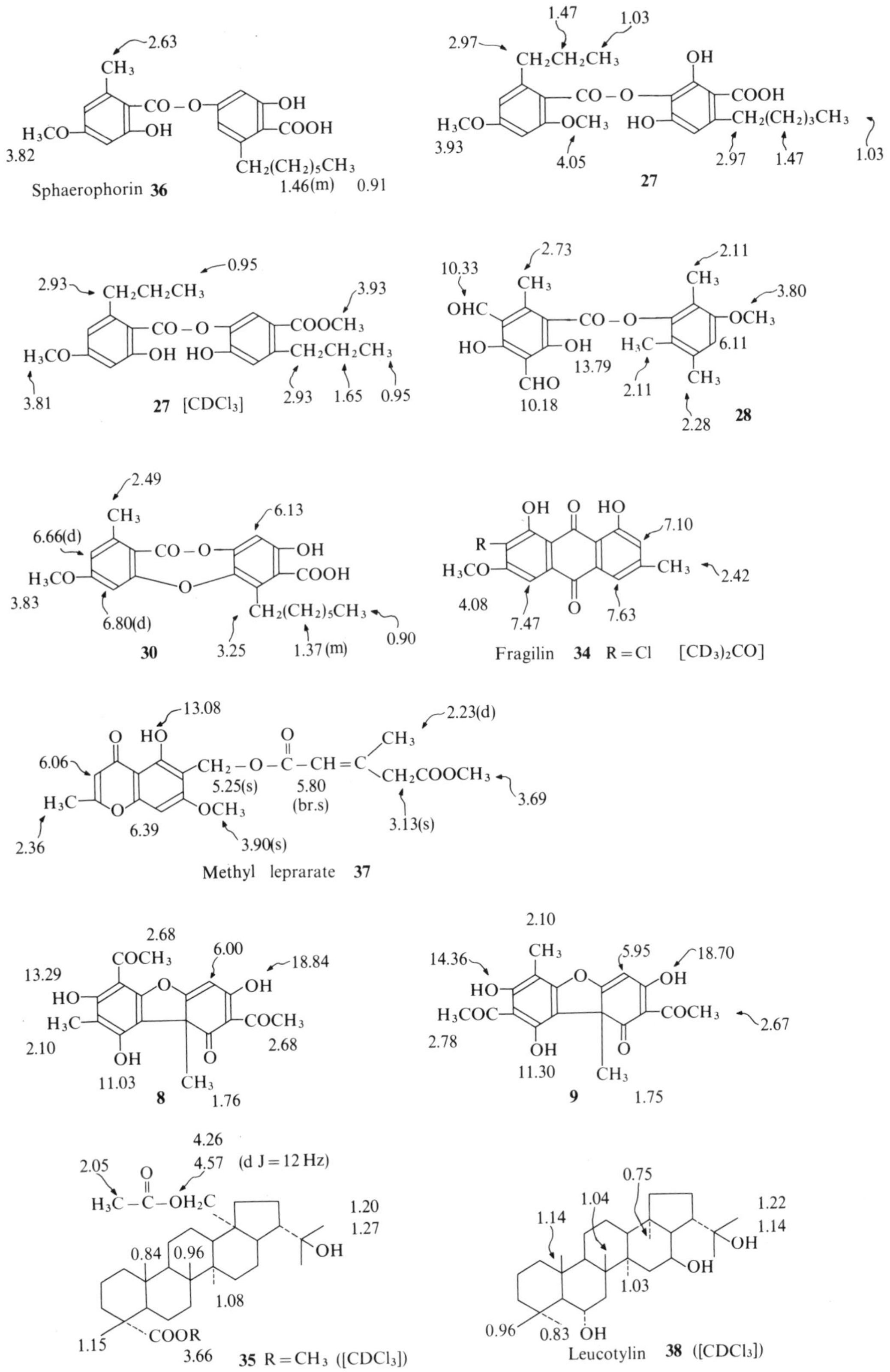

2.63
CH_3
CO – O
OH
COOH
H_3CO
OH
3.82
$CH_2(CH_2)_5CH_3$
1.46(m) 0.91
Sphaerophorin **36**

1.47 1.03
2.97
$CH_2CH_2CH_3$
OH
CO – O
COOH
H_3CO
OCH_3
HO
$CH_2(CH_2)_3CH_3$
3.93
4.05
2.97 1.47 1.03
27

0.95
2.93
$CH_2CH_2CH_3$
3.93
CO – O
$COOCH_3$
H_3CO
OH HO
$CH_2CH_2CH_3$
3.81
27 [$CDCl_3$]
2.93 1.65 0.95

2.73
2.11
10.33
CH_3
CH_3
3.80
OHC
CO – O
OCH_3
HO
OH
H_3C
6.11
13.79
CHO
2.11
CH_3
10.18
2.28
28

2.49
CH_3
6.13
6.66(d)
CO – O
OH
COOH
H_3CO
O
3.83
6.80(d)
$CH_2(CH_2)_5CH_3$
0.90
30
3.25
1.37(m)

OH O HO
7.10
R
H_3CO
CH_3
2.42
4.08
O
7.47
7.63
Fragilin **34** R = Cl [$(CD_3)_2CO$]

13.08
O HO
O
CH_3
2.23(d)
6.06
$CH_2 – O – C – CH = C$
CH_2COOCH_3
5.25(s) 5.80 (br.s)
3.69
H_3C
O
OCH_3
6.39
3.13(s)
2.36
3.90(s)
Methyl leprarate **37**

2.68
$COCH_3$
6.00
18.84
13.29
HO
O
OH
H_3C
$COCH_3$
2.10
OH
O
2.68
11.03
CH_3
8
1.76

2.10
CH_3
5.95
18.70
14.36
HO
O
OH
H_3COC
$COCH_3$
2.67
2.78
OH
O
11.30
CH_3
9
1.75

4.26
4.57 (d J = 12 Hz)
2.05
O
$H_3C – C – OH_2C$
1.20
1.27
OH
0.84 0.96
1.08
1.15
COOR
3.66
35 R = CH_3 ([$CDCl_3$])

0.75
1.04
1.22
1.14
1.14
OH
OH
1.03
0.96
0.83 OH
Leucotylin **38** ([$CDCl_3$])

Scheme 4.

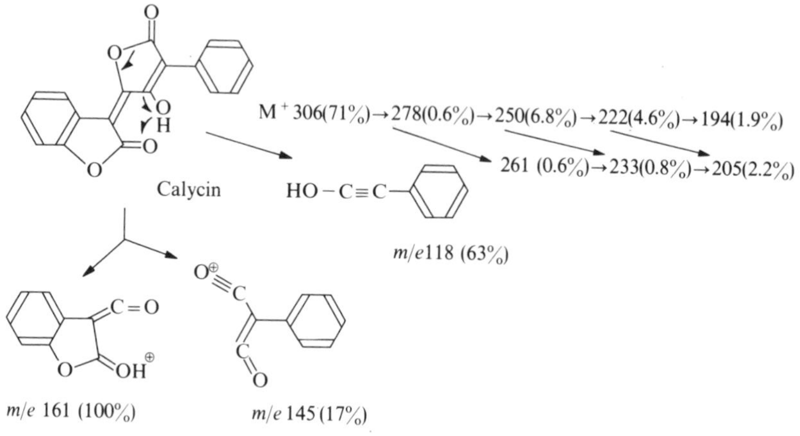

Calycin

M^+ 306(71%) → 278(0.6%) → 250(6.8%) → 222(4.6%) → 194(1.9%)

261 (0.6%) → 233(0.8%) → 205(2.2%)

HO−C≡C−Ph

m/e 118 (63%)

m/e 161 (100%)

m/e 145 (17%)

Scheme 5.

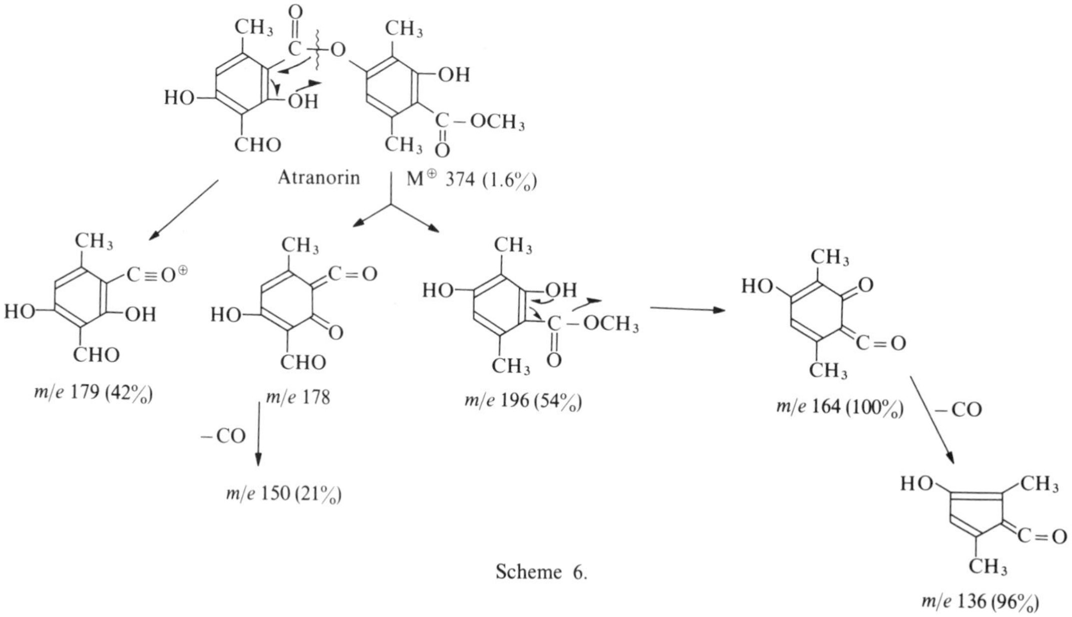

Atranorin

$M^{\oplus}$ 374 (1.6%)

m/e 179 (42%)

m/e 178

−CO

m/e 150 (21%)

m/e 196 (54%)

m/e 164 (100%)

−CO

m/e 136 (96%)

Scheme 6.

metry is now one of the most powerful tools for determination of structures of natural products. The nature, disposition and stereochemical correlation of the functional groups and nonfunctional protons as well can be determined by this method. The classical Zeisel's methoxyl determination method has now been replaced by the calculation of protons of methoxyl signals in the NMR spectra. The number of phenolic, enolic and alcoholic hydroxyls can be determined by the NMR signals of those functional groups or by the acetoxyl signals of the acetylated compound. The analysis of spin-spin coupling pattern and the coupling constant gives reliable evidence for the location and stereochemical disposition of protons in the molecule. The figures given in Scheme 4 indicate the chemical shift (δ ppm) observed in the NMR spectra.

6.4 Mass Spectrometry

The mass spectrometry of lichen substances has extensively been studied[35] resulting in the fragmentation patterns of depsides (Scheme 5), depsidones (Scheme 6), depsone (Scheme 7), dibenzofuranes and diphenylbutadienes (pulvinic acid derivatives, cf. Scheme 7). Shibata *et al.*[16] empolyed mass spectral analysis to investigate the constituents of the lichen. *Acroscyphus sphaeropholoides*, collected in the Himalayan

[35] *S. Huneck, C. Djerassi, D. Becher, M. Barber, M. von Ardenne, K. Steinfelder, R. Tümmler*, Tetrahedron *24*, 2707 (1968).

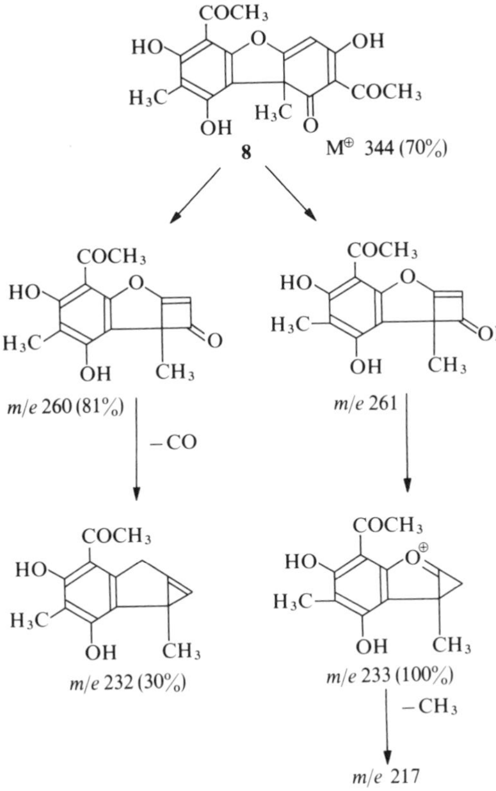

Scheme 7.

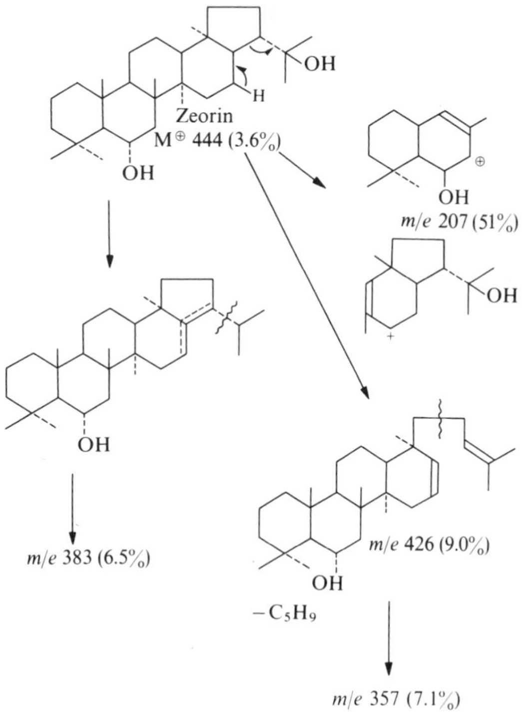

Scheme 8.

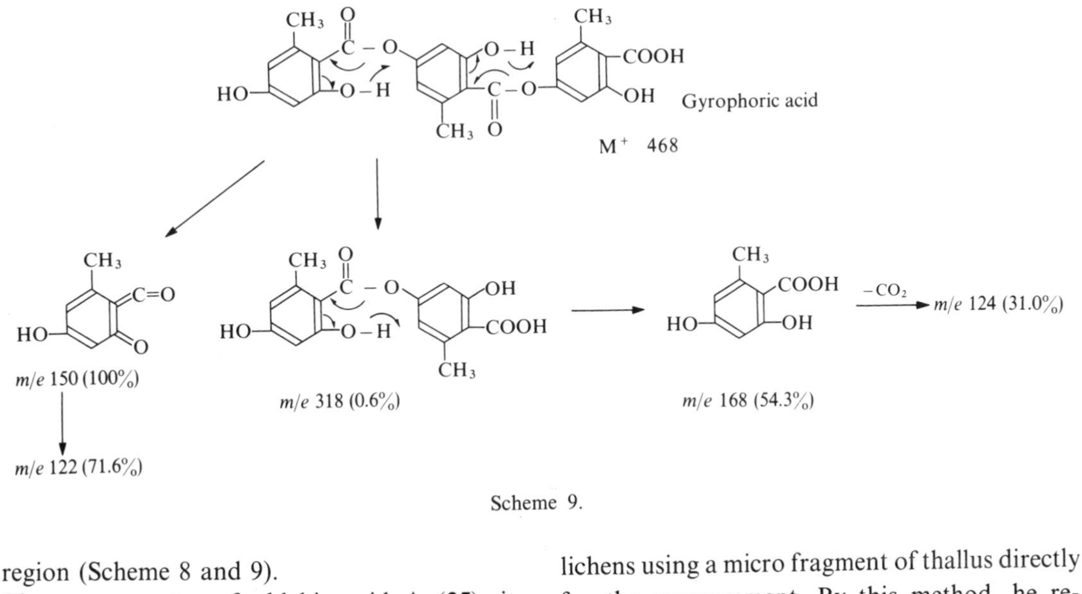

Scheme 9.

region (Scheme 8 and 9).

The mass spectra of phlebic acid A (**35**), its methyl ester and the $LiAlH_4$ reduction product of the methyl ester were compared to show the presence of a carboxyl group on the A-ring[24] (Scheme 10).

Lichen Mass Spectrometry (LMS): J. Santesson[36] carried out mass spectrometry of lichens using a micro fragment of thallus directly for the measurement. By this method, he reported[37] an extensive survey of the anthraquinones in 230 species of *Caloplaca* from which emodin, parietin, fallacinal, fallacinol,

[36] *J. Santesson*, Ark. för Kemi *30*, 363 (1969).

[37] *J. Santesson*, Phytochemistry *9*, 2149 (1970).

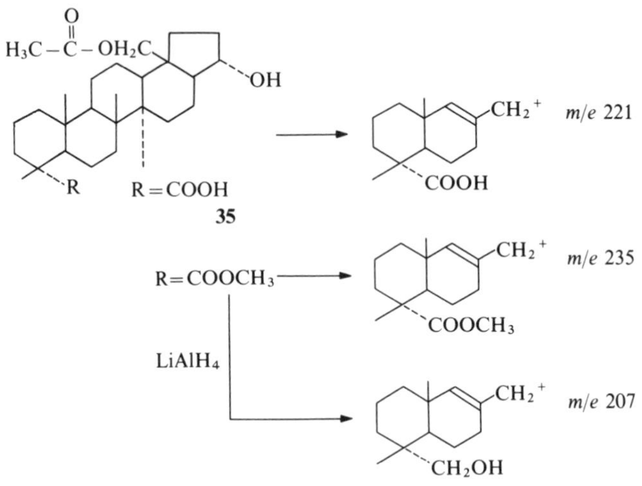

Scheme 10.

parietinic acid, xanthorin, 2-chloroemodin, fragilin and 1-*O*-methylfragilin have been identified to classify the lichens of *Caloplaca* genus into thirteen chemical groups.

6.5 X-Ray Crystallography

X-ray crystallography is now available for the investigation of the complex structures of natural products and some lichen substances have been studied successfully by this method. For the X-ray diffraction experiment, it is usually necessary to prepare a suitable single crystal of a derivative of the natural compound containing a heavy atom. Depending on the functional group of the compound, direct bromination, bromoacetylation, *p*-bromobenzoylation and *p*-bromophenacylation are employed for that purpose. Using the X-ray diffraction method, the structures of nephroarctin (**28**)[38], phlebic acid A (**35**, R = H)[24b], zeorin (**23**, R = H)[39], leucotylin (**23**, R = OH)[40], (+)rugulosin (**15**)[41] and eumitrin C (**19**)[19] have been established. The establishment of the absolute configuration or stereochemical structure of a certain compound by X-ray crystallography can help solve the structures of other related natural products which may be correlated with the original one by chemical or physical methods.

7 Synthesis of Lichen Substances

Synthesis of lichen depsides:[42] A phenolic component (S) is ethoxycarbonylated and converted into the acid chloride which is condensed with the other phenolic component (A) having an aldehyde group (Scheme 11).

After condensation, the free hydroxyls in the A-part are fully ethoxycarbonylated and then the aldehyde group is oxidized to carboxyl. The ethoxycarbonyl is finally removed by the action of ammonia.

Synthesis of lichen depsides by direct condensation[43,44]: Using *N,N'*-dicyclohexyl-carbodiimide (DCC) or trifluoroacetic anhydride (TFAA), phenylcarboxylic acid (S) is condensed with the phenolic component (A) (Scheme 12).

Depsidones are regarded to be biosynthesized by the intramolecular oxidative coupling of the corresponding depsides. This has been demonstrated chemically by the intramolecular oxidative coupling of a synthetic depside using active manganese dioxide in boiling chloroform to yield nor-diploicin (**39** in Scheme 13).[43]

A biogenetical chemical synthesis of (±) usnic acid (**8**) was achieved by the oxidative coupling of 2 moles of methylphloroacetophenone using

[38] *M. Nuno, Y. Kuwada, K. Kamiya*, Chem. Commun. 1101 (1968).

[39] *I. Yosioka, T. Nakanishi, H. Yamauchi, O. Kitagawa, T. R. Fujiwara, K. Tomita*, XIVth Symp. Chem. Nat. Prod., Symp. papers p. 335 (Fukuoka, Oct. 1970).

[40] *T. Nakanishi, T. Fujiwara, K. Tomita*, Tetrahedron Lett. 1491 (1968).

[41] *N. Kobayashi, Y. Iitaka, U. Sankawa, Y. Ogihara, S. Shibata*, Tetrahedron Lett. 6133 (1968); *N. Kobayashi, Y. Iitaka, S. Shibata*, Acta Crystallogr. *B 26*, 188 (1970).

[42] *Y. Asahina, A. Hashimoto*, Yakugaku-zasshi (J. Pharm. Soc. Japan) *58*, 221 (1938); *Y. Asahina, I. Yosioka*, Ber. *70*, 1823 (1937).

[43] *C. J. Brown, D. E. Clark, W. D. Ollis, P. L. Veal*, Proc. Chem. Soc. 393 (1960).

[44] *S. Neelakanan, R. Padmasani, T. R. Seshadri*, Tetrahedron *21*, 3531 (1965).

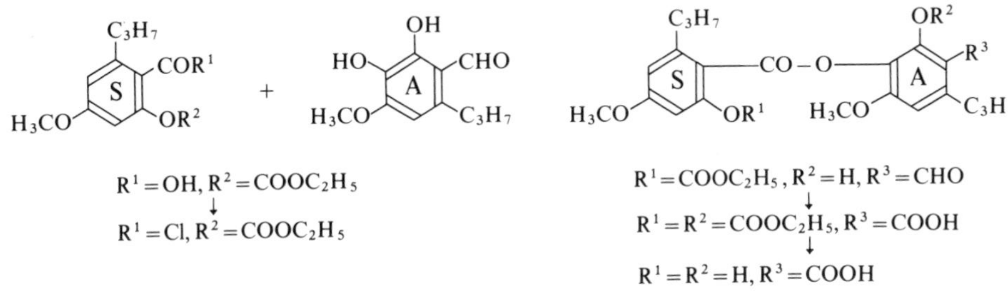

$R^1 = OH, R^2 = COOC_2H_5$ → $R^1 = Cl, R^2 = COOC_2H_5$

$R^1 = COOC_2H_5, R^2 = H, R^3 = CHO$ → $R^1 = R^2 = COOC_2H_5, R^3 = COOH$ → $R^1 = R^2 = H, R^3 = COOH$

Scheme 11.

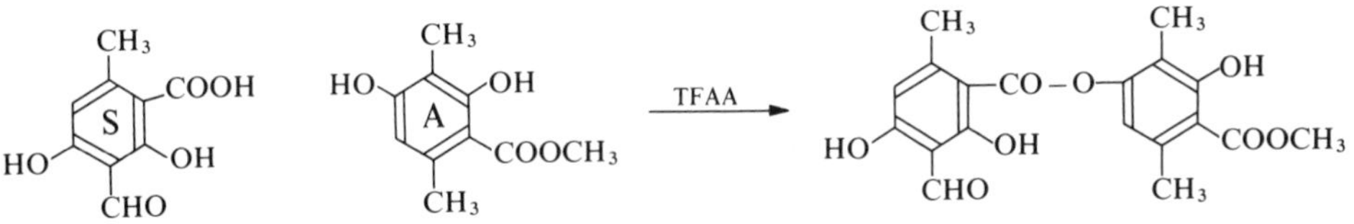

Scheme 12.

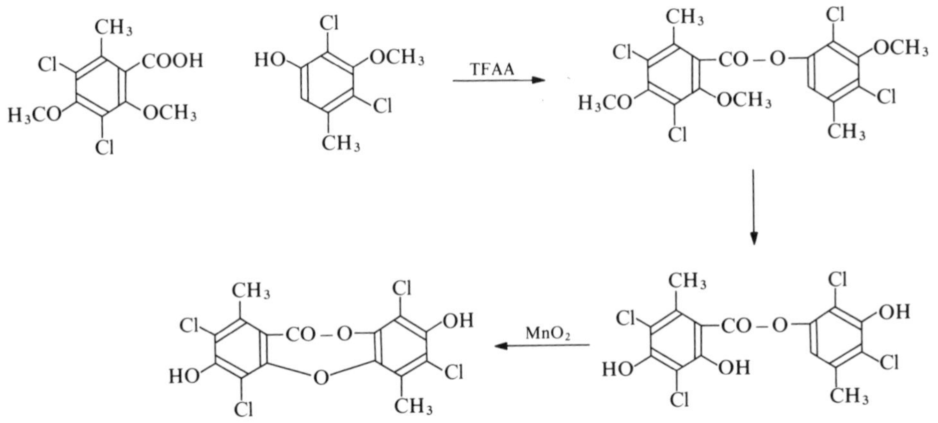

39

Scheme 13.

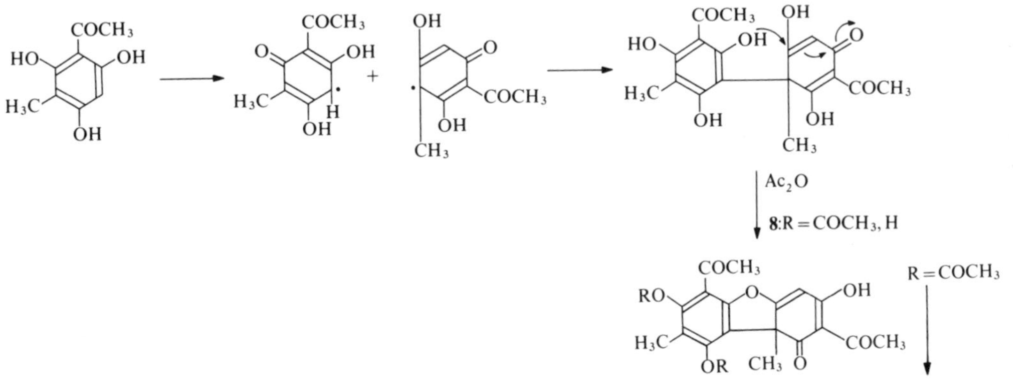

Ac_2O

8: $R = COCH_3, H$

$R = COCH_3$ → $R = H$

Scheme 14.

potassium ferricyanide[45], or peroxidase and hydrogen peroxide[46,47] as the reagents (Scheme 14). A purely chemical synthesis of a lichen dibenzofurane, strepsilin dimethyl ether[48], has been reported to confirm the structure **40**.

[45] *D. H. R. Barton, A. M. Deflorin, O. E. Edwards*, J. Chem. Soc. 530 (1956).

[46] *A. Penttila, H. M. Fales*, Chem. Commun. 656 (1966).

[47] *H. Taguchi, U. Sankawa, S. Shibata*, Chem. Pharm. Bull. (Tokyo) *17*, 2054, 2061 (1969).

[48] *J. D. Brewer, J. A. Elix*, Tetrahedron Lett. 4139 (1969).

40

8 Separation and Cultivation of Symbionts of Lichens

The phycobionts which are commonly found in lichens are *Nostoc* (blue-green alga) or *Trebouxia* (green alga), sometimes *Trentepohlia* and *Coccomyxa* (green algae). Although these lichen phycobionts are essentially the same as free-living algae, the lichenized fungi (mycobionts) are mostly different from free-living fugi. For the physiological investigations of lichens, isolation and cultivation of lichen symbionts have been studied by some lichenologists[49,50]. But still little is known about the biochemical metabolism of isolated symbionts.

The phycobiont can be separated from lichen by the micromanipulator method or by the fragmentation method. The mycobiont is isolated mostly by the micromanipulator method or the test tube method. The spore isolation by the test tube method, which is generally used for the separation of mycobionts with less danger of contamination, is carried out in the following way: a fruiting body (apothecium) of lichen is supported on a small agar globe and placed under a sterilized agar slant in a test tube. Spores ejected from the apothecium may be caught on the surface of agar slant, part of which is cut and transferred to an other organic medium slant in a test tube for cultivation. The growth rates of mycobionts are much slower than those of free-living fungi and the optimal temperature of cultivation is lower, ca. 20°.

Using a ^{14}C-tracer technique, Smith *et al.*[51] studied the carbohydrate metabolism of the isolated lichen symbionts to find that *Nostoc* produces D-glucose which is converted into D-mannitol in the mycobiont, while *Trebouxia* releases D-ribitol which is transformed into D-mannitol and D-arabitol in the mycobiont. Komiya and Shibata[52] proved that D-ribitol produced by the cultivated phycobiont of *Ramalina crassa* and *R. subbreviuscula* is converted into D-mannitol and D-arabitol in the isolated mycobiont of the same lichens. The identification of polyols was performed successfully by gas-liquid chromatography after trifluoroacetylation of the metabolites. A pulvinic acid derivative[53] and roccellic acid, eugenitol, eugenitin and rupicolon[54] were obtained from the isolated and cultivated mycobionts of lichens. Komiya and Shibata[55] obtained crystalline (+)usnic acid (**8**) from the cultured mycobionts of *Ramalina crassa* and *R. yasudae*. Although depsides and depsidones have not been isolated as crystals from the separated mycobionts, it is almost certain that almost all the lichen substances, including the water soluble

[49] cf. *V. Ahmadjian*, The Lichen Symbiosis, Blaisdell Publ. Co., Waltham, Toroto, London 1967.

[50] *E. A. Thomas*, Beitr. Kryptogammenflora Schweiz Vol. 9, p. 1–208 (1939).

[51] *D. H. Lewis, D. C. Smith*, New Physiologist *66*, 143 (1967); *D. H. S. Richardson, D. C. Smith*, New Physiologist *67*, 469; *69*, 69 (1968); Nature *214*, 879 (1967).

[52] *T. Komiya, S. Shibata*, Phytochemistry *10*, 695 (1971).

[53] *K. Mosbach*, Acta Chem. Scand. *21*, 2331 (1967).

[54] *C. H. Fox, S. Huneck*, Phytochemistry *8*, 1301 (1969).

[55] *T. Komiya, S. Shibata*, Chem. Pharm. Bull. (Tokyo) *17*, 1305 (1969).

polysaccharide components, are the metabolites of fungal symbionts of lichens.

9 Biological Activities of Lichen Substances

The *in vitro* antimicrobial activities of some lichen substances have been reported[56]. In particular, usnic acid[57-59] and didymic acid show pronounced activities against Gram positive bacteria and tubercular bacilli *in vitro*[57]. A direct antitumor activity was observed in polyporic acid of *Sticta coronata*[60], while an indirect, host-mediated antitumor activity against implanted sarcoma 180 in mice was demonstrated with the water-soluble polysaccharides of lichens, lichenan, pustulan and some other glycans[61].

(+)Usnic acid, evernic acid, atranorin and perlatoric acid were shown to be immunologically active in allergic contact dermatitis caused by exposure to lichens[62].

[56] cf. *S. Shibata*, Especial Compounds of Lichens, Handb. d. Pflanzenphysiologie, Bd. X, p. 612, Springer Verlag, Berlin 1958.

[57] *S. Shibata, Y. Miura*, J. Japan. Med. *1*, 518 (1948); *2*, 22 (1949).

[58] Lääke Oy Pharm. Manufact. Turku, Finland.

[59] *V. P. Savicz, M. A. Litvinov, E. N. Moiseeva*, Planta Med. *8*, 191 (1960).

[60] *J. F. Burton, B. F. Cain*, Nature *184*, 1326 (1959); *B. F. Cain*, J. Chem. Soc. 936 (1961).

[61] *S. Shibata, Y. Nishikawa, M. Tanaka, F. Fukuoka, M. Nakanishi*, Z. Krebsforsch. *71*, 102 (1968); *S. Shibata, Y. Nishikawa, Cheng Fu Mei, F. Fukuoka, M. Nakanishi*, Gann *59*, 156 (1968); *Y. Nishikawa, T. Takeda, S. Shibata, F. Fukuoka*, Chem. Pharm. Bull. (Tokyo) *17*, 1910 (1969); *T. Takeda, Y. Nishikawa, S. Shibata*, Chem. Pharm. Bull. (Tokyo) *18*, 1074 (1970); *Y. Nishikawa, M. Tanaka, S. Shibata, F. Fukuoka*, Chem. Pharm. Bull. (Tokyo) *18*, 1431 (1970).

[62] *J. C. Mitchell, S. Shibata*, J. Invest. Dermatology *52*, 517 (1969).

10 Bibliography

[63] *W. Zopf*, Die Flechtenstoffe, Gustav Fischer, Jena 1907.

[64] *Y. Asahina, S. Shibata*, Chemistry of Lichen Substances, Japan Soc. Promotion of Science, Tokyo 1954.

[65] *S. Shibata*, Especial Compounds of Lichens, Handb. d. Pflanzenphysiologie Bd. X. p. 560–623, Springer Verlag, Berlin 1958.

[66] *S. Shibata*, Lichen Substances, Moderne Methoden d. Pflanzenanalyse, Bd. VI. p. 155–193, Springer Verlag, Berlin 1963.

[67] *S. Shibata*, Biogenetical and Chemotaxonomical Aspects of Lichen Substances, Beiträge zur Biochemie u. Physiologie von Naturstoffen, Festschrift Kurt Mothes zum 65. Geburtstag, p. 451–465, Gustav Fischer, Jena 1965.

[68] *S. Neelakantan*, Recent Developments in the Chemistry of Lichen Substances, Advancing Frontiers in the Chemistry of Natural Products, p. 35–84, Hindstan Publ. Corp. Delhi 1965.

[64] *M. E. Hale, Jr.*, The Biology of Lichens, Edward Arnold Ltd. London 1967.

[70] *S. Huneck*, Lichen Substance, Progress in Phytochemistry, Vol. I, p. 224–346, Interscience Publ., London, New York, Sydney 1968.

[71] *C. F. Culberson*, Chemical and Botanical Guide to Lichen Products, The University North Carolina Press, Chapel Hill 1969; *C. F. Culberson*, Supplements for Chemical and Botanical Guide to Lichen Products, The University of North Carolina Press, Chapel Hill 1971.

6.2 Biosynthesis of Isoprenoids

Shigeo Nozoe and Akihiko Kawaguchi
Faculty of Pharmaceutical Sciences, Tohoku University, Aobayama, Sendai, 980 Japan

1 Introduction

The term "isoprenoids" is used as a general designation for a group of substances which are essentially biosynthesized from a common precursor, mevalonic acid. Isoprenoids are widely distributed in a great variety of organisms and some of these have important functions within living organisms while others exibit interesting physiological activity. There are a number of such isoprenoid compounds which act as male and female sex hormones, plant hormones and sex hormones of microorganisms. It has been revealed that long-chain isoprenyl alcohols play an important role in the biosynthesis of some of the polymers of bacterial cell-walls[1,2].

Studies on the biosynthesis of these compounds are becoming numerous and the mechanisms of such biosynthetic reactions are gradually being elucidated to a certain extent by experiments using intact biological systems as well as cell-free

[1] *Y. Higashi, J. L. Strominger, C. C. Sweeley*, Proc. Nat. Acad. Sci. U.S. *57*, 1878 (1967).

[2] *A. Wright, M. Dankert, P. Fennessey, P. W. Robbins*, Proc. Nat. Acad. Sci. U.S. *57*, 1798 (1967).

systems. As many enzymic reactions take part in the biosynthesis of isoprenoids, the studies on the mechanism of each individual step of the biosynthesis have been carried out by using both crude and purified enzyme systems.

2 Methodology of Biosynthetic Research

2.1 Biological Systems

Various biological systems are available for studies on the biosynthesis of natural products: the biological system in the intact state, callus from tissue culture or organ culture, the cell-free system obtained by disruption of cell-walls or membranes and the purified enzyme obtained by fractionation of the cell-free preparation. When the intact living organism is used for the experiments, a labeled substrate, considered to be a precursor, is added to the diet of animals or added to the culture medium of microorganisms or applied to the roots, stems or leaves of plants. Then the products are isolated from the organisms and conversion rates are measured. Most knowledge on the biosynthesis of natural products has been obtained by these techniques. However, there are many difficulties and limitations in actual practice. To elucidate the detailed mechanisms of such reactions, intermediates products must be isolated; the pools of intermediates are, however usually small in the intact state. In addition, some labeled intermediates are not used by living cells owing to the lack of transport across cell-membranes or cell-walls. To solve these difficulties, cell-free extracts obtained by disruption of cells are used for some investigations. The advantage of using a cell-free enzyme system is that it makes possible the use of a substrate which can not permeate cell membranes and that some intermediates can accumulate with the aid of inhibitors or in the absence of coenzymes which are required for a specific enzyme. The cell-free preparation can further be fractionated by centrifugation, precipitation with ammonium sulfate, dialysis or gel filtration.

Some examples of biosynthetic studies utilizing various enzymes (in the purified or crude state) such as prenyl transferase (farnesyl pyrophosphate synthetase), squalene synthetase, 2,3-oxidosqualene-sterol cyclase and the cyclizing enzymes of various terpenoids will be described in Section **3**.

2.2 Preparation of the Isotopically Labeled Substance

Since tracer experiments represent one of the most convenient techniques for studies on the biosynthesis of isoprenoids, the preparation of labeled substrates is an important problem. Labeled substrates are prepared either by a purely chemical synthesis or by use of a biological system or a combination of these. Many of the fundamental radioactive substrates, such as ^{14}C-labeled mevalonic acid, position- and stereospecifically labeled ^{3}H-mevalonic acid and ^{14}C-isopentenyl pyrophosphate, are now commercially available. These labeled substrates are used either individually or as doubly- or triply-labeled substrates. Chemical syntheses of acetic acid with chiral methyl group, R-[$^{2}H_1$, $^{3}H_1$] acetate and S-[$^{2}H_1$, $^{3}H_1$]acetate, were reported and the method to determine the chirality of the acetic acid was established by J. W. Cornforth *et al*[91]. Mevalonate, chiral at C-6 methyl, *i.e.*, [(6R)-6-$^{2}H_1$, $^{3}H_1$]MVA was also synthesized and was used in the experiment on the sterol biosynthesis[92].

Cell-free preparations from mammalian liver are frequently used for the preparation of radioactive substrates, such as farnesyl pyrophosphate, squalene and so on[3]. Among the chemical methods of preparation, $^{3}H_2$-gas (Wilzbach's method of hydrogenation etc.), lithium aluminum tritide, and sodium borotritide are used and ^{14}C-labeled Grignard reagents or a Wittig reagent (methylenetriphenylphosphorane etc.) are often used.

The use of ^{13}C-labeled precursor for the elucidation of biosynthetic pathways has become a well established techniques. A number of reports, including some reviews dealing with this topic have recently appeared[93]. The commercially available [1-^{13}C], [2-^{13}C], and [1,2-$^{13}C_2$]-acetate are the most frequently used substrate for the biosynthetic experiment of isoprenoid. More complex precursors, such as [2-^{13}C], [4-^{13}C], [3,4-$^{13}C_2$] and [4,5-$^{13}C_2$]-mevalonate have been synthesized chemically and utilized in the biosynthetic studies of sesquiterpenes.

---

[3] *R. B. Clayton*, Methods in Enzymology, *S. P. Colowick, N. O. Kaplan* (Eds.), Vol. 15, p. 305–309, Academic Press, New York 1969.

3 Examples of Studies on Biosynthesis

In studying the biosynthesis of isoprenoids, the reactions leading to the final natural product can be classified into three stages. The first process is the formation of a polyisoprene chain, in the form of a pyrophosphate ester of polyprenyl alcohol, or a hydrocarbon chain (squalene and phytoene) from mevalonic acid; the second is the formation of a basic carbon skeleton from this polyisoprene chain by cyclization; the third is the final formation of various natural products by a secondary modification via oxidation, reduction or alkylation of the substance formed in the second process. The first process seems to be common to many organisms wherever isoprenoids are produced, while the second and the third processes are usually specific to each species. Cyclizations of the isoprene chain are usually initiated by solvolytic elimination of the pyrophosphoryl group or by protonation of the double bond or epoxide in the isoprene chain. This cyclization reaction is terminated by deprotonation or introduction of a hydroxyl ion after a variety of processes involving concerted ring closure, hydride- or alkyl- group shifts.

3.1 Biosynthesis of Isoprene Chains

Mevalonic acid (**1**) and isopentenyl pyrophosphate (**2**) have been identified as the common biosynthetic precursors of isoprenoids, respectively in 1956[4] and 1958[5]. Later, the route(s) from mevalonic acid or isopentenyl pyrophosphate to polyprenyl pyrophosphate has been studied in detail by experiments using cell-free systems or partially purified enzymes and the reaction mechanisms including the stereochemistry in each step have been elucidated to a considerable extent[6]. Formation of farnesyl pyrophosphate (**5**) from isopentenyl pyrophosphate has been demonstrated by experiments using an enzyme prepared from liver tissue[7], pumpkin fruit[8], yeast[9] etc. These enzyme preparations are reported to have the activities for condensation of isopentenyl pyrophosphate (**2**) with dimethylallyl pyrophosphate (**3**) and with geranyl pyrophosphate (**4**) to give farnesyl pyrophosphate (**5**). This enzyme has been obtained from avian liver in a stable crystalline form[94]. The protein is a dimer of molecular weight of *ca.* 86000. According to the study on substrate specificity for this enzyme[10–12], it has been found

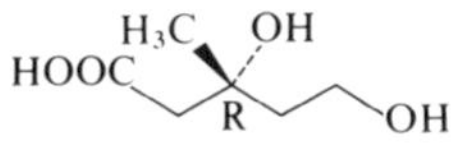

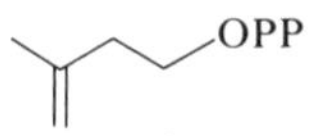

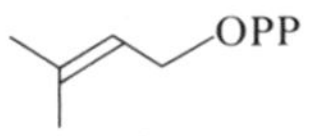

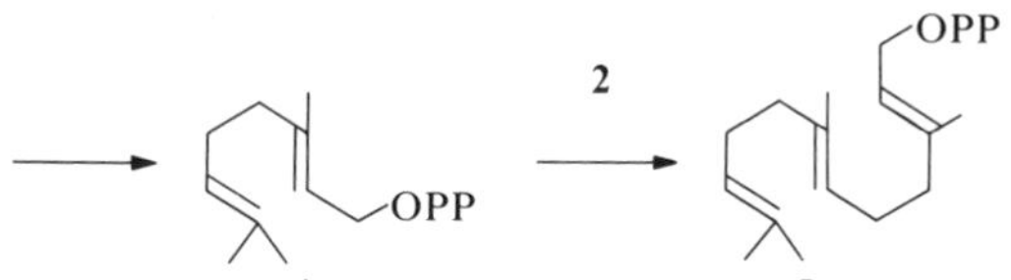

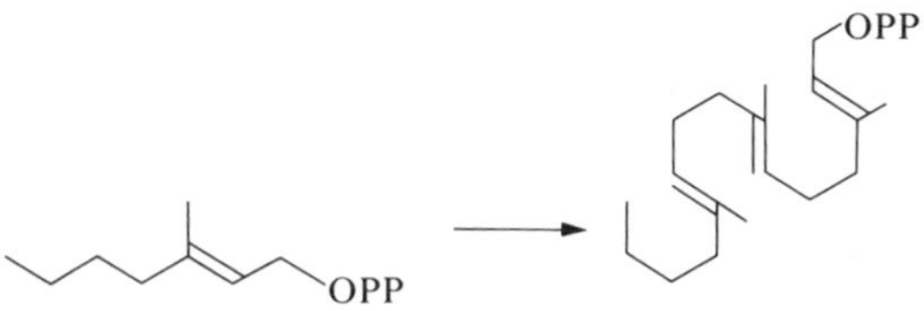

[4] *P. A. Tavormina, M. H. Gibbs, J. W. Huff*, J. Am. Chem. Soc. *78*, 6210 (1956).

[5] *S. Chaykin, J. A. Law, A. H. Phillips, T. T. Tchen, K. Bloch*, Proc. Nat. Acad. Sci. U.S. *44*, 738 (1958).

[6] *J. W. Cornforth, et al.*, Proc. Roy. Soc. *163*, 436 (1966); J. Biol. Chem. *241*, 3970 (1966).

[7] *P. W. Holloway, G. Popjak*, Biochem. J. *104*, 57 (1967).

[8] *K. Ogura, T. Nishino, S. Seto*, J. Biochem. (Tokyo) *64*, 197 (1968).

[9] *F. Lynen, B. W. Arganoff, H. Eggerer, U. Henning, E. M. Möslein*, Angew. Chem. *71*, 657 (1959).

[10] *G. Popjak, J. L. Rabinowitz, J. M. Baron*, Biochem. J. *113*, 861 (1969).

[11] *K. Ogura, T. Nishino, T. Koyama, S. Seto*, J. Am. Chem. Soc. *92*, 6036 (1970).

[12] *T. Nishino, K. Ogura, S. Seto*, J. Am. Chem. Soc. *93*, 794 (1971).

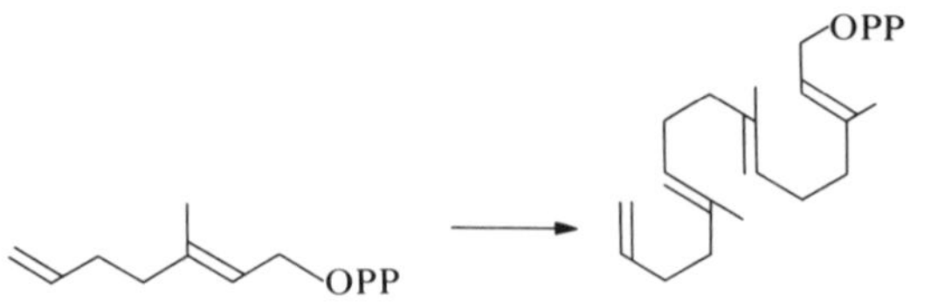

OPP

8

OPP

9

2

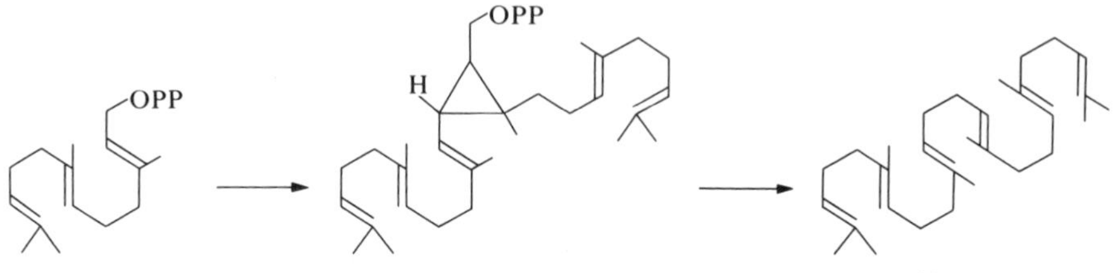

OPP

5

OPP H

10

11

OPP

R_1 R_2

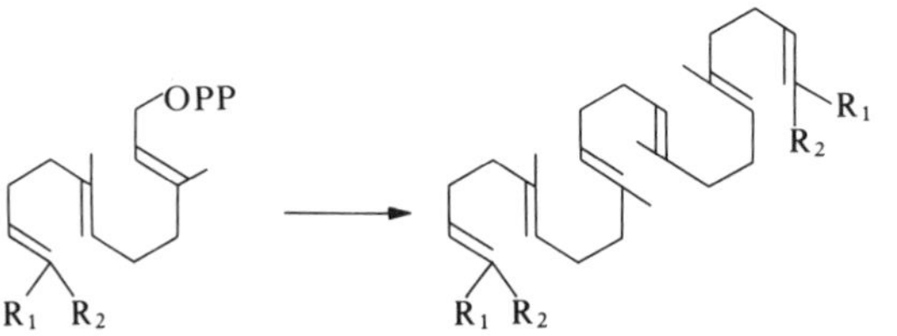

12: $R_1 = Et$, $R_2 = Me$
13: $R_1 = Me$, $R_2 = Et$

R_1 R_2

14: $R_1 = Et$, $R_2 = Me$
15: $R_1 = Me$, $R_2 = Me$

that the farnesyl pyrophosphate synthetase has a similar activity with artificial substrates, such as compound **6** and compound **8** to afford trishomofarnesyl pyrophosphate (**7**) and 16,16′-bisnorgeranylgeranyl pyrophosphate (**9**) respectively.

The enzymes that catalyze the formation of geranylgeranyl pyrophosphate and heptaprenyl pyrophosphate (C_{35}-OPP) were extracted from plant seeds[13] and *Micrococcus lysodeikticus*[14,15]. All the enzymes which catalyze the formation of polyprenyl pyrophosphate were found to be present in the supernatant fraction from the high speed centrifugation (100,000 × *g*) of the cell-free extracts.

Another important precursor of the isoprene chain is the hydrocarbon, squalene (**11**), which is a precursor for sterols and triterpenes. Squalene is known to be formed by the reductive condensation of two molecules of farnesyl pyrophosphate (**5**)[16]. Treatment of the microsomal fraction of yeast cells with farnesyl pyrophosphate in the absence of a reducing agent, NADPH, was found to produce an intermediate product, presqualene alcohol pyrophosphate (**10**)[17]. This is found to be converted into squalene by the treatment of the microsomal fraction in the presence of NADPH and Mg^{++}. The structure of presqualene alcohol has been elucidated by chemical and physicochemical means and further confirmed by chemical synthesis[18-20]. According to experiments on the substrate specificity of squalene synthetase, homologs of squalene, such as the compounds **14** and **15**, are formed respectively from corresponding homologs of farnesyl pyrophosphate (**12** and **13**) by the action of pig liver microsomal enzyme[21]. Recent investigations dealing with a

[13] *M. O. Oster, C. A. West*, Arch. Biochem. Biophys. *127*, 112 (1968).

[14] *A. A. Kandutsch, H. Paulus, E. Levin, K. Bloch*, J. Biol. Chem. *239*, 2507 (1964).

[15] *C. M. Allen, W. Alworth, A. Macrae, K. Bloch*, J. Biol. Chem. *242*, 1895 (1967).

[16] *G. Popjak, J. W. Cornforth*, Biochem. J. *101*, 553 (1966).

[17] *H. C. Rilling*, J. Biol. Chem. *241*, 3233 (1966).

[18] *H. C. Rilling, W. W. Epstein*, J. Am. Chem. Soc. *91*, 1041 (1969).

[19] *W. W. Epstein, H. C. Rilling*, J. Biol. Chem. *245*, 4597 (1970).

[20] *L. J. Altmann, R. C. Kowerski, H. C. Rilling*, J. Am. Chem. Soc. *93*, 1783 (1971).

mechanisms of squalene biosynthesis using various substrate analogues provided more detailed informations as to this enzymic conversion[95]. In the synthesis of carotenoids, the condensation reaction of two geranylgeranyl pyrophosphates to phytoene takes place, but the detailed reaction mechanism remains to be solved.

Rubber differs from the isoprene chain described above in the respect that the geometrical configuration about the double bond is all *cis*. Rubber can be synthesized *in vitro* from mevalonic acid and isopentenyl pyrophosphate by incubation with latex of *Hevea brasiliensis*[22]. Another group of isoprenoids whose double bonds are in the *cis* configuration comprises the polyprenols (isoprenoid alcohols). With a few exceptions, some of the internal isoprene residues are *trans* and the rest are *cis*[23].

3.2 Cyclization of the Isoprene Chain and Secondary Modifications

Isoprenoids can be divided into several groups: i) terpenoids formed from polyprenyl pyrophosphate, ii) steroids and triterpenes formed from squalene, iii) carotenoids from phytoene and iv) substances formed by coupling with materials of other biosynthetic origin. In this section, recent investigations on the biosynthesis of each of these types of substances will be described.

3.2.1 Terpenoids

3.2.1.1 Monoterpenes

Monoterpenes are presumed to be formed by cyclization of geranyl pyrophosphate (**4**), but experimental evidences of such a cyclization are quite few. Labeled geranyl pyrophosphate or geraniol given to *Menyanthes trifoliata* is incorporated into loganin (**16**)[24-27] (or loganic acid), which is a precursor of the non-tryptophan portion of indole alkaloids. However, there is still no experimental proof how the five-membered ring in **16** is formed from geranyl pyrophosphate.

According to the experiments on the biosynthesis of thujone (**17**)[28] and camphor (**18**)[29,30] in higher plants, the radioactivity from the administered 2-^{14}C-mevalonic acid is localized only in the C-6 position. This has been explained to be due to i) a large pool of dimethylallyl pyrophosphate in plants, ii) dimethylallyl pyrophosphate not originating from mevalonic acid, or iii) compartmentation effects in higher plants. Such a phenomenon has also been observed in the biosynthesis of the sesquiterpene, coriamyrtin[31]. On the other hand, a different result has been obtained by experiments which confirm equal distribution of the radioactivity both at the C-4 and C-8 positions of linalool (**19**) biosynthesized by *Cinnamomum camphora* from 2-^{14}C-mevalonic acid[32]. These results show an example of the complexities of experiments using the whole plant in a live condition.

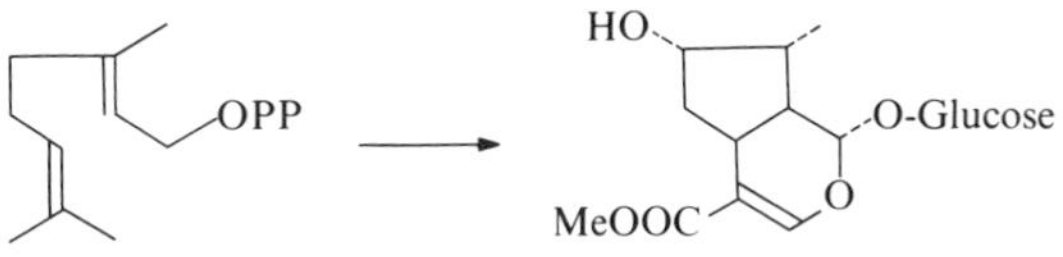

OPP → HO, O-Glucose, MeOOC, O

4 **16**

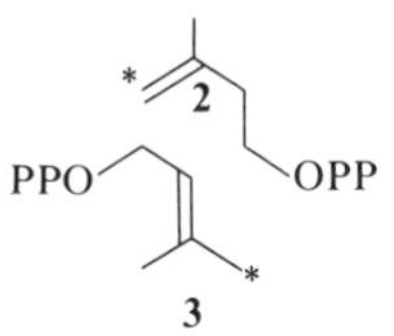

PPO, OPP, 2, *

3

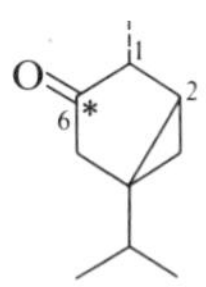

O, 1, 2, 6, *   **17**   1, O, 2, 6, *   **18**   OH, 2, 1, 4, 7, 8, *   **19**

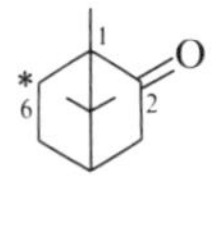

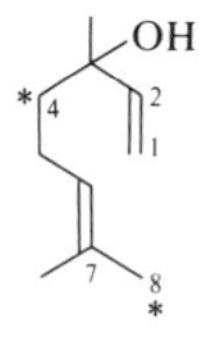

---

[21] *K. Ogura, T. Koyama, S. Seto*, J. Am. Chem. Soc. *94*, 307 (1972).

[22] *B. L. Archer, D. Barnard, E. G. Cockbain, J. W. Cornforth, R. H. Cornforth, G. Popjak*, Proc. Roy. Soc. *163*, 519 (1966).

[23] *F. W. Hemming*, Natural Substances Formed Biologically from Mevalonic Acid, *T. W. Goodwin* (Ed.), p. 105, Academic Press, New York 1970.

[24] *A. R. Battersby, R. S. Kapil, R. Southgate*, Chem. Commun. 131 (1968).

[25] *P. Loew, D. Arigoni*, Chem. Commun. 137 (1968).

[26] *A. R. Battersby, et al.*, Chem. Commun. 826, 827 (1970).

[27] *S. Escher, P. Leow, D. Arigoni*, Chem. Commun. 823 (1970).

[28] *J. P. Kutney*, Bioorg. Chem. *1*, 194 (1971) and references cited therein.

[29] *D. V. Banthorpe, K. W. Turnbull*, Chem. Commun. 177 (1966).

[30] *D. V. Banthorpe, J. Mann, K. W. Turnbull*, J. Chem. Soc. (*C*), 2689 (1970).

[31] *D. V. Banthrope, D. Baxendale*, Chem. Commun. 1553 (1968); J. Chem. Soc. (*C*), 2694 (1970).

[32] *T. Suga, T. Shishibori, M. Bukeo*, Phytochemistry *10*, 2725 (1971).

3.2.1.2 Sesquiterpenes

Sesquiterpenes are considered to be formed by cyclization of farnesyl pyrophosphate (**5**), which is initiated by the solvolytic removal of the pyrophosphate group, or in certain cases, by protonation of the double bond.

The pool-size of a first cyclic intermediate, which usually is hydrocarbon or monoalcohol derivative, is so small that it hardly be recognized. In certain cases, however, the hydrocarbon precursors of the sesquiterpene metabolites have been isolated; for instance, sativene for helminthosporal, cuprenene for helicobasidin, hirsutene[96] for coriolins, trichodiene for tichothecenes, bergamotene[97] for fumagillin or ovalicin and protoilludene[116] for illudin biosyntheses. An experimental example for the formation of a carbon skeleton is found in the enzymic formation of germacrene-C (**20**), with a 10 membered ring structure, which is considered to be the precursor of a large number of bicyclic sesquiterpenes, by using a cell-free preparation from immature seeds of *Kadzura japonica*[33]. A supernatant fraction obtained by centrifugation (100,000 × *g*) of the cell-free preparation of fungal cells produces γ-monocyclofarnesol (**21**), which is formed by protonation of the terminal double bond of farnesyl pyrophosphate[34]. Trichodiene (**25**), a precursor hydrocarbon of trichothecin (**26**)[35], was found to be formed from farnesyl pyrophosphate with a cell-free system prepared from disrupted cell of *Trichothecium roseum*[98]. Farnesyl- or geranyl-pyrophosphate added to cultures of *T. roseum* is also incorporated into trichothecin (**26**)[36,99]. A comparison of the structure of trichodiene (**25**) with that of cuprenene (**27**), a precursor hydrocarbon of helicobasidin (**28**), suggests that they are respectively formed by successive 1,2-methyl shifts and by simple deprotonation of the common cationic intermediate **23**. Although the cation **23** was thought to be formed by protonation of γ-bisabolene (**24**), experiments using double-labeled mevalonic acid (2-^{14}C-, (4R)-4-^{3}H-) indicate that two H_R atoms at C-4 of mevalonic acid which are known to be located on the double bonds in the farnesyl pyrophosphate (H asterisked in **5**) are retained in helicobasidin (**28**)[37,38]. This result suggests that the hydrogen atom at C-6 in the cation **22** must shift to the carbon of five membered ring (possibly at C-2) during the cyclization process giving rise to the cation, **23**. Such a long-range hydride shift has also been observed in the formation of the tricyclic sesterterpene, ophiobolin (**46**) from geranylfarnesyl pyrophosphate as described later. Hydride shift or shifts have also been found to be involved in other sesquiterpene biosyntheses such as those of illudin[100], longifolene[101], culmorin[102], germacrene[33], dendrobine[103].

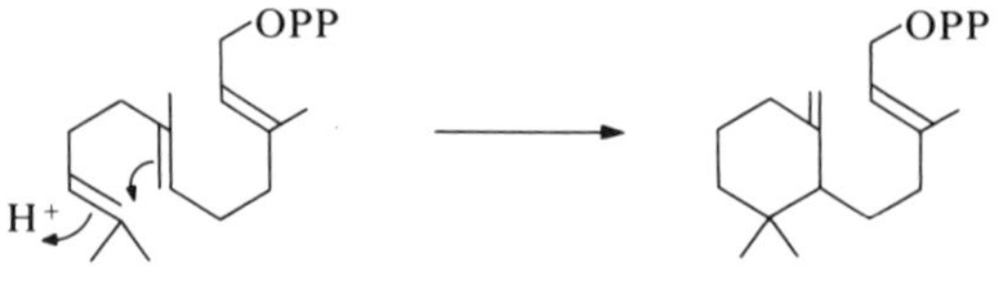

Many of the sesquiterpenes are oxidized after the cyclization of farnesyl pyrophosphate. The majority of them retain the basic carbon skeleton but it is sometimes modified by the cleavage of carbon-carbon linkages or by skeletal rearrangements. For example, helminthosporal (**30**) and coriamyrtin (**32**) or tutin are now known to be formed respectively from sativene (**29**)[39,40] and copaborneol (**31**)[41,42] by oxidative cleavage of C-C bond as indicated. A phytotoxic compound, fomannosin (**35**) is thought to be derived from the compound having the protoilludane skeleton by oxidative bond fission. Co-occurence of fomannosin (**35**) and the compound **33** in *Fomitopsis insularis* supports this hypothesis[116]. This wood-rotting mushroom also produces compound **34**

---

[33] *K. Morikawa, Y. Hirose, S. Nozoe*, Tetrahedron Lett. 1131 (1971).

[34] *K. T. Suzuki, N. Suzuki, S. Nozoe*, Chem. Commun. 527 (1971).

[35] *S. Nozoe, Y. Machida*, Tetrahedron Lett. 2671 (1970).

[36] *J. R. Hanson*, Chem. Ind. 1643 (1967); Chem. Commun. 511 (1970).

[37] *S. Nozoe, M. Morisaki, H. Matsumoto*, Chem. Commun. 926 (1970).

[38] *P. M. Adams, J. R. Hanson*, Chem. Commun. 1414 (1971).

[39] *P. de Mayo, J. R. Robinson, E. Y. Spencer, R. W. White*, Experientia *18*, 359 (1962).

[40] *P. de Mayo, R. E. Williams*, J. Am. Chem. Soc. *87*, 3275 (1965).

[41] *M. Biollaz, D. Arigoni*, Chem. Commun. 633 (1969).

[42] *A. Corbella, P. Garifoldi, G. Jommi, S. Scolastico*, Chem. Commun. 634 (1969).

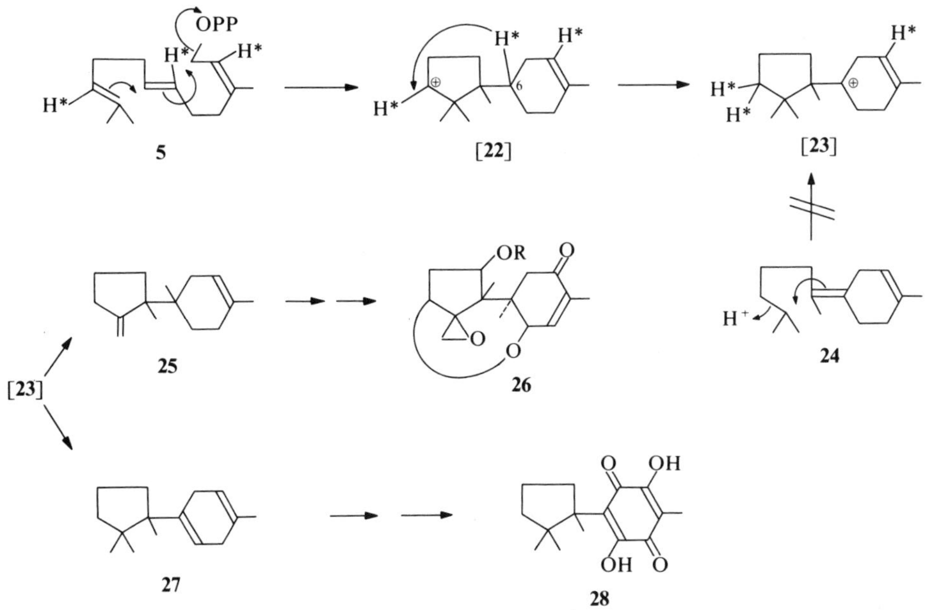
OPP
H*
H*
H*
5
H*
H*
H*
6
[22]
H*
H*
H*
[23]
OR
O
O
O
25
26
[23]
H⁺
24
O
OH
27
OH
O
28

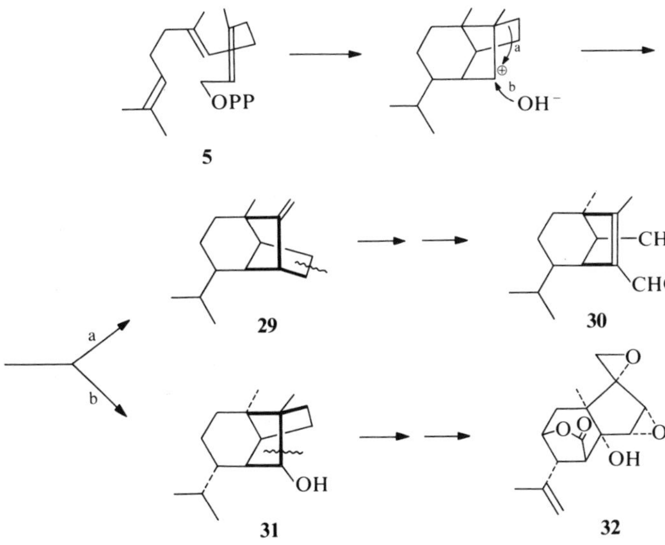
OPP
5
a
b
OH⁻
29
CHO
CHO
30
a
b
OH
31
O
O
O
O
OH
32

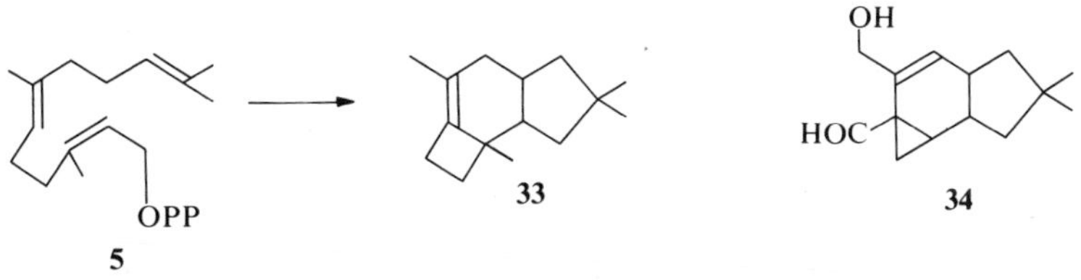
OPP
5
33
OH
HOC
34

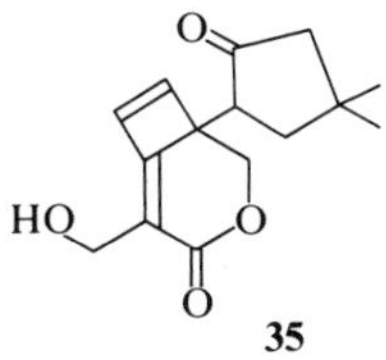
O
HO
O
O

35

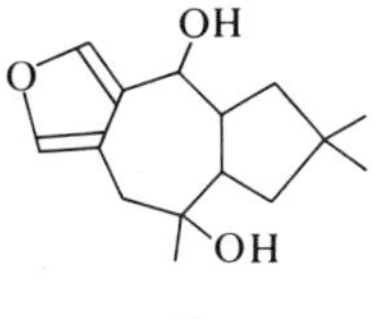
OH
O
OH

36

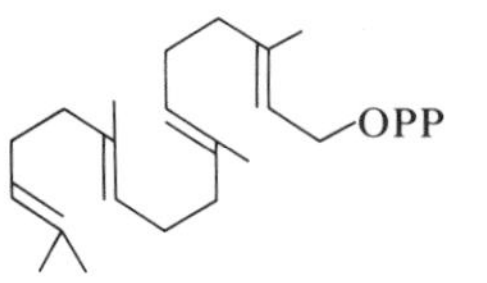

and compound **36**. The conversion of compound **34** into compound **36** has been verified by addition to a medium of this fungus[43].

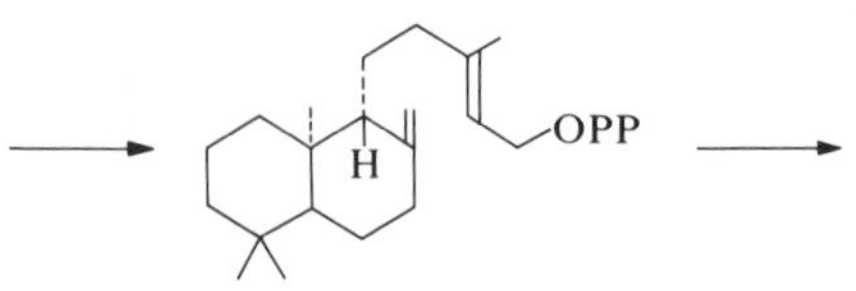

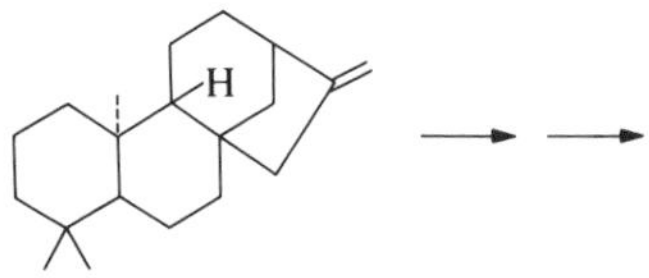

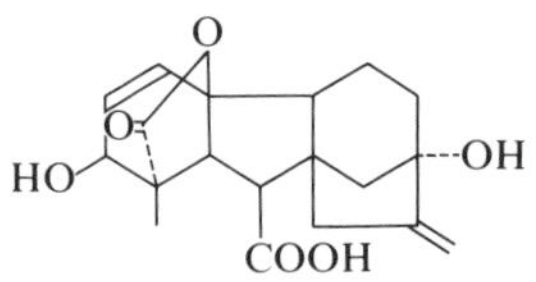

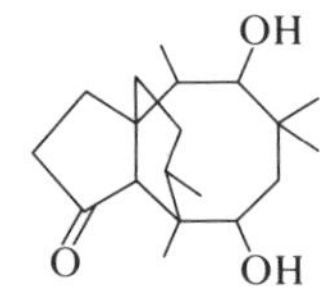

Use of ^{13}C-labeled precursor in the biosynthetic experiments provide useful information as to the pathway including cyclization, skeletal rearrangement, and oxidative fission of C-C bond and so on. Sesquiterpene metabolites studied with the use of ^{13}C-NMR techniques include trichothecolone[104], helicobasidin[105], coriolin[106], ovalicin[107], fomannosin[108], cyclonerodiol[109], phytuberin[110], paniculide[111] and capsidiol[112] (some of these by the double labeling method).

3.2.1.3 Diterpenes

The biosynthesis of the plant growth hormone, gibberellic acid (**40**), has been studied in detail and the hydrocarbon, kaurene (**39**), has been proved[45] as an intermediate in its biosynthesis by adding labeled kaurene to a culture of *Gibberella fujikuroi*. Recent work using ^{13}C-labeled precursor, 2-^{13}C-MVA, confirmed the accepted labeling patterns[113]. The route from geranylgeranyl pyrophosphate (**37**) to kaurene (**39**) has been studied with cell-free preparations from the endosperm of immature seeds of *Echynocystis macrocarpa*[46] and mycelia of the fungus *Gibberella fujikuroi*[47]. Copalyl pyrophosphate (**38**) was found to be an intermediate from geranylgeranyl pyrophosphate to kaurene. It has been found that (RS)-14,15-oxidogeranylgeranyl pyrophosphate was cyclized to a mixture of 3α- and 3β-hydroxykaurene by soluble enzyme preparation obtained from immature *E. macrocarpa*[114]. A cell-free preparation from seedlings of castor beans (*Ricinus communis* L.) has been found to produce kaurene and several diterpene hydrocarbons[48]. Enzymic formation of pleuromutilin (**41**) from geranylgeranyl pyrophosphate (**37**) using a cell-free system prepared by sonication of cells of the mushroom, *Pleurotus mutilus* P., has been reported[49].

3.2.1.4 Sesterterpenes

Terpenes with 25 carbon atoms, *i.e.*, sesterterpenes, have been isolated from fungi[50,51] and insect wax[52,53]. Sesterterpenes were assumed to be derived from geranylfarnesyl pyrophosphate (**42**) and this view is supported by isolation of geranylnerolidol (**44**) and geranylfarnesol (**43**)

[43] *S. Nozoe, S. Urano*, private communication.

[44] *S. Nozoe, S. Urano, H. Matsumoto*, Tetrahedron Lett. 2153 (1971).

[45] *B. E. Cross, R. H. B. Galt, J. R. Hanson*, J. Chem. Soc. 295 (1964).

[46] *C. D. Utter, C. A. West*, J. Biol. Chem. *242*, 3285 (1967).

[47] *I. Shechter, C. A. West*, J. Biol. Chem. *244*, 3200 (1969).

[48] *D. R. Robinson, C. A. West*, Biochemistry *9*, 80 (1969).

[49] *D. Arigoni*, Chem. of Natural Products *5*, 331 (1968).

[50] *S. Nozoe, M. Morisaki, K. Tsuda, Y. Iitaka, N. Takahashi, S. Tamura, K. Ishibashi, M. Shirasaka*, J. Am. Chem. Soc. *87*, 4968 (1965).

[51] *L. Canonica, A. Ficchi, M. Galli Kienle, A. Scala*, Tetrahedron Lett. 1329 (1966).

[52] *Y. Iitaka, I. Watanabe, I. T. Harrison, S. Harrison*, J. Am. Chem. Soc. *90*, 1092 (1968).

[53] *T. Rios, S. Perez C., F. Colunga*, Chem. Ind. 1184 (1965).

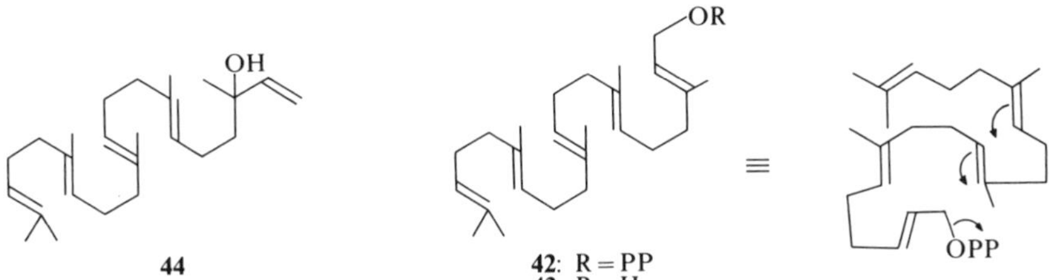

44

42: R = PP
43: R = H

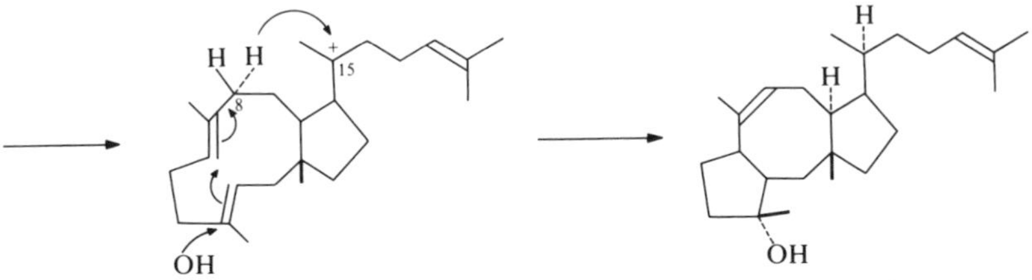

45

46

from such sesterterpene producing organisms[54,55]. The tricyclic sesterterpene, ophiobolin-F (**46**) was found to be formed from geranylfarnesyl pyrophosphate (**42**) by using a cell-free system prepared by disruption of *Cochliobolus heterostrophus*[56]. The mechanism of the cyclization is thought to be as depicted below and some experimental evidences have been reported[57,58]. The stereospecific 1,5-hydride shift[59] from C-8α to C-15 position in the cationic intermediate **45** initiates a further cyclization of the 11-membered intermediate to a tricyclic system.

3.2.2 Steroids and Triterpenes

Lanosterol (**48**), the first cyclic precursor in the biosynthesis of steroidal compounds, is now firmly established as being formed by the cyclization of 2,3-oxidosqualene (**47**)[60]. Investigations concerning substrate specificity of the enzyme catalyzing this cyclization have been carried out in detail and the minimum structural requirement of the substrate for cyclization was demonstrated[61]. An example of enzymic cyclization of a chemically synthesized substrate, 20,21-dehydro-2,3-oxidosqualene (**49**), indicated the formation of a compound having a protostane skeleton[62], such as **50**. Substances with a protostane skeleton are found in nature and cyclization of 2,3-oxidosqualene into fusidic acid (**51**) was observed[63]. A cell-free preparation of *Cephalosporium caerulens* producing helvolic acid converted 2,3-oxidosqualene into both lanosterol (**48**) and a protosterol type compound (**52**) which is presumed to be an intermediate of helvolic acid[64].

Many plant sterols differ from cholesterol, which is considered to be a typical animal sterol, with respect to having additional C_1 or C_2 residues at C-24 and/or a double bond at C-22. In animals the first cyclic precursor of cholesterol is lanosterol, but in higher plants existence of lanosterol is extremely restricted. Its place seems to be taken by cycloartenol (**53**)[65]. Cell-free preparations from *Ochromonas malhamensis* convert 2,3-oxidosqualene into cycloartenol but not into

[54] *S. Nozoe, M. Morisaki, K. Fukushima, S. Okuda*, Tetrahedron Lett. 4457 (1968).

[55] *T. Rios, S. Perez C.*, Chem. Commun. 214 (1968).

[56] *S. Nozoe, M. Morisaki*, Chem. Commun. 1319 (1969).

[57] *L. Canonica, S. Fiecchi, M. Galli Kienle, B. M. Ranzi, A. Scala*, Tetrahedron Lett. 337 (1967); 3035 (1966).

[58] *S. Nozoe, M. Morisaki, K. Tsuda, S. Okuda*, Tetrahedron Lett. 3365 (1967); 2347 (1968).

[59] *L. Canonica, A. Fiecchi, M. Galli Kienle, B. M. Ranzi, A. Scala*, Tetrahedron Lett. 275 (1968).

[60] *E. E. van Tamelen*, Accounts Chem. Res. *1*, 111 (1968).

[61] *E. E. van Tamelen*, Special Lecture presented at the XXIII Intl. Congr. of Pure and Applied Chemistry, Vol. 5, p. 85, Butterworth, London 1971.

[62] *E. J. Corey, K. Lin, H. Yamamoto*, J. Am. Chem. Soc. *91*, 2132 (1969).

[63] *W. O. Godtfredsen, H. Lorck, E. E. van Tamelen, J. D. Willett, R. B. Clayton*, J. Am. Chem. Soc. *90*, 208 (1968).

[64] *A. Kawaguchi, S. Okuda*, private communication.

[65] *R. H. Kemp, A. S. A. Hamman, L. J. Goad, T. W. Goodwin*, Phytochemistry *7*, 447 (1968).

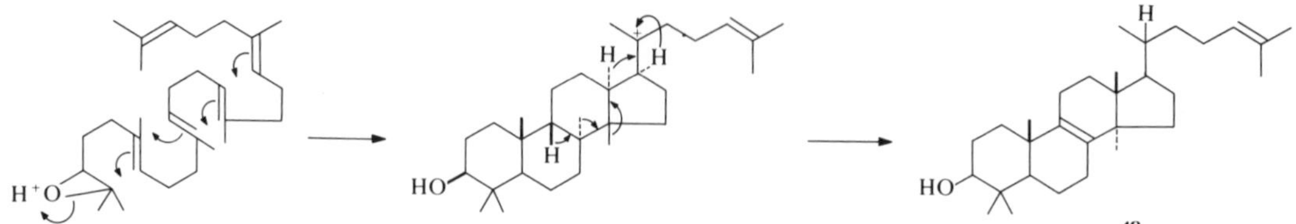

47 48

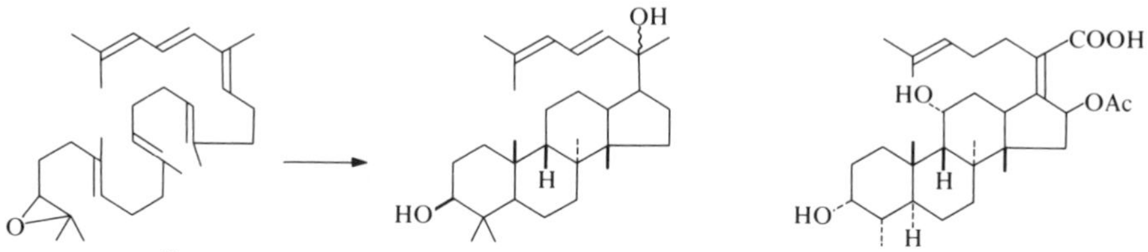

49 50 51

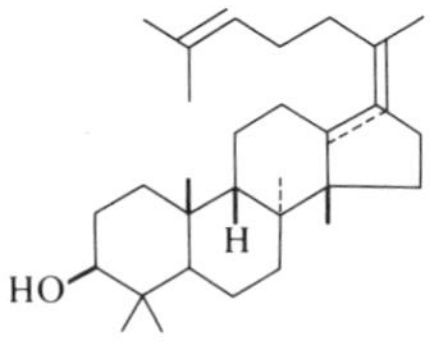

52 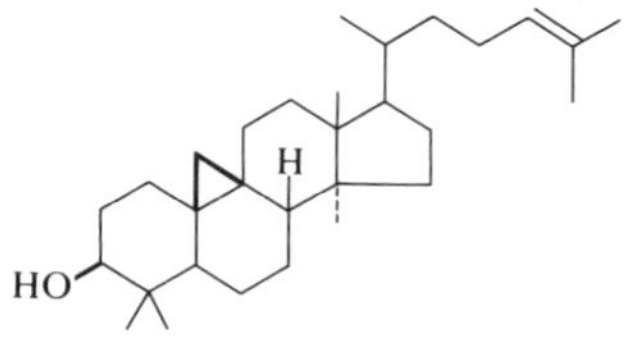 53

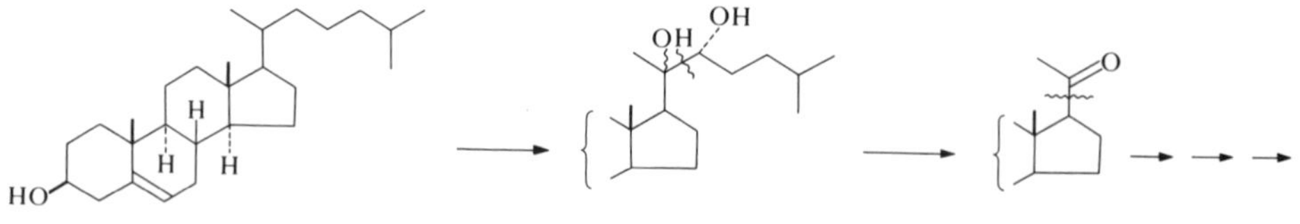

54

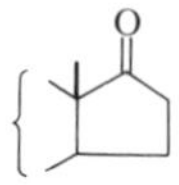

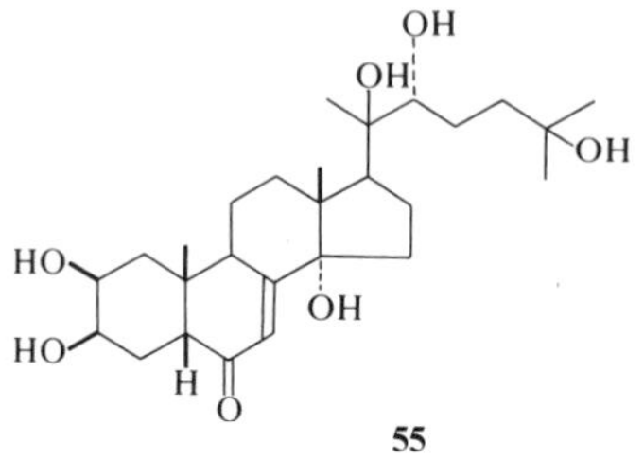

55

lanosterol[66]. Cycloartenol is a product of direct cyclization and is not produced by isomerization of lanosterol[67].

Conversion of lanosterol (**48**) into cholesterol (**54**) has been examined in detail by using a homogenate or crude liver enzyme system, and the mechanism of each step, including the stereochemistry of the reaction, has been clarified to a considerable extent[68].

[66] *H. H. Rees, L. J. Goad, T. W. Goodwin*, Tetrahedron Lett. 723 (1968); Biochem. Biophys. Acta *176*, 892 (1969).

[67] *L. J. Goad, T. W. Goodwin*, Eur. J. Biochem. *7*, 502 (1969).

Cholesterol is further metabolized to bile acids and steroid hormones. Much information about this metabolism has been obtained with mammals and a number of extensive reviews have appeared[69,70].

[68] *L. J. Goad*, Natural Substances Formed Biologically from Mevalonic Acid, *T. W. Goodwin* (Ed.), Academic Press, New York 1970.

[69] *H. Danielsson, T. T. Tchen, in* Metabolic Pathways, *D. M. Greenberg* (Ed.), Vol. 2, p. 132, Academic Press, New York 1968.

[70] *L. T. Samuels, K. B. Eil-Nes, in* Metabolic Pathways, *D. M. Greenberg* (Ed.), Vol. 2, p. 169, Academic Press, New York 1968.

The insect moulting hormone, ecdysone (**55**), which came to light in recent years, is also formed by the oxidative modification of cholesterol[71]. Owing to the inability of insects to synthesize sterols[72], insects feeding on plants may retain the enzymes involved in the conversion of phytosterols to cholesterol. This dealkylation is thought to proceed *via* the epoxide from a chemical analog[73].

The triterpene alcohol, β-amyrin (**56**), was shown to be cyclized from 2,3-oxidosqualene product (**47**) by using a cell-free preparation from germinating peas[74,75]. Enzymic cyclization of squalene (**11**) into tetrahymanol (**57**) was also demonstrated by using a cell-free preparation as well as by the growing cell system of the protozoon, *Tetrahymena pyriformis*[76]. Squalene has also been shown to cyclize into hopane-type triterpenes with a cell-free system derived from an *Acetobacter*[115].

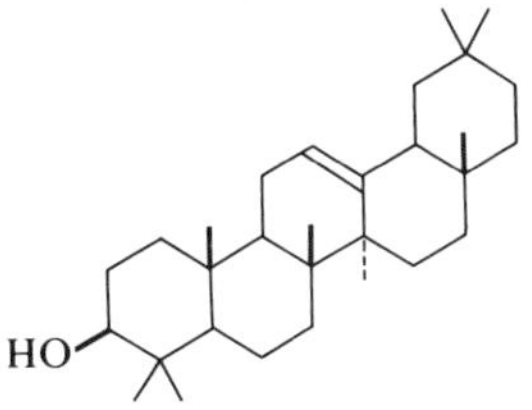

56 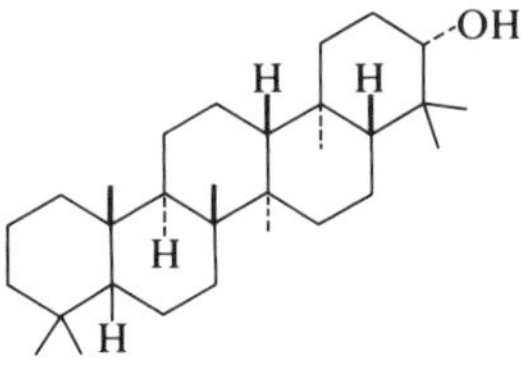 **57**

3.2.3 Carotenoids

Carotenoids are biosynthesized through an intermediate, phytoene (**58**), formed by the condensation of two molecules of geranylgeranyl pyrophosphate (**37**)[77]. Phytoene (**58**) is converted into many carotenoid compounds by a sequential dehydrogenation, cyclization of the terminal ring and/or oxygenation[78]. There are some important substances formed by the oxidative degradation of carotenoids. Trisporic acid (**59**), the sexual hormone of fungi, is known to be biosynthesized from β-carotene, *via* retinal (**60**)[79]. Other substances thought to be formed by oxidative degradation of carotenoids are abscissic acid (**61**)[80] and the allenic compound **62** which is a component of a grasshopper secretion[81].

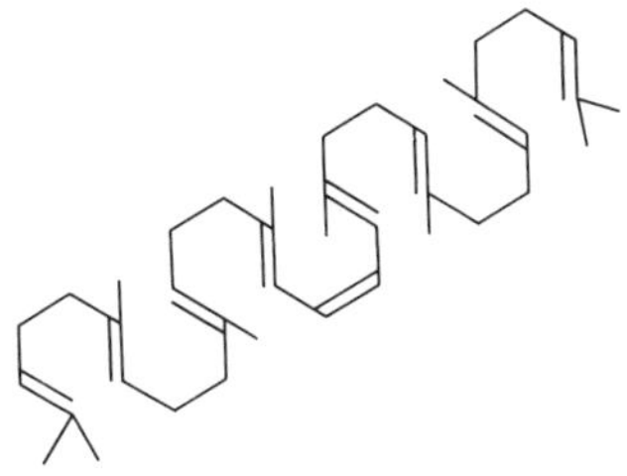

58

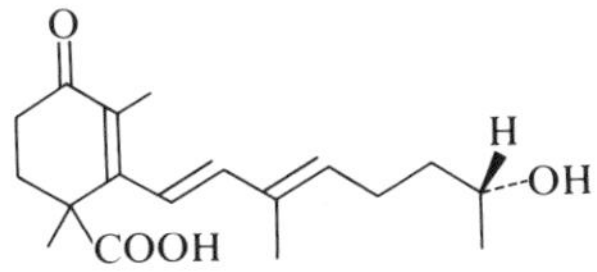

59

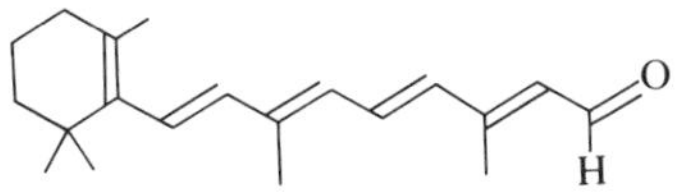

60

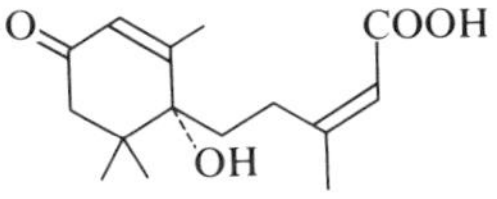

61 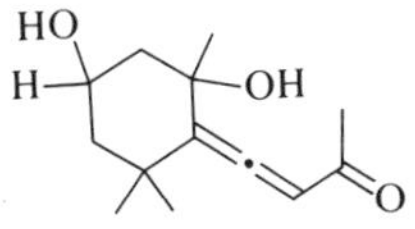 **62**

3.2.4 Prenylphenols and Prenylquinones

There are many indications that the side chains of various prenylquinones and prenylphenols arise from mevalonic acid. Tetrahydrocannabinol (**63**), the active component of hashish, is an example of a diprenylphenol and is found together with cannabigerol (**64**)[82], can-

---

[71] *J. B. Siddall, A. D. Cross, J. H. Fried*, J. Am. Chem. Soc. *88*, 862 (1966).

[72] *R. B. Clayton*, J. Lipid Res. *5*, 3 (1964).

[73] *N. Ikekawa, et al.*, Chem. Commun. 1498 (1971).

[74] *E. J. Corey, P. R. O. de Montellano*, J. Am. Chem. Soc. *89*, 3362 (1967).

[75] *E. Capstack, Jr., N. Rosin, G. A. Blondin, W. R. Nes*, J. Biol. Chem. *240*, 3258 (1965).

[76] *E. Caspi, et al.*, J. Am. Chem. Soc. *90*, 3563, 3564 (1968); Biochem. J. *109*, 931 (1968).

[77] *R. J. H. Williams, G. Britton, J. M. Charlton, T. W. Goodwin*, Biochem. J. *104*, 767 (1967).

[78] *T. W. Goodwin*, Biochem. J. *123*, 293 (1971).

[79] *D. J. Austin, J. D. Bu'Lock, D. Drake*, Experientia, 348 (1970).

[80] *K. Ohkuma, J. L. Lyon, F. T. Addicott, O. E. Smith*, Science *147*, 1952 (1965).

[81] *J. Meinwald, K. Erickson, M. Hartshorn, Y. C. Meinwald, T. Eisner*, Tetrahedron Lett. 2959 (1968).

[82] *Y. Shoyama, T. Yamauchi, I. Nishioka*, Chem. Pharm. Bull. (Tokyo) *18*, 1327 (1970) and references cited therein.

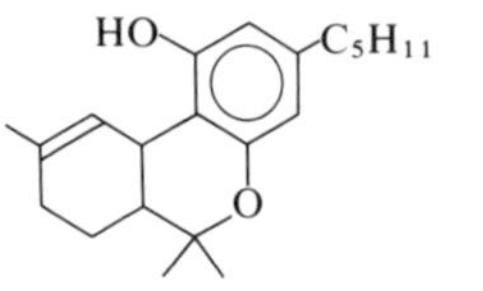

63

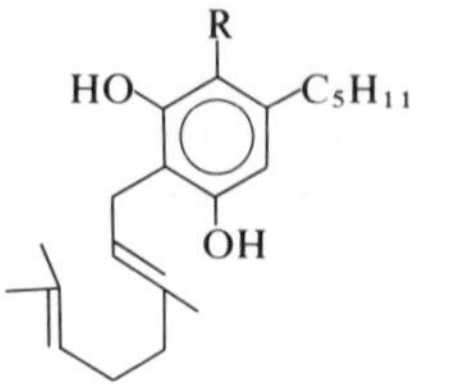

64: R = H
65: R = COOH

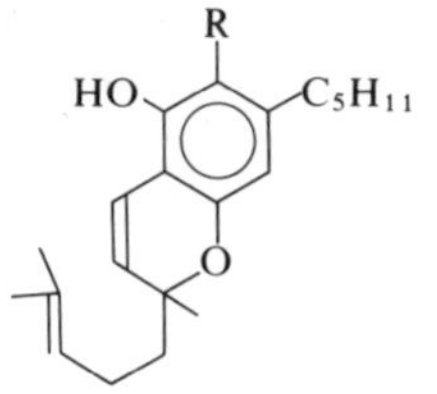

66: R = H
67: R = COOH

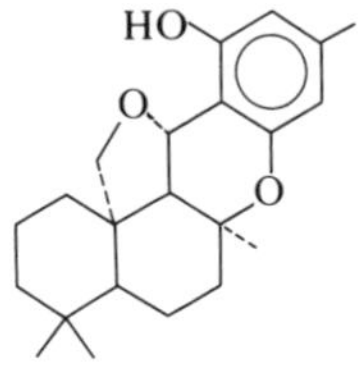

68

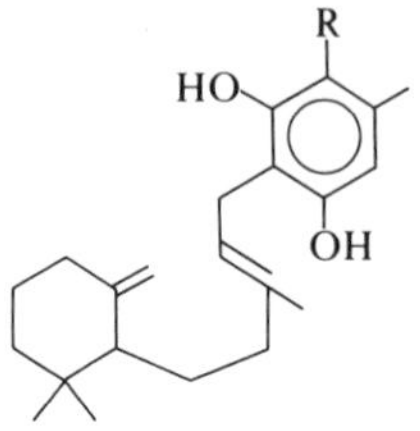

69: R = H
70: R = COOH

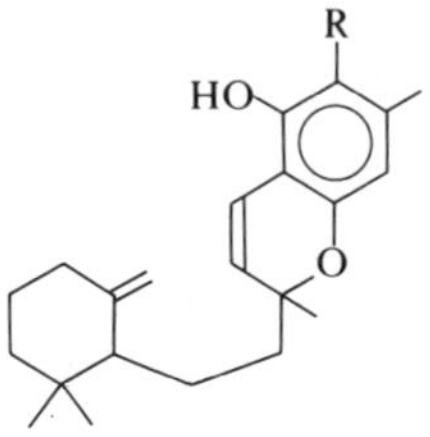

71: R = H
72: R = COOH

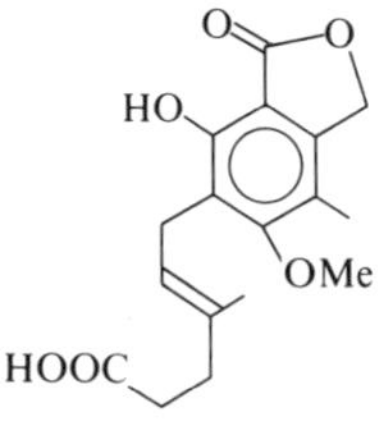

73

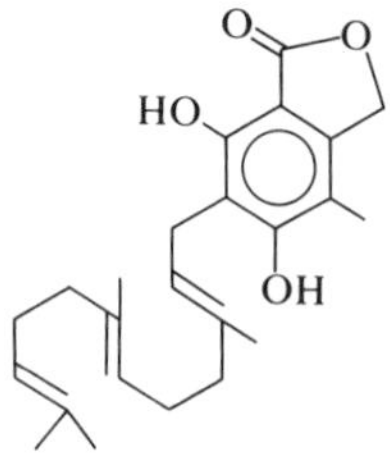

74

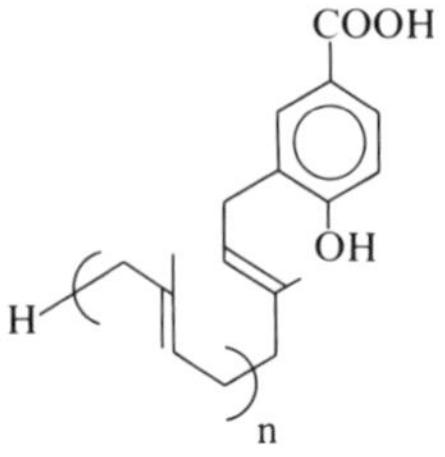

75

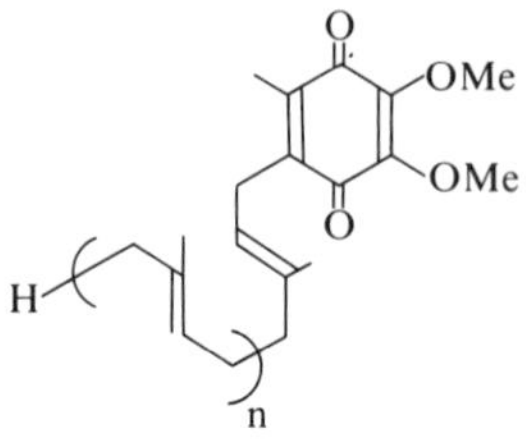

76

nabichromene (**66**) and their carboxylic acid derivatives (**65, 67**). The antibiotic siccanin (**68**)[83] is a triprenylphenol closely related biosynthetically with tetrahydrocannabinol (**63**), since a series of compounds corresponding to the congeners of **63** were found in *Helminthosporium siccans* D., *i.e.*, presiccanochromene (**69**)[84], siccanochromene (**71**) and the corresponding carboxylic acid derivatives **70** and **72**[85]. In the crude metabolite of the fungi, a diprenyl derivative has been found as a minor component[86]. The cell-free preparation from *Helminthosporium siccans* produces siccanochromen-A (**71**) from mevalonic acid or farnesyl pyrophosphate in the presence of orsellinic acid, but in the absence of orsellinic acid γ-monocyclofarnesol (**21**) is accumulated[87]. Siccanochromene-A (**71**) was found to be synthesized *via* presiccanochromenic acid (**70**) and siccanochromenic acid (**72**) in an experiment using a cell-free preparation[88]. Conversion

[83] *K. Hirai, K. T. Suzuki, S. Nozoe*, Tetrahedron *27*, 6057 (1971).

[84] *S. Nozoe, K. T. Suzuki*, Tetrahedron Lett. 3643 (1968).

[85] *S. Nozoe, K. T. Suzuki*, Tetrahedron Lett. 2457 (1969).

[86] *S. Nozoe, T. Sato, T. Yatsunami, K. T. Suzuki*, private communication.

[87] *K. T. Suzuki, N. Suzuki, S. Nozoe*, Chem. Commun. 527 (1971).

[88] *K. T. Suzuki, S. Nozoe*, private communication.

of siccanochromene-A to siccanin has been proved by an experiment using a growing cell system of the fungi[87]. Mycophenolic acid (**73**) is now known to be a degradation product of the triprenylphenol, **74**[89]. Prenylation of the aromatic compound was demonstrated by means of a feeding experiment using appropriate substrates. Coenzyme Q (**76**), a polyprenylquinone, is known to be formed from the polyprenyl phenol **75** which, in turn, arises by condensation of *p*-hydroxybenzoic acid and polyprenyl pyrophosphate[90].

---

[89] *L. Canonica, et al.*, Chem. Commun. 1357 (1970); Chem. Commun. 257 (1971).

[90] *K. Folkers, et al.*, J. Am. Chem. Soc. *88*, 4754 (1966); *90*, 5587 (1968); Biochem. Biophys. Res. Commun. *28*, 324 (1967).

Recommended Literature for Further Reading

[91] *J. W. Cornforth, J. W. Redmond, H. Eggerer, W. Buckel, C. Gutschow*, Eur. J. Biochem. *14*, 1 (1970); *D. Arigoni, et al.*, Chem. Commun. 921 (1975) and references cited therein.

[92] *K. H. Clifford, G. T. Phillips*, Chem. Commun. 419 (1975).

[93] *U. Séquin, A. I. Scott*, Science *186*, 101 (1975); *M. Tanabe, in* Biosynthesis, Vol. 2, p. 241, *T. A. Geissmann* (Ed.), Chemical Society (Specialist Periodical Report), London 1973; *M. Tanabe, in* Biosynthesis Vol. 3, p. 247, 1975.

[94] *B. C. Reed, H. C. Rilling*, Biochemistry, *14*, 50 (1975); For reaction mechanism, see *C. D. Poulter, D. M. Satterwhite, H. C. Rilling*, J. Am. Chem. Soc. *98*, 3376 (1976).

[95] *E. J. Corey, R. P. Volante*, J. Am. Chem. Soc. *98*, 1291 (1976); *P. R. O. de Montellano, R. Castillo, W. Vinson, J. S. Wei*, J. Am. Chem. Soc. *98*, 2018 (1976); J. Am. Chem. Soc. 3021 (1976).

[96] *S. Nozoe, J. Furukawa, U. Sankawa, S. Shibata*, Tetrahedron Lett. 195 (1976).

[97] *S. Nozoe, H. Kobayashi, N. Morisaki*, Tetrahedron Lett. 4625 (1976); *D. E. Cane, G. G. S. King*, Tetrahedron Lett. 4737 (1976).

[98] *R. Evans, A. H. Holtom, J. R. Hanson*, Chem. Commun. 465 (1973); *R. Evans, J. R. Hanson*, J. Chem. Soc., Perkin I, 326 (1976).

[99] *R. Evans, J. R. Hanson, T. Marten*, J. Chem. Soc., Perkin I, 1212 (1976).

[100] *J. R. Hanson, T. Marten, R. Nyfeler*, J. Chem. Soc., Perkin I, 876 (1976).

[101] *F. Dorn, P. Bernasconi, D. Arigoni*, Chimia *29*, 25 (1975).

[102] *J. R. Hanson, R. Nyfeler*, Chem. Commun. 824 (1975).

[103] *A. Corbella, P. Gariboldi, G. Jommi, M. Sisti*, Chem. Commun. 288 (1975).

[104] *J. R. Hanson, T. Marten, M. Siverns*, J. Chem. Soc., Perkin I, 1033 (1974).

[105] *M. Tanabe, K. T. Suzuki, W. C. Jankowski*, Tetrahedron Lett. 4723 (1973).

[106] *M. Tanabe, K. T. Suzuki, W. C. Jankowski*, Tetrahedron Lett. 2271 (1974); *T. C. Feline, G. Mellow, R. B. Jones, L. Phillips*, Chem. Commun. 63 (1974).

[107] *D. E. Cane, R. H. Levin*, J. Am. Chem. Soc. *98*, 1183 (1976).

[108] *D. E. Cane, R. B. Nachbar*, Tetrahedron Lett. 2097 (1976).

[109] *R. Evans, J. R. Hanson, R. Nyfeler*, Chem. Commun. 814 (1975).

[110] *A. Stoessl, E. W. B. Ward*, Tetrahedron Lett. 3271 (1976).

[111] *K. H. Overton, et al.* Chem. Commun. 105 (1976).

[112] *F. C. Baker, C. J. Brooks, S. A. Hutchinson*, Chem. Commun. 293 (1975).

[113] *R. Evans, J. R. Hanson, M. Siverns*, J. Chem. Soc., Perkin I, 1514 (1975).

[114] *C. A. West, R. M. Coates*, J. Am. Chem. Soc. *98*, 4659 (1976).

[115] *C. Anding, M. Rohmer, G. Ourisson*, J. Am. Chem. Soc. *98*, 74 (1976).

[116] *S. Nozoe, H. Kobayashi, S. Urano, J. Furukawa*, Tetrahedron Lett. 1381 (1977).

Subject Index

Abascissic acid 233
Acansterol 13
Acanthasterol 6
Acetate-malonate pathway 134
2-Acetylflexuosin A 65
N-Acethylvincoside 123
10-Acetoxy-18-hydroxy-2,7-dollabelladine 83
3β-Acetoxynorerythrosauamine 72
Achillin 61
Aconine 132
Aconitine 103, 132
Acrostalic acid 74
Acrostalidic acid 74
Acutimine 103
2-Acylindoles 102
S-Adenosylmethionine 102
Adrenalcorticosteroid hormones 3
Aflatoxins 136, 140
Ajaconine 132
Ajimaliccine 122, 128
Ajimaline 102, 129
Akummicine 122
Albiocolide 52
GA_{12}-Aldehyde 78
Aldosterone 3
Alkaloids 101–132
– / amaryllidaceae 102
– / aporphine 102
– / benzophenanthridine 102
– / classification 102
– / indole 102
– / ipecaunha 102, 123
– / isoquinoline 102
– / quinolene 102
– / quinolizidine 102
– / steroid 11, 102
– / tropane 102
Allenes 171
Allomelanines 197
Allylic chloride 38
Alternilin 65
Amaralin 65
Amberboin 61
Ambrosin 63
Ambrosiol 63
Amitenone 160
β-Amyrin 233
Anabasine 102
Anagyrine 125
Anhydrocoronopilin 63
Anhydroivaxillarin 60
Anibin 138
Annelids 159
Annonalide 70
Annotonine 102, 103
Antheridiogen-An 78
Antheridiol 5, 14, 15
Anthraquinones 210
9,10-Anthraquinone 162
Aphid 164
Aphidicolene 80
Aphidicolin 80
Aphinin 164
Aplysin-20 66
Aplysioviolin 150
Apoatropine 124
Aquatic fungus 5, 14
Arbiglovin 60
Arborescin 60
Arctiopicrin 52
Arecaidine 125
Arecoline 125
Argemonine 102
Aristolactone 53
Arnifolin 64
Aromaticin 64
Aromatin 64
Artabsin 60
Artemisiifolin 53
3-Arylcoumarins 190
4-Arylcoumarins 179
Aryliodine reagents 34
Asperdiol 86
Aspidosperma 121, 123
Aspidospermine 102, 130
Asterosterol 6
Atisine 102, 132
Atranorin 218, 223
Atropine 124
Aureothin 137, 140
Aurones 179
Austricin 61
Autumnolide 65
Averufin 163
Axivalin 60
Azo-steroid 41
Baccatin IV, VI, VII 86
Bachtrachotoxinin A 12
Bacteriochlorin 153, 158
Bacteriochlorophyll 156
Badkhysin 60
Baeyer-Villiger oxidation 96
Bahia I, II 61
Balchanolide 52
Bamford-Stevens reaction 41
Barbatusin 72
Batrachotoxin 5, 11, 12, 103
Belladine 102
Benzofuran 184
Benzophenanthridine alkaloids 102
Benzylisoquinolines 102
Berberine 102, 117, 126
Bianthronyl-10 211
Bigelovin 64
Biladienes 157
Biladiene a, c 158
Bilanes 157
b-Bilenes 157
Bile pigments 142
Biliproteins 149

Bilirubin
– / conjugates 145
Bilirubin IXα 145
Biliverdin IXα 144
Bioquinones 159
γ-Bisabolene 228
Bisbenzylisoquinolines 102
16,16′-Bisnorgeranylgeranyl pyrophosphate 226
Blue-pigment 149
Boldine 126
Bovinone 160
Brevilin A 64
B-Ring diene 41
Bufotalin 4, 139
Bufatoxin 4
Bulbocapnine 126
Buldulin 64
α-Bulnesene 58
β-Bulnesene 58
Bulnesol 58
Burcine 129
Caffeine 125
C-Calebassine 131
Calocephalin 61
Calopropene 13
Calycanthine 103
Calycin 218
Calysterol 13
Campesterol 10
Camphor 227
Camptothecin 103, 131
Cannabichromene 233
Capaurimine 113
(±)-Capaurimine 113
Caperatic acid 208
Caracurine 103
Carajurin 166
Cardiac active principles 4
Cardiac aglycones 4
Carotenoids 168, 233
Catalases 153, 154
Catharanthine 122
Caulerpol 86
Cembrene-A 82
Cephaeline 127
Cernuine 102
Cevadine 132
Chalcones 179
Chamissonin 53
Chelidonine 102, 117, 118, 126
Chemotaxonomy 98
Chenodesoxycholic acid 2
Chihuahuin 52
Chimaphilin 161, 166
Chimonanthine 121
Chlorohaemins 154
Chloroiride 36
Chlorophyll 153–156, 158
Chlorophyllase 155
Cholestane 2
Cholesterol 2, 7, 8, 232
Cholic acid 2
Chromone 136, 182, 185, 186
Chrysanthemin A, B 60
Chrysophanol 165
Chrysotalunin 162
Cinchona alkaloids 102
Cinchonidine 131
Cinchonine 131
Clerodendrin A 68
Cnicin 52
Cobalt(II) chelate 157
Cocaine 124
Codeine 114, 118, 127
Coenzyme Q 235
Colchicine 102, 119, 121, 124
Colchicine alkaloids 102
Coleon D, E 72
Conacytone 72
Conchosin A, B 63
Concinndiol 66
Conessine 102
Confertiflorin 63
Confertin 64
Copaborneol 228
Copalyl pyrophosphate 230
Coprinin 160
Cordeauxione 162
Coriamyrtin 228
Coronaridine 122
Coronopilin 63
Corrins 158
Cortexone 10
Corticoid side chain 38, 41
Corticosteroids 26, 34
Cortisol 3
Cortisone 3, 26, 34
Corynanthe 121, 123
Corynantheine 102, 122
Corytuberine 119
Costunolide 52
Cotarine 102
Cotarnine 126
Coumarins 139, 179, 183–185, 189
Crinitol 86
Crinine 102
Crinoids 163
Crotofolin A 84
Cryptoaustotine 102
Cryptochlorophaeic acid 216
Cularine 102
Cularine alkaloids 102
Cumambrin A, B 61
Cumanin 65
Cuprenene 228
Curcumenol 58
Curcumol 58
Curidione 51
Currin 153
Curzerenone 56
Cyathin A_3 84
Cycloartenol 7, 10, 231
– / biosynthesis 8
Cyclobutatusin 72
Cyclogranisolide 15
Cyclohexenones 26

Cyclopropane 6, 12, 13
Cyclopropyl group 9
Cynaropicrin 60
Cyperaquinone 165
Cytochrome 153, 154, 156
Cytochrome C 157
Cytochrome models 158
Cytodeuteroporphyrin 156
Dalbergiones 160
Damsin 63
Damsinic acid 63
Daphniphylline 103
Daunomycin 163
Deacetylmatricarin 61
Decarboxytrichosiderin B, C 197
Decipidiene-triol 84
7-Dehydrochlolesterol 4
Dehydrocostus lactone 60
11,13-Dehydroopodin 64
20,21-Dehydro-2,3-oxidosqualene 231
dl-16-Dehydroprogesterone 26
1-Dehydroxybaccatin IV 86
Delphisine 81
Demecolcine 124
23-Demethylgorgosterol 6
Dendrobine 103
– /-stereochemistry 105
Dendrobinediol 104
Dendrobium nobile L 103
Deoxylapachol 161
Deoxyelephantopin 53
Deoxyloganine 123
Deoxymikanolide 53
Depside 184, 209
Depsidone 184, 209
Desacetylconfertiflorin 63
Deserpidine 129
Desmosterol 10
Desoxycholic acid 2
Deuteroporphyrin IX 156
Dibenzofuran 184
– / derivatives 209
Dibenzopyrocoline alkaloids 102
Dichotine 103
Dictyol A, B 83
Didymic acid 209, 223
Digitogenin 4
Digitonin 4, 88
Dihydrogorgosterol 12
Dihydromexicanin 65
Dihydromikanolide 53
Dihydropseudoivalin 61
Dihydroscandenolide 53
5,6-Dihydroxyindole 192
5,6-Dihydroxyindole-2-carboxylic acid 191
3β,7β-Dihydroxykaurenolide 75
1,8-Dihydroxynaphthalene 199
α-Diketones 34
Dimethylally pyrophosphate 225
2,2-Dimethylchromanols 184
Dioscin 88
Dioscorine 124
Diosgenin 31
Diospyrin 161
5(2*H*)-dipyrrylmethanone 142
Diterpenoids 66, 230
Dolastane 84
Dolatriol 83
Dollabelladiene 84
Dopa 191
Dopachrome 191
Dopaquinone 191
3,8,13-Duratriene-1,3-diol 81
Eburnamine 102
Ecdysone 7, 233
α-Ecdysone 5, 10
β-Ecdysone 10
Ecdysones
– / synthesis 31
Echinoderms 159
Echitamine 103, 129
Elemanal 56
α-Elemene 56
(+)β-Elemene 56
γ-Elemene 56
δ-Elemene 56
Elemol 56
δ-Elemenol 56
Epi-δ-elemenol 56
α-Elemenone 56
β-Elemenone 56
γ-Elemenone 56
Elephantin 53
Elephantopin 53
Elsinochrome A 165
Elsinochromes 164
Embelin 160
Emetine 123, 127
Emodin 165, 211
Endocrocin 210
Enol-butenolides 172
Ephedrine 124
10-Epieupatoroxin 61
Epishyobunone 56
Epitulipinolide 52
Epoxynephthenol acetate 82
Equilenin 3, 22, 25
Ergocornine 131
Ergonovine 131
Ergosterol 3, 31, 36
Ergosteroldiene 36
Ergot alkaloids 102
Ergotamine 102, 131
Ervsothiovine 127
Erysodine 102
Erysothiopine 127
Erythrina alkaloids 102
α-Erythroidine 102
β-Erythroidine 127
Eserine 102
Estafiatin 61
Esterone 3
Estradiol 3
Estriol 3
Estrogen deficiency 3
Estrone 22

– / synthesis 26
Eumelanins 191
Eumitrin A, B, C 212
Eupachlorin 60
Eupachloroxin 60
Euparotin 61
Eupatoriopicrin 52
Eupatoroxin 61
Eupatundin 61
Evernic acid 223
Farnesol 7
Farnesyl pyrophosphate 225
Fastigilin A, B, C 64
Ferulin 60
Flavone 183, 185
Flavonoids 137, 179, 182–185, 190
Flavoobscurin A, B_1, B_2 211
Flexuosin A, B 64, 65
Floribundic acid 69
α-Fluoro-ketone 38
Fly repellant 18
Foliol 75
Fomannosin 228
G.L-Formamide 81
Franserin 63
Frog venon 5, 11
Fucoxanthin 172
Fukujusonorone 21
Fungal polyacetylenes 175
Furanodiene 54
Furanodienone 54
Furocoumarins 189
Furonaphthoquinone 211
Furoquinoilines 102
Furostane bisglycoside 91
Furostane glycosides 90
Fusidic acid 10, 231
Gafrinin 52
Gailardilin 65
Gaillardin 61
Galanthamine 102, 119, 128
Galipine 131
Gallopheomelanins 194, 195
Geigerin 61
Geigerinin 65
Geissoschizine 122
Geldanamycin 165
Gelsemine 103, 130
Gemini ketones 158
Genin 4
Genuin aglycone 96
Genuine sapogenins 94
Geranylfarnesol 230
Geranylfarnesyl pyrophosphate 231, 230
Geranylgeranyl pyrophosphate 230, 233
Geranylnerolidol 230
Geranyl pyrophosphate 225
Gereserine 128
Germacranolides 53
Germacrene A, B, C, D 51
Germacrone 51
Germafuranolides 54
Germine 132
Gibberellic acid 78, 230
Globicin 60
Glucuronide linkage
– / selective cleavage in saponin 100
Gnidicin 85
Gnididin 85
Gnidimacrin 85
Gnidimacrin 20-palmitate 85
Gniditrin 85
Gorgosterol 6, 12
Grayanic acid 216
Grosshemin 60
Guai-3,7-diene 58
α-Guaiene 58
Guaiol 58
Guaioxide 58
γ-Gurjunene 58
Gutierolide 68
Gypsoside A 94
Haemanthamine 119
Haemin 154, 156
Haemoglobin 153, 154
Haemoproteids 153
Haems 154
Hakomori method 97
Half stercobilin 148
Half-stercobilinogens 148
Hallachrome 163
Hamalol 128
Harmaline 102, 128
Harmine 128
Hedycaryol 51
Helenalin 64
Helenium lactone 61
Heliangin 52
Helicobasidin 228
Helminthosporal 228
Heteratisine 102
Hispidin 138
Holotoxin A, B 100
Homobatrachotoxin 11
dl-D-Homotestosterone 26
Hormone
– / female sex 3
– / insect moulting 5, 10
– / sex 5
Huratoxin 86
Husbanonine 102
(–)-Hydrastine 117
Hydroquinine 131
8-Hydroxyachillin 61
Hydroxybalchanolide 52
Hydroxycostunolide 52
3-Hydroxydamsin 63
14-Hydroxyguai-1,3,5,9,11-pentaenyl stearate 58
ent-11α-Hydroxykauren-15α-yl acetate 75
7β-Hydroxy-(–)-kaur-16-en-19-oic acid 78
Hydroxypelenolide 52
1α-Hydroxyvitamin D_3 41
Hygrine 102
Hymenin 63
Hymenolin 63
Hypericin 164

Hysterin 63
Iboga 121, 123
Ibogaine 121, 130
Ibogamine 102
Icetexone 72
Imidazole alkaloids 102
Indole alkaloids 102
Ingol 86
Inunolide 53
Ipecacuanha alkaloids 102, 123
Iriediol 85
Irieol A 85
Isabelin 11, 13
Islandicin 165
Isoacrostalidic acid 74
Isobatrachotoxin 11
Isoboldine 119
Isocericenine 56
Isocoumarins 134
Iso-curcumenol 58
Isoflavones 190
Isoflavonoid 137, 179, 183, 185, 190
Isofucosterol 10
Isofuranodienone 54
Isogermafurene 56
Isohelenalin 19
Isolinearol 75
Isomesobilirhodin 148
Isomesobiliviolin 148
G.L-isonitrile 81
Isopelletierine 125
Isopentenylcoumarins 189
Isopentenyl pyrophosphate 225
4-Isopropyl-2-pyridone 105
Isoquinoline alkaloids 102
Isoshyobunone 56
Isosideritol 76
G.L-Isothiocyanate 81
Isousnic acid 209
Isovincoside 123
Isoxazole 26, 31
Ivaxillarin 60
Jacquinelin 61
Jamine 103
Jatrophone 85
Jatrorrhizine 126
Javanicin 165
Jervine 4, 102
Juglone 161, 166
Julichromes 181
Julichrome $Q_{2.3}$ 162
Jurineolide 52
Jurubine 90, 91
Kaurene 230
Kessane 58
Kessanol 58
Kessanyl acetate 58
Kessoglycol diacetate 58
α-Kessyl acetate 58
α-Kessyl alcohol 58
Kesting-Craven 159
Kirenol 70
Klyne rule 97
Kojic acid 139
Kuhn method 97
Laccaic acids 163
Lachnanthofluorone 166
Lactucin 61
Lanosterol 2, 7, 10, 231, 232
Lapachol 161, 165
α-Lapachone 161
β-Lapachone 161
Lasiocoryin 69
Laudanosine 102
Lawsone 161, 166
Lead tetrafluoride 38
Leprapinic acid 216
Leucodin 61
Leucothol A, B, D 80
Leucotylin 220
α-Levantenoide 67
β-Levantenoide 67
Lichen substances 208–223
Lichexanthone 216
Lignins 180, 189
Liguloxide 58
Liguloxidol 58
Ligustrin 61
Linalool 227
Linaridial 69
Linderadine 54
Linderalactone 54
Linderane 54
Linifolin A, B 22
Lithium dialkylcopper 41
Lithocholic acid 2
Litseaculane 54
Litsealactone 54
LL-Z 1220 140
Lobelamine 125
Lobelanidine 125
Lobeline 125
Lobophytolide 82
Loganine 123, 227
Lumirhodopsin 203
Lumisterol 3
Lunacrine 102
Lupinine 102
Lycodine 102
Lyconnotine 102
Lycorenine 102
Lycorine 102, 119, 128
Lycotonine 102
Lycopodine 102
Lycopodine alkaloids 102
Lycoxanthol 72
Lygnans 179, 181, 184, 189
Lysergic acid 121, 131
Macrocyclic tetrapyrrole pigments 157
Macrolides 173
Malonyl CoA 134
Mansonones 161, 165
Marin coelenterate 12
Marine sterols 23-demethylgorgosterol 6
Marker's degradation 96, 97
Mascaroside 75

Matricarin 61
Matricin 60
Matrine 102
Melanins 190
Melodinus scandens Forst 108
Meloscine 108, 111
Merochlorophaeic acid 216
Mescaline 124
Mesobilirhodin 148
Mesobiliviolin 148
Mesoporphyrin IX 156
Mesostercobilin IXα 148
Mesourobilin 147
Metarhodopsin I, II, III 203
Methoxynepetaefolin 69
m-Methoxytetralone 23
2-Methylanthraquinone 161
Methyl D-homoestrone 26
7-Methyljuglone 161, 165
O-Methynorbelladine 119
4-*O*-Methylphysodic acid 216
Mevalonic acid 7, 208, 225
Mexicanin A, C, D, E, I, H 22, 23
Mikanolide 53
Miscandenin 56
Mitomycin C 165
Mitomycins 164
Mitraphylline 102
Moffatt oxidation 12
Mokko lactone 60
Mollisin 165
Momilactone-A, B, C 70
Monocrotaline 102
γ-Monocyclofarnesol 234, 228
Monotamine 102
Monoterpenes 227
Morphine 102, 113, 114, 118, 119, 127
Morphine alkaloids 102
Mukulol 82
Muscarine 124
Mycophenolic acid 235
Myoglobin 153, 154

Narceine 124
Narcotine 102, 117, 118, 126
Nardol 58
Natural pigments 133–205
Neocembrene-A 82
Neoflavonoids 179, 190
Neolignan 189
Neolin 81
Neolinderalactone 54, 55
Neolinderane 54
Neosericenine 54
Neothiobinupharidine 103
Neoxanthin 172
Nephroarctin 216, 220
Nephthenol 82
Neutral saponin 88
Nicandrenone 5, 17, 18
Nicotine 125
Nitrone 41
Nitrosyl fluoride 38
Nologenin 93
Nolonin 93
Nonaromatic steroids 26
19-Norcholestanol 6
Nor-diplocin 220
Norlaudanosoline 116
18-Norsteroid 21
19-Norsteroid 25, 31
Noscapine 126
Obtusaquinone 166
Occidenol 56
Ochotensimine 102
Octaethylporphin complexes 158
Odoratin 65
Onopordopicrine 52
Oocyan 144
Ophiobolin 228
Ophiobolin-F 231
Opsin 203, 204
Oxepin ring 139
2,3-Oxidosqualene 10, 213, 233
a-Oxobilanes 157
b-Oxobilanes 157
Oxodendrobine 103
P 487_1 204
P 499_1 201–204
P 500_1 202
P 502_1 201, 202, 204
P 523_2 201, 202
Pachydictyol A 84
Paniculoside I, II, III 76
Papaverine 126
Parietib 216
Parillin 91
Parthenin 63
1-Epi-parthenin 63
Partheniol 58
Parthenolide 52
Paucin 64
Pelenolide a, b 52
Pemptoporphyrin 157
Peridinin 172
Perivine 129
Perlatoric acid 223
Peruvinin 64
Pharbitic acid 78
Phenolic oxidative coupling 116
Pheomelanins 193
Pheophorbides 155
Phlebiakauranol 75
Phlebianorkauranol 75
Phlebiarubrone 160, 166
Phlebic acid A 216, 219
Phorcabiline 145
Photo-Claisen rearragements 185
Photoepimerization 41
Photo-Fries rearrangements 185
Photo induced reactions 185
Phthalide isoquinotines 102
Phycobilins 149
Phycobiliverdin 149
Phycobiliviolin 149
Phycocyanins 149
Phycocyanobilin 149

Phycoerythobilin 149
Phycoerythrins 149
Physalin A, B 16
Physostigmine 128
Phytochrome 150
Phytoene 233
Phytosteroids 7, 10
α-Picolines 26
Picrolichenic acid 209, 216
Pilocarpine 125
Piloquinone 164, 165
Pimaradiene-3β-ol 70
Piperidine alkaloid 102
Pododacric acid 72
Pogostol 58
Polyacetylenes 175–178
Polydalin 53
Polyenes 168–174
Polyketide 134, 165, 208
Polyphenolic compounds 179
Polyporic acid 160, 161, 212
Porphins 153, 157, 158
Porphobilinogens 157
Porphyrin C 157
Porphyropsin 202
Porpurin 162
Portentol 216
Portulal 68
Preakuammicine 122
Precalciferol 31
Preisocalamendiol 51
Prelumirhodopsin 203
Prenylphenols 233
Prenylquinones 233
Presiccanochromene 234
Presiccanochromenic acid 234
Presqualene alcohol 9
Primin 159, 160
Prironoquinonoids 135
Pristimerin 166
Proaporphine 119
Proaporphine alkaloids 102
Prochiral hydrogen 7
Procurcumenol 58
Progesterone 3, 26, 30
Pronuciferine 102
Prostratin 86
Protoberberine alkaloids 102
Protobilirubin IXα 145
Protobiliverdin IXα 144
Protocetraric acid 216
Protohaem 154
Protolichesteric acid 208
Protopine 102
Protopine alkaloids 102
Protoporphyrin IX dimethyl ester 155
Protoveratrine 132
Pseudobatrachotoxin 11
Pseudoivalin 61
Pseudoneolinderane 54
Pseudopurpurin 166
Pterobilin 145
Pterocarpanes 190
Pterocarpins 183
Pulchellin 23
Pulvilloric acid 166
Purine alkaloids 102
Pycnamine 102
Pyran compounds 134
Pyranocoumarins 189
Pyrethrosin 53
Pyridine alkaloids 102
α-Pyrone 134
γ-Pyrone 134
2-Pyrones 184
Pyronone 134
Pyrroles 153, 156, 157
2,3,5-Pyrroletricarboxylic acid 193
Pyrrolidine alkaloids 102
Pyrrolizidine alkaloids 102
Pyrromethanes 157
Pyrromethenes 157
5(1*H*)-pyrromethenone 142
Pyrromycin 163, 165
ε-Pyrromycinone 163
η-Pyrromycinone 163
Quercetinase 187
Quinidine 131
Quinine 102, 131
o-Quinodimethane 26
Quinolizidine alkaloids 102
Quinolone alkaloids 102
Quinon 183
Quinone-methides 166
Quinonoids 208
Quinotine alkaloids 102
Rapanone 160
Rebaudioside A, B 76
Rescinnamine 129
Reserpine 103, 129
Reticuline 116, 118, 119
Retigeric acid A, B 213, 216
Retinal 203, 233
11-*cis*-Retinal 201, 204
all-*trans*-Retinal 204
Retinol 203
11-*cis*-Retinol 204
11-*cis*-Retinolester 204
all-*trans*-Retinolester 204
Retroprogesterone 31
Retrotesterone 31
Rhodoaphin-be 164
Rhodocomatulin 163
Rhodopsin 202–204
Rotenoid 179, 183, 184, 190
Rubidomycin 163
Rugulosin 211, 220
Rutecarpine 121
Salonitenolide 52
Salonitolide 53
Salsolin 63, 126
Salutaridinol-I 119
Salviol 72
Sapogenin 4
Saponic properties 98
Saponines 88–100

– / antimicrobial activities 98
– / basic steroid 88, 94
– / recovery from complexes 95
– / triterpenoids 88, 93
Sarpagine 102
Sarpedobiline 145
Sarsaparilloside 90
Sativene 228
Saussurea lactone 56
Scabiolide 53
Scandenolide 53
Schizopeltic acid 216
Scopolamine 124
(−)-Scoulerine 118
Scrobiculin 216
Sea cucumber 100
Sea urchins 162
Secologanine 123
Seco C/D steroid 16
Securinine 107, 108, 125
Selagine 102
Sericenic acid 54
Sericenine 54
Serotonine 102
Serpentine 121, 122
Serratinine 102
Sesquiterpenoids 9, 14, 228
Shanorellin 160
Sheep's wool fat 2
Shikimate
– / biogenesis 166
– / pathway 137
Shikimic acid 208
Shikimic acid pathway 139, 188
Shiromodiol 51
Shiromool 51
Shyobunone 56
Siccanin 234
Siccanochromene 234
Siccanochromenic acid 234
Sideritol 76
Sieverin 60
Sieversinin 60
Sinularin 86
Sinulariolide 86
β-Sitosterol 10
Skyrin 162, 211
Smith degradation 96
α-Solamargine 95
Solanaceae 17
Solanidane 94
Salanum alkaloids 88, 94
Solorinic acid 163
Solstitialin 61
Solstitialin acetate 61
Sparteine 102, 125
Spathulin 65
Sphaerococcenol A 85
Spirobenzyl isoquinolines 102
Spirographisporphyrin 157
Spirosolane 88, 94
Spirostanol glycosides 90
Sponge 6, 13
Squalene 2, 7, 226, 233
Starfish saponins 99
Stemarin 80
Stemmadenine 122
Stemodin 80
Stemodinone 80
Stemolide 72
Steptovaricins 165
Stercobilin IXα 148
Stercobilinogen 148
Steroids 1–41
Steroid alkaloids 11, 102
Steroid saponins 88
Stevin 65
Stigmasterol 31
Stilbenes 179, 190
Stizolin 53
Strophanthidin 4
Strychnine 102, 129
Stylatulide 86
Tabernathine 130
Tabersonine 122
Tachysterol 3
Tamaulipin A, B 52
Tannins 180, 189
Tanshinones 165
Taondiol 81
Taxodone 166
Taxol 85
Tazettine 102
Teaflavin 187
Temisin 56
Tenebrio molitor 10
Tenulin 64
Terpenoids 43–86, 227
Testosterone 3
Tetradehydrocorrin-complex 158
Tetrahydrocannabinol 233, 234
S(−)-Tetrahydropalmatine 108
Tetrahydrosecurinine 108
Tetrahymanol 233
Tetrandrine 127
Tetraneurin A, B, C, D, E, F 63
Tetrangomycin 164
Tetrapyrrole ligand systems 158
Tetrapyrrole pigments 142–159
Teucvin 69
Thebaine 118
Thelephoric acid 160, 212
Theobromine 125
Theophylline 125
Thermarol 70
Thujone 227
Thurberilin 64
Tocopherols 184
Toluquinone 159
Tomatidine 103
Torilin 58
Tosylhydrazone 41
C-Toxiferine I 130
Triazole 36
Trichodiene 228
Trichosiderin A, B, C, E, F 195–197

Trichothecin 228
ent-7α, 16α,17-Trihydroxykauran-19-oic acid 75
Tripdiolide 72
Triphenylphosphine chlororhodium 36
Triptolide 72
Trishomofarnesyl pyrophosphate 226
Trisporic acid 233
Triterpenes 231
Triterpenoid saponins 93
Tropacocaine 124
Tropane alkaloids 102
Tropine 102
α-Truxillic acid 185
β-Truxillic acid 185
Tryptamine 121
Tulipinolide 52
Tuberostemonine 103
Tubocurarine 127
Tylocrebrine 128
Umbelliferone 139
d-Urobilin 147
i-Urobilin 147
Urobilinogen 147
Urospermal A, B 52
Usnic acid 208, 2099, 214, 220 222, 223
Uteroverdin 144
Uvedalin 53
Vakognavine 76
Veatchine 102
Veratramine 132
Vernodalin 56
Vernolepin 56
Vernolide 52
Vernomenin 56
Vernomygdin 53
Vilangin 160
Villalstonine 103
Volucrisporin 166
Vinblastine 102, 130
Vincaleukoblastine 102
Vincamine 130
Vincoside 123
Vincristine 103
Vindoline 121, 122
Vinylurobilin 148
Virginolide 60
Vitamin B_{12} 154, 157
Vitamin B_{12s} 157
Vitamin D 3, 4, 31
Vobasine 102
Vulpinic acid 213
Westphalen rearrangement 41
Wieland-Mischler keton 26
Withaferin 19
Withaferin A 5
Withanolide 17, 19, 21
Withaphysalin A, B 17
Xanthones 179, 182, 183, 186, 189
Xanthophylls 168
Xylindein 164
Yangonin 137, 138
Yohimbine 102, 129
Yononin 90
Yuzurimine 103
Zaluzanin A, B, C, D 60
Zaylanane 54
Zederone 54
Zedoarone 56
Zeorin 213, 218, 220
Zexbrevin 52
Zeylanicine 54
Zeylanidine 54
Zeylanine 54
Zygadenine 132

DATE DUE			
Chemistry Dept			